Springer Proceedings in Mathematics & Statistics

Volume 261

Springer Proceedings in Mathematics & Statistics

This book series features volumes composed of selected contributions from workshops and conferences in all areas of current research in mathematics and statistics, including operation research and optimization. In addition to an overall evaluation of the interest, scientific quality, and timeliness of each proposal at the hands of the publisher, individual contributions are all refereed to the high quality standards of leading journals in the field. Thus, this series provides the research community with well-edited, authoritative reports on developments in the most exciting areas of mathematical and statistical research today.

More information about this series at http://www.springer.com/series/10533

Masanao Ozawa · Jeremy Butterfield
Hans Halvorson · Miklós Rédei
Yuichiro Kitajima · Francesco Buscemi
Editors

Reality and Measurement in Algebraic Quantum Theory

NWW 2015, Nagoya, Japan, March 9–13

Springer

Editors
Masanao Ozawa
Graduate School of Informatics
Nagoya University
Nagoya, Japan

Jeremy Butterfield
Trinity College
Cambridge University
Cambridge, UK

Hans Halvorson
Department of Philosophy
Princeton University
Princeton, NJ, USA

Miklós Rédei
Department of Philosophy, Logic and Scientific Method
London School of Economics and Political Science
London, UK

Yuichiro Kitajima
College of Industrial Technology
Nihon University
Narashino, Japan

Francesco Buscemi
Graduate School of Informatics
Nagoya University
Nagoya, Japan

ISSN 2194-1009 ISSN 2194-1017 (electronic)
Springer Proceedings in Mathematics & Statistics
ISBN 978-981-13-2486-4 ISBN 978-981-13-2487-1 (eBook)
https://doi.org/10.1007/978-981-13-2487-1

Library of Congress Control Number: 2018954022

Mathematics Subject Classification (2010): 03G12, 03G30, 06C15, 18B25, 81P10, 81P15, 81P16, 81P40, 81P45, 81T05, 82C10, 94A15, 94A17

This Springer imprint is published by the registered company Springer Nature Singapore Pte Ltd.
The registered company address is: 152 Beach Road, #21-01/04 Gateway East, Singapore 189721, Singapore

Preface

This volume contains papers based on presentations at the "Nagoya Winter Workshop 2015: Reality and Measurement in Algebraic Quantum Theory (NWW 2015)", held in Nagoya, Japan, from March 9 to 13, 2015. The foundations of quantum theory have been a source of mysteries, puzzles, and confusions, and have encouraged innovations in mathematical languages to describe, analyze, and delineate this wonderland. Both ontological and epistemological questions about quantum reality and measurement have been placed in the center of the mysteries explored originally by Bohr, Heisenberg, Einstein, and Schrödinger. This volume describes how those traditional problems are nowadays explored from the most advanced perspectives formed by emerging new paradigms such as algebraic methods in quantum field theory, new research fields on quantum information and computation, and category theoretical descriptions of physical theories as well as by the advanced experimental techniques applicable to those foundational problems traditionally only discussed through ingenious thought experiments. Specifically, it includes new research results in quantum information theory, quantum measurement theory, information thermodynamics, operator algebraic and category theoretical foundations of quantum theory, and the interplay between experimental and theoretical investigations on the uncertainty principle. This book is intended for a broad audience of mathematicians, theoretical and experimental physicists, and philosophers of science.

NWW 2015 is the sixth workshop of the Nagoya Winter Workshop Series on Quantum Information, Measurement, and Foundations, held annually in February/March from 2010 in the Higashiyama Campus, Nagoya University, Nagoya, Japan, aiming to achieve advances in quantum information, quantum measurement, and quantum foundations. The first five workshops were organized by Francesco Buscemi (Nagoya U.) and Masanao Ozawa (Nagoya U.). NWW 2015 was a special member of the series emphasized on topics related to the research project "Reality and Measurement in Algebraic Quantum Theory" led by Yuichiro Kitajima (Nihon U.) and M. Ozawa, supported by the Templeton Foundation (ID 37751) from October 2012 to June 2015. NWW 2015 was organized by Y. Kitajima, F. Buscemi, and M. Ozawa collaborating with the guest organizers,

Jeremy Butterfield (U. of Cambridge), Hans Halvorson (Princeton U.), and Miklós Rédei (LSE, London).

We are grateful to the John Templeton Foundation (Grant ID 35771) and the Japan Society for the Promotion of Science (JSPS) (Grant No. JP26247016) for financial supports. We thank Nagoya University for providing the workshop venue, the Noyori Conference Hall, which contributed to the pleasant and stimulating atmosphere during the workshop. We thank the publisher, Springer Japan, represented by Ms. Chino Hasebe (Executive Editor), Mr. Masayuki Nakamura (Editorial Department), and Mr. Yoshio Saito (Editorial Department), for assistance in this publication. Last but not least, we thank the local secretaries, Ms. Yoko Tashiro and Ms. Mizuho Katsuse, who made the workshop run smoothly and efficiently.

Nagoya, Japan
June 2018

Masanao Ozawa
Jeremy Butterfield
Hans Halvorson
Miklós Rédei
Yuichiro Kitajima
Francesco Buscemi

Contents

Part I
Quantum Reality

A Generalisation of Stone Duality to Orthomodular Lattices

Sarah Cannon and Andreas Döring

Abstract With each orthomodular lattice L we associate a spectral presheaf $\underline{\Sigma}^L$, generalising the Stone space of a Boolean algebra, and show that (a) the assignment $L \mapsto \underline{\Sigma}^L$ is contravariantly functorial, (b) $\underline{\Sigma}^L$ is a complete invariant of L, and (c) for complete orthomodular lattices there is a generalisation of Stone representation in the sense that L is mapped into the clopen subobjects of the spectral presheaf $\underline{\Sigma}^L$. The clopen subobjects form a complete bi-Heyting algebra, and by taking suitable equivalence classes of clopen subobjects, one can regain a complete orthomodular lattice isomorphic to L. We interpret our results in the light of quantum logic and in the light of the topos approach to quantum theory.

Keywords Orthomodular lattice · Stone space · Stone duality · Invariant · Spectrum · Functor · Bi-Heyting algebra

1 Introduction

Classical dualities and the lack of dualities for nondistributive/ noncommutative algebras. Stone duality [1] is one of the classical dualities. It relates a kind of algebras (Boolean algebras) to a kind of topological spaces (Stone spaces). There are many variants and generalisations of Stone duality [2], which are all similar in spirit. Another important classical duality is Gelfand duality, relating C^*-algebras and locally compact Hausdorff spaces. (Again, there are a number of variants.) The classical dualities always have on one side some kind of distributive or commutative

S. Cannon
College of Computing, Georgia Institute of Technology,
North Avenue, Atlanta, GA 30332, USA
e-mail: sarah.cannon@gatech.edu

A. Döring (✉)
Independent researcher, Frankfurt, Germany
e-mail: andreas.doering@posteo.de

M. Ozawa et al. (eds.), *Reality and Measurement in Algebraic Quantum Theory*,
Springer Proceedings in Mathematics & Statistics 261,
https://doi.org/10.1007/978-981-13-2487-1_1

algebras, organised into a category, and on the other side a corresponding kind of topological spaces, forming another category, related by a dual equivalence.

Yet, in quantum theory and in a vast number of mathematical situations, *non*-distributive and *non*commutative algebras are of interest. For these, there mostly are no general functorial correspondences or dualities with suitable (generalised) spaces known. In fact, much of the difficulty consists in determining what kind of dual spaces would be suitable. These generalised spaces would have to be noncommutative spaces, in a sense to be made precise. Of course, the vast field of noncommutative geometry has as one of its starting points the assumption that there should be spaces corresponding to noncommutative algebras, and that there is much to be gained from using geometric methods when dealing with noncommutative algebras. This is doubtlessly true and has led to many deep and beautiful results, but since concrete noncommutative spaces are often lacking, noncommutative geometry is mostly done as algebra and only implicitly deals with spaces and geometric objects.

The spectral presheaf of an orthomodular lattice as a dual space. In this article, we will go another route: we will provide a new, concrete kind of dual space for any orthomodular lattice L. Here, orthomodular lattices (OMLs) are seen as a natural, generally nondistributive generalisation of Boolean algebras.

The dual space that we will assign to an OML will be a presheaf, which means it is not a single set (equipped with a topology), but a 'diagram' of sets (in fact, topological spaces), canonically linked together by continuous functions. More specifically, the spectral presheaf $\underline{\Sigma}^L$ of an orthomodular lattice L consists of the Stone spaces of all the Boolean subalgebras of L, organised into a presheaf over the partially ordered set of these Boolean subalgebras. This seemingly simple-minded construction raises the question whether one does not lose too much information: is it possible to encode an orthomodular lattice, as a nondistributive structure, by considering the (Stone spaces of) its Boolean, distributive parts only? Maybe surprisingly, the answer is in the affirmative. One of our main results shows that two orthomodular lattices L and M are isomorphic if and only if their spectral presheaves $\underline{\Sigma}^L$ and $\underline{\Sigma}^M$ are isomorphic (Theorem 3.18). In order to show this, we first have to develop the necessary categorical background in some detail, including the notion of morphisms between presheaves over different base categories, and a dual notion of copresheaves and their morphisms. Among other things, we show that the assignment $L \to \underline{\Sigma}^L$ is contravariantly functorial.

We also provide a certain generalisation of Stone representation to complete orthomodular lattices. Recall that every Boolean algebra B is isomorphic to the concrete Boolean algebra of clopen subsets of the Stone space Σ_B of B. In a similar fashion, every complete orthomodular lattice L can be represented within the clopen subobjects of its spectral presheaf $\underline{\Sigma}^L$. The representing map is called daseinisation. The clopen subobjects of $\underline{\Sigma}^L$ form a complete bi-Heyting algebra, and by using the adjoint of daseinisation, we can form suitable equivalence classes such that the set of equivalence classes becomes a complete OML that is canonically isomorphic to L. This is the content of our second main result (Theorem 4.19).

The topos approach and physical interpretation. This work is of course inspired by the so-called topos approach to quantum theory [3–8]. A spectral presheaf was first defined by Isham, Hamilton, and Butterfield for the noncommutative von Neumann algebra $\mathcal{B}(\mathcal{H})$ of all bounded operators on a Hilbert space $\mathcal{H}$ [9] and was later generalised to arbitrary von Neumann algebras [10]. In the topos approach, the spectral presheaf plays the role of a generalised state space for a quantum system, providing a topological-geometric perspective that is not available ordinarily. Just as in classical physics, propositions about the values of physical quantities are represented by (clopen) sub'sets' of the quantum state space. This led to the development of a new form of logic for quantum systems, based upon the internal logic of the topos of presheaves in which the spectral presheaf lies [4, 11, 12]. In [13, 14], the second author considered the question if the spectral presheaf determines a von Neumann algebra up to isomorphism (it does not, but it determines the algebra up to Jordan-$*$-isomorphism).

Orthomodular lattices are key structures in quantum logic [15, 16], where they represent algebras of propositions about a quantum system. The lattice operations are interpreted logically as conjunction and disjunction, while the orthocomplement is interpreted as negation. We provide a topological-geometric underpinning of this kind of quantum logic by providing a concrete dual space for every orthomodular lattice. Moreover, we represent the elements of a complete OML by clopen subsets (technically, subobjects) of this dual space. The fact that the clopen subsets form a complete bi-Heyting algebra and not a complete OML may seem to be a disadvantage at first sight, but in fact it is a great improvement over standard quantum logic, since many conceptual problems are avoided. For example, there is a material implication. Moreover, one can use the adjoint of daseinisation to map back to the complete OML. We will briefly discuss some of the interpretational advantages of the bi-Heyting algebra representation in Sect. 4.4.

Overview and organisation. This article is largely self-contained. Section 2 provides some mathematical background on orthomodular lattices, Stone duality etc. and some preliminary results, in particular concerning the Boolean substructure of an orthomodular lattice. In Sect. 3, we introduce the spectral presheaf of an orthomodular lattice (Sect. 3.1) and consider maps between spectral presheaves (Sect. 3.2). There is some detailed discussion of categories of presheaves over varying base categories and with values in another category $\mathcal{D}$ (Sect. 3.3), as well as of copresheaves with values in $\mathcal{C}$ (Sect. 3.4). A dual equivalence between $\mathcal{C}$ and $\mathcal{D}$ lifts to a dual equivalence between **Copresh**$(\mathcal{C})$ and **Presh**$(\mathcal{D})$, and we apply this to Stone duality in particular (Sect. 3.5). These results are then employed to show that two orthomodular lattices are isomorphic if and only if their spectral presheaves are isomorphic, and the isomorphisms can be constructed explicitly from each other (Sects. 3.6, 3.7; Theorem 3.18). We provide some interpretation, including physical interpretation, of this result in (Sect. 3.8) and also show the analogous result for complete orthomodular lattices (Sect. 3.9; Theorem 3.29). In Sect. 4, we are concerned with the representation of complete OMLs. We define clopen subobjects of the spectral presheaf (Sect. 4.1), show that they form a complete bi-Heyting algebra $\mathrm{Sub_{cl}}\,\underline{\Sigma}^L$ (Sect. 4.2), and then introduce the map called daseinisation that takes elements of a complete OML L to clopen subobjects of its spectral presheaf $\underline{\Sigma}^L$. We interpret this as a representation of L within $\mathrm{Sub_{cl}}\,\underline{\Sigma}^L$ (Sect. 4.3) and give some physical interpretation in (Sect. 4.4). The adjoint of daseinisation is introduced and some of its properties are discussed in (Sect. 4.5). We then show that using this adjoint, one can form equivalence classes of clopen subobjects such that the set E of equivalence classes becomes a complete OML isomorphic to L in a natural way, which means that we have a generalisation of Stone representation to complete orthomodular lattices (Sect. 4.6; Theorem 4.19). Section 5 concludes with a list of some open problems.

2 Background and Preliminary Results

We assume familiarity with some basics of order and lattice theory such as the definitions of partially ordered sets (posets), meets (greatest lower bounds), joins (least upper bounds), lattices, and complete lattices [17, 18]. Additionally, some familiarity with category theory is assumed, including the definitions (but no advanced properties) of presheaves, copresheaves, dual equivalences, and topoi; see, e.g., [19–24].

Throughout, we will denote the category of posets and monotone maps between them as **Pos**, the category of sets and functions between them as **Set**, and the category of Boolean algebras and Boolean algebra homomorphisms as **BA**.

2.1 *Ortholattices and Orthomodular Lattices*

Our results focus on orthomodular lattices, which we now define. Good references are [16, 25].

Definition 2.1 An *orthocomplementation function* on a lattice L is a map $a \mapsto a'$ for each lattice element a, satisfying

1. $a' \vee a = 1, a' \wedge a = 0$ (Complement Law),
2. $a'' = a$ (Involution Law),
3. If $a \leq b$, then $b' \leq a'$ (Order-Reversing).

Definition 2.2 An *orthocomplemented lattice*, also called an *ortholattice*, is a bounded lattice with an orthocomplementation function.

Definition 2.3 An *orthomodular lattice (OML)* L is an ortholattice such that for any $x, y \in L$ with $x \leq y$, it holds that $x \vee (x' \wedge y) = y$. This is the *orthomodularity property*.

Figure 1 depicts four small ortholattices. Ortholattices (i), (iii), and (iv) have a unique orthocomplementation function, as shown. The second has three valid orthocomplementation functions; the orthocomplement of a could be any of b, c, or d.

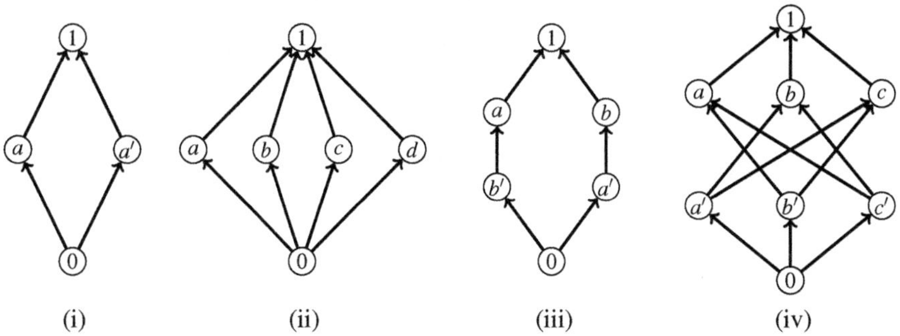

Fig. 1 Four valid ortholattices. An arrow $a \to b$ means $a \leq b$

Of these ortholattices, (i), (ii), and (iv) are orthomodular lattices. In (iii), elements b' and a satisfy $b' \leq a$, but

$$b' \vee (b \wedge a) = b' \vee 0 = b' \neq a. \tag{1}$$

Another example of an OML is the lattice of subspaces of any inner product space, with the orthogonal complement operation on these subspaces as the orthocomplementation function. The closed subspaces of a separable Hilbert space form a complete orthomodular lattice; such lattices are at the heart of Birkhoff-von Neumann style quantum logic [26], where the closed subspaces represent propositions about the values of physical quantities of a quantum system. More generally, the projections in any von Neumann algebra $\mathscr{N}$ form a complete OML $\mathscr{P}(\mathscr{N})$.

We will refer to an orthocomplement-preserving lattice homomorphism between two OMLs as an orthomodular lattice homomorphism. Orthomodular lattices and orthomodular lattice homomorphisms form a category **OML**. It will be useful to note that De Morgan's laws, which are an important property of Boolean algebras, hold in the more general case for all ortholattices (and thus all orthomodular lattices).

2.2 *Distributive Substructure of an Orthomodular Lattice*

We will consider Boolean sublattices of orthomodular lattices (OMLs).

Definition 2.4 A *Boolean sublattice*, also called a *Boolean subalgebra*, of an orthomodular lattice L is a complemented distributive sublattice with complements given by the orthocomplementation function of L.

Lemma 2.5 *Every element a of an orthomodular lattice L is in some Boolean subalgebra of L.*

For $a \neq 0, 1$, one Boolean subalgebra containing a is the four-element sublattice $\{0, a, a', 1\}$.

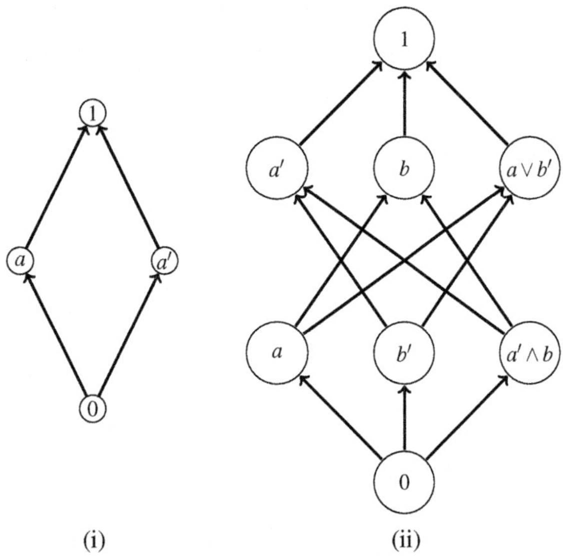

Fig. 2 Boolean sublattices of an orthomodular lattice L containing (i) element a, and (ii) elements a and b with $a \leq b$

Proposition 2.6 *Let L be an ortholattice. L is orthomodular if and only if for all elements $a, b \in L$ with $a \leq b$ there is a Boolean subalgebra of L containing both a and b.*

Proof The forward implication can be found in [17]; concretely, for $a, b \neq 0, 1$, a Boolean sublattice of L containing a and b is displayed in Fig. 2. For the converse, assume that for all $a \leq b$ in L there is some Boolean subalgebra of L containing both a and b. Then elements a and b and their complements satisfy distributivity, meaning

$$a \vee (a' \wedge b) = (a \vee a') \wedge (a \vee b) = 1 \wedge b = b. \tag{2}$$

This is the orthomodularity condition, so as it holds for all $a \leq b$ then L is orthomodular.

This is the reason we consider orthomodular lattices instead of ortholattices, as Proposition 2.6 plays a key role in the proofs of Lemma 2.12 and Proposition 2.13 and subsequently in Theorem 3.17, which is the main result of Sect. 3.

2.2.1 The Context Category $\mathscr{B}(L)$

Definition 2.7 For an orthomodular lattice L, let $\mathscr{B}(L)$ denote the *poset of Boolean sublattices of* L, where the partial order on $\mathscr{B}(L)$ is given by inclusion. $\mathscr{B}(L)$ is also called the *context category* of L.

Seen as a category, the poset $\mathscr{B}(L)$ has a unique arrow from Boolean subalgebra B' to Boolean subalgebra B whenever $B' \subseteq B$. This arrow will be denoted $i_{B',B}$, and simply indicates that $B' \subseteq B$.

Additionally, whenever $B' \subseteq B$, one can define an inclusion map between Boolean subalgebras $inc_{B',B} : B' \to B$ given by $inc_{B',B}(b) = b$ for all $b \in B'$. As B' is closed under meets, joins, and orthocomplements, it follows that $inc_{B',B}$ is a Boolean algebra homomorphism, that is, a morphism in category **BA**.

Let $\varphi : L \to M$ be an orthomodular lattice homomorphism. If B is a Boolean subalgebra of L, then

$$\varphi|_B : B \longrightarrow \varphi[B]$$

is a morphism of Boolean (sub)algebras, since the image $\varphi[B]$ clearly is a Boolean subalgebra of M. Hence, every morphism $\varphi : L \mapsto M$ of OMLs induces a morphism between their context categories:

$$\tilde{\varphi} : \mathscr{B}(L) \longrightarrow \mathscr{B}(M) \tag{3}$$

$$B \longmapsto \varphi[B]. \tag{4}$$

If $\varphi : L \to M$ is an isomorphism of OMLs, then clearly $\varphi|_B : B \to \varphi[B]$ is an isomorphism of Boolean algebras. Summing up,

Proposition 2.8 *There is a functor from* $\mathscr{B}$: **OML** $\to$ **Pos** *sending each orthomodular lattice* L *to its context category* $\mathscr{B}(L)$ *and each homomorphism* $\varphi : L \to M$ *of OMLs to the corresponding morphism* $\tilde{\varphi} : \mathscr{B}(L) \to \mathscr{B}(M)$.

Since functors preserve isomorphisms, we have

Lemma 2.9 *If* $\varphi : L \to M$ *is an isomorphism of orthomodular lattices, then* $\tilde{\varphi} : \mathscr{B}(L) \to \mathscr{B}(M)$ *is an order isomorphism in* **Pos**.

2.2.2 The Partial Orthomodular Lattice L_{part}

The Boolean sublattices of an OML can also be used to generate a second structure, called the partial orthomodular lattice associated with L.

Definition 2.10 Let L be an OML. The *partial orthomodular lattice* L_{part} *associated with* L has the same elements and orthocomplements as L, as well as lattice operations $\vee$ and $\wedge$ inherited from L but only defined for finite families of elements $(a_i)_{i \in I}$ in L such that there is some $B \in \mathscr{B}(L)$ that contains a_i for all $i \in I$. Such families of elements are called *compatible families*.

Definition 2.11 A *morphism of partial orthomodular lattices* is a function $p : L_{part} \to M_{part}$ that preserves orthocomplements and existing finite meets and joins.

The following lemma depends critically on orthomodularity:

Lemma 2.12 *If $a \leq b$ in orthomodular lattice L and $p : L_{part} \to M_{part}$ is a partial orthomodular lattice homomorphism, then $p(a) \leq p(b)$.*

Proof Suppose $a, b \in L$ and $a \leq b$. By Proposition 2.6, there is some Boolean subalgebra of L that contains both a and b. This means that the meet $a \wedge b = a$ is defined in L_{part}, and thus is preserved by any partial orthomodular lattice homomorphism p:

$$p(a) = p(a \wedge b) = p(a) \wedge p(b). \tag{5}$$

From this it follows that $p(a) \leq p(b)$.

Partial orthomodular lattices associated with OMLs and partial orthomodular lattice homomorphisms form a category **POML**. The motivation for considering partial OMLs comes from the 'Bohrification' construction that can be applied to an orthomodular lattice, as will be explained in Sect. 3; L_{part} can be seen as a topos-external description of the Bohrification $\overline{\mathscr{L}}$ of L, which is an object in the topos $\mathbf{Set}^{\mathscr{B}(L)}$ of (covariant) functors from the context category $\mathscr{B}(L)$ to **Set**.

Proposition 2.13 *Let L and M be OMLs, and L_{part} and M_{part} their associated partial OMLs. There is a bijective correspondence between isomorphisms $L \to M$ in* **OML** *and isomorphisms $L_{part} \to M_{part}$ in* **POML**.

Proof Let $\varphi : L \to M$ be an isomorphism in **OML**. As a homomorphism between orthomodular lattices, it preserves orthocomplements and finite meets and joins. In particular, it preserves all meets and joins that are defined in L_{part}, meaning that it induces a homomorphism $\varphi : L_{part} \to M_{part}$. As $\varphi : L \to M$ is an isomorphism, so is $\varphi : L_{part} \to M_{part}$.

Conversely, let $p : L_{part} \to M_{part}$ be an isomorphism of partial ortholattices in **POML**. Let $(a_i)_{i \in I}$ be any finite family of elements in L; our goal is to show that

$$p\left(\bigvee_{i \in I} a_i\right) = \bigvee_{i \in I} p(a_i), \tag{6}$$

which implies that p preserves all joins, not just those joins that are defined in L_{part}. The same result for meets then follows by taking orthocomplements.

First, suppose that there is some Boolean subalgebra B of L such that $a_i \in B$ for all $i \in I$. Thus $\bigvee_{i \in I} a_i$ is defined in L_{part}, and as partial orthomodular lattice isomorphism p preserves all joins that are defined in L_{part},

$$p\left(\bigvee_{i \in I} a_i\right) = \bigvee_{i \in I} p(a_i). \tag{7}$$

Now, assume there is no $B \in \mathscr{B}(L)$ such that $a_i \in B$ for all $i \in I$. Consider the element $\bigvee_{i \in I} a_i$ of L. Note that for each i, $a_i \leq \bigvee_{i \in I} a_i$, meaning that by Lemma 2.12, $p(a_i) \leq p\left(\bigvee_{i \in I} a_i\right)$. As this is true for all i, it follows that

$$\bigvee_{i\in I} p(a_i) \leq p\left(\bigvee_{i\in I} a_i\right). \tag{8}$$

Now, let $p^{-1} : M_{part} \to L_{part}$ be the inverse of partial orthomodular lattice isomorphism p, which is also a partial orthomodular lattice isomorphism. For all i, $p(a_i) \leq \bigvee_{i\in I} p(a_i)$. Again by Lemma 2.12, p^{-1} preserves inequalities, so this equation becomes

$$a_i = p^{-1}(p(a_i)) \leq p^{-1}\left(\bigvee_{i\in I} p(a_i)\right). \tag{9}$$

As this is true for all $i \in I$, it follows that

$$\bigvee_{i\in I} a_i \leq p^{-1}\left(\bigvee_{i\in I} p(a_i)\right). \tag{10}$$

Applying p to the above equation and again invoking Lemma 2.12, this becomes

$$p\left(\bigvee_{i\in I} a_i\right) \leq p\left(p^{-1}\left(\bigvee_{i\in I} p(a_i)\right)\right) = \bigvee_{i\in I} p(a_i) \tag{11}$$

Equations 8 and 11 together imply

$$p\left(\bigvee_{i\in I} a_i\right) = \bigvee_{i\in I} p(a_i), \tag{12}$$

showing that p preserves all joins in L, not only those joins which are defined in L_{part}.

Showing that p preserves all meets in L follows easily. Let $(a_i)_{i\in I}$ be any family of elements in L. Then $(a_i')_{i\in I}$ is also a family of elements in L, and we know

$$p\left(\bigvee_{i\in I} a_i'\right) = \bigvee_{i\in I} p(a_i'). \tag{13}$$

Recall that orthocomplementation is preserved by p and satisfies De Morgan's laws. Then,

$$p\left(\bigwedge_{i\in I} a_i\right) = p\left(\left[\bigvee_{i\in I} a_i'\right]'\right) = \left[p\left(\bigvee_{i\in I} a_i'\right)\right]' = \left[\bigvee_{i\in I} p(a_i')\right]' \tag{14}$$

$$= \bigwedge_{i\in I} [p(a_i')]' = \bigwedge_{i\in I} p(a_i'') = \bigwedge_{i\in I} p(a_i). \tag{15}$$

Thus, as p preserves all meets and joins in L, as well as all orthocomplements, p is in fact an isomorphism of OMLs, $p : L \to M$.

As $p : L_{part} \to M_{part}$ and $p : L \to M$ are the same on every element of L, and $\varphi : L \to M$ and the induced $\varphi : L_{part} \to M_{part}$ are the same on every element of L, then there is a bijective correspondence between isomorphisms $\varphi : L \to M$ and isomorphisms $p : L_{part} \to M_{part}$.

Note it is in the construction of an isomorphism of OMLs from an isomorphism of partial OMLs that the orthomodularity condition (in the form of Lemma 2.12) is essential. This result does not hold for arbitrary ortholattices, and is the reason we consider orthomodular lattices instead.

2.2.3 Example

We now consider a small OML L^*, and examine $\mathscr{B}(L^*)$ and L^*_{part}. Let L^* be as in Fig. 3. Consider the Boolean sublattices of L^*. The two-element Boolean lattice $B_0 = \{0, 1\}$ is a sublattice of L^*. The four element Boolean lattice (Fig. 1i) appears as a sublattice of L five times, as B_a, B_b, B_c, B_d, and B_e. The eight element Boolean lattice (Fig. 1iv) appears twice, as $B_{a,b,c}$ and $B_{c,d,e}$. This yields the context category shown in Fig. 3.

The partial orthomodular lattice L^*_{part} has the same elements as L^* but meets and joins only defined for compatible elements. Table 1 lists all pairs of elements in L^* that do not have a well-defined meet or join. For L^*, larger families of elements are compatible precisely when they contain none of the pairs in Table 1, though this is not the case in general. To see this, consider L^* with additional elements f and f' such that $B_{e,f,a}$ is a Boolean sublattice. Then, for elements a, c, and e, all pairwise meets and joins are defined but not the meet or join of all three elements.

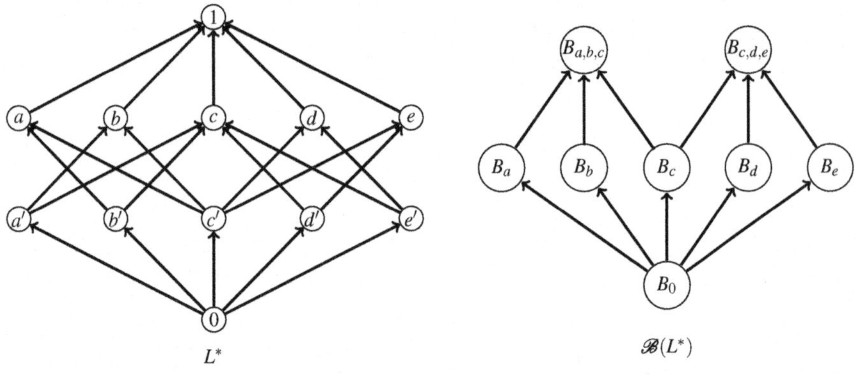

Fig. 3 An orthomodular lattice L^* with twelve elements and its context category $\mathscr{B}(L^*)$

Table 1 Pairs of elements that are not compatible in L^*; that is, pairwise meets and joins between these elements are not defined in L^*_{part}

a, d	a', d	b, d	b', d
a, d'	a', d'	b, d'	b', d'
a, e	a', e	b, e	b', e
a, e'	a', e'	b, e'	b', e'

2.3 *Stone Duality*

There is a well-known duality between Boolean algebras and Stone spaces. We recall the main definitions and fix the notation for later use, see also, e.g., [27].

Definition 2.14 A *Stone space* is a compact totally disconnected Hausdorff space.

There is a category **Stone** whose objects are Stones spaces and whose arrows are continuous functions between these topological spaces. Let $\{0, 1\}$ denote the two element Boolean algebra consisting of only a bottom element 0 and a top element 1.

Definition 2.15 The *Stone space* of a Boolean algebra B is the topological space Σ_B with set of elements

$$\Sigma_B = \{\lambda : B \to \{0, 1\} \mid \lambda \text{ is a Boolean algebra homomorphism,} \tag{16}$$

$$\text{also called a state or an ultrafilter in } B\} \tag{17}$$

and topology generated by a basis of, for all $b \in B$, the sets

$$U_b := \{\lambda \in \Sigma_B : \lambda(b) = 1\}. \tag{18}$$

We use the notation Σ_B instead of the more common Ω_B (or just Ω), since we will generalise the Stone space Σ_B to the spectral presheaf $\underline{\Sigma}^L$ of an orthomodular lattice L, and the notation $\underline{\Sigma}$ (or $\underline{\Sigma}^{\mathcal{N}}$) for the spectral presheaf of a von Neumann algebra $\mathcal{N}$ is already established. Moreover, the spectral presheaf is an object in a topos, and the subobject classifier in a topos is traditionally denoted Ω, which could lead to confusion.

Each $\lambda \in \Sigma_B$ is also called a *state of* B, and states correspond bijectively to *ultrafilters*: given λ, the set

$$F_\lambda := \{a \in B \mid \lambda(a) = 1\}$$

is an ultrafilter in B.

We can construct a contravariant functor Σ : **BA** $\to$ **Stone** from the category of Boolean algebras and Boolean algebra homomorphisms to the category of Stone spaces and continuous functions, given

(i) on objects: for each $B \in Ob(\mathbf{BA})$, let $\Sigma(B) := \Sigma_B$, the Stone space of B,
(ii) on arrows: for each morphism $(\phi : B' \to B) \in Arr(\mathbf{BA})$ of Boolean algebras, let

$$\Sigma(\phi) : \Sigma(B) \longrightarrow \Sigma(B') \tag{19}$$

$$\lambda \longmapsto \lambda \circ \phi. \tag{20}$$

Furthermore, to each Stone space X we can associate a canonical Boolean algebra. Let cl_X denote the set of subsets of X that are simultaneously closed and open, i.e. *clopen*. With meets given by intersections and joins given by unions, this is a Boolean algebra. Additionally considering morphisms, we obtain a functor $cl : \mathbf{Stone} \to \mathbf{BA}$, given

(i) on objects: for all $X \in Ob(\mathbf{Stone})$, let $cl(X) := cl_X$,
(ii) on arrows: for all $(f : X \to X') \in Arr(\mathbf{Stone})$, let

$$cl(f) : cl_{X'} \longrightarrow cl_X \tag{21}$$

$$S \longmapsto f^{(-1)}(S), \tag{22}$$

where $f^{(-1)}$ denotes the inverse image function of f.

Throughout, we will use the notation f^{-1} to denote function inverses and $f^{(-1)}$ to denote inverse image functions.

If we replace each clopen subset $S \subseteq X$ by its characteristic function $\chi_S : X \to \{0, 1\}$, we can write $cl(f)(\chi_S) = \chi_S \circ f$, which makes the morphism part of the functor $cl : \mathbf{Stone} \to \mathbf{BA}$ formally identical to the morphism part of $\Sigma : \mathbf{BA} \to \mathbf{Stone}$.

The two functors give rise to a dual equivalence between the categories **BA** and **Stone**:

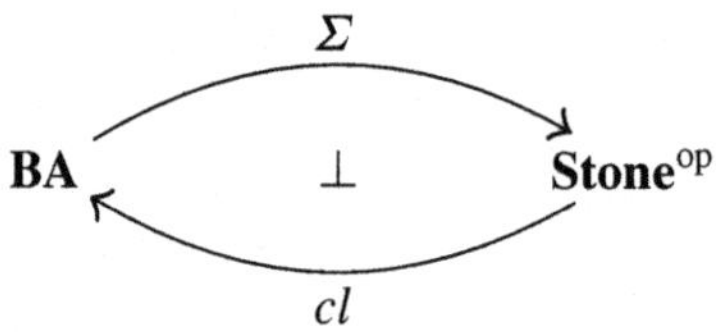

That is, there are natural isomorphisms $Bo : Id_{\mathbf{BA}} \to cl \circ \Sigma$ in **BA** and $St : Id_{\mathbf{Stone}} \to \Sigma \circ cl$ in **Stone**. In particular, the components of these isomorphisms are given as follows:

$$Bo_B : B \to cl(\Sigma_B) \tag{23}$$

$$b \mapsto U_b = \{\lambda \in \Sigma_B \mid \lambda(b) = 1\} \tag{24}$$

$$St_X : X \to \Sigma(cl(X)) \tag{25}$$

$$x \mapsto \lambda_x \tag{26}$$

where $\lambda_x : cl(X) \to \{0, 1\}$ is given by

$$\lambda_x(S) = \begin{cases} 1 : x \in S \\ 0 : x \notin S \end{cases} \tag{27}$$

Later, it will be of use to know the explicit components of $Bo^{-1} : cl \circ \Sigma \to Id_{\mathbf{BA}}$. Each component Bo_B^{-1} is a map from $cl(\Sigma_B)$ to B. Let S be any clopen subset in $cl(\Sigma_B)$. As S is closed and the subset of a compact space Σ_B, S is compact. As S is open, it can be written as the union of basic open sets. Compactness implies that this open cover has a finite subcover of basic open sets, which are of the form $U_b = \{\lambda \in \Sigma_B \mid \lambda(b) = 1\}$. That is, for some finite index set $J \subseteq B$,

$$S = \bigcup_{b \in J} U_b.$$

Let

$$s_* = \bigvee_{b \in J} b \in B. \tag{28}$$

Then, the action of Bo_B^{-1} is as follows.

$$Bo_B^{-1} : cl(\Sigma_B) \longrightarrow B \tag{29}$$

$$S \longmapsto s_*. \tag{30}$$

2.4 Complete Orthomodular Lattices and Their Boolean Substructure

All of the concepts defined above for orthomodular lattices also hold for complete orthomodular lattices (cOMLs). Let **cOML** denote the category of complete orthomodular lattices, and **cBA** the subcategory of complete Boolean algebras. Morphisms in both categories preserve all meets, all joins, and orthocomplements.

The following two results are immediate, and the Boolean algebras stated to exist are the same as in Fig. 2:

Proposition 2.16 *Every element a of complete orthomodular lattice L is in some complete Boolean subalgebra of L.*

Proposition 2.17 *In a complete orthomodular lattice L, for any elements $a, b \in L$ satisfying $a \leq b$ there is complete Boolean subalgebra of L containing both a and b.*

We define the complete analogue of the context category $\mathcal{B}(L)$:

Definition 2.18 The *complete context category* of a complete orthomodular lattice L, denoted $\mathscr{B}_c(L)$, is the poset of complete Boolean subalgebras of L, ordered by inclusion.

As before, when we consider the poset $\mathscr{B}_c(L)$ as a category, arrows will be denoted in the form $i_{B',B} : B' \hookrightarrow B$. We will usually drop the 'complete' and just call $\mathscr{B}_c(L)$ the context category of L.

Any morphism $\varphi : L \to M$ of cOMLs induces an order-preserving map $\tilde{\varphi} : \mathscr{B}_c(L) \to \mathscr{B}_c(M)$ between the context categories, where on each complete Boolean subalgebra B of L,

$$\tilde{\varphi}(B) := \varphi[B].$$

Clearly, $\varphi|_B : B \to \varphi[B]$ is a morphism of complete Boolean algebras.

Summing up, there is a functor $c\mathscr{B} : \mathbf{cOML} \to \mathbf{Pos}$, given

(i) on objects: for each $L \in \mathbf{cOML}$, let $c\mathscr{B}_L := \mathscr{B}_c(L)$, the (complete) context category of L,
(ii) on arrows: for each morphism $\varphi : L \to M$ of cOMLs, let

$$c\mathscr{B}(\varphi) := \tilde{\varphi} : \mathscr{B}_c(L) \longrightarrow \mathscr{B}_c(M) \tag{31}$$
$$B \longmapsto \varphi[B]. \tag{32}$$

There is also a complete version of the partial Boolean algebra L_{part} associated with an orthomodular lattice L:

Definition 2.19 Let L be a complete orthomodular lattice. The *partial complete orthomodular algebra* L^c_{part} *associated with* L has the same elements and orthocomplements as L, and has lattice operations $\bigvee$ and $\bigwedge$ inherited from L but only defined for (possibly infinite) families of elements $(a_i)_{i\in I}$ in L such that there is a $B \in \mathscr{B}_c(L)$ that contains a_i for all $i \in I$. Such families of elements are called *compatible families*.

Definition 2.20 A *morphism of partial complete orthomodular algebras* is a function $p : L^c_{part} \to M^c_{part}$ that preserves orthocomplements and existing meets and joins.

There is a category **pcOML** of partial cOMLs and morphisms of partial cOMLs between them. The complete versions of Lemma 2.12 and Proposition 2.13 are

Lemma 2.21 *If $a \leq b$ in complete orthomodular lattice L and $p : L_{part} \to M_{part}$ is a morphism of partial cOMLs, then $p(a) \leq p(b)$.*

Proposition 2.22 *Let L and M be complete orthomodular lattices, and L_{part} and M_{part} their associated partial complete orthomodular lattices. There is a bijective correspondence between isomorphisms $L \to M$ in* **cOML** *and isomorphisms $L_{part} \to M_{part}$ in* **pcOML**.

As before, Lemmas 2.21 and 2.22 depend on orthomodularity and do not hold for arbitrary complete ortholattices.

2.5 Stonean Spaces and Stone Duality for Complete Boolean Algebras

Just as there is a duality between Boolean algebras and Stone spaces, there is a duality between complete Boolean algebras and Stonean spaces.

Definition 2.23 A *Stonean space* is an extremely disconnected compact Hausdorff space.

In an extremely disconnected topological space, the closure of every open subspace is open and the interior of every closed subspace is closed. Recall that a Stone space is a totally disconnected compact Hausdorff space. As 'extremely disconnected' is a stronger condition than 'totally disconnected,' all Stonean spaces are also Stone spaces but not vice versa. The following lemmas characterise the relation between Stonean spaces and complete Boolean algebras.

Proposition 2.24 ([2]) *A Boolean algebra is complete if and only if its Stone space is Stonean.*

Proposition 2.25 ([28]) *The clopen subsets of a Stonean space form a complete Boolean algebra. Complementation is given by set-theoretic complementation, and meets and joins for a family of clopen subsets* $\{S_i \mid i \in I\}$ *are given by:*

$$\bigvee_{i \in I} S_i = \mathrm{cls}\Big(\bigcup_{i \in I} S_i\Big) \tag{33}$$

$$\bigwedge_{i \in I} S_i = \mathrm{int}\Big(\bigcap_{i \in I} S_i\Big) \tag{34}$$

Here cls *denotes the closure and* int *the interior of a subset with respect to the Stone topology.*

The correspondence between complete Boolean algebras and Stonean spaces can be extended to a dual equivalence of categories. There is a category **Stonean**, whose objects are Stonean spaces and whose morphisms are continuous open maps.

Proposition 2.26 ([29]) *There is a dual equivalence of categories between* **cBA** *and* **Stonean***:*

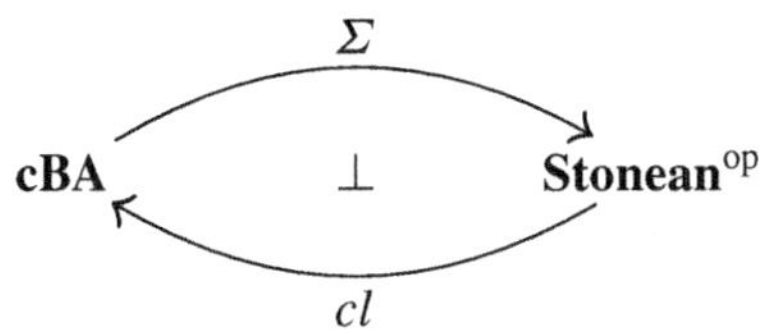

This duality is witnessed by the natural isomorphisms $Bo : Id_{\mathbf{cBA}} \to cl \circ \Sigma$ and $St : Id_{\mathbf{Stonean}} \to \Sigma \circ cl$ (where we use the same notation as in Stone duality for OMLs and BAs). Propositions 2.24 and 2.25, above, are consequences of this dual equivalence, but the references listed above provide explicit proofs that give more intuition as to why such results are true.

As corollaries of Proposition 2.26, we also have the following facts that will later be essential for extending the isomorphism result of Theorem 3.18 to complete orthomodular lattices.

Fact 2.27 *For every $B \in Ob(\mathbf{cBA})$, the component Bo_B of the natural isomorphism Bo is an isomorphism of cBAs.*

Proof $Bo_B : B \to cl(\Sigma_B)$ is an arrow in **cBA**.

Fact 2.28 *If $\eta : X \to Y$ is any continuous open map between Stonean spaces, then $cl(\eta)$ is a morphism of cBAs.*

Proof $cl(\eta) : cl(Y) \to cl(X)$ is an arrow in **cBA**.

2.6 Galois Connections and the Adjoint Functor Theorem for Posets

We briefly recall the definition of Galois connections and the adjoint functor theorem for posets (in fact, for complete lattices) for later use, see also, e.g., [18].

Definition 2.29 Let P and Q be posets. A pair of monotone maps $f : P \to Q$ and $g : Q \to P$ is a *Galois connection between P and Q* if, for all $p \in P$ and all $q \in Q$,

$$f(p) \leq q \text{ iff } p \leq g(q). \tag{35}$$

A Galois connection is written (f, g), where f is called the lower adjoint (or left adjoint) of g, and g is called the upper adjoint (or right adjoint) of f.

Proposition 2.30 (Adjoint functor theorem for posets) *Let P and Q be complete lattices and $f : P \to Q$ a monotone map. Then,*

1. *f preserves arbitrary joins if and only if f has an upper adjoint g, meaning (f, g) is a Galois connection. For all $q \in Q$, this map g is given by*

$$g(q) = \bigvee\{p \in P \mid f(p) \leq q\}. \tag{36}$$

2. *f preserves arbitrary meets if and only if f has a lower adjoint h, meaning (h, f) is a Galois connection. For all $q \in Q$, this map h is given by*

$$h(q) = \bigwedge\{p \in P \mid q \leq f(p)\}. \tag{37}$$

There are more general versions of this theorem, but the above form is what we will need.

Galois connections have several interesting properties that will be of use to us.

Proposition 2.31 *Let P and Q be complete lattices and $f : P \to Q$ and $g : Q \to P$ such that (f, q) is a Galois connection. The following hold:*

1. *f preserves arbitrary joins,*
2. *g preserves arbitrary meets,*
3. *For all $p \in P$, $p \leq (g \circ f)(p)$,*
4. *For all $q \in Q$, $(f \circ g)(q) \leq q$,*
5. *For all $p \in P$, $(f \circ g \circ f)(p) = f(p)$,*
6. *For all $q \in Q$, $(g \circ f \circ g)(q) = g(q)$.*

3 The Spectral Presheaf of an Orthomodular Lattice

We now define and examine the spectral presheaf of an orthomodular lattice, the main focus of this work.

3.1 Definition

A spectral presheaf was originally defined for von Neumann algebras as part of an alternate topos-based formulation of quantum mechanics. However, one can also define the spectral presheaf of an orthomodular lattice as follows.

Definition 3.1 Let L be an orthomodular lattice with context category $\mathscr{B}(L)$. The *spectral presheaf* $\underline{\Sigma}^L$ *of* L is the contravariant, **Set**-valued functor with domain $\mathscr{B}(L)$ given

(i) on objects: for all $B \in Ob(\mathscr{B}(L))$, let $\underline{\Sigma}^L_B := \Sigma_B$, the Stone space of B. Here, $\underline{\Sigma}^L_B$ denotes the component of $\underline{\Sigma}^L$ at B.
(ii) on arrows: for all $(i_{B'B} : B' \hookrightarrow B) \in Arr(\mathscr{B}(L))$, let

$$\underline{\Sigma}^L(i_{B'B}) : \underline{\Sigma}^L_B \longrightarrow \underline{\Sigma}^L_{B'} \tag{38}$$

$$\lambda \longmapsto \lambda|_{B'}. \tag{39}$$

Here, $\lambda|_{B'}$ denotes the restriction of λ to the subalgebra B'.

The spectral presheaf $\underline{\Sigma}^L$ of an OML L is an object in the functor category $\mathbf{Set}^{\mathscr{B}(L)^{\mathrm{op}}}$ of contravariant, **Set**-valued functors with domain $\mathscr{B}(L)$. The category $\mathbf{Set}^{\mathscr{B}(L)^{\mathrm{op}}}$ is a topos. In fact, we will shortly also consider another category in which $\underline{\Sigma}^L$ is an object, namely the category of **Stone**-valued presheaves. The advantage of considering $\underline{\Sigma}^L$ as a **Stone**-valued presheaf is that the components of $\underline{\Sigma}^L$ are explicitly seen as topological spaces in **Stone** rather than simply as sets, and the restriction maps $\underline{\Sigma}^L(i_{B'B})$ are continuous functions (in fact, surjective, continuous and open functions).

3.1.1 Example

Consider the orthomodular lattice L^* from Sect. 2.2.3. This lattice and its context category appear in Fig. 3. The spectral presheaf of L^* is a functor from $\mathcal{B}(L)$ to **Set**. Each Boolean subalgebra B of L^* is mapped to its Stone space Σ_B.

We now consider the action of the spectral presheaf on an inclusion map in $\mathcal{B}(L)$. We know that $B_a \subseteq B_{a,b,c}$, meaning there is an arrow $i_{B_a,B_{a,b,c}}$ corresponding to this in $\mathcal{B}(L)$. Note the Stone space of B_a has two elements, called λ_a and $\lambda_{a'}$, where $\lambda_a(a) = 1$ and $\lambda_{a'}(a) = 0$. Additionally, the Stone space of $B_{a,b,c}$ has three elements $\lambda_{a,b}$, $\lambda_{a,c}$, and $\lambda_{b,c}$, where the subscripts denote the two elements out of a, b, and c that are mapped to 1, while the third is mapped to 0; this completely determines the functions' values on all of $B_{a,b,c}$. Then, $\underline{\Sigma(L^*)}(i_{B_a,B_{a,b,c}})$ is a map r from $\Sigma_{B_{a,b,c}}$ to Σ_{B_a} whose action on elements of $\Sigma_{B_{a,b,c}}$ simply restricts the domains of the homomorphisms to B_a:

$$r(\lambda_{a,b}) = \lambda_a \tag{40}$$

$$r(\lambda_{a,c}) = \lambda_a \tag{41}$$

$$r(\lambda_{b,c}) = \lambda_{a'} \tag{42}$$

Note that as the inverse image of any open set of Σ_{B_a} is open in $\Sigma_{B_{a,b,c}}$, then this map r is in fact a continuous map when Σ_{B_a} and $\Sigma_{B_{a,b,c}}$ are considered as topological spaces rather than simply as sets. The images of other inclusion arrows under the spectral presheaf of L^* can be determined similarly and are also continuous maps between topological spaces.

3.2 *Maps Between Spectral Presheaves*

The next obvious step is to consider maps between spectral presheaves of orthomodular lattices. Specifically, if L and M are orthomodular lattices and $\varphi : L \to M$ is a morphism of OMLs, then we want to define some map Φ, determined by φ, from $\underline{\Sigma}^M$ to $\underline{\Sigma}^L$. This is done in two steps, below. The first step transforms $\underline{\Sigma}^M$ into a contravariant functor from $\mathcal{B}(L)$ to **Set**, while the second step then gives a natural transformation within $\mathbf{Set}^{\mathcal{B}(L)^{\mathrm{op}}}$ from this new functor to $\underline{\Sigma}^L$. In particular, such a map will be used to show that $L \cong M$ if and only if $\underline{\Sigma}^L \cong \underline{\Sigma}^M$, the goal of this section. This result implies that the spectral presheaf $\underline{\Sigma}^L$ determines up to isomorphism the orthomodular lattice L it comes from.

Step 1. Let $\varphi : L \to M$ be a morphism of OMLs. Recall from Sect. 2.2 that $\varphi : L \to M$ induces a monotone map $\tilde{\varphi} : \mathcal{B}(L) \to \mathcal{B}(M)$ between the context categories. This map $\tilde{\varphi}$ then induces a map between functor categories (topoi) $\tilde{\varphi}^* : \mathbf{Set}^{\mathcal{B}(M)^{\mathrm{op}}} \to \mathbf{Set}^{\mathcal{B}(L)^{\mathrm{op}}}$, given by 'pullback', that is, precomposition: for each $\underline{P} \in Ob(\mathbf{Set}^{\mathcal{B}(M)^{\mathrm{op}}})$, let

$$\tilde{\varphi}^*(\underline{P}) := \underline{P} \circ \tilde{\varphi},$$

which is the presheaf over $\mathscr{B}(L)$ with components

$$\forall B \in \mathscr{B}(L) : (\tilde{\varphi}^*(\underline{P}))_B = \underline{P}_{\tilde{\varphi}(B)}.$$

Those familiar with topos theory [20, 21, 23] will recognise the map $\tilde{\varphi}^*$ as the inverse image part the essential geometric morphism induced by the functor $\tilde{\varphi} : \mathscr{B}(L) \to \mathscr{B}(M)$ between the base categories of the topoi $\mathbf{Set}^{\mathscr{B}(L)^{\mathrm{op}}}$ and $\mathbf{Set}^{\mathscr{B}(M)^{\mathrm{op}}}$.

Thus, $\tilde{\varphi}^*$ maps $\underline{\Sigma}^M$ to some functor from $\mathscr{B}(L)$ to $\mathbf{Set}$, which is not necessarily $\underline{\Sigma}^L$. However, since a map from $\underline{\Sigma}^M$ to $\underline{\Sigma}^L$ is desired, it is now necessary to define a way to transform $\tilde{\varphi}^*(\underline{\Sigma}^M)$ to $\underline{\Sigma}^L$ within the functor category $\mathbf{Set}^{\mathscr{B}(L)^{\mathrm{op}}}$. This is done via a natural transformation as follows.

Step 2. Let $B \in \mathscr{B}(L)$. A morphism $\varphi : L \to M$ of OMLs induces a Boolean algebra homomorphism $\varphi|_B : B \to \tilde{\varphi}(B)$. By Stone duality, this corresponds to a unique morphism $\Sigma_{\tilde{\varphi}(B)} \to \Sigma_B$ of Stone spaces, sending λ to $\lambda \circ \varphi|_B$. Note that $\Sigma_{\tilde{\varphi}(B)}$ is the component of $\tilde{\varphi}^*(\underline{\Sigma}^M)$ at $B \in \mathscr{B}(L)$, and Σ_B is the component of $\underline{\Sigma}^L$ at B. Hence, for each $B \in \mathscr{B}(L)$ we have a map

$$\zeta_{\varphi,B} : \tilde{\varphi}^*(\underline{\Sigma}^M)_B = \underline{\Sigma}^M_{\tilde{\varphi}(B)} \longrightarrow \underline{\Sigma}^L_B = \Sigma_B \tag{43}$$

$$\lambda \longmapsto \lambda \circ \varphi|_B. \tag{44}$$

Lemma 3.2 *The maps $\zeta_{\varphi,B}$, where $B \in \mathscr{B}(L)$, are the components of a natural transformation between functors in* $\mathbf{Set}^{\mathscr{B}(L)^{\mathrm{op}}}$*:*

$$\zeta_\varphi : \tilde{\varphi}^*(\underline{\Sigma}^M) \longrightarrow \underline{\Sigma}^L. \tag{45}$$

Proof Recall

$$\tilde{\varphi}^*(\underline{\Sigma}^M)_B = \underline{\Sigma}^M_{\tilde{\varphi}(B)} = \Sigma_{\tilde{\varphi}(B)} \tag{46}$$

$$\tilde{\varphi}^*(\underline{\Sigma}^M)(i_{B',B}) = \underline{\Sigma}^M(i_{\tilde{\varphi}(B'),\tilde{\varphi}(B)}) = r_{\tilde{\varphi}(B),\tilde{\varphi}(B')} \tag{47}$$

For $B', B \in \mathscr{B}(L)$, where $i_{B',B}$ is an inclusion arrow, to show ζ_φ is a natural transformation it is necessary to show that the following diagram commutes:

$$\begin{array}{ccc}
\underline{\Sigma}^M_{\tilde{\varphi}(B')} & \xleftarrow{\underline{\Sigma}^M(i_{\tilde{\varphi}(B'),\tilde{\varphi}(B)})} & \underline{\Sigma}^M_{\tilde{\varphi}(B)} \\
\Big\downarrow \zeta_{\varphi,B'} & & \Big\downarrow \zeta_{\varphi,B} \\
\underline{\Sigma}^L_{B'} & \xleftarrow{\underline{\Sigma}^L(i_{B',B})} & \underline{\Sigma}^L_B
\end{array}$$

Let $\lambda : \tilde{\varphi}(B) \to \{0, 1\}$ be any element of $\underline{\Sigma}^M_{\tilde{\varphi}(B)}$. Then,

$$\left[\zeta_{\varphi,B'} \circ \underline{\Sigma}^M(i_{\tilde{\varphi}(B'),\tilde{\varphi}(B)})\right](\lambda) = \zeta_{\varphi,B'}(\lambda|_{\tilde{\varphi}(B')}) \tag{48}$$
$$= \lambda|_{\tilde{\varphi}(B')} \circ \varphi|_{B'} \tag{49}$$
$$= (\lambda \circ \varphi)|_{B'}, \tag{50}$$
$$\left[\underline{\Sigma}^L(i_{B',B}) \circ \zeta_{\varphi,B}\right](\lambda) = \underline{\Sigma}^L(i_{B',B})(\lambda \circ \varphi|_B) \tag{51}$$
$$= (\lambda \circ \varphi|_B)|_{B'} \tag{52}$$
$$= (\lambda \circ \varphi)|_{B'}. \tag{53}$$

Thus, the diagram commutes and ζ_φ is a natural transformation.

The two maps $\tilde{\varphi}^*$ and ζ_φ defined above can be combined to give, for any homomorphism $\varphi : L \to M$, a map from $\underline{\Sigma}^M$ to $\underline{\Sigma}^L$, written $\Phi = \langle \tilde{\varphi}^*, \zeta_\varphi \rangle$. As $\tilde{\varphi}^*$ is completely determined by $\tilde{\varphi}$ (as is ζ_φ), this can also equivalently be written $\Phi = \langle \tilde{\varphi}, \zeta_\varphi \rangle$. Note that the process described above is not a standard composition $\zeta_\varphi \circ \tilde{\varphi}^*$, as these two maps are not within the same category; $\tilde{\varphi}^*$ is a map between topoi $\mathbf{Set}^{\mathcal{B}(M)^{\mathrm{op}}}$ and $\mathbf{Set}^{\mathcal{B}(L)^{\mathrm{op}}}$, while ζ_φ is a natural transformation within $\mathbf{Set}^{\mathcal{B}(L)^{\mathrm{op}}}$.

So far, we have shown that every morphism $\varphi : L \to M$ of OMLs induces a morphism $\langle \tilde{\varphi}, \zeta_\varphi \rangle : \underline{\Sigma}^M \to \underline{\Sigma}^L$ in the 'opposite' direction between their spectral presheaves. In order to understand this properly as a contravariant functor, we will show this is an example of a more general construction and define a suitable category of presheaves over varying base categories and their morphisms.

3.3 *The Category of $\mathcal{D}$-Valued Presheaves*

The rather unintuitive definition of a map between spectral presheaves, above, can in fact be understood best as an arrow in a suitable category **Presh**(**Stone**). We now define and explore such presheaf categories. This subsection and the next considerably expand some work done by the second author in [13].

First, we develop some general theory of presheaf categories over varying base categories with values in a category $\mathcal{D}$. Since the base categories of such presheaves are not the same in general, the morphisms between the presheaves are not just natural transformations.

Let $H : \mathcal{K} \to \mathcal{J}$ be a functor between small categories. For clarity, the action of H on an object K of $\mathcal{K}$ will be written as $H(K)$ rather than H_K. For any category $\mathcal{L}$, H induces a "pullback" map H^*, analogous to $\tilde{\varphi}^*$, above, from $\mathcal{L}^{\mathcal{J}}$ to $\mathcal{L}^{\mathcal{K}}$ which acts by precomposing by H. That is, on objects $R \in \mathcal{L}^{\mathcal{J}}$,

$$H^*R = R \circ H : \mathcal{K} \to \mathcal{L}. \tag{54}$$

Specifically, for any $K \in \mathcal{K}$,

$$(H^*R)_K = (R \circ H)_K = R_{H(K)}. \tag{55}$$

This is captured by the following commutative diagram for each $R \in \mathscr{L}^{\mathscr{J}}$:

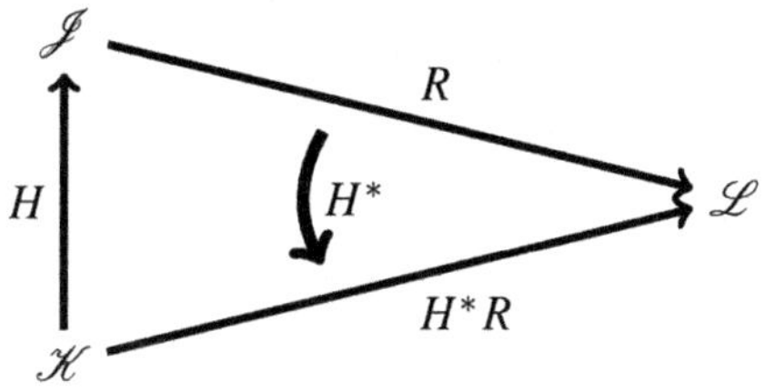

We can additionally show H^* satisfies the even stronger property of being a functor from $\mathscr{L}^{\mathscr{J}}$ to $\mathscr{L}^{\mathscr{K}}$ by defining its action on arrows of $\mathscr{L}^{\mathscr{J}}$ as well. An arrow in $\mathscr{L}^{\mathscr{J}}$ is a natural transformation $\tau : R \to R'$, for $R, R' : \mathscr{J} \to \mathscr{L}$. Applying H^* produces a natural transformation $H^*\tau : H^*R \to H^*R'$ in $\mathscr{L}^{\mathscr{K}}$, where for each $K \in \mathscr{K}$,

$$(H^*\tau)_K = \tau_{H(K)}. \tag{56}$$

Checking the necessary diagram shows that $H^*\tau$ is a valid natural transformation precisely because τ is.

Proposition 3.3 $H^* : \mathscr{L}^{\mathscr{J}} \to \mathscr{L}^{\mathscr{K}}$ *is a functor.*

Proof One can verify, using the definition of H^*, that it preserves identity arrows and composition.

The following elementary facts about H^* follow from the definition of H^* and will be useful in later proofs.

Fact 3.4 *For* $H : \mathscr{J}' \to \mathscr{J}$ *and induced functor* $H^* : \mathscr{L}^{\mathscr{J}} \to \mathscr{L}^{\mathscr{J}'}$, $\tilde{H} : \mathscr{J}'' \to \mathscr{J}'$ *and induced functor* $\tilde{H}^* : \mathscr{L}^{\mathscr{J}'} \to \mathscr{L}^{\mathscr{J}''}$,

$$(H \circ \tilde{H})^* = \tilde{H}^* \circ H^*. \tag{57}$$

Fact 3.5 *Suppose* $H : \mathscr{K} \to \mathscr{J}$, $R : \mathscr{J} \to \mathscr{L}$, *and* $S : \mathscr{L} \to \mathscr{M}$. *Then*

$$H^*(S \circ R) = S \circ (H^*R). \tag{58}$$

That is, the following diagram commutes:

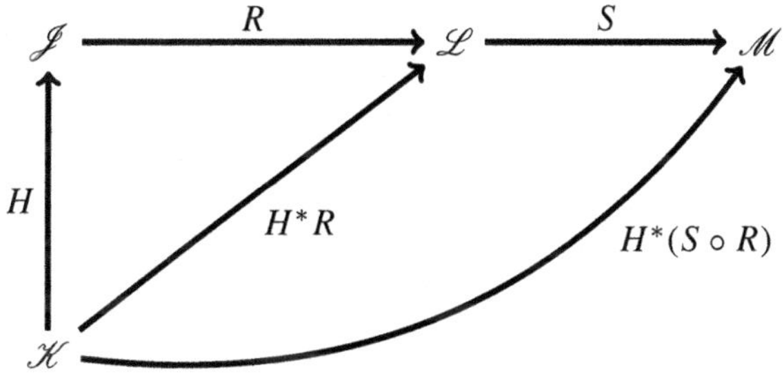

Fact 3.6 *Let* $Id : \mathcal{J} \to \mathcal{J}$ *be the identity functor on category* $\mathcal{J}$*. Let* $R, R' \in \mathcal{L}^{\mathcal{J}}$*, and let* $\eta : R \to R'$ *be a natural transformation. Then* $Id^* R = R$ *and* $Id^*\eta = \eta : R \to R'$*.*

We now proceed to use the functor H^* to define a presheaf category.

Definition 3.7 The category **Presh**($\mathcal{D}$) of $\mathcal{D}$-valued presheaves has as its objects functors (presheaves) of the form $\underline{P} : \mathcal{J} \to \mathcal{D}^{op}$, where $\mathcal{J}$ is a small category. Arrows are pairs

$$\langle H, \eta \rangle : (\underline{P}' : \mathcal{J}' \to \mathcal{D}^{op}) \to (\underline{P} : \mathcal{J} \to \mathcal{D}^{op}), \tag{59}$$

where $H : \mathcal{J} \to \mathcal{J}'$ is a functor and $\eta : H^*\underline{P}' \to \underline{P}$ is a natural transformation in $(\mathcal{D}^{op})^{\mathcal{J}}$:

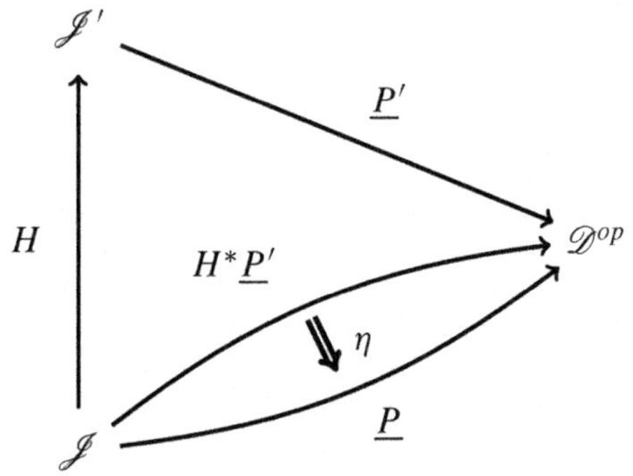

Let $\underline{P}_i : \mathcal{J}_i \to \mathcal{D}^{op}$, for $i = 1, 2, 3$, be functors. Given two arrows $\langle \tilde{H}, \tilde{\eta} \rangle : \underline{P}_3 \to \underline{P}_2$ and $\langle H, \eta \rangle : \underline{P}_2 \to \underline{P}_1$, the composition $\langle H, \eta \rangle \circ \langle \tilde{H}, \tilde{\eta} \rangle : \underline{P}_3 \to \underline{P}_1$ is given by

$$\langle H, \eta \rangle \circ \langle \tilde{H}, \tilde{\eta} \rangle = \langle \tilde{H} \circ H, \eta \circ H^*\tilde{\eta} \rangle, \tag{60}$$

where $\eta \circ H^*\tilde{\eta}$ denotes vertical composition of natural transformations. The intuition behind this definition of composition can be seen in the following diagram.

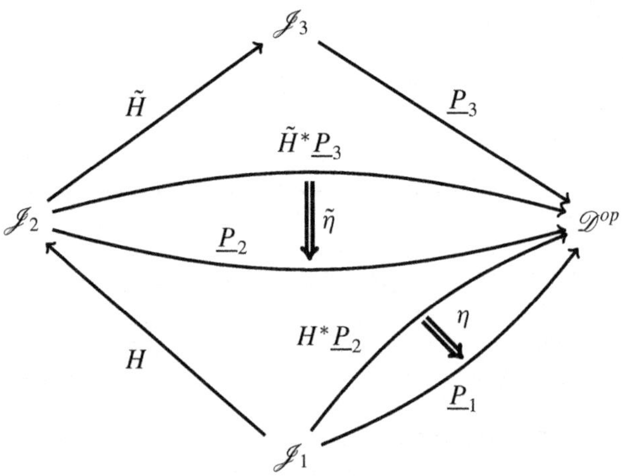

Lemma 3.8 **Presh**$(\mathscr{D})$ *is a category.*

Proof First, it is necessary to show that composition as given above is well-defined, that is, that $\langle H, \eta\rangle \circ \langle \tilde{H}, \tilde{\eta}\rangle$ is a valid arrow from $\underline{P}_3$ to $\underline{P}_1$. Consider the diagram above. Clearly $\tilde{H} \circ H$ is a functor from $\mathscr{J}_1$ to $\mathscr{J}_3$, as required. Then, the natural transformation $\eta \circ H^*\tilde{\eta}$ is from $H^*(\tilde{H}^*\underline{P}_3)$ to $H^*\underline{P}_2$ to $\underline{P}_1$ in $(\mathscr{D}^{op})^{\mathscr{J}_1}$. As $H^* \circ \tilde{H}^* = (\tilde{H} \circ H)^*$ by Fact 3.4, it follows that $\eta \circ H^*\tilde{\eta} : (\tilde{H} \circ H)^*\underline{P}_3 \to \underline{P}_1$, as required.

It is also necessary to show that this composition is associative, which will be done algebraically. Suppose $\underline{P}_4 : \mathscr{J}_4 \to \mathscr{D}^{op}$ is a presheaf and $\hat{H} : \mathscr{J}_3 \to \mathscr{J}_4$ is a functor, and that $\langle \hat{H}, \hat{\eta}\rangle$ is an arrow from $\underline{P}_4$ to $\underline{P}_3$. Then, by the definition of composition, the functoriality of H^*, the associativity of functors and natural transformations, and Fact 3.4,

$$\left(\langle H, \eta\rangle \circ \langle \tilde{H}, \tilde{\eta}\rangle\right) \circ \langle \hat{H}, \hat{\eta}\rangle = \langle \tilde{H} \circ H, \eta \circ H^*\tilde{\eta}\rangle \circ \langle \hat{H}, \hat{\eta}\rangle \tag{61}$$

$$= \langle \hat{H} \circ \left(\tilde{H} \circ H\right), \left(\eta \circ H^*\tilde{\eta}\right) \circ (\tilde{H} \circ H)^*\hat{\eta}\rangle \tag{62}$$

$$= \langle \hat{H} \circ \left(\tilde{H} \circ H\right), \eta \circ \left(H^*\tilde{\eta} \circ (H^* \circ \tilde{H}^*)\hat{\eta}\right)\rangle \tag{63}$$

$$= \langle \left(\hat{H} \circ \tilde{H}\right) \circ H, \eta \circ H^* \left(\tilde{\eta} \circ \tilde{H}^*\hat{\eta}\right)\rangle \tag{64}$$

$$= \langle H, \eta\rangle \circ \langle \hat{H} \circ \tilde{H}, \tilde{\eta} \circ \tilde{H}^*\hat{\eta}\rangle \tag{65}$$

$$= \langle H, \eta\rangle \circ \left(\langle \tilde{H}, \tilde{\eta}\rangle \circ \langle \hat{H}, \hat{\eta}\rangle\right) \tag{66}$$

Finally, it remains only to show that every object $\underline{P} : \mathscr{J} \to \mathscr{D}^{op}$ of **Presh**$(\mathscr{D})$ has an identity arrow. If $Id_J : \mathscr{J} \to \mathscr{J}$ is the identity functor on J and $id_{\underline{P}} : \underline{P} \to \underline{P}$ is the identity natural transformation on $\underline{P}$, then $\langle Id_J, id_{\underline{P}}\rangle$ is the appropriate identity arrow on $\underline{P}$, which can be easily verified using the definitions above. Thus, **Presh**$(\mathscr{D})$ is a valid category.

It is possible to view spectral presheaves and spectral presheaf maps as defined in the previous subsection as a subcategory of **Presh**(**Set**). Specifically, it is the subcategory with objects and arrows determined as follows.

$$\text{Objects: } \{\underline{\Sigma}^L : \mathscr{B}(L) \to \textbf{Set} \mid L \text{ is an orthomodular lattice.}\} \tag{67}$$

$$\text{Morphisms: } \{\langle \tilde{\varphi}, \zeta_\varphi\rangle \mid \varphi \text{ is an orthomodular lattice homomorphism.}\} \tag{68}$$

The latter is an arrow in **Presh**(**Set**), depicted here:

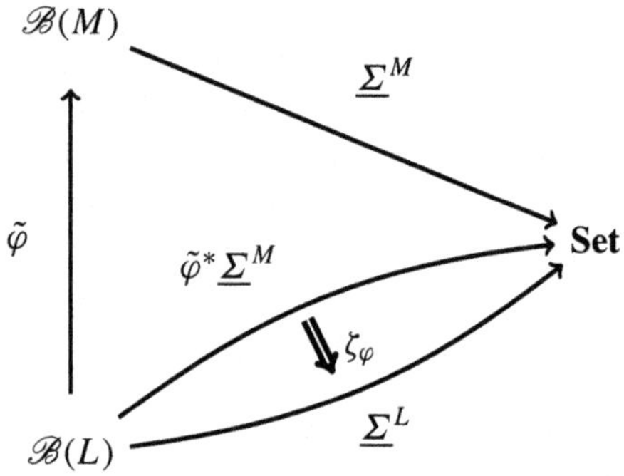

In fact, this subcategory is the image of a functor; there is a contravariant functor $SP : \mathbf{OML} \to \mathbf{Presh}(Set)$ which acts as follows for all orthomodular lattices L and all orthomodular lattice homomorphisms $\varphi : L \to M$:

$$SP(L) = \underline{\Sigma}^L \tag{69}$$

$$SP(\varphi) = \langle \tilde{\varphi}, \zeta_\varphi \rangle : \underline{\Sigma}^M \to \underline{\Sigma}^L. \tag{70}$$

Proposition 3.9 *SP is a functor.*

Proof First, we must check that SP preserves identities. Suppose $i : L \to L$ is the identity orthomodular lattice homomorphism on L. Then, $\tilde{i} : \mathscr{B}(L) \to \mathscr{B}(L)$ is also clearly the identity functor on category $\mathscr{B}(L)$. Furthermore, ζ_i has components given by

$$\zeta_{i,B} : \underline{\Sigma}^L_B \to \underline{\Sigma}^L_B \tag{71}$$

$$\lambda \mapsto \lambda \circ i = \lambda \tag{72}$$

Thus, as each $\zeta_{i,B}$ is just the identity map on $\underline{\Sigma}^L_B$ in **Set**, it follows that ζ_i is the identity natural transformation on $\underline{\Sigma}^L$. Thus, $\langle \tilde{i}, \zeta_i \rangle$ is the identity arrow of $\underline{\Sigma}^L$ in category **Presh**(**Set**).

Next, it is necessary to show that SP preserves composition. Suppose $\varphi : L \to M$ and $\rho : M \to N$ are orthomodular lattice homomorphisms. Recalling that SP is contravariant, we wish to show that $SP(\rho \circ \varphi) = SP(\varphi) \circ SP(\rho)$. Consider the following diagram, which depicts arrows $SP(\varphi) : \underline{\Sigma}^M \to \underline{\Sigma}^L$ and $SP(\rho) : \underline{\Sigma}^N \to \underline{\Sigma}^M$ in **Presh**(**Set**):

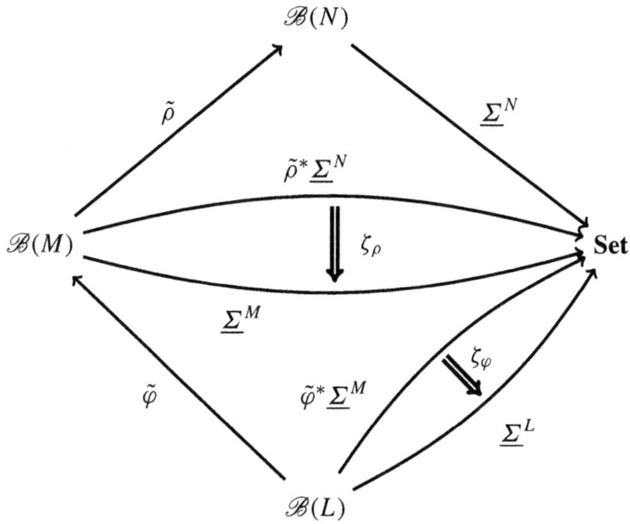

Recall the definition of composition in **Presh**(**Set**):

$$SP(\varphi) \circ SP(\rho) = \langle \tilde{\varphi}, \zeta_\varphi \rangle \circ \langle \tilde{\rho}, \zeta_\rho \rangle = \langle \tilde{\rho} \circ \tilde{\varphi}, \zeta_\varphi \circ \tilde{\varphi}^* \zeta_\rho \rangle \tag{73}$$

Note also that the map from $\mathscr{B}(L)$ to $\mathscr{B}(N)$ induced by the composition $\rho \circ \varphi$ is precisely $\tilde{\rho} \circ \tilde{\varphi}$, which follows from the definition in Sect. 2.2 of such induced maps. Thus,

$$SP(\rho \circ \varphi) = \langle \tilde{\rho} \circ \tilde{\varphi}, \zeta_{\rho \circ \varphi} \rangle$$

It simply remains to show that the natural transformations $\zeta_\varphi \circ \tilde{\varphi}^* \zeta_\rho$ and $\zeta_{\rho\circ\varphi}$ from presheaf $\tilde{\varphi}^* \tilde{\rho}^* \underline{\Sigma}^N$ to presheaf $\underline{\Sigma}^L$ in $\mathbf{Set}^{\mathscr{B}(L)^{\mathrm{op}}}$ are equal. Consider any element $B \in \mathscr{B}(L)$. Recall, from Fact 3.4 and previous definitions, that

$$(\tilde{\varphi}^* \tilde{\rho}^* \underline{\Sigma}^N)_B = ((\tilde{\rho} \circ \tilde{\varphi})^* \underline{\Sigma}^N)_B = \underline{\Sigma}^N_{(\tilde{\rho}\circ\tilde{\varphi})(B)} = \Sigma_{(\tilde{\rho}\circ\tilde{\varphi})(B)}. \tag{74}$$

The action of the component at B of natural transformation $\zeta_{\rho\circ\varphi}$ is, by the definition of ζ,

$$\zeta_{\rho\circ\varphi} : \Sigma_{(\tilde{\rho}\circ\tilde{\varphi})(B)} \to \Sigma_B \tag{75}$$

$$\lambda \mapsto \lambda \circ (\rho \circ \varphi)|_B \tag{76}$$

Now consider natural transformation $\zeta_\varphi \circ \tilde{\varphi}^* \zeta_\rho$.

$$\left(\zeta_\varphi \circ \tilde{\varphi}^* \zeta_\rho\right)_B = \zeta_{\varphi,B} \circ (\tilde{\varphi}^* \zeta_\rho)_B = \zeta_{\varphi,B} \circ \zeta_{\rho,\tilde{\varphi}(B)}$$

The action of this composition is given as follows.

$$\zeta_{\varphi,B} \circ \zeta_{\rho,\tilde{\varphi}(B)} : \underline{\Sigma}^N_{(\tilde{\rho}\circ\tilde{\varphi})(B)} \rightarrow \underline{\Sigma}^M_{\tilde{\varphi}(B)} \rightarrow \underline{\Sigma}^L_B \tag{77}$$

$$\lambda \mapsto \lambda \circ \rho|_{\tilde{\varphi}(B)} \mapsto \lambda \circ \rho|_{\tilde{\varphi}(B)} \circ \varphi|_B \tag{78}$$

$$= \lambda \circ (\rho \circ \varphi)|_B \tag{79}$$

As the two natural transformations we are considering have the same component for every $B \in \mathscr{B}(L)$, then they must be the same natural transformation, implying SP preserves composition and is a functor.

Thus, the image in **Presh**(**Set**) of functor SP, consisting of the spectral presheaves of orthomodular lattices and the spectral presheaf maps between them, is a category. Of note, the functor SP is neither full nor faithful.

3.4 The Category of $\mathscr{C}$-Valued Copresheaves

Dual to the notion of a presheaf is that of a copresheaf. This definition yields another category **Copresh**($\mathscr{C}$) as follows.

Definition 3.10 Let $\mathscr{C}$ be a category. The category **Copresh**($\mathscr{C}$) of $\mathscr{C}$-valued copresheaves has as its objects functors (copresheaves) of the form $\overline{Q} : \mathscr{J} \rightarrow \mathscr{C}$, where $\mathscr{J}$ is a small category. Arrows are pairs

$$\langle I, \theta \rangle : (\overline{Q} : \mathscr{J} \rightarrow \mathscr{C}) \rightarrow (\overline{Q}' : \mathscr{J}' \rightarrow \mathscr{C}), \tag{80}$$

where $I : \mathscr{J} \rightarrow \mathscr{J}'$ is a functor and $\theta : \overline{Q} \rightarrow I^*\overline{Q}'$ is a natural transformation in $\mathscr{C}^{\mathscr{J}}$:

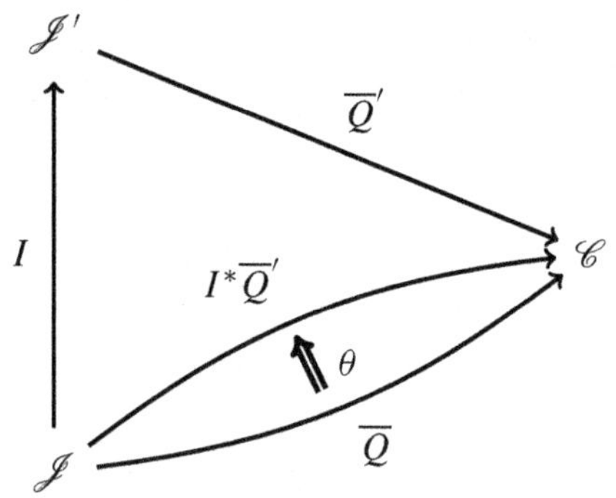

Let $\overline{Q}_i : \mathscr{J}_i \rightarrow \mathscr{C}$, for $i = 1, 2, 3$, be functors. Given two arrows $\langle I, \theta \rangle : \overline{Q}_1 \rightarrow \overline{Q}_2$ and $\langle \tilde{I}, \tilde{\theta} \rangle : \overline{Q}_2 \rightarrow \overline{Q}_3$, the composition $\langle \tilde{I}, \tilde{\theta} \rangle \circ \langle I, \theta \rangle : \overline{Q}_1 \rightarrow \overline{Q}_3$ is given by

$$\langle \tilde{I}, \tilde{\theta} \rangle \circ \langle I, \theta \rangle = \langle \tilde{I} \circ I, (I^*\tilde{\theta}) \circ \theta \rangle, \tag{81}$$

where $(I^*\tilde{\theta}) \circ \theta$ denotes vertical composition of natural transformations within functor category $\mathscr{C}^{\mathscr{I}_1}$. The intuition behind this definition of composition can be seen in the following diagram.

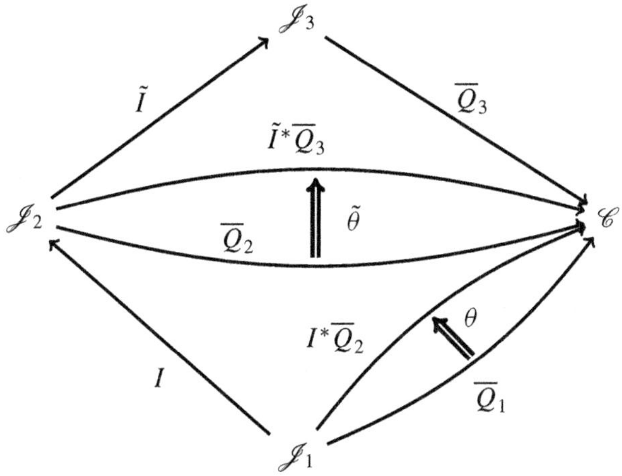

Just as with category **Presh**($\mathscr{D}$), it follows that **Copresh**($\mathscr{C}$) is a well-defined category, though this proof is omitted due to its similarities to the proof above.

3.5 Dual Equivalences and Stone Duality

3.5.1 Lifting Dual Equivalences to Presheaf and Copresheaf Categories

Having defined the categories of $\mathscr{D}$-valued presheaves and $\mathscr{D}$-valued copresheaves and their morphisms, we now turn to the question of how such categories relate if $\mathscr{C}$ and $\mathscr{D}$ are dually equivalent. In [13], the following result was proven:

Lemma 3.11 *Let $\mathscr{C}$, $\mathscr{D}$ be two categories that are dually equivalent,*

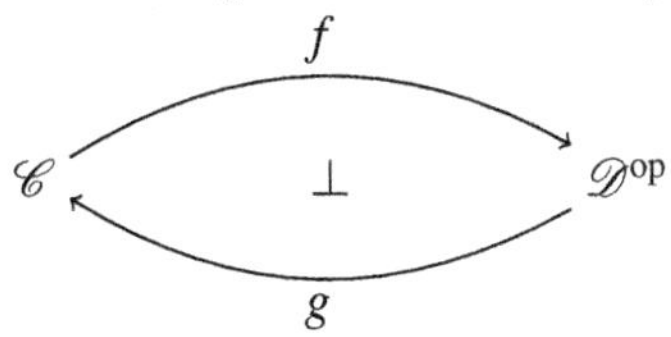

Then there is a dual equivalence

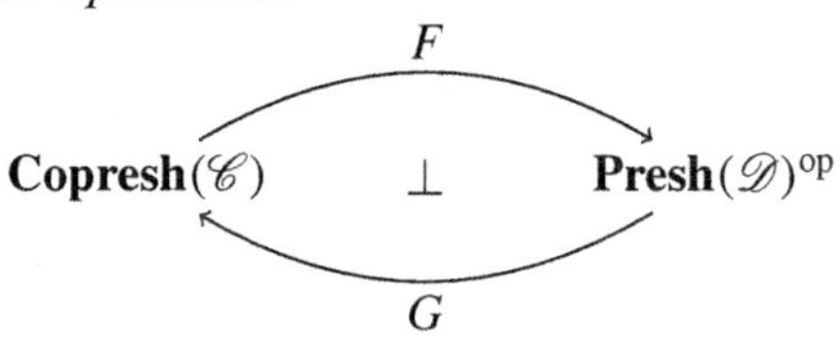

The actions of the functors F and G are defined in the proof of the above theorem in the following way. First, consider G : **Presh**$(\mathscr{D}) \to$ **Copresh**$(\mathscr{C})$. If $\underline{P} : \mathscr{J} \to \mathscr{D}^{op}$ is an object of **Presh**$(\mathscr{D})$, then $G(\underline{P}) : \mathscr{J} \to \mathscr{C}$ is the (covariant) functor $g \circ \underline{P}$. That is, for all objects J and arrows $a : J' \to J$ in $\mathscr{J}$,

$$G(\underline{P})_J = (g \circ \underline{P})_J = g(\underline{P}_J) \in Ob(\mathscr{C}) \tag{82}$$

$$G(\underline{P})(a) = (g \circ \underline{P})(a) \in Morph(\mathscr{C}) \tag{83}$$

It is now time to consider the action of G on morphisms on **Presh**$(\mathscr{D})$. Let

$$\langle H, \eta \rangle : (\underline{P}' : \mathscr{J}' \to \mathscr{D}^{op}) \to (\underline{P} : \mathscr{J} \to \mathscr{D}^{op}) \tag{84}$$

be an arrow in **Presh**$(\mathscr{D})$. Then, as G is contravariant, $G(\langle H, \eta \rangle)$ is an arrow in **Copresh**$(\mathscr{C})$ from $G(\underline{P}) = g \circ \underline{P}$ to $G(\underline{P}') = g \circ \underline{P}'$. Specifically,

$$G(\langle H, \eta \rangle) = \langle H, g(\eta) \rangle, \tag{85}$$

where $g(\eta) : g \circ \underline{P} \to H^*(g \circ \underline{P}')$ is a natural transformation with components

$$(g(\eta))_J = g(\eta_J) : (g \circ \underline{P})_J \to (g \circ H^* \underline{P}')_J. \tag{86}$$

Because g is a contravariant functor, components $g(\eta)_J$ are arrows in the opposite direction of components η_J. The following diagram is not a commutative diagram, but is intended to give some visual intuition behind the definitions above and why $\langle H, g(\eta) \rangle : G(\underline{P}) \to G(\underline{P}')$ is in fact a morphism in **Copresh**$(\mathscr{C})$.

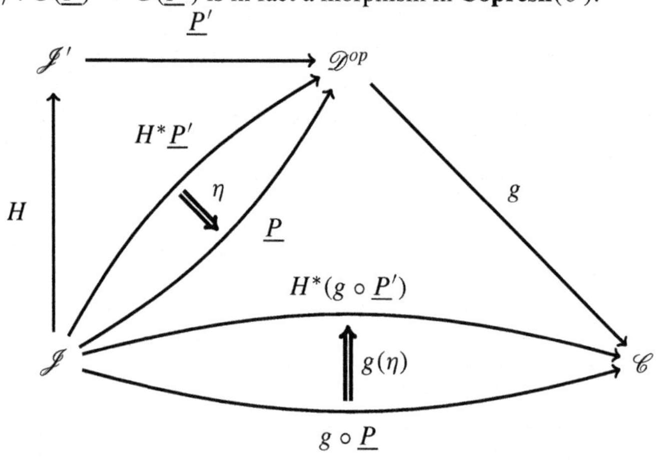

In [13], the action of contravariant functor F : **Copresh**$(\mathscr{C}) \to$ **Presh**$(\mathscr{D})$ is defined as follows. On an object $\overline{Q} : \mathscr{J} \to \mathscr{C}$ of **Copresh**$(\mathscr{C})$, F acts as post-composition by $f : \mathscr{C} \to \mathscr{D}^{op}$. That is,

$$F(\overline{Q}) = f \circ \overline{Q} : \mathscr{J} \to \mathscr{C} \to \mathscr{D}^{op}. \tag{87}$$

On morphisms $\langle I, \theta\rangle : (\overline{Q} : \mathcal{J} \to \mathcal{C}) \to (\overline{Q}' : \mathcal{J}' \to \mathcal{C})$ in **Copresh**$(\mathcal{C})$, contravariant functor F acts as follows:

$$F(\langle I, \theta\rangle) = \langle I, f(\theta)\rangle, \tag{88}$$

where $f(\theta) : I^*(F(\overline{Q}')) \to F(\overline{Q})$ is a natural transformation with components, for each $J \in \mathcal{J}$, given by

$$f(\theta)_J = f(\theta_J) : (f \circ I^*\overline{Q}')_J \to (f \circ \overline{Q})_J \tag{89}$$

As the functor $f : \mathcal{C} \to \mathcal{D}^{op}$ is contravariant, the natural transformations $f(\theta)$ and θ are in opposite directions. The following is again not a commutative diagram, but captures the intuition behind this definition of F.

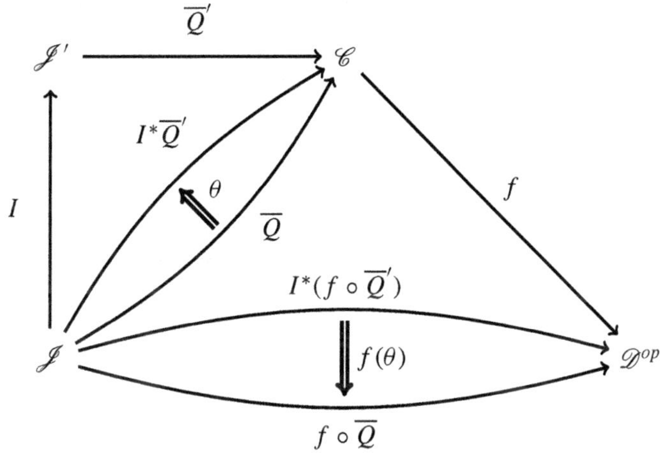

3.5.2 Stone Duality, the Spectral Presheaf, and the Bohrification of an OML

Recall there is a dual equivalence between the category **BA** of Boolean algebras and the category **Stone** of Stone spaces, given by functors Σ : **BA** $\to$ **Stone**^{op} and cl : **Stone**$^{\text{op}} \to$ **BA**. By Lemma 3.11, there is then a duality

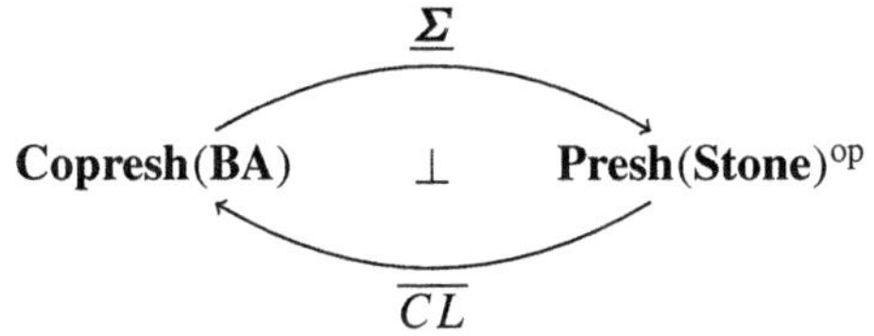

We now define the actions of $\overline{CL}$ and $\underline{\Sigma}$ on so-called Bohrifications in the category **Copresh**(**BA**) and spectral presheaves in the category **Presh**(**Stone**). The Bohrification of a unital C^*-algebra was introduced by Heunen, Landsman, and Spitters in [8]. Our construction for orthomodular lattices is analogous, it is the tautological inclusion copresheaf:

Definition 3.12 For an orthomodular lattice L, the *Bohrification* $\overline{\mathscr{L}}$ of L is the copresheaf from $\mathscr{B}(L)$ to **BA** given by:

$$\text{On objects: } \overline{\mathscr{L}}_B = B \tag{90}$$

$$\text{On morphisms: } \overline{\mathscr{L}}(i_{B',B}) = inc_{B',B}, \text{ the inclusion homomorphism} \tag{91}$$

Recall $i_{B',B}$ denotes the arrow in poset $\mathscr{B}(L)$ from B' to B which signifies that $B' \subseteq B$, while $inc_{B',B}$ denotes the Boolean algebra homomorphism $B' \hookrightarrow B$ that maps each element in B' to the same element of B.

Functor $\underline{\boldsymbol{\Sigma}}$: We are interested in the action of the functor $\underline{\boldsymbol{\Sigma}}$ on Bohrifications of orthomodular lattices and maps between them. First consider the action of $\underline{\boldsymbol{\Sigma}}$ on the Bohrification $\overline{\mathscr{L}}$ of an orthomodular lattice L, which is an object in **Copresh**(**BA**). $\underline{\boldsymbol{\Sigma}}$ acts by postcomposition with Σ, that is,

$$\underline{\boldsymbol{\Sigma}}(\overline{\mathscr{L}}) = \Sigma \circ \overline{\mathscr{L}} : \mathscr{B}(L) \to \mathbf{BA} \to \mathbf{Stone} \tag{92}$$

Specifically, on objects B of $\mathscr{B}(L)$, the functor $\underline{\boldsymbol{\Sigma}}(\overline{\mathscr{L}})$ in **Presh**(**Stone**) acts as follows: for all $B \in \mathscr{B}(L)$,

$$\underline{\boldsymbol{\Sigma}}(\overline{\mathscr{L}})_B = (\Sigma \circ \overline{\mathscr{L}})_B = \Sigma(\overline{\mathscr{L}}_B) = \Sigma_B. \tag{93}$$

On arrows $i_{B',B}$ in $\mathscr{B}(L)$, this functor $\underline{\boldsymbol{\Sigma}}(\overline{\mathscr{L}})$ has the following action:

$$\underline{\boldsymbol{\Sigma}}(\overline{\mathscr{L}})(i_{B',B}) = (\Sigma \circ \overline{\mathscr{L}})(i_{B',B}) = \Sigma(inc_{B',B}) = r_{B,B'}, \tag{94}$$

where r denotes the restriction map, that is, precomposition with the inclusion map. Note that as $\Sigma \circ \overline{\mathscr{L}}$ is a presheaf from $\mathscr{B}(L)$ to **Stone** with the same action on both objects and arrows of $\mathscr{B}(L)$ as $\underline{\Sigma}^L$, then in fact $\Sigma \circ \overline{\mathscr{L}} = \underline{\Sigma}^L$. That is,

$$\underline{\boldsymbol{\Sigma}}(\overline{\mathscr{L}}) = \underline{\Sigma}^L. \tag{95}$$

Now consider the action of functor $\underline{\boldsymbol{\Sigma}}$ on morphisms between Bohrifications, that is, on arrows $\langle I, \theta \rangle : \overline{\mathscr{L}} \to \overline{\mathscr{M}}$ in **Copresh**(**BA**). By Eq. 88,

$$\underline{\boldsymbol{\Sigma}}(\langle I, \theta \rangle) = \langle I, \Sigma(\theta) \rangle, \tag{96}$$

where $\Sigma(\theta)$ is the natural transformation with components $\Sigma(\theta)_B = \Sigma(\theta_B)$ for all $B \in \mathscr{B}(L)$.

Functor $\overline{CL}$: We are interested in the action of the functor $\overline{CL}$ on a spectral presheaf $\underline{\Sigma}^L \in$ **Presh**(**Stone**), for some orthomodular lattice L. $\overline{CL}$ acts on $\underline{\Sigma}^L$ as postcomposition with cl : **Stone** $\to$ **BA**, yielding $cl \circ \underline{\Sigma}^L$, a functor with domain $\mathscr{B}(L)$ in **Copresh**(**BA**). The functor $\overline{CL}(\underline{\Sigma}^L)$ acts on objects $B \in \mathscr{B}(L)$ by

$$\overline{CL}(\underline{\Sigma}^L)_B = (cl \circ \underline{\Sigma}^L)_B = cl(\underline{\Sigma}^L_B) = cl(\Sigma_B), \tag{97}$$

where $cl(\Sigma_B)$ is the Boolean algebra of clopen subsets of Σ_B. On arrows $i_{B,B'} : B' \to B$ in $\mathscr{B}(L)$,

$$\overline{CL}(\underline{\Sigma}^L)(i_{B',B}) = (cl \circ \underline{\Sigma}^L)(i_{B',B}) = cl(r_{B,B'}) : cl(\Sigma_{B'}) \to cl(\Sigma_B). \tag{98}$$

Recall functor cl maps a morphism to its inverse image morphism, denoted by exponent (-1). For any clopen subset S of $\Sigma_{B'}$, the map $cl(r_{B,B'})$ acts as

$$cl(r_{B,B'})(S) = r^{(-1)}_{B,B'}(S) = \{\lambda \in \Sigma_B : \lambda|_{B'} \in S\}, \tag{99}$$

which is a clopen subset of Σ_B.

Now, consider how the map $\overline{CL}$ acts on spectral presheaf morphisms $\langle \tilde{\varphi}, \zeta_\varphi \rangle : \underline{\Sigma}^M \to \underline{\Sigma}^L$ in **Presh**(**Stone**). From Eq. 85,

$$\overline{CL}(\langle \tilde{\varphi}, \zeta_\varphi \rangle) = \langle \tilde{\varphi}, cl(\zeta_\varphi) \rangle \tag{100}$$

where $cl(\zeta_\varphi)$ is a natural transformation between functors in $\mathbf{BA}^{\mathscr{B}(L)}$, from functor $\overline{CL}(\underline{\Sigma}^L) = cl \circ \underline{\Sigma}^L$ to functor $cl \circ \tilde{\varphi}^*(\underline{\Sigma}^M)$. Map $cl(\zeta_\varphi)$ has components for each $B \in \mathscr{B}(L)$ that map from $cl(\underline{\Sigma}^L_B) = cl(\Sigma_B)$ to $cl((\tilde{\varphi}^*\underline{\Sigma}^M)_B) = cl(\Sigma_{\tilde{\varphi}(B)})$, given by:

$$cl(\zeta_\varphi)_B = cl(\zeta_{\varphi,B}) = \zeta^{(-1)}_{\varphi,B} : cl(\Sigma_B) \to cl(\Sigma_{\tilde{\varphi}(B)}). \tag{101}$$

Again, here the exponent denotes inverse image, rather than inverse. Specifically, the action of $cl(\zeta_\varphi)_B$ on a clopen subset S of $\underline{\Sigma}^L_B$ is given by

$$cl(\zeta_\varphi)_B(S) = \zeta^{(-1)}_{\varphi,B}(S) = \{\lambda \in \Sigma_{\tilde{\varphi}(B)} : \zeta_{\varphi,B}(\lambda) \in S\} = \{\lambda \in \Sigma_{\tilde{\varphi}(B)} : \lambda \circ \varphi|_B \in S\}. \tag{102}$$

3.6 *Concrete Isomorphisms Between Spectral Presheaves and Bohrifications*

Now that the action of the functors $\underline{\Sigma}$ and $\overline{CL}$ has been defined, we explore the relationship between spectral presheaves in **Presh**(**Stone**) and Bohrifications in **Copresh**(**BA**) further. From Lemma 3.11 and Stone duality, it is not hard to see that if L and M are orthomodular lattices, with spectral presheaves $\underline{\Sigma}^L$ and $\underline{\Sigma}^M$ and Bohrifications $\overline{\mathscr{L}}$ and $\overline{\mathscr{M}}$, then there is an isomorphism $\underline{\Sigma}^M \to \underline{\Sigma}^L$ in **Presh**(**Stone**) if and only if there is an isomorphism $\overline{\mathscr{L}} \to \overline{\mathscr{M}}$ in **Copresh**(**BA**). Our goal in this subsection is to construct such isomorphisms from each other explicitly. This is done in Theorem 3.15 below. The concrete form will be useful later in the proof of Theorem 3.18, one of the main results.

We first show $\overline{\mathscr{L}}$ and $\overline{CL}(\underline{\Sigma}^L) = cl \circ \underline{\Sigma}^L$ are naturally isomorphic in the functor category $\mathbf{BA}^{\mathscr{B}(L)}$. For each $B \in \mathscr{B}(L)$, this requires an isomorphism from $\overline{\mathscr{L}}_B$ to $(cl \circ \underline{\Sigma}^L)_B$. Recall

$$\overline{\mathscr{L}}_B = B \text{ and } (cl \circ \underline{\Sigma}^L)_B = cl(\underline{\Sigma}^L_B) = cl(\Sigma_B). \tag{103}$$

The dual equivalence between **BA** and **Stone** given in Sect. 2.3 is witnessed by a natural isomorphism $Bo : Id_{\mathbf{BA}} \to cl \circ \Sigma$ with components $Bo_B : B \to cl(\Sigma_B)$. Using those components of Bo corresponding to $B \in \mathscr{B}(L)$ gives a map $\{Bo_B\}_{B \in \mathscr{B}(L)} : \overline{\mathscr{L}} \to cl \circ \underline{\Sigma}^L$, which we now show comprise a natural isomorphism as desired.

Lemma 3.13 *The map* $\{Bo_B\}_{B \in \mathscr{B}(L)} : \overline{\mathscr{L}} \to cl \circ \underline{\Sigma}^L$ *is a natural isomorphism. That is, these two functors are naturally isomorphic in the functor category* $\mathbf{BA}^{\mathscr{B}(L)}$.

Proof First it is necessary to show that this map is a natural transformation, that is, that the following diagram commutes for every $B', B \in \mathscr{B}(L)$ such that $B' \subseteq B$:

$$\begin{array}{ccc}
\overline{\mathscr{L}}_{B'} & \xrightarrow{\overline{\mathscr{L}}(i_{B',B})} & \overline{\mathscr{L}}_B \\
{\scriptstyle Bo_{B'}}\downarrow & & \downarrow{\scriptstyle Bo_B} \\
(cl \circ \underline{\Sigma}^L)_{B'} & \xrightarrow{(cl \circ \underline{\Sigma}^L)(i_{B',B})} & (cl \circ \underline{\Sigma}^L)_B
\end{array}$$

Recall that

$$(cl \circ \underline{\Sigma}^L)_B = cl(\underline{\Sigma}^L_B) = cl(\Sigma_B) = (cl \circ \Sigma)_B. \tag{104}$$

Additionally, note that

$$(cl \circ \underline{\Sigma}^L)(i_{B',B}) = cl(\underline{\Sigma}^L(i_{B',B})) = cl(r_{B,B'}) = cl(\Sigma(inc_{B',B})) = (cl \circ \Sigma)(inc_{B',B}). \tag{105}$$

Thus, also applying the definition of $\overline{\mathscr{L}}$, the above diagram can be rewritten as

$$\begin{array}{ccc}
B' & \xrightarrow{inc_{B',B}} & B \\
{\scriptstyle Bo_{B'}}\downarrow & & \downarrow{\scriptstyle Bo_B} \\
(cl \circ \Sigma)_{B'} & \xrightarrow{(cl \circ \Sigma)(inc_{B',B})} & (cl \circ \Sigma)_B
\end{array}$$

The above diagram commutes because $inc_{B',B} : B' \to B$ is a morphism in category **BA** and because $Bo : Id_{\mathbf{BA}} \to cl \circ \Sigma$ is a natural transformation. Thus, the collection $\{Bo_B\}_{B \in \mathscr{B}(L)} : \overline{\mathscr{L}} \to cl \circ \underline{\Sigma}^L$ is a valid natural transformation. As each arrow Bo_B is an isomorphism then it is in fact a natural isomorphism.

Natural isomorphism $\{Bo_B\}_{B \in \mathscr{B}(L)}$ will now simply be written in a slight abuse of notation as Bo, and we will remember it only has components for all $B \in \mathscr{B}(L)$.

While the above lemma presents an interesting result, it will be more useful to know that the functors $\overline{\mathscr{L}}$ and $\overline{CL}(\underline{\Sigma}^L) = cl \circ \underline{\Sigma}^L$ are isomorphic in category **Copresh(BA)** rather than just naturally isomorphic in $\mathbf{BA}^{\mathscr{B}(L)}$.

Lemma 3.14 *The morphism* $\langle Id_{\mathscr{B}(L)}, Bo\rangle : \overline{\mathscr{L}} \to cl \circ \underline{\Sigma}^L$ *is an isomorphism in* **Copresh(BA)**.

Proof For natural isomorphism $Bo = \{Bo_B\}_{B \in \mathscr{B}(L)}$ there exists some inverse natural isomorphism which we denote by $Bo^{-1} : cl \circ \underline{\Sigma} \to \overline{\mathscr{L}}$. We now use Fact 3.6 to show that morphism $\langle Id_{\mathscr{B}(L)}, Bo^{-1}\rangle : cl \circ \underline{\Sigma}^L \to \overline{\mathscr{L}}$ is an inverse to morphism $\langle Id_{\mathscr{B}(L)}, Bo\rangle$ in **Copresh(BA)**:

$$\begin{aligned} \langle Id_{\mathscr{B}(L)}, Bo\rangle \circ \langle Id_{\mathscr{B}(L)}, Bo^{-1}\rangle &= \langle Id_{\mathscr{B}(L)} \circ Id_{\mathscr{B}(L)}, (Id^*_{\mathscr{B}(L)}Bo) \circ Bo^{-1}\rangle && (106)\\ &= \langle Id_{\mathscr{B}(L)}, Bo \circ Bo^{-1}\rangle && (107)\\ &= \langle Id_{\mathscr{B}(L)}, Id_{cl \circ \underline{\Sigma}^L}\rangle && (108) \end{aligned}$$

$$\begin{aligned} \langle Id_{\mathscr{B}(L)}, Bo^{-1}\rangle \circ \langle Id_{\mathscr{B}(L)}, Bo\rangle &= \langle Id_{\mathscr{B}(L)} \circ Id_{\mathscr{B}(L)}, (Id^*_{\mathscr{B}(L)}Bo^{-1}) \circ Bo\rangle && (109)\\ &= \langle Id_{\mathscr{B}(L)}, Bo^{-1} \circ Bo\rangle && (110)\\ &= \langle Id_{\mathscr{B}(L)}, Id_{\overline{\mathscr{L}}}\rangle && (111) \end{aligned}$$

Thus, $\langle Id_{\mathscr{B}(L)}, Bo\rangle$ is an isomorphism in **Copresh(BA)**, meaning $\overline{\mathscr{L}}$ and $cl \circ \underline{\Sigma}^L$ are isomorphic in this category of copresheaves.

Theorem 3.15 *Let L and M be orthomodular lattices,* $\underline{\Sigma}^L$ *and* $\underline{\Sigma}^M$ *their spectral presheaves, and* $\overline{\mathscr{L}}$ *and* $\overline{\mathscr{M}}$ *their Bohrifications. There is an isomorphism* $\underline{\Sigma}^M \to \underline{\Sigma}^L$ *in the category* **Presh(Stone)** *if and only if there is an isomorphism* $\overline{\mathscr{L}} \to \overline{\mathscr{M}}$ *in the category* **Copresh(BA)**, *and these isomorphisms can be explicitly constructed from each other.*

Proof Suppose there is an isomorphism $\langle H, \eta\rangle : \underline{\Sigma}^M \to \underline{\Sigma}^L$ in **Presh(Stone)**. Then, as functors preserve isomorphisms, there is an isomorphism in **Copresh(BA)** given by

$$\overline{CL}(\langle H, \eta\rangle) : \overline{CL}(\underline{\Sigma}^L) \to \overline{CL}(\underline{\Sigma}^M), \qquad (112)$$

or equivalently,

$$\langle H, cl(\eta)\rangle : cl \circ \underline{\Sigma}^L \to cl \circ \underline{\Sigma}^M, \tag{113}$$

where $cl(\eta)$ is the natural transformation with components, for all $B \in \mathscr{B}(L)$, given by $cl(\eta)_B = cl(\eta_B) = \eta_B^{(-1)}$, where the exponent (-1) denotes the inverse image function. By the previous lemma, there are isomorphisms in **Copresh(BA)**

$$\langle Id_{\mathscr{B}(L)}, Bo\rangle : \overline{\mathscr{L}} \to cl \circ \underline{\Sigma}^L \tag{114}$$

$$\langle Id_{\mathscr{B}(M)}, Bo^{-1}\rangle : cl \circ \underline{\Sigma}^M \to \overline{\mathscr{M}} \tag{115}$$

Composing these two isomorphisms on either side of isomorphism $\langle H, cl(\eta)\rangle$ gives an isomorphism from $\overline{\mathscr{L}}$ to $\overline{\mathscr{M}}$, as desired. Specifically, this composition evaluates as follows:

$$\langle Id_{\mathscr{B}(M)}, Bo^{-1}\rangle \circ \langle H, cl(\eta)\rangle \circ \langle Id_{\mathscr{B}(L)}, Bo\rangle \tag{116}$$

$$= \langle Id_{\mathscr{B}(M)}, Bo^{-1}\rangle \circ \langle H \circ Id_{\mathscr{B}(L)}, (Id^*_{\mathscr{B}(L)}cl(\eta)) \circ Bo\rangle \tag{117}$$

$$= \langle Id_{\mathscr{B}(M)} \circ H \circ Id_{\mathscr{B}(L)}, (H^* Bo^{-1}) \circ (Id^*_{\mathscr{B}(L)}cl(\eta)) \circ Bo\rangle \tag{118}$$

$$= \langle H, (H^* Bo^{-1}) \circ cl(\eta) \circ Bo\rangle \tag{119}$$

Some visual intuition is provided below:

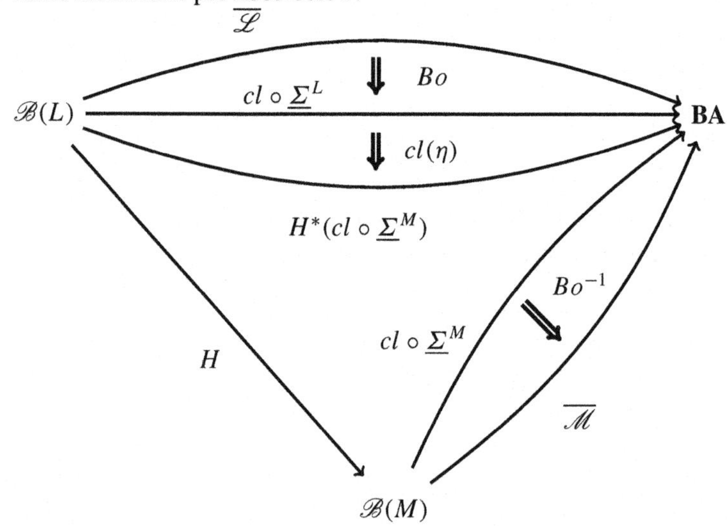

We conclude whenever there is an isomorphism $\langle H, \eta\rangle : \underline{\Sigma}^M \to \underline{\Sigma}^L$ in **Presh(Stone)**, then $\langle H, (H^* Bo^{-1}) \circ cl(\eta) \circ Bo\rangle : \overline{\mathscr{L}} \to \overline{\mathscr{M}}$ is an isomorphism in **Copresh(BA)**, completing the first half of this proof.

Now, suppose that there is an isomorphism $\langle I, \theta\rangle : \overline{\mathscr{L}} \to \overline{\mathscr{M}}$ in **Copresh**(**BA**). Recall $\underline{\boldsymbol{\Sigma}}$: **Copresh**(**BA**) $\to$ **Presh**(**Stone**) that is dual to $\overline{CL}$. As functors preserve isomorphisms, there is an isomorphism in **Presh**(**Stone**) from $\underline{\boldsymbol{\Sigma}}(\overline{\mathscr{M}})$ to $\underline{\boldsymbol{\Sigma}}(\overline{\mathscr{L}})$, given by

$$\underline{\boldsymbol{\Sigma}}(\langle I, \theta\rangle) = \langle I, \Sigma(\theta)\rangle, \tag{120}$$

where $\Sigma(\theta)$ is the natural transformation with components $\Sigma(\theta)_B = \Sigma(\theta_B)$ for all B in $\mathscr{B}(L)$. Recalling from (95) that

$$\underline{\boldsymbol{\Sigma}}(\overline{\mathscr{M}}) = \Sigma \circ \overline{\mathscr{M}} = \underline{\Sigma}^M \text{ and } \underline{\boldsymbol{\Sigma}}(\overline{\mathscr{L}}) = \Sigma \circ \overline{\mathscr{L}} = \underline{\Sigma}^L, \tag{121}$$

it follows that $\langle I, \Sigma(\theta)\rangle$ is an isomorphism in **Presh**(**Stone**) from $\underline{\Sigma}^M$ to $\underline{\Sigma}^L$, as desired.

3.7 The Spectral Presheaf of an OML Is a Complete Invariant

We now prove our first main result: two orthomodular lattices are isomorphic if and only if their spectral presheaves are isomorphic, hence the spectral presheaf is a complete invariant of an OML.

The proof is separated into the following two theorems.

Theorem 3.16 *Let L and M be orthomodular lattices. If $\varphi : L \to M$ is an isomorphism in* **OML**, *then there is an isomorphism $\langle \tilde{\varphi}, \zeta_\varphi\rangle : \underline{\Sigma}^M \to \underline{\Sigma}^L$ in* **Presh**(**Stone**), *where the natural transformation ζ_φ has components $\zeta_{\varphi,B} = \Sigma(\varphi|_B)$ for all B in $\mathscr{B}(L)$.*

Proof Suppose $\varphi : L \to M$ is an isomorphism of orthomodular lattices, with inverse $\psi = \varphi^{-1} : M \to L$. Then, by Lemma 2.9, $\tilde{\varphi} : \mathscr{B}(L) \to \mathscr{B}(M)$ is an order isomorphism of posets, with inverse $\tilde{\psi}$. Additionally, for each $B \in \mathscr{B}(L)$, $\varphi|_B : B \to \varphi[B]$ is an isomorphism of Boolean algebras, with inverse $\psi|_{\tilde{\varphi}(B)}$.

By Stone duality, applying functor Σ to Boolean algebra isomorphism $\varphi|_B : B \to \tilde{\varphi}(B)$ yields a continuous isomorphism $\Sigma(\varphi|_B) : \Sigma_{\tilde{\varphi}(B)} \to \Sigma_B$ in **Stone**. As $\tilde{\varphi}(B) \in \mathscr{B}(M)$, then

$$\Sigma_{\tilde{\varphi}(B)} = \underline{\Sigma}^M_{\tilde{\varphi}(B)} = (\underline{\Sigma}^M \circ \tilde{\varphi})_B = (\tilde{\varphi}^* \underline{\Sigma}^M)_B. \tag{122}$$

Additionally, as $B \in \mathscr{B}(L)$, then $\Sigma_B = \underline{\Sigma}^L_B$. Thus, $\Sigma(\varphi|_B)$ is in fact a Stone space isomorphism from $(\tilde{\varphi}^* \underline{\Sigma}^M)_B$ to $\underline{\Sigma}^L_B$. Let isomorphism $\Sigma(\varphi|_B)$ be denoted

$$\Sigma(\varphi|_B) := \zeta_{\varphi,B} : (\tilde{\varphi}^* \underline{\Sigma}^M)_B \to \underline{\Sigma}^L_B. \tag{123}$$

Note this coincides exactly with the definition of $\zeta_{\varphi,B}$ given in Step 2 of Sect. 3.2, where the action of isomorphism $\zeta_{\varphi,B}$ on a homomorphism $\lambda : \tilde{\varphi}(B) \to \{0, 1\}$ is given by precomposition with $\varphi|_B$. The components $(\zeta_{\varphi,B})_{B\in\mathcal{B}(L)}$ thus form a natural isomorphism from $\tilde{\varphi}^*\underline{\Sigma}^M$ to $\underline{\Sigma}^L$, because as we proved in Lemma 3.2, for every $B' \subseteq B$ in $\mathcal{B}(L)$ the following diagram commutes:

$$\begin{array}{ccc}
\underline{\Sigma}^M_{\tilde{\varphi}(B')} & \xleftarrow[= r_{\tilde{\varphi}(B),\tilde{\varphi}(B')}]{\underline{\Sigma}^M(i_{\tilde{\varphi}(B'),\tilde{\varphi}(B)})} & \underline{\Sigma}^M_{\tilde{\varphi}(B)} \\
\downarrow \zeta_{\varphi,B'} & & \downarrow \zeta_{\varphi,B} \\
\underline{\Sigma}^L_{B'} & \xleftarrow[= r_{B',B}]{\underline{\Sigma}^L(i_{B',B})} & \underline{\Sigma}^L_B
\end{array}$$

Since $\tilde{\varphi} : \mathcal{B}(L) \to \mathcal{B}(M)$ is an isomorphism and $\zeta_\varphi : \tilde{\varphi}^*\underline{\Sigma}^M \to \underline{\Sigma}^L$ is a natural isomorphism, then the composite

$$\langle \tilde{\varphi}, \zeta_\varphi \rangle : \underline{\Sigma}^M \to \underline{\Sigma}^L \tag{124}$$

is an arrow in **Presh**(**Stone**), depicted here:

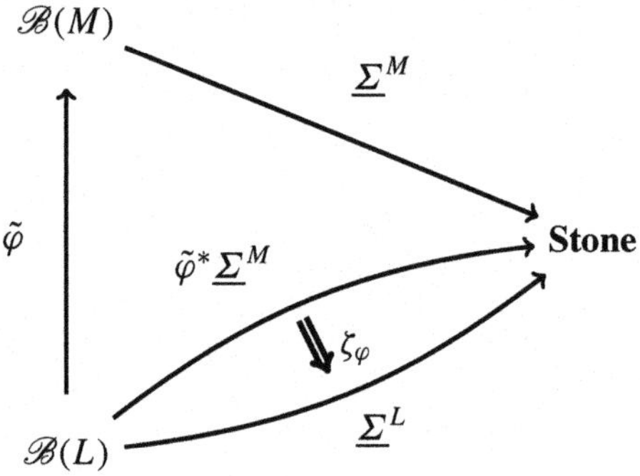

It only remains to show that this arrow has an inverse, that is, that it is an isomorphism in **Presh**(**Stone**). Recall that $\tilde{\psi} : \mathcal{B}(M) \to \mathcal{B}(L)$ is the inverse of $\tilde{\varphi}$, and consider the arrow

$$\langle \tilde{\psi}, \tilde{\psi}^*(\zeta_\varphi^{-1}) \rangle : \underline{\Sigma}^L \to \underline{\Sigma}^M. \tag{125}$$

This arrow is depicted in the following diagram:

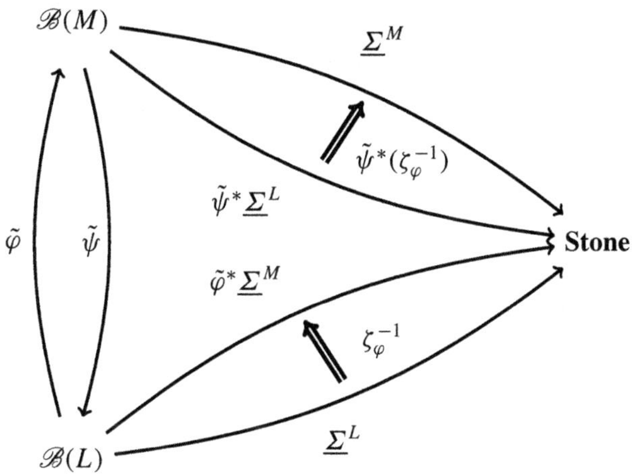

That both compositions of arrow $\langle\tilde{\varphi}, \zeta_\varphi\rangle$ with its inverse give the identity morphism is now checked algebraically.

$$\langle\tilde{\psi}, \tilde{\psi}^*(\zeta_\varphi^{-1})\rangle \circ \langle\tilde{\varphi}, \zeta_\varphi\rangle = \langle\tilde{\varphi} \circ \tilde{\psi}, \tilde{\psi}^*(\zeta_\varphi^{-1}) \circ \tilde{\psi}^*\zeta_\varphi\rangle \tag{126}$$
$$= \langle Id_{\mathscr{B}(M)}, \tilde{\psi}^*(Id_{\tilde{\varphi}^*\underline{\Sigma}^M})\rangle \tag{127}$$
$$= \langle Id_{\mathscr{B}(M)}, Id_{\tilde{\psi}^*\tilde{\varphi}^*\underline{\Sigma}^M}\rangle \tag{128}$$
$$= \langle Id_{\mathscr{B}(M)}, Id_{\underline{\Sigma}^M}\rangle. \tag{129}$$

$$\langle\tilde{\varphi}, \zeta_\varphi\rangle \circ \langle\tilde{\psi}, \tilde{\psi}^*(\zeta_\varphi^{-1})\rangle = \langle\tilde{\psi} \circ \tilde{\varphi}, \zeta_\varphi \circ \tilde{\varphi}^*(\tilde{\psi}^*(\zeta_\varphi^{-1}))\rangle \tag{130}$$
$$= \langle Id_{\mathscr{B}(L)}, \zeta_\varphi \circ (\tilde{\psi} \circ \tilde{\varphi})^*(\zeta_\varphi^{-1}))\rangle \tag{131}$$
$$= \langle Id_{\mathscr{B}(L)}, \zeta_\varphi \circ (Id_{\mathscr{B}(L)})^*(\zeta_\varphi^{-1})\rangle \tag{132}$$
$$= \langle Id_{\mathscr{B}(L)}, \zeta_\varphi \circ \zeta_\varphi^{-1}\rangle \tag{133}$$
$$= \langle Id_{\mathscr{B}(L)}, Id_{\underline{\Sigma}^L}\rangle. \tag{134}$$

Thus, $\langle\tilde{\varphi}, \zeta_\varphi\rangle : \underline{\Sigma}^M \to \underline{\Sigma}^L$ is an isomorphism in **Presh(Stone)**, as desired.

In order to prove the next result, recall from Sect. 2.2.2 the definition of a partial orthomodular lattice, which captures all aspects of lattice structure within each boolean subalgebra of L, as well as capturing inclusion relations between Boolean subalgebras.

Theorem 3.17 *Let L and M be orthomodular lattices. If there is an isomorphism $\langle H, \eta\rangle : \underline{\Sigma}^M \to \underline{\Sigma}^L$ in* **Presh(Stone)***, then there is an isomorphism from L to M in* **OML** *that can be explicitly constructed from $\langle H, \eta\rangle$.*

Proof Let $\langle H, \eta \rangle : \underline{\Sigma}^M \to \underline{\Sigma}^L$ be an isomorphism between spectral presheaves of orthomodular lattices. Note $H : \mathcal{B}(L) \to \mathcal{B}(M)$ is necessarily an isomorphism with inverse $H^{-1} : \mathcal{B}(M) \to \mathcal{B}(L)$. By Theorem 3.15, there exists a isomorphism from $\overline{\mathcal{L}}$ to $\overline{\mathcal{M}}$ in **Copresh(BA)**, specifically,

$$\langle H, (H^* Bo^{-1}) \circ cl(\eta) \circ Bo \rangle : \overline{\mathcal{L}} \to \overline{\mathcal{M}}. \tag{135}$$

For simplicity, define

$$\rho := (H^* Bo^{-1}) \circ cl(\eta) \circ Bo : \overline{\mathcal{L}} \to H^* \overline{\mathcal{M}}. \tag{136}$$

This natural transformation ρ has components for each $B \in \mathcal{B}(L)$ that map from $\overline{\mathcal{L}}_B = B$ to $(H^*\overline{\mathcal{M}})_B = \overline{\mathcal{M}}_{H(B)} = H(B)$, where $H(B)$ is an element of $\mathcal{B}(M)$, that is, a Boolean subalgebra of M:

$$\rho_B : B \to H(B). \tag{137}$$

By the construction of ρ in the proof of Theorem 3.15, each component ρ_B is a Boolean algebra isomorphism.

Suppose that $B', B \in \mathcal{B}(L)$ with $B' \subseteq B$, that is, $i_{B',B}$ is an arrow in $\mathcal{B}(L)$. Recall that $\overline{\mathcal{L}}(i_{B',B}) = inc_{B',B}$, the inclusion Boolean algebra homomorphism from B' to B. Additionally, $H(i_{B',B})$ is an arrow in $\mathcal{B}(M)$ from $H(B')$ to $H(B)$; as poset categories have at most one arrow with a given domain and codomain, it must be that $H(i_{B',B}) = i_{H(B'),H(B)}$. Then,

$$(\overline{\mathcal{M}} \circ H)(i_{B',B}) = inc_{H(B'),H(B)}. \tag{138}$$

The naturality of ρ then means that the following diagram commutes:

$$\begin{array}{ccc} B' & \xrightarrow{\rho_{B'}} & H(B') \\ \downarrow{\scriptstyle inc_{B',B}} & & \downarrow{\scriptstyle inc_{H(B'),H(B)}} \\ B & \xrightarrow{\rho_B} & H(B) \end{array}$$

Let $a \in L$ such that $a \in B, B'$. Then

$$\rho_B(a) = (\rho_B \circ inc_{B',B})(a) = (inc_{H(B'),H(B)} \circ \rho_{B'})(a) = \rho_{B'}(a). \tag{139}$$

From this, it follows that if element a is in any two Boolean subalgebras B_1, B_2 of L (not necessarily related by containment), then

$$\rho_{B_1}(a) = \rho_{B_1 \cap B_2}(a) = \rho_{B_2}(a). \tag{140}$$

As every element of L is in at least one Boolean subalgebra, this yields a well-defined map as follows:

$$\varphi : L_{part} \to M_{part} \tag{141}$$
$$a \mapsto \rho_B(a), \text{ where } B \in \mathscr{B}(L) \text{ is any Boolean subalgebra containing} a \tag{142}$$

This map φ is a partial orthomodular lattice homomorphism because it preserves all defined meets and joins, i.e. those within some Boolean subalgebra, as well as orthocomplementation. It remains to check that φ is an isomorphism of partial orthomodular lattices.

As ρ is a natural isomorphism, each component ρ_B is an isomorphism of Boolean algebras and has an inverse $\rho_B^{-1} : H(B) \to B$; note the subscript in ρ_B^{-1} reflects its codomain. Just as above, for any $m \in M$ and any $B_1, B_2 \in \mathscr{B}(M)$ that contain m, it can be shown that $\rho_{H^{-1}(B_1)}^{-1}(m) = \rho_{H^{-1}(B_2)}^{-1}(m)$. Thus, as any $m \in M$ is in at least one $B \in \mathscr{B}(M)$, it is possible to define a partial orthomodular lattice homomorphism

$$\psi : M_{part} \to L_{part} \tag{143}$$
$$m \mapsto \rho_{H^{-1}(B)}^{-1}(m), \text{ where } B \in \mathscr{B}(M) \text{ is any Boolean subalgebra containing } m \tag{144}$$

One can now verify that ψ is an inverse to φ. Let $a \in L$, and let $B \in \mathscr{B}(L)$ contain a. Then,

$$(\psi \circ \varphi)(a) = (\rho_B^{-1} \circ \rho_B)(a) = Id_B(a) = a \tag{145}$$

Similarly, for any $m \in M$ contained in some Boolean algebra $B \in \mathscr{B}(M)$,

$$(\varphi \circ \psi)(m) = (\rho_{H^{-1}(B)} \circ \rho_{H^{-1}(B)}^{-1})(m) = Id_B(m) = m \tag{146}$$

Thus ψ is an inverse to φ, meaning φ is a partial orthomodular lattice isomorphism. By Proposition 2.13, φ preserves all meets and joins, not just those within Boolean subalgebras, and as it also already preserves orthocomplementation this means that $\varphi : L \to M$ is an isomorphism of orthomodular lattices.

Specifically, for any element $a \in L$, the action of φ on a as constructed in the proof above is given as follows. Let $B \in \mathscr{B}(L)$ be any Boolean subalgebra containing a. Then,

$$\varphi(a) = \rho_B(a) = ((H^* Bo^{-1}) \circ cl(\eta) \circ Bo)_B(a) = ((H^* Bo_B^{-1}) \circ cl(\eta)_B \circ Bo_B)(a) \tag{147}$$

$$= (Bo^{-1}_{H(B)} \circ cl(\eta_B) \circ Bo_B)(a). \tag{148}$$

Recall that $Bo_B : B \to cl(\Sigma(B))$ is the component at B of the natural transformation that witnesses Stone duality; $Bo^{-1}_{H(B)}$ is the component at $H(B) \in \mathcal{B}(M)$ of the inverse of this same natural transformation; and cl : **Stone** $\to$ **BA** is one functor of the dual equivalence between **BA** and **Stone**. Specific actions of these maps are given in Sect. 2.3. In practice, to calculate $\varphi(a) = \rho_B(a)$ it is simplest to choose $B = B_a = \{0, a, a', 1\}$, the Boolean algebra with four elements, as we will do in the later proofs of Theorems 3.19 and 3.20.

Theorem 3.18 *Two orthomodular lattices L and M are isomorphic in* **OML** *if and only if their spectral preserves $\underline{\Sigma}^L$ and $\underline{\Sigma}^M$ are isomorphic in* **Presh(Stone)**.

Proof Theorems 3.16 and 3.17.

We give some interpretation of this result in Sect. 3.8, but first we present an even stronger result. For an orthomodular lattice isomorphism $\varphi : L \to M$, denote the spectral presheaf isomorphism constructed in the proof of Theorem 3.16 by $SP(\varphi) : \underline{\Sigma}^M \to \underline{\Sigma}^L$. For a spectral presheaf isomorphism $\langle H, \eta \rangle : \underline{\Sigma}^M \to \underline{\Sigma}^L$, denote the orthomodular lattice isomorphism constructed in the proof of Theorem 3.17 by $OML(\langle H, \eta \rangle) : L \to M$.

Theorem 3.19 *For all orthomodular lattice isomorphisms $\varphi : L \to M$,*

$$OML(SP(\varphi)) = \varphi. \tag{149}$$

Proof Consider an orthomodular lattice isomorphism $\varphi : L \to M$. Then, $\langle \tilde{\varphi}, \zeta_\varphi \rangle$ is an isomorphism in **Presh(Stone)**, where ζ_φ is a natural isomorphism with components given by

$$\zeta_{\varphi,B} : \Sigma_{\tilde{\varphi}(B)} \to \Sigma_B \tag{150}$$

$$\lambda \mapsto \lambda \circ \varphi|_B \tag{151}$$

Each component $\zeta_{\varphi,B}$ is an isomorphism of Stone spaces. To construct $OML(\langle \tilde{\varphi}, \zeta_\varphi \rangle)$, consider the natural isomorphism in **Copresh(BA)**

$$\rho = (\tilde{\varphi}^* Bo^{-1}) \circ cl(\zeta_\varphi) \circ Bo : \overline{\mathcal{L}} \to \tilde{\varphi}^* \overline{\mathcal{M}}. \tag{152}$$

Each component of this natural isomorphism is a Boolean algebra isomorphism from B to $\tilde{\varphi}(B)$ given by

$$\rho_B = \left((\tilde{\varphi}^* Bo^{-1}) \circ cl(\zeta_\varphi) \circ Bo\right)_B \tag{153}$$

$$= (\tilde{\varphi}^* Bo^{-1})_B \circ cl(\zeta_\varphi)_B \circ Bo_B \tag{154}$$

$$= Bo^{-1}_{\tilde{\varphi}(B)} \circ \zeta^{(-1)}_{\varphi,B} \circ Bo_B \tag{155}$$

Let $a \in L$, and consider the Boolean algebra $B_a \subseteq L$ with elements $\{0, a, a', 1\}$. $OML(\langle\tilde{\varphi}, \zeta_\varphi\rangle)$ is the homomorphism from L to M whose action on element a is $\rho_{B_a}(a)$, which we will now calculate. The Stone space of B_a has two elements λ_a and $\lambda_{a'}$, where $\lambda_a(a) = 1$, $\lambda_a(a') = 0$, and $\lambda_{a'}(a) = 0$, $\lambda_{a'}(a) = 0$. Thus,

$$Bo_{B_a}(a) = \{\lambda \in \Sigma_{B_a} \mid \lambda(a) = 1\} = \{\lambda_a\}. \tag{156}$$

As $\varphi|_B$ is a Boolean algebra isomorphism, then $\tilde{\varphi}(B)$ is the four-element Boolean algebra with elements $\{0, \varphi(a), \varphi(a)', 1\}$, which we will denote $B_{\varphi(a)}$. The Stone space Σ_{B_a} has two elements, which we denote $\lambda_{\varphi(a)}$ and $\lambda_{\varphi(a)'}$, where $\lambda_{\varphi(a)}(\varphi(a)) = 1$, $\lambda_{\varphi(a)}(\varphi(a')) = 0$ and $\lambda_{\varphi(a)'}(\varphi(a)) = 0$, $\lambda_{\varphi(a')}(\varphi(a)) = 0$. Then,

$$\zeta_{\varphi,B}^{(-1)}(Bo_B(a)) = \zeta_{\varphi,B}^{(-1)}(\{\lambda_a\}) \tag{157}$$

$$= \{\lambda \in \Sigma_{\tilde{\varphi}(B_a)} \mid (\lambda \circ \varphi|_{B_a})(a) = 1\} \tag{158}$$

$$= \{\lambda_{\varphi(a)}\}. \tag{159}$$

In order to calculate $\rho_{B_a}(a) = Bo^{-1}_{\tilde{\varphi}(B)}(\{\lambda_{\varphi(a)}\})$, recall the definition for the components of Bo^{-1} given at the end of Sect. 2.3: write $S = \bigcup_{b\in J} U_b$ as a finite union of basic open sets for some index set J, then $Bo^{-1}(S) = \bigvee_{b\in J} b$. As $\{\lambda_{\varphi(a)}\} = U_{\varphi(a)}$ is itself a basic open set, then $Bo^{-1}_{\tilde{\varphi}(B)}(\{\lambda_{\varphi(a)}\}) = \varphi(a)$. Thus,

$$\rho_{B_a}(a) = \left(Bo^{-1}_{\tilde{\varphi}(B)} \circ \zeta_{\varphi,B}^{(-1)} \circ Bo_B\right)(a) = Bo^{-1}_{\tilde{\varphi}(B)}(\{\lambda_{\varphi(a)}\}) = \varphi(a) \tag{160}$$

Thus, $OML(\langle\tilde{\varphi}, \zeta_\varphi\rangle)$ is the orthomodular lattice homomorphism from L to M mapping a to $\rho_{B_a}(a) = \varphi(a)$, meaning that $\varphi = OML(\langle\tilde{\varphi}, \zeta_\varphi\rangle) = (OML \circ SP)(\varphi)$.

Theorem 3.20 *Let* $\langle H, \eta\rangle : \underline{\Sigma}^M \to \underline{\Sigma}^L$ *be an isomorphism in* **Presh**(**Stone**) *between the spectral presheaves of two orthomodular lattices* M *and* L. *Then*

$$SP(OML(\langle H, \eta\rangle)) = \langle H, \eta\rangle. \tag{161}$$

Proof Consider an isomorphism $\langle H, \eta\rangle : \underline{\Sigma}^M \to \underline{\Sigma}^L$ in **Presh**(**Stone**). To construct $OML(\langle H, \eta\rangle) : L \to M$, consider natural isomorphism in **Copresh**(**BA**):

$$\rho = (H^* Bo^{-1}) \circ cl(\eta) \circ Bo : \overline{\mathscr{L}} \to H^*\overline{\mathscr{M}}. \tag{162}$$

Each component of this natural isomorphism is a Boolean algebra isomorphism from B to $H(B)$ given by

$$\rho_B = Bo^{-1}_{H(B)} \circ \eta_B^{(-1)} \circ Bo_B : B \to H(B). \tag{163}$$

Let $a \in L$; the four-element Boolean B_a contains a. We first want to calculate

$$OML(\langle H, \eta\rangle)(a) = \rho_{B_a}(a). \tag{164}$$

Note that $H(B_a)$ is also a four-element Boolean algebra because H is an order isomorphism of posets; as there is no $B \in \mathcal{B}(L)$ such that $\{0, 1\} \subset B \subset B_a$, then this also holds true for $H(B_a)$ in $\mathcal{B}(M)$. We name its elements $\{0, h(a), h(a)', 1\}$. Note that we are not defining some function $h : L \to M$, but rather simply using function notation to indicate that the elements of $H(B_a)$ depend on the chosen element a. We now calculate

$$\begin{aligned}(\eta_{B_a}^{(-1)} \circ Bo_{B_a})(a) &= \eta_{B_a}^{(-1)}(\{\lambda_a\}) & (165)\\ &= \{\lambda \in \Sigma_{H(B_a)} \mid \eta_{B_a}(\lambda) = \lambda_a\} & (166)\\ &= \{\lambda \in \Sigma_{H(B_a)} \mid \eta_{B_a}(\lambda)(a) = 1\} & (167)\end{aligned}$$

As $\eta_{B_a} : \Sigma_{H(B_a)} \to \Sigma_{B_a}$ is an isomorphism of Stone spaces, it must be that exactly one of the two elements $\sigma = \lambda_{h(a)}$ or $\sigma = \lambda_{h(a)'}$ of $\Sigma_{H(B_a)}$ satisfies $\eta_{B_a}(\sigma)(a) = 1$. If $(\eta_{B_a}^{(-1)} \circ Bo_{B_a})(a) = \{\lambda_{h(a)}\} = U_{h(a)}$, then applying $Bo_{H(B_a)}^{-1}$ yields $h(a)$, while the other case yields $h(a)'$. Thus,

$$\rho_{B_a}(a) = \begin{cases} h(a) \ : \eta_{B_a}(\lambda_{h(a)}) = \lambda_a \\ h'(a) : \eta_{B_a}(\lambda_{h(a)'}) = \lambda_a \end{cases} \tag{168}$$

Thus, $OML(\langle H, \eta\rangle)$ is a homomorphism $\varphi : L \to M$ given by $\varphi(a) = \rho_{B_a}(a)$ as above.

We now want to show that $SP(\varphi) = \langle H, \eta\rangle$. First, consider $\tilde{\varphi}$, and let B be any element of $\mathcal{B}(L)$. We want to show that $\tilde{\varphi}(B) = H(B)$. First, let $a \in L$ and consider the four-element Boolean subalgebra $B_a = \{0, a, a', 1\}$. Recall that $H(B_a)$ has four elements which we call $\{0, h(a), h'(a), 1\}$, and note that either $\varphi(a) = h(a)$ and $\varphi(a') = h(a)'$, or $\varphi(a) = h(a)'$ and $\varphi(a') = h(a)$. In either case,

$$\tilde{\varphi}(B_a) = \{\varphi(x) \mid x \in B_a\} = \{0, h(a), h(a)', 1\} = H(B_a). \tag{169}$$

Now, let B be an arbitrary Boolean subalgebra of L. Let $\varphi(a)$ be any element in $\tilde{\varphi}(B)$, where a is some element of B. Then, $\varphi(a) \in \tilde{\varphi}(B_a) = H(B_a)$. As $B_a \subseteq B$, then $H(B_a) \subseteq H(B)$, meaning $\varphi(a) \in H(B)$ and thus $\tilde{\varphi}(B) \subseteq H(B)$.

Conversely, let $h \in H(B)$. Then $B_h = \{0, h, h', 1\} \subseteq H(B)$, implying that

$$H^{-1}(B_h) \subseteq H^{-1}(H(B)) = B. \tag{170}$$

$H^{-1}(B_h)$ is a four-element Boolean subalgebra of B because H is an order isomorphism, so because $\tilde{\varphi}$ and H are the same on four-element Boolean subalgebras then

$$\tilde{\varphi}(H^{-1}(B_h)) = H(H^{-1}(B_h)) = B_h. \tag{171}$$

Thus $h \in B_h$ is equal to some element $\varphi(a)$ in

$$\tilde{\varphi}(H^{-1}(B_h)) = \{\varphi(a) \mid a \in H^{-1}(B_h) \subseteq B\}. \tag{172}$$

As a is thus also an element of B, then $h \in \tilde{\varphi}(B)$ implying $H(B) \subseteq \tilde{\varphi}(B)$ and thus $H(B) = \tilde{\varphi}(B)$ for all $B \in \mathscr{B}(L)$, so $\tilde{\varphi} = H$.

It only remains to show that $\zeta_\varphi = \eta$, i.e for all $B \in \mathscr{B}(L)$, $\zeta_{\phi,B} = \eta_B$. Recall $\zeta_{\varphi,B}$ and η_B are both isomorphisms from $\Sigma_{\tilde{\varphi}(B)} = \Sigma_{H(B)}$ to Σ_B. Fix $\lambda \in \Sigma_{\tilde{\varphi}(B)} = \Sigma_{H(B)}$ and fix $a \in B$; we want to show that $\zeta_{\varphi,B}(\lambda)(a) = \eta_B(\lambda)(a)$.

As described in the proof of Theorem 3.16, component $\zeta_{\varphi,B}$ acts on an element $\lambda \in \Sigma_B$ by precomposing by $\varphi|_B$:

$$\zeta_{\varphi,B} : \Sigma_{\tilde{\varphi}(B)} \to \Sigma_B \tag{173}$$

$$\lambda \mapsto \lambda \circ \varphi|_B \tag{174}$$

Thus,

$$\zeta_{\varphi,B}(\lambda)(a) = \lambda(\varphi(a)). \tag{175}$$

As η is a natural transformation, then as B_a is a Boolean algebra contained in B, the following diagram commutes:

$$\begin{array}{ccc}
\Sigma_{\tilde{\varphi}(B_a)} & \xleftarrow{\; r_{\tilde{\varphi}(B),\tilde{\varphi}(B_a)} \;} & \Sigma_{\tilde{\varphi}(B)} \\
\Big\downarrow \eta_{B_a} & & \Big\downarrow \eta_B \\
\Sigma_{B_a} & \xleftarrow{\; r_{B,B_a} \;} & \Sigma_B
\end{array}$$

In particular, this implies that

$$\eta_B(\lambda)(a) = \eta_B(\lambda)|_{B_a}(a) = \eta_{B_a}(\lambda|_{\tilde{\varphi}(B_a)})(a). \tag{176}$$

Recall:

$$\varphi(a) = \begin{cases} h(a) \ : \eta_{B_a}(\lambda_{h(a)}) = \lambda_a \Leftrightarrow \eta_{B_a}(\lambda_{h(a)})(a) = 1 \\ h(a)' : \eta_{B_a}(\lambda_{h(a)'}) = \lambda_a \Leftrightarrow \eta_{B_a}(\lambda_{h(a)})(a) = 0 \end{cases} \tag{177}$$

Specifically, for any $\lambda|_{\tilde{\varphi}(B)} \in \Sigma_{H(B)} = \{\lambda_{h(a)}, \lambda_{h(a)'}\}$, whether $\lambda|_{\tilde{\varphi}(B)} = \lambda_{h(a)}$ or $\lambda|_{\tilde{\varphi}(B)} = \lambda_{h(a)'}$, an exhaustive check shows

$$\lambda(\varphi(a)) = \lambda|_{\tilde{\varphi}(B_a)}(\varphi(a)) = \eta_{B_a}(\lambda|_{\tilde{\varphi}(B)})(a). \tag{178}$$

Combining this with Eqs. 175 and 176,

$$\zeta_{\varphi,B}(\lambda)(a) = \lambda(\varphi(a)) = \eta_{B_a}(\lambda|_{\tilde{\varphi}(B)})(a) = \eta_B(\lambda)(a) \tag{179}$$

Thus, $\zeta_\varphi = \eta$, meaning

$$SP(OML(\langle H, \eta \rangle)) = \langle H, \eta \rangle. \tag{180}$$

Theorem 3.21 *There are bijections SP and OML between orthomodular lattice isomorphisms $\varphi : L \to M$ and spectral presheaf isomorphisms $\langle H, \eta \rangle : \underline{\Sigma}^M \to \underline{\Sigma}^L$.*

Proof Theorems 3.19 and 3.20.

3.8 Interpretation of the Results so Far

Mathematical aspects. Theorem 3.18 is of some mathematical interest. While the duality between Stone spaces and Boolean algebras has been well-known for many years, we are not familiar with any attempts to generalise this duality to general orthomodular lattices. The spectral presheaf of an orthomodular lattice provides a new notion of 'dual space' for an orthomodular lattice, given by a functor whose image is, rather than a single Stone space, a collection of Stone spaces linked together into a presheaf by continuous restriction maps. Theorem 3.18 implies that the assignment of the spectral presheaf $\underline{\Sigma}^L$ to an OML L (implicitly) preserves all the structure of an orthomodular lattice, as one would require in a duality type situation.

In the case where the orthomodular lattice L is in fact a Boolean algebra, the spectral presheaf does not quite reduce to the Stone space of the Boolean algebra, as our construction of the spectral presheaf considers all Boolean subalgebras of L while Stone duality does not. This is necessary to avoid certain no-go theorems about extending classical dualities [30]. Yet, for a Boolean algebra B the poset of contexts has a unique top element, which is B itself, and the component of the spectral presheaf $\underline{\Sigma}^B$ at B is the Stone space of B. In this sense, for Boolean algebras the spectral presheaf is very close to the Stone space.

Theorem 3.18 shows that the spectral presheaf of an orthomodular lattice is a complete invariant, hence determines the orthomodular lattice up to isomorphism and vice versa. This is stronger than the corresponding result for von Neumann algebras, where a spectral presheaf determines a von Neumann algebra only up to Jordan $*$-isomorphism rather than up to isomorphism [13].

Relation with earlier results by Harding and Navara. In [31], Harding and Navara prove that an isomorphism of context categories yields an isomorphism of orthomodular lattices, though this isomorphism is only unique when the orthomodular lattices have no maximal four-element Boolean subalgebras. We considered not just the context category but rather a functor on the context category; an isomorphism between spectral presheaves $\langle H, \eta \rangle$ consists of not only an isomorphism H between context categories but also a natural isomorphism η. The additional data of η enables the proof of Theorem 3.21, that there is a concrete bijection between orthomodular lattice isomorphisms and spectral presheaf isomorphisms. Additionally, Theorem 3.17 provides a way to construct an isomorphism of orthomodular lattices from an isomorphism of their spectral presheaves by only considering four-element Boolean subalgebras; it is precisely when considering maximal four-element Boolean subalgebras that the process employed by [31] fails to construct a unique isomorphism.

Quantum logic and physical interpretation. Many considerations in physics can fundamentally be phrased in terms of *propositions*. Such propositions are of the form "the physical quantity A has a value in the (Borel) set $\Delta \subseteq \mathbb{R}$", short "$A \,\varepsilon\, \Delta$". Of course, the truth value of such a proposition depends on the state of the system.[1]

In classical physics, a proposition such as "$A \,\varepsilon\, \Delta$" is represented by a (Borel) subset of the state space $\mathscr{S}$ of the system. If $f_A : \mathscr{S} \to \mathbb{R}$ is the (Borel) function representing the physical quantity A, then the subset $f_A^{-1}(\Delta)$ of $\mathscr{S}$ contains all the states for which A has a value in Δ: if $s \in f_A^{-1}(\Delta)$, then $f_A(s) \in \Delta$. Hence, $f_A^{-1}(\Delta)$ represents the proposition "$A \,\varepsilon\, \Delta$". The Borel subsets of the state space $\mathscr{S}$ form a σ-complete Boolean algebra.

For quantum theory, such a state space picture is lacking. Instead, one uses the closed subspaces of Hilbert space as representatives of propositions. The closed subspaces form a complete orthomodular lattice, and this is the motivation to also consider more general orthomodular lattices as algebras modeling propositions in quantum theory and quantum logic.

The spectral presheaf $\underline{\Sigma}^L$ plays the role of a state space for the quantum system described by an OML L, akin to the classical state space $\mathscr{S}$. Our results so far show that the spectral presheaf is a complete invariant of an OML, which implies that instead of modeling quantum logic with the OML, one can model quantum logic based on the spectral presheaf without losing any information. To do so concretely, we will not just need the spectral presheaf $\underline{\Sigma}^L$ (this is like having the state space of a classical system only), but also a representation of the OML L, that is, of the propositions, by suitable subsets of the quantum state space (this is like having the algebra of Borel subsets of the form $f_A^{-1}(\Delta)$). The representation of the OML L by subsets—technically, subobjects—of $\underline{\Sigma}^L$ should generalise the well-known Stone representation for Boolean algebras. For a concrete generalisation of the Stone representation theorem to complete orthomodular lattices, see Sect. 4 and in particular Theorem 4.19 below.

3.9 The Spectral Presheaf of a Complete OML Is a Complete Invariant

We finally treat the case of *complete* orthomodular lattices (cOMLs). Note that the isomorphism result of Theorem 3.18 doesn't immediately apply to complete OMLs.

[1]In quantum theory, for most states a given proposition "$A \,\varepsilon\, \Delta$" is neither true nor false, but can only be assigned a probability, which is usually interpreted as the probability that when measuring the physical quantity A in the given state, a measurement outcome in Δ is obtained.

This is because the isomorphism between orthomodular lattices L and M constructed from an isomorphism from L_{part} to M_{part} in the proof of Theorem 3.17 is not necessarily a morphism of complete orthomodular lattices, that is, it may only preserve finite meets and joins, not arbitrary meets and joins.

Luckily, the extra effort needed in proving the results for complete OMLs is very moderate, and with a little care the proofs carry over virtually unchanged, with just the additional 'complete' in the right places. For this reason, we will not give all the details here.

Recall from Sect. 2.5 that the clopen subsets of the Stone space of a complete Boolean algebra form a complete Boolean algebra, and that there is a duality between complete Boolean algebras and *Stonean* spaces, which are extremely disconnected compact Hausdorff spaces. The appropriate kind of morphisms between these spaces are continuous open maps.

We first define the spectral presheaf of a complete OML:

Definition 3.22 Let L be a cOML, and let $\mathcal{B}_c(L)$ be its context category consisting of the *complete* Boolean subalgebras of L. The *spectral presheaf* $\underline{\Sigma}^L$ *of* L is the contravariant functor over $\mathcal{B}_c(L)$ given

(i) on objects: for all $B \in \mathcal{B}_c(L)$, let $\underline{\Sigma}^L_B := \Sigma(B)$, the Stonean space of B,
(ii) on arrows: for all inclusions $i_{B'B} : B' \to B$, let

$$\underline{\Sigma}^L(i_{B'B}) : \underline{\Sigma}^L_B \longrightarrow \underline{\Sigma}^L_{B'} \tag{181}$$

$$\lambda \longmapsto \lambda|_{B'}. \tag{182}$$

If we consider the Stonean spaces $\underline{\Sigma}^L_B$, $B \in \mathcal{B}_c(L)$, with their topology (and not just as mere sets), the restriction maps $\underline{\Sigma}^L(i_{B'B})$ are surjective, continuous, closed, and open. In particular, they are open since the inclusion $B' \hookrightarrow B$ is a morphism of cBAs, and the restriction $\underline{\Sigma}^L(i_{B'B})$ is the dual map to this inclusion, so by Proposition 2.26, $\underline{\Sigma}^L(i_{B'B})$ is a morphism in **Stonean**. Hence, the spectral presheaf $\underline{\Sigma}^L$ of a cOML L is an object in the category **Presh**(**Stonean**) of presheaves with values in Stonean spaces.

Definition 3.23 Let L be a cOML, and let $\mathcal{B}_c(L)$ be its context category. The *Bohrification* $\overline{\mathcal{L}}$ *of* L is the covariant functor over $\mathcal{B}_c(L)$ given

(i) on objects: for each $B \in \mathcal{B}_c(L)$, let $\overline{\mathcal{L}}_B := B$, the complete Boolean algebra B itself,
(ii) on arrows: for each inclusion arrow $i_{B'B}$, let $\overline{\mathcal{L}}(i_{B'B}) : B' \hookrightarrow B$ be the inclusion homomorphism of cBAs.

The Bohrification $\overline{\mathcal{L}}$ of a cOML L is an object in the category **Copresh**(**cBA**) of copresheaves with values in complete Boolean algebras.

We consider the action of the functors Σ and $\overline{CL}$ of the dual equivalence between **Copresh**(**cBA**) and **Presh**(**Stonean**) on Bohrifications and spectral presheaves, respectively. A brief check shows that they are just as for the general case of arbitrary orthomodular lattices in Sect. 3.5:

$$\Sigma(\overline{\mathscr{L}}) = \underline{\Sigma}^L \tag{183}$$

$$\Sigma(\langle I, \theta\rangle) = \langle I, \Sigma(\theta)\rangle \tag{184}$$

$$CL(\underline{\Sigma}^L) = cl \circ \underline{\Sigma}^L \tag{185}$$

$$CL(\langle\tilde{\varphi}, \zeta_\varphi\rangle) = \langle\tilde{\varphi}, \zeta_\varphi^{(-1)}\rangle \tag{186}$$

The lemmas and theorems of Sects. 3.6 and 3.7 also have analogous versions for the complete case:

Lemma 3.24 *The map* $\{Bo_B\}_{B\in\mathscr{B}_c(L)} : \overline{\mathscr{L}} \to cl \circ \underline{\Sigma}^L$ *is a natural isomorphism. That is, these two functors are isomorphic in the functor category* $\mathbf{cBA}^{\mathscr{B}_c(L)}$.

The natural isomorphism $\{Bo_B\}_{B\in\mathscr{B}_c(L)}$ will now simply be written in a slight abuse of notation as Bo for the sake of simplicity.

Lemma 3.25 *The morphism* $\langle Id_{\mathscr{B}_c(L)}, Bo\rangle : \overline{\mathscr{L}} \to cl \circ \underline{\Sigma}^L$ *is an isomorphism in* **Copresh(cBA)**.

Theorem 3.26 *Let L and M be complete orthomodular lattices,* $\underline{\Sigma}^L$ *and* $\underline{\Sigma}^M$ *their spectral presheaves, and* $\overline{\mathscr{L}}$ *and* $\overline{\mathscr{M}}$ *their Bohrifications. Then there is an isomorphism* $\underline{\Sigma}^M \to \underline{\Sigma}^L$ *in* **Presh(Stonean)** *if and only if there is an isomorphism* $\overline{\mathscr{L}} \to \overline{\mathscr{M}}$ *in* **Copresh(cBA)**, *and these isomorphisms can be explicitly constructed from each other.*

If $\langle H, \eta\rangle$ is an isomorphism between the spectral presheaves of cOMLs L and M, then the corresponding isomorphism from $\overline{\mathscr{L}}$ to $\overline{\mathscr{M}}$ in **Copresh(cBA)** is:

$$\rho := \langle Id_{\mathscr{B}(M)}, Bo^{-1}\rangle \circ \langle H, cl(\eta)\rangle \circ \langle Id_{\mathscr{B}(L)}, Bo\rangle = \langle H, (H^* Bo^{-1}) \circ cl(\eta) \circ Bo\rangle.$$

In particular, each component of the natural isomorphism $(H^* Bo^{-1}) \circ cl(\eta) \circ B$ is an isomorphism in **cBA**. This follows from Proposition 2.26, and in particular, Facts 2.27 and 2.28. That this isomorphism (renamed ρ in the proof of Theorem 3.17 as it is above) preserves arbitrary meets and joins is essential for being able to construct an isomorphism of cOMLs from an isomorphism of spectral presheaves in Theorem 3.28 below.

Theorem 3.27 *Let L and M be complete orthomodular lattices. If* $\varphi : L \to M$ *is an isomorphism in* **cOML**, *then there is an isomorphism* $\langle\tilde{\varphi}, \zeta_\varphi\rangle : \underline{\Sigma}^M \to \underline{\Sigma}^L$ *in* **Presh(Stonean)**, *where the natural transformation* ζ_φ *has components* $\zeta_{\varphi,B} = \Sigma(\varphi|_B)$ *for all B in* $\mathscr{B}_c(L)$.

Theorem 3.28 *Let L and M be complete orthomodular lattices. If there is an isomorphism* $\langle H, \eta\rangle : \underline{\Sigma}^M \to \underline{\Sigma}^L$ *in* **Presh(Stonean)**, *then there is an isomorphism from L to M in* **cOML** *that can be constructed explicitly from* $\langle H, \eta\rangle$.

The proof of this theorem for cOMLs is made possible by the fact that ρ is an isomorphism of *complete* Boolean algebras and thus induces an isomorphism of partial *complete* orthomodular lattices, which in turn determines a (unique) isomorphism of cOMLs. Summing up, we have:

Theorem 3.29 *Two complete orthomodular lattices L and M are isomorphic in* **cOML** *if and only if their spectral presheaves $\underline{\Sigma}^L$ and $\underline{\Sigma}^M$ are isomorphic in* **Presh**(**Stonean**).

4 Representing a Complete Orthomodular Lattice

The goal of this section is to find a 'representation' of a complete orthomodular lattice by clopen subobjects of its spectral presheaf, in analogy to the Stone representation of a Boolean algebra by clopen subsets of its Stone space.

In Sect. 4.1, we define and describe the clopen subobjects of the spectral presheaf of a complete orthomodular lattice, and in Sect. 4.2 we show that they form a complete bi-Heyting algebra. In Sect. 4.3, we define a map called 'daseinisation' from a complete orthomodular lattice to the clopen subobjects of its spectral presheaf. If we interpret the elements of the cOML as propositions about (the values of physical quantities of) a quantum system, then this map can be seen as a 'translation' of the quantum propositions into clopen subobjects of a generalised phase space. In Sects. 4.5 and 4.6, we use the adjoint of the daseinisation map to relate the lattice structure of the clopen subobjects of the spectral presheaf to the lattice structure of the original orthomodular lattice.

4.1 Clopen Subobjects of the Spectral Presheaf

For the remainder of this section, we assume L is a complete orthomodular lattice and $\underline{\Sigma}^L$ is its spectral presheaf.

Definition 4.1 Let $F : \mathscr{C} \to \mathbf{Set}$ be a functor. A functor $G : \mathscr{C} \to \mathbf{Set}$ is a *subfunctor of* F if for all $C \in Ob(\mathscr{C})$, $G_C \subseteq F_C$ and for all $a : C \to D$ in $Arr(\mathscr{C})$, $G(a)$ is the restriction of $F(a)$ to domain G_C and codomain G_D.

Note that this implies $G(a)(G_C) \subseteq G_D$.

Definition 4.2 A *subobject* of $\underline{\Sigma}^L$ is a subfunctor $\underline{S} : \mathscr{B}_c(L)^{op} \to \mathbf{Set}$ of $\underline{\Sigma}^L$.

This is the same definition of a subobject of $\underline{\Sigma}^L$ as in the topos sense. That is, recalling the definition of a subobject in a topos, subfunctors of $\underline{\Sigma}^L$ correspond precisely to monic arrows with codomain $\underline{\Sigma}^L$ in the functor category $\mathbf{Set}^{\mathscr{B}(L)^{\mathrm{op}}}$ (see e.g. [23]).

Definition 4.3 A subobject $\underline{S}$ of $\underline{\Sigma}^L$ is *clopen* if for all $B \in \mathscr{B}_c(L)$, the component $\underline{S}_B$ is a clopen subset of $\underline{\Sigma}^L_B$. The set of clopen subobjects of $\underline{\Sigma}^L$ will be denoted $\mathrm{Sub_{cl}}\,\underline{\Sigma}^L$.

There is an obvious partial order on $\mathrm{Sub_{cl}}\,\underline{\Sigma}^L$: let $\underline{S}$ and $\underline{T}$ be clopen subobjects in $\mathrm{Sub_{cl}}\,\underline{\Sigma}^L$. Then define

$$\underline{S} \leq \underline{T} \quad :\Longleftrightarrow \quad \forall B \in \mathscr{B}_c(L) : \underline{S}_B \subseteq \underline{T}_B.$$

With respect to this partial order, all meets and joins exist. Let $(\underline{S}_i)_{i\in I} \subseteq \mathrm{Sub_{cl}}\,\underline{\Sigma}^L$ be an arbitrary family of clopen subobjects. Their meet resp. join is given by

$$\forall B \in \mathscr{B}_c(L) : \left(\bigwedge_{i\in I} \underline{S}_i\right)_B = \mathrm{int} \bigcap_{i\in I} \underline{S}_{i;B}, \tag{187}$$

$$\left(\bigvee_{i\in I} \underline{S}_i\right)_B = \mathrm{cls} \bigcup_{i\in I} \underline{S}_{i;B}, \tag{188}$$

where $\underline{S}_{i;B}$ denotes the component of $\underline{S}_i$ at B. The interior resp. closure are taken with respect to the Stone topology. Since the Stone spaces of the complete Boolean (sub)algebras are Stonean, i.e. extremely disconnected, we obtain clopen subsets at each stage $B \in \mathscr{B}_c(L)$, and the meet and the join of clopen subobjects are clopen subobjects again as required.

Hence, $\mathrm{Sub_{cl}}\,\underline{\Sigma}^L$ is a complete lattice. It is also distributive, since meets and joins are defined stagewise, at each $B \in \mathscr{B}_c(L)$ separately. At $B \in \mathscr{B}_c(L)$, the meet and the join are the meet and join of clopen subsets of the Stonean space $\underline{\Sigma}^L_B$ of B, which form a complete Boolean algebra (which of course is distributive). It is easy to show that in $\mathrm{Sub_{cl}}\,\underline{\Sigma}^L$, finite meets distribute over arbitrary joins and finite joins distribute over arbitrary meets.

4.2 The Clopen Subobjects Form a Complete Bi-Heyting Algebra

It was shown in [12] that $\mathrm{Sub_{cl}}\,\underline{\Sigma}^L$ is a complete bi-Heyting algebra. For general information on bi-Heyting algebras, see [32]. For convenience of the reader, we briefly recall the definitions and main results.

Definition 4.4 A *Heyting algebra* is a bounded lattice H such that for all elements $a, b \in H$, there is a greatest element $x \in H$ such that $a \wedge x \leq b$. Such an element x is called the *relative pseudocomplement of a with respect to b* or the *Heyting implication from a to b* and is denoted $a \Rightarrow b$. The *pseudocomplement* of a, also called the *Heyting negation of a*, is the element $\neg a := (a \Rightarrow 0)$.

In the above definition, the element $\neg a$ is called a pseudocomplement of a because $a \wedge \neg a = 0$ but it is not necessarily true that $a \vee \neg a = 1$.

Definition 4.5 A Heyting algebra is *complete* if it is complete as a lattice.

In a complete Heyting algebra, finite meets distribute over arbitrary joins [32].

One can also define the dual notion of a co-Heyting algebra, also called a Brouwer algebra.

Definition 4.6 A *co-Heyting algebra* (or *Brouwer algebra*) is a bounded lattice H such that for all elements $a, b \in H$, there is a least element $x \in H$ such that $a \leq b \vee x$. Such an element x is called the *co-Heyting implication from a to b*, and is denoted $a \Leftarrow b$. The *co-Heyting negation of a* is the element $\sim a := (1 \Leftarrow a)$.

Dually to the negation in a Heyting algebra, the co-Heyting negation satisfies $a \vee \sim a = 1$ but it might not necessarily be true that $a \wedge \sim a = 0$.

Definition 4.7 A co-Heyting algebra is *complete* if it is complete as a lattice.

In a complete co-Heyting algebra, finite joins distribute over arbitrary meets [32].

Definition 4.8 A *bi-Heyting algebra* is a bounded lattice that is both a Heyting algebra and a co-Heyting algebra. A bi-Heyting algebra is *complete* if it is complete as a lattice.

A bi-Heyting algebra is distributive, but generalises a Boolean algebra by splitting up the notion of complementation into two separate notions, Heyting negation and co-Heyting negation. Heyting negation is intuitionistic, satisfying $a \wedge \neg a = 0$ but not necessarily $a \vee \neg a = 1$; logically, this means that the law of excluded middle need not hold. The co-Heyting negation is paraconsistent, satisfying $a \vee \sim a = 1$ but not necessarily $a \wedge \sim a = 0$; logically, this means that the law of noncontradiction need not hold.

Proposition 4.9 $\mathrm{Sub_{cl}}\, \underline{\Sigma}^L$ *is a complete bi-Heyting algebra.*

Proof We already saw that $\mathrm{Sub_{cl}}\, \underline{\Sigma}^L$ is a complete distributive lattice. It remains to show that there are both a Heyting and a co-Heyting structure on $\mathrm{Sub_{cl}}\, \underline{\Sigma}$.

The map

$$\underline{S} \wedge (-) : \mathrm{Sub_{cl}}\, \underline{\Sigma}^L \longrightarrow \mathrm{Sub_{cl}}\, \underline{\Sigma}^L \tag{189}$$

$$\underline{T} \longmapsto \underline{S} \wedge \underline{T} \tag{190}$$

is monotone and preserves arbitrary joins. Hence, by Proposition 2.30, this map has an upper adjoint which is denoted $\underline{S} \to (-)$ and is given by:

$$\underline{S} \Rightarrow (-) : \mathrm{Sub_{cl}}\, \underline{\Sigma}^L \longrightarrow \mathrm{Sub_{cl}}\, \underline{\Sigma}^L \tag{191}$$

$$\underline{T} \longmapsto (\underline{S} \Rightarrow \underline{T}) := \bigvee \{\underline{R} \in \mathrm{Sub_{cl}}\, \underline{\Sigma}^L \mid \underline{S} \wedge \underline{R} \leq \underline{T}\} \tag{192}$$

Additionally, by Proposition 2.31, this map satisfies

$$\underline{S} \wedge (\underline{S} \Rightarrow \underline{T}) \leq \underline{T}. \tag{193}$$

The map $\underline{S} \Rightarrow (-) : \mathrm{Sub_{cl}}\, \underline{\Sigma}^L \to \mathrm{Sub_{cl}}\, \underline{\Sigma}^L$ gives a well-defined Heyting implication in the complete distributive lattice $\mathrm{Sub_{cl}}\, \underline{\Sigma}^L$, with $\underline{S}$ varying over $\mathrm{Sub_{cl}}\, \underline{\Sigma}^L$. Thus, $\mathrm{Sub_{cl}}\, \underline{\Sigma}^L$ is a Heyting algebra. It is complete as $\mathrm{Sub_{cl}}\, \underline{\Sigma}^L$ is a complete lattice. The Heyting negation of this algebra will be denoted $\neg$, and is given by

$$\neg : \mathrm{Sub}_{\mathrm{cl}}\,\underline{\Sigma}^L \longrightarrow \mathrm{Sub}_{\mathrm{cl}}\,\underline{\Sigma}^L \tag{194}$$
$$\underline{S} \longmapsto \neg\underline{S} := (\underline{S} \Rightarrow \underline{0}). \tag{195}$$

Here $\underline{0}$ is the clopen subobject of $\underline{\Sigma}^L$ with $\underline{0}_B = \emptyset$ for all $B \in \mathscr{B}_c(L)$, the bottom element of $\mathrm{Sub}_{\mathrm{cl}}\,\underline{\Sigma}^L$.

Analogously, the following monotone map preserves arbitrary meets:

$$\underline{S} \vee (-) : \mathrm{Sub}_{\mathrm{cl}}\,\underline{\Sigma}^L \longrightarrow \mathrm{Sub}_{\mathrm{cl}}\,\underline{\Sigma}^L \tag{196}$$
$$\underline{T} \longmapsto \underline{S} \vee \underline{T} \tag{197}$$

Thus, by Proposition 2.30, it has a lower adjoint which we will call $(-) \Leftarrow \underline{S}$ given by:

$$(-) \Leftarrow \underline{S} : \mathrm{Sub}_{\mathrm{cl}}\,\underline{\Sigma}^L \longrightarrow \mathrm{Sub}_{\mathrm{cl}}\,\underline{\Sigma}^L \tag{198}$$
$$\underline{T} \longmapsto (\underline{T} \Leftarrow \underline{S}) := \bigwedge\{\underline{R} \in \mathrm{Sub}_{\mathrm{cl}}\,\underline{\Sigma}^L \mid \underline{T} \leq \underline{S} \vee \underline{R}\} \tag{199}$$

By Proposition 2.31, this map satisfies

$$\underline{T} \leq \underline{S} \vee (\underline{T} \Leftarrow \underline{S}) \tag{200}$$

It is clear by the definition of this map and Eq. 200 that this gives a co-Heyting implication for $\mathrm{Sub}_{\mathrm{cl}}\,\underline{\Sigma}^L$ (where $\underline{S}$ varies over $\mathrm{Sub}_{\mathrm{cl}}\,\underline{\Sigma}^L$), demonstrating that $\mathrm{Sub}_{\mathrm{cl}}\,\underline{\Sigma}^L$ is a complete co-Heyting algebra and thus a complete bi-Heyting algebra. The co-Heyting negation is given by

$$\sim : \mathrm{Sub}_{\mathrm{cl}}\,\underline{\Sigma}^L \longrightarrow \mathrm{Sub}_{\mathrm{cl}}\,\underline{\Sigma}^L \tag{201}$$
$$\underline{S} \longmapsto \sim\underline{S} := (\underline{\Sigma}^L \Leftarrow \underline{S}). \tag{202}$$

4.3 Daseinisation as Representation of a Complete OML

In this subsection we define a map from a complete orthomodular lattice L to the complete bi-Heyting algebra $\mathrm{Sub}_{\mathrm{cl}}\,\underline{\Sigma}^L$, called the *daseinisation map*. This can be interpreted as an approximation map, which for each element a of L 'brings into existence' an approximation of a as a subspace of each of the Stonean spaces Σ_B for $B \in \mathscr{B}_c(L)$ (hence the name). Daseinisation was first defined in a quantum theory context in [4] and discussed in detail in [11, 33] for the projection lattice of a von Neumann algebra. Here, we give a streamlined presentation and generalise to arbitrary complete OMLs.

Let L be a complete orthomodular lattice, let $a \in L$, and let $B \in \mathscr{B}_c(L)$ be a complete Boolean subalgebra of L, not necessarily containing a. Then, we define

$$\delta^o_B(a) := \bigwedge \{b \in B \mid b \geq a\}, \tag{203}$$

the smallest element of B that is greater than or equal to a. If $a \in B$, then $\delta^o_B(a) = a$. Note that the superscript of o denotes that this is outer daseinisation, that is, approximating element a in B from above. It is precisely at this step that completeness of orthomodular lattice L is required to define daseinisation; we need to know that the infinite meet in the definition of $\delta^o_B(a)$ exists.

Note that the inclusion map

$$B \hookrightarrow L \tag{204}$$
$$a \mapsto a \tag{205}$$

is a morphism of complete OMLs and hence preserves all meets in particular, so it has a lower adjoint, which is precisely

$$\delta^o_B : L \longrightarrow B \tag{206}$$
$$a \longmapsto \delta^o_B(a). \tag{207}$$

By Stone duality, we have an isomorphism between the complete Boolean algebra B and the clopen subobjects of its Stone space Σ_B, which is Stonean because B is complete. From Sect. 2.3, this isomorphism is given by

$$Bo_B : B \longrightarrow cl(\Sigma_B) = cl(\underline{\Sigma}^L_B) \tag{208}$$
$$b \longmapsto \{\lambda \in \underline{\Sigma}^L_B \mid \lambda(b) = 1\} \tag{209}$$

Recall cl is the functor which maps a Stone space to its Boolean algebra of clopen subsets. Here, Bo_B is an isomorphism of complete Boolean algebras. In particular, the element $\delta^o_B(a)$ of B corresponds to the clopen subset of $\underline{\Sigma}^L_B$ given by:

$$\underline{\delta}^o{}_B(a) := Bo_B(\delta^o_B(a)) = \{\lambda \in \underline{\Sigma}^L_B \mid \lambda(\delta^o_B(a)) = 1\}. \tag{210}$$

(The reason for using the notation with underlining will become clear shortly.)

Suppose that $B' \subseteq B$ in $\mathcal{B}_c(L)$. Clearly, it holds that $\delta^o_B(a) \leq \delta^o_{B'}(a)$. Then,

$$\lambda \in \underline{\delta}^o{}_B(a) \Leftrightarrow \lambda(\delta^o_B(a)) = 1 \tag{211}$$
$$\Rightarrow \lambda(\delta^o_{B'}(a)) = 1 \tag{212}$$
$$\Leftrightarrow \lambda|_{B'}(\delta^o_{B'}(a)) = 1 \tag{213}$$
$$\Leftrightarrow \lambda|_{B'} \in \underline{\delta}^o{}_{B'}(a) \tag{214}$$

We conjecture that this result can be strengthened to show that $\lambda \in \underline{\delta}^o{}_B(a)$ if and only if $\lambda|_{B'} \in \underline{\delta}^o{}_{B'}(a)$, but such a result is not necessary for our purposes so we do not pursue this line of investigation. Note that this result implies that for every inclusion arrow $i_{B',B}$ in $\mathcal{B}_c(L)$, the restriction of $\underline{\Sigma}^L(i_{B',B}) = r_{B,B'}$ to domain $\underline{\delta}^o{}_B(a) \subseteq \underline{\Sigma}^L_B$

has codomain contained in $\underline{\delta^o}_{B'}(a) \subseteq \underline{\Sigma}^L_{B'}$. This means that the functor from $\mathcal{B}_c(L)$ to **Set** which sends B to $\underline{\delta^o}_B(a)$ is a valid subfunctor of $\underline{\Sigma}^L$; we will call this functor $\underline{\delta^o}(a)$. Clearly this functor

$$\underline{\delta^o}(a) := \left(\underline{\delta^o}_B(a)\right)_{B \in \mathcal{B}_c(L)} \tag{215}$$

is thus also a subobject of the spectral presheaf. It is a clopen subobject because for each $B \in \mathcal{B}_c(L)$, the subset $\underline{\delta^o}_B(a)$ of Σ_B is clopen.

We are now ready to define the daseinisation map for complete orthomodular lattice L and discuss its properties.

Definition 4.10 The map

$$\underline{\delta^o} : L \longrightarrow \mathrm{Sub_{cl}}\, \underline{\Sigma}^L \tag{216}$$

$$a \longmapsto \underline{\delta^o}(a) \tag{217}$$

from the complete orthomodular lattice L to the complete bi-Heyting algebra $\mathrm{Sub_{cl}}\, \underline{\Sigma}^L$ is called *outer daseinisation*, or more simply *daseinisation.*

The daseinisation map can be seen as a process by which an element a in the complete orthomodular lattice L is approximated in each classical context B and subsequently each Stone space Σ_B, ultimately yielding a clopen subobject of $\underline{\Sigma}^L$. Returning to the notion of an orthomodular lattice as a quantum logic whose elements are propositions, for each classical context B the daseinisation process first associates to proposition a the strongest proposition within B that must be true if proposition a is true, which above we called $\delta^o_B(a)$. The next step of daseinisation associates to each of these strongest propositions the collection of local valuations (elements of the Stone space of B, i.e., Boolean algebra homomorphisms from B to $\{0, 1\}$) for which the proposition holds, which we called $\underline{\delta^o}_B(a)$. These sets of local valuations are clopen and are linked together by the restriction maps to create a clopen subobject $\underline{\delta^o}(a)$. This analysis shows that the daseinisation process associates to each quantum proposition a subobject of the spectral presheaf of the complete orthomodular lattice to which it belongs, just as a classical proposition corresponds to a subset of the state space of the classical system.

Lemma 4.11 *The daseinisation map* $\underline{\delta^o} : L \to \mathrm{Sub_{cl}}\, \underline{\Sigma}^L$ *has the following properties:*

1. $\underline{\delta^o}(0) = \underline{0}$, $\underline{\delta^o}(1) = \underline{\Sigma}^L$,
2. $\underline{\delta^o}$ *is monotone, that is,* $a \leq b$ *in* L *implies* $\underline{\delta^o}(a) \leq \underline{\delta^o}(b)$ *in* $\mathrm{Sub_{cl}}\, \underline{\Sigma}^L$,
3. $\underline{\delta^o}$ *is injective, but not surjective,*
4. $\underline{\delta^o}$ *preserves all joins.*

Proof (1) is obvious form the definition of $\underline{\delta^o}$; for all $B \in \mathcal{B}_c(L)$, $\delta^o_B(0) = 0$ and $\underline{\delta^o}_B(0) = \emptyset$. Similarly, $\delta^o_B(1) = 1$ and $\underline{\delta^o}_B(1) = \Sigma_B = \underline{\Sigma}^L_B$.

(2) also follows directly from the definition of $\underline{\delta^o}$. If $a \leq b$, then $\delta^o_B(a) \leq \delta^o_B(b)$ and $\underline{\delta^o}_B(a) \subseteq \underline{\delta^o}_B(b)$ for all $B \in \mathcal{B}_c(L)$, meaning that $\underline{\delta^o}(a) \leq \underline{\delta^o}(b)$ in $\mathrm{Sub_{cl}}\, \underline{\Sigma}^L$.

For (3), let a and b be distinct elements of L. Then,

$$\bigwedge_{B\in\mathscr{B}_c(L)} \delta^o_B(a) = a \neq b = \bigwedge_{B\in\mathscr{B}_c(L)} \delta^o_B(b) \tag{218}$$

This implies that there must be some $B \in \mathscr{B}_c(L)$ such that $\delta^o_B(a) \neq \delta^o_B(b)$. As Bo_B is a complete Boolean algebra isomorphism, it follows for this B that

$$\underline{\delta}^o{}_B(a) = Bo_B(\delta^o_B(a)) \neq Bo_B(\delta^o_B(a)) = \underline{\delta}^o{}_B(b) \tag{219}$$

As $\underline{\delta}^o(a)$ and $\underline{\delta}^o(b)$ differ at this component, then they are not the same subobject of $\underline{\Sigma}^L$. Thus, $\underline{\delta}^o$ is injective. On the other hand, $\underline{\delta}^o$ clearly is not surjective, since not every clopen subobject of $\underline{\Sigma}^L$ is of the form $\underline{\delta}^o(a)$ for some $a \in L$.

For (4), note that joins are colimits, which are calculated stagewise, at each $B \in \mathscr{B}_c(L)$ separately. We saw that for each B, the map $\delta^o_B : L \to B, a \mapsto \delta^o_B(a)$, is the lower adjoint of the inclusion map $B \hookrightarrow L$, so it preserves all colimits. The map cl that takes $\delta^o_B(a)$ to $\underline{\delta}^o{}_B(a)$ is an isomorphism of complete Boolean algebras, so it preserves all joins.

Stone duality provides a representation of every complete Boolean algebra B by a concrete complete Boolean algebra, viz. the algebra of clopen subsets of the Stone space Σ_B,

$$B \longrightarrow cl(\Sigma_B). \tag{220}$$

In analogy, and as a generalisation, daseinisation can be interpreted as providing a 'representation' of every complete orthomodular lattice L by a concrete algebra of clopen subobjects of a generalised Stone space, viz. the spectral presheaf $\underline{\Sigma}^L$,

$$L \longrightarrow \mathrm{Sub_{cl}}\,\underline{\Sigma}^L. \tag{221}$$

We saw that $\underline{\Sigma}^L$ is a complete invariant of L (Theorem 3.18) and generalises the Stone space Σ_B of a complete Boolean algebra B in a straightforward manner. Of course, the algebra $\mathrm{Sub_{cl}}\,\underline{\Sigma}^L$ in which we are 'representing' the cOML L is not a cOML itself, but is a complete bi-Heyting algebra. The representation provided by $\underline{\delta}^o$ preserves top and bottom element, the order and all joins. Moreover, the representation is faithful, since $\underline{\delta}^o$ is an injective map (Lemma 4.11).

In Sect. 4.5, we will show that daseinisation has an adjoint, which will then allow us to regain the cOML L from the complete bi-Heyting algebra $\mathrm{Sub_{cl}}\,\underline{\Sigma}^L$ in (Sect. 4.6), further strengthening the analogy with Stone representation. Bur first, we will give some physical interpretation of the results so far.

4.4 Some Physical Interpretation

The representation $\underline{\delta}^o : L \to \underline{\Sigma}^L$ of a complete orthomodular lattice L is structurally similar to the Stone representation of a Boolean algebra, and the interpretation of the spectral presheaf as a state space for the quantum system is vindicated.

Yet, the fact that we map from L into a complete bi-Heyting algebra (and not into another complete OML) may seem to be a drawback at first sight. We will give some brief arguments why, on the contrary, the bi-Heyting algebra picture provides many advantages compared to standard quantum logic [15, 16]. Some more discussion can be found in [4, 11, 12].

Distributivity and existence of a material implication. One key problem of standard quantum logic is the lack of a material implication. In a bi-Heyting algebra, the Heyting implication, $\Rightarrow$, plays the role of a material implication and hence is given as part of the structure. The existence of the Heyting implication depends on the distributivity of the underlying lattice (see the argument after Eq. (189)). The fact that the lattice in which we represent propositions about our quantum system is distributive has further advantages. The behaviour of meets and joins has a clear interpretation, and situations such as the 'quantum breakfast' do not pose any interpretational issues.[2]

Availability of higher-order logic. By daseinisation $\underline{\delta}^o : L \to \underline{\Sigma}^L$ we do not just map into a bi-Heyting algebra, but this algebra is given by the (clopen) subobjects of a presheaf, which is an object in the presheaf topos $\mathbf{Set}^{\mathcal{B}(L)^{\mathrm{op}}}$. The topos comes equipped with an internal higher-order intuitionistic logic [20, 21, 23], which can now be employed for quantum theory. This is largely a task for the future.

Superposition without linearity. One characteristic feature of quantum theory is the existence of superposition. In a given (vector) state $|\psi\rangle$, the disjunction of two propositions P, Q can be true while neither of the propositions is true. For example, if p, q are one-dimensional subspaces that represent the propositions P, Q and the one-dimensional subspace $\mathbb{C}\,|\psi\rangle$ lies in the plane spanned by p and q (without being equal to either p or q), then the state $|\psi\rangle$ makes the proposition (P or Q) true without making P true or Q true. We see that superposition relates to the behaviour of joins, given by spans of subspaces of a linear space. Interestingly, our representation $L \to \underline{\Sigma}^L$ preserves all joins and hence preserves that structural aspect of orthomodular lattices which comes from superposition. This is true despite the fact that different from Hilbert space the spectral presheaf, which is the underlying (generalised) space, is not a linear space.

Good interpretation of all conjunctions. Every quantum system has incompatible physical quantities that cannot be measured simultaneously. In fact, a context is usually understood to be a subset of physical quantities that can be measured simultaneously. Accordingly, certain propositions of the form "$A\,\varepsilon\,\Delta$ and $B\,\varepsilon\,\Gamma$" about the

---

[2]For those not familiar with this example: if you go to the quantum hotel and they offer you eggs and (bacon or sausage), you cannot expect to get (eggs and bacon) or (eggs and sausage) due to nondistributivity of 'and' and 'or'. As a formula, $e \wedge (b \vee s) \neq (e \wedge s) \vee (e \wedge s)$ in general in an orthomodular lattice.

values of incompatible physical quantities should be meaningless, since there is no possible experiment that could measure A and B simultaneously. In an orthomodular lattice, any pair of elements has a meet, so there are many meets that have no good physical interpretation.

In our bi-Heyting algebra $\mathrm{Sub_{cl}}\,\underline{\Sigma}^L$, all meets exist as well, but nonetheless we avoid the interpretational problem described above: meets in $\mathrm{Sub_{cl}}\,\underline{\Sigma}^L$ are taken stagewise, in each context separately. (Here, a context is a collection of compatible propositions, forming a Boolean algebra.) If we start from two incompatible propositions, we first apply a process of coarse-graining. Each proposition is approximated by a weaker proposition in every context (see Eq. (203)). The meet is then taken only between compatible propositions, each of which is a weakening of the original proposition. For example, if we consider a context B that contains an element p that represents the proposition "$A \,\varepsilon\, \Delta$", then $\delta^o_B(p) = p$, so the 'approximation' to p within the context B is p itself, as expected. If q is another element of the OML that represents the (incompatible) proposition "$B \,\varepsilon\, \Gamma$", then q is not contained in the context B, so $\delta^o_B(q) \gneq q$, and the approximation of q within B represents a properly weaker proposition than the original one. The meet at B is the meet $p \wedge \delta^o_B(q)$, and analogously for all other contexts, including those contexts $\tilde{B}$ that contain q (where p has to be properly approximated). In this way, we avoid taking any meets of incompatible elements.

Additional paraconsistent fragment of the logic. Apart from the Heyting algebra aspect, which provides an intuitionistic logic for every quantum system, there is also a co-Heyting algebra aspect. Logically, this represents a paraconsistent logic. Some properties of the Heyting and co-Heyting structure are discussed in [12]. A bi-Heyting algebra can be seen as a fairly modest generalisation of a Boolean algebra. The concept of negation becomes split into a Heyting negation (pseudocomplement) for which the law of the excluded middle does not hold (i.e., $a \wedge \neg a = 0$, but $a \vee \neg a \leq 1$), and a co-Heyting negation, for which the law of non-contradiction does not hold (i.e., $a \vee \sim a = 1$, but $a \wedge \sim a \geq 0$). The latter property is a direct consequence of coarse-graining and is not problematic interpretationally.

Summing up, our representation $\underline{\delta}^o : L \to \underline{\Sigma}^L$ translates from standard quantum logic to a new, distributive form of logic for quantum systems that has many good interpretational properties. Daseinisation 'creates' distributivity and splits negation into two concepts, relating to the Heyting and the co-Heyting fragment, respectively.

In the following subsections, we will show that daseinisation has an adjoint that can be used to map back from the complete bi-Heyting algebra $\mathrm{Sub_{cl}}\,\underline{\Sigma}^L$ to the complete OML L. This gives an even closer link between our new form of quantum logic and standard quantum logic formulated in OMLs.

4.5 The Adjoint of Daseinisation

$\underline{\delta}^o$ is a join-preserving map between two complete lattices, so by Proposition 2.30, $\underline{\delta}^o$ has a meet-preserving upper adjoint $\varepsilon : \mathrm{Sub}_{\mathrm{cl}}\,\underline{\Sigma}^L \to L$. This map ε is defined by:

$$\varepsilon : \mathrm{Sub}_{\mathrm{cl}}\,\underline{\Sigma}^L \longrightarrow L \tag{222}$$

$$\underline{S} \longmapsto \bigvee\{a \in L \mid \underline{\delta}^o(a) \leq \underline{S}\}. \tag{223}$$

The following lemma, adapted from an unpublished result by Carmen Constantin, provides more insight into this map ε.

Lemma 4.12 *Let L be a complete orthomodular lattice, with spectral presheaf $\underline{\Sigma}^L$. The upper adjoint ε of $\underline{\delta}^o$ is given by*

$$\varepsilon : \mathrm{Sub}_{\mathrm{cl}}\,\underline{\Sigma}^L \to L \tag{224}$$

$$\underline{S} \mapsto \bigwedge_{B \in \mathscr{B}_c(L)} Bo_B^{-1}(\underline{S}_B) \tag{225}$$

Proof Suppose that a is some lower bound for the set $\{Bo_B^{-1}(\underline{S}_B) \mid B \in \mathscr{B}_c(L)\}$. That is, for each $B \in \mathscr{B}_c(L)$,

$$a \leq Bo_B^{-1}(\underline{S}_B). \tag{226}$$

As $Bo_B^{-1}(\underline{S}_B)$ is an element of B that is greater than or equal to a and $\delta^o_B(a)$ is the least element of B that is greater than or equal to a, then

$$a \leq Bo_B^{-1}(\underline{S}_B) \tag{227}$$

$$\Leftrightarrow \quad \delta^o_B(a) \leq Bo_B^{-1}(\underline{S}_B) \tag{228}$$

$$\Leftrightarrow \quad Bo_B(\delta^o_B(a)) \subseteq Bo_B(Bo_B^{-1}(\underline{S}_B)) \tag{229}$$

$$\Leftrightarrow \quad \underline{\delta}^o{}_B(a) \subseteq \underline{S}_B \tag{230}$$

This exactly characterises the lower bounds a of the set $\{Bo_B^{-1}(\underline{S}_B) \mid B \in \mathscr{B}_c(L)\}$. That is,

$$\{a \in L \mid a \leq Bo_B^{-1}(\underline{S}_B)\ \forall\ B \in \mathscr{B}_c(L)\} = \{a \in L \mid \underline{\delta}^o{}_B(a) \subseteq \underline{S}_B\ \forall\ B \in \mathscr{B}_c(L)\} \tag{231}$$

$$= \{a \in L \mid \underline{\delta}^o(a) \leq \underline{S}\}. \tag{232}$$

In a complete lattice, joins can be written in terms of meets. That is,

$$\bigwedge_{B\in\mathcal{B}_c(L)} Bo_B^{-1}(\underline{S}_B) = \bigvee\{a\in L \mid a\leq Bo_B^{-1}(\underline{S}_B)\ \forall\ B\in\mathcal{B}_c(L)\} \tag{233}$$

$$= \bigvee\{a\in L \mid \underline{\delta}^o(a)\leq \underline{S}\} \tag{234}$$

$$= \varepsilon(\underline{S}). \tag{235}$$

The previous lemma implies the following result, which is stronger than could be expected for an arbitrary Galois connection:

Lemma 4.13 $\varepsilon\circ\underline{\delta}^o = id_L$.

Proof Let $a\in L$. Then,

$$(\varepsilon\circ\underline{\delta}^o)(a) = \varepsilon(\underline{\delta}^o(a)) = \bigwedge_{B\in\mathcal{B}_c(L)} Bo_B^{-1}(\underline{\delta}^o{}_B(a)) \tag{236}$$

$$= \bigwedge_{B\in\mathcal{B}_c(L)} Bo_B^{-1}(Bo_B(\delta^o_B(a))) \tag{237}$$

$$= \bigwedge_{B\in\mathcal{B}_c(L)} \delta^o_B(a) \tag{238}$$

$$= a. \tag{239}$$

From the general properties of a Galois connection, it also follows that

$$\underline{\delta}^o\circ\varepsilon\leq id_{\mathrm{Sub_{cl}}\,\underline{\Sigma}^L}. \tag{240}$$

Lemma 4.14 *The map* $\varepsilon:\mathrm{Sub_{cl}}\,\underline{\Sigma}^L\to L$ *has the following properties:*

1. $\varepsilon(\underline{0})=0$, $\varepsilon(\underline{\Sigma}^L)=1$,
2. ε *is monotone,*
3. ε *is surjective, but not injective,*
4. ε *preserves all meets.*

Proof (1) and (2) are obvious. For (3), note that if $a\in L$, then $a=(\varepsilon\circ\underline{\delta}^o)(a)$ by Lemma 4.13, so a is in the image of ε.

(4) holds since ε is an upper adjoint, which preserves all limits, which are meets here.

4.6 *Regaining a cOML from the Algebra of Clopen Subobjects*

It is clear that a cOML L and the complete bi-Heyting algebra $\mathrm{Sub_{cl}}\,\underline{\Sigma}^L$ cannot be isomorphic as lattices in general, because L is not necessarily distributive but

$\mathrm{Sub}_{\mathrm{cl}}\,\underline{\Sigma}^L$ is. Additionally, $\mathrm{Sub}_{\mathrm{cl}}\,\underline{\Sigma}^L$ contains significantly more elements in general than L. However, we will show that by forming certain equivalence classes within $\mathrm{Sub}_{\mathrm{cl}}\,\underline{\Sigma}^L$, we obtain a complete OML that is isomorphic to L. Of course, Lemma 4.13 already gives a clear hint that it is possible to reconstruct L from $\mathrm{Sub}_{\mathrm{cl}}\,\underline{\Sigma}^L$, and we will make this explicit now.

We can use the map ε to define an equivalence relation on $\mathrm{Sub}_{\mathrm{cl}}\,\underline{\Sigma}^L$:

Definition 4.15 For $\underline{S}, \underline{T}$ in $\mathrm{Sub}_{\mathrm{cl}}\,\underline{\Sigma}^L$, define $\underline{S} \sim \underline{T}$ if and only if $\varepsilon(\underline{S}) = \varepsilon(\underline{T})$.

This is clearly a well-defined equivalence relation. We will write $[\underline{S}]$ for the equivalence class of $\underline{S}$. Let

$$E := \left\{ [\underline{S}] \mid \underline{S} \in \mathrm{Sub}_{\mathrm{cl}}\,\underline{\Sigma}^L \right\} \tag{241}$$

be the set of equivalence classes. We observe right away that E is a partially ordered set in a canonical manner: define

$$[\underline{S}] \leq [\underline{T}] :\Leftrightarrow \varepsilon(\underline{S}) \leq \varepsilon(\underline{T}). \tag{242}$$

Then $[\underline{\emptyset}]$ is the bottom element and $[\underline{\Sigma}^L]$ is the top element.

Lemma 4.16 *Let $[\underline{S}] \in E$. Then $[(\underline{\delta}^o \circ \varepsilon)(\underline{S})] = [\underline{S}]$ and $(\underline{\delta}^o \circ \varepsilon)(\underline{S})$ is the smallest representative of $[\underline{S}]$.*

Proof By Proposition 2.31, we have $\varepsilon(\underline{\delta}^o \circ \varepsilon)(\underline{S}) = \varepsilon(\underline{S})$, so $[(\underline{\delta}^o \circ \varepsilon)(\underline{S})] = [\underline{S}]$. Moreover, if $\underline{T}$ is a representative of $[\underline{S}]$, then $\varepsilon(\underline{T}) = \varepsilon(\underline{S})$ and since $\underline{\delta}^o \circ \varepsilon \leq id_{\mathrm{Sub}_{\mathrm{cl}}\,\underline{\Sigma}^L}$,

$$\underline{T} \geq (\underline{\delta}^o \circ \varepsilon)(\underline{T}) = (\underline{\delta}^o \circ \varepsilon)(\underline{S}). \tag{243}$$

Lemma 4.17 *There is a bijective set map from E to the set underlying the complete orthomodular lattice L, given by*

$$g : E \longrightarrow L \tag{244}$$

$$[\underline{S}] \longmapsto \varepsilon(\underline{S}) \tag{245}$$

Proof Clearly g is well defined, as if $[\underline{S}] = [\underline{T}]$ then $g([\underline{S}]) = \varepsilon(\underline{S}) = \varepsilon(\underline{T}) = g([\underline{T}])$ by definition. Consider the function

$$f : L \to E \tag{246}$$

$$a \mapsto [\underline{\delta}^o(a)] \tag{247}$$

We will now show that f is an inverse to g, meaning E and L are isomorphic as sets. First, let $a \in L$. Then, by Lemma 4.13,

$$(g \circ f)(a) = g\left([\underline{\delta}^o(a)]\right) = \varepsilon\left(\underline{\delta}^o(a)\right) = a. \tag{248}$$

Now, let $\underline{S} \in \mathrm{Sub_{cl}}\,\underline{\Sigma}^L$. Then,

$$(f \circ g)\left([\underline{S}]\right) = f\left(\varepsilon(\underline{S})\right) = \left[\underline{\delta}^o\left(\varepsilon(\underline{S})\right)\right] = [\underline{S}], \tag{249}$$

where we used Lemma 4.16 in the last step. Thus, as both compositions of f and g are the identity, then $g : E \to L$ is a set bijection, and $f = g^{-1}$.

We can now use g (and g^{-1}) to equip E with the structure of a complete OML canonically: define the order by

$$[\underline{S}] \leq [\underline{T}] :\Leftrightarrow g([\underline{S}]) \leq g([\underline{T}]). \tag{250}$$

Since $g([\underline{S}]) \leq g([\underline{T}]) \Leftrightarrow \varepsilon(\underline{S}) \leq \varepsilon(\underline{T})$, this is exactly the order we had defined on E before. Since g is a bijection (and an order isomorphism, as we now know), all meets and joins in E with respect to this order exist and correspond to those in L by construction. Moreover, following Eva [34], one defines an orthocomplementation on E by

$$' : E \longrightarrow E \tag{251}$$

$$[\underline{S}] \longmapsto [\underline{S}]' := g^{-1}(g([\underline{S}])^{\perp}) = [\underline{\delta}^o(\varepsilon(\underline{S})^{\perp})]. \tag{252}$$

This makes E into a cOML that is isomorphic to L. The maps g and g^{-1} are isomorphisms of cOMLs.

We want to relate the lattice structure on $E = \mathrm{Sub_{cl}}\,\underline{\Sigma}^L/\sim$ more directly to the lattice structure on the bi-Heyting algebra $\mathrm{Sub_{cl}}\,\underline{\Sigma}^L$ of clopen subsets. The meets in E can be written in terms of the meets in $\mathrm{Sub_{cl}}\,\underline{\Sigma}^L$ as follows:

Lemma 4.18 *For all families $\left([\underline{S}_i]\right)_{i\in I}$ of elements of E, where $\underline{S}_i \in \mathrm{Sub_{cl}}\,\underline{\Sigma}^L$,*

$$\bigwedge_{i\in I}[\underline{S}_i] = \left[\bigwedge_{i\in I}\underline{S}_i\right]. \tag{253}$$

Proof We have

$$\bigwedge_{i\in I}[\underline{S}_i] = g^{-1}(\bigwedge_{i\in I} g([\underline{S}_i])) = [\underline{\delta}^o(\bigwedge_{i\in I}\varepsilon(\underline{S}_i))] = [\underline{\delta}^o(\varepsilon(\bigwedge_{i\in I}\underline{S}_i)] = [\bigwedge_{i\in I}\underline{S}_i], \tag{254}$$

where we applied Lemma 4.16 in the last step.

As in any complete lattice, the joins in E can be written in terms of the meets. For all families $\left([\underline{S}_i]\right)_{i\in I}$ of elements of E, where $\underline{S}_i \in \mathrm{Sub_{cl}}\,\underline{\Sigma}^L$,

$$\bigvee_{i\in I}[\underline{S}_i] := \bigwedge\{[\underline{T}] \mid [\underline{S}_i] \leq [\underline{T}] \;\; \forall\; i \in I\}. \tag{255}$$

Note that in general,

$$\bigvee_{i \in I} [\underline{S}_i] \neq \left[\bigvee_{i \in I} \underline{S}_i \right]. \tag{256}$$

Summing up, we have the following generalisation of the Stone representation theorem to complete orthomodular lattices:

Theorem 4.19 *For every complete orthomodular lattice L, there exists a map*

$$\underline{\delta}^o : L \longrightarrow \text{Sub}_{\text{cl}}\, \underline{\Sigma}^L \tag{257}$$
$$a \longmapsto \underline{\delta}^o(a) \tag{258}$$

called daseinisation into the complete bi-Heyting algebra of clopen subobjects of the spectral presheaf $\underline{\Sigma}^L$ of L. This map is injective, preserves top and bottom elements, the order and all joins.

The adjoint of $\underline{\delta}^o$ is a map

$$\varepsilon : \text{Sub}_{\text{cl}}\, \underline{\Sigma}^L \longrightarrow L \tag{259}$$
$$\underline{S} \longmapsto \bigwedge_{B \in \mathscr{B}_c(L)} Bo_B^{-1}(\underline{S}_B) \tag{260}$$

taking clopen subobjects to elements of the cOML L. The map ε is surjective, preserves top and bottom elements, the order and all meets. The quotient $E = \text{Sub}_{\text{cl}}\, \underline{\Sigma}^L / \sim$, where $\underline{S} \sim \underline{T}$ if and only if $\varepsilon(\underline{S}) = \varepsilon(\underline{T})$, is canonically isomorphic to L as a complete orthomodular lattice.

5 Conclusion

We conclude with a list of some open problems:

- How does the complement in $E = \text{Sub}_{\text{cl}}\, \underline{\Sigma}^L / \sim$, given by $[\underline{S}]' = [\underline{\delta}^o(\varepsilon(\underline{S})^{\perp})]$, relate to the Heyting and co-Heyting negation in $\text{Sub}_{\text{cl}}\, \underline{\Sigma}^L$?
- Can the representation suggested in Sect. 4 be generalised from complete OMLs to all OMLs?
- Is there a characterisation of those posets that can show up as context categories of orthomodular lattices?
- How can we employ the logic of the presheaf topos $\mathbf{Set}^{\mathscr{B}(L)^{\text{op}}}$ to discuss higher-order aspects of the new logic for quantum systems?
- How does the work presented here relate to Quantum Set Theory and OML-valued models?[3]

[3]On this topic, there is some ongoing work with Masanao Ozawa and Benjamin Eva.

In a more general perspective, one can ask which kind of nondistributive or noncommutative structures allow us to associate a spectral presheaf with them and prove duality type or (partial) representation results? A necessary precondition seems to be that the nondistributive or noncommutative structure under consideration has distributive or commutative parts which each have a dual space given by one of the classical dualities. For example, it is conceivable that compact Lie groups are amenable to methods similar to those developed in this article. A context in a compact Lie group would be a Lie-commuting compact subgroup. By Pontryagin duality, we obtain dual spaces that can be fit together into a spectral presheaf, and one can consider the question if this is a complete invariant of the compact Lie group. As a more direct generalisation of the algebras considered in this article, one could use the duality between spatial frames and sober spaces as a starting point. We hope to come back to these problems in the future.

Acknowledgements We are very grateful to Masanao Ozawa for giving us the opportunity to contribute an article (without page limit!) to the proceedings of the *Nagoya Winter Workshop* 2015. Moreover, AD thanks Masanao for his continued generosity and friendship over the years and for many discussions. We also thank Chris Isham, Rui Soares Barbosa, Carmen Constantin, Boris Zilber, and Benjamin Eva for discussions.

References

1. M.H. Stone. The theory of representations for Boolean algebras. *Trans. Amer. Math. Soc.*, 40:37–111, 1936.
2. Peter T. Johnstone. *Stone Spaces*, volume 3 of *Cambridge studies in advanced mathematics*. Cambridge University Press, Cambridge, 1982.
3. Andreas Döring and Christopher J. Isham. A topos foundation for theories of physics: I. Formal languages for physics. *J. Math. Phys.*, 49, Issue 5:053515, 2008.
4. Andreas Döring and Christopher J. Isham. A topos foundation for theories of physics: II. Daseinisation and the liberation of quantum theory. *J. Math. Phys.*, 49, Issue 5:053516, 2008.
5. Andreas Döring and Christopher J. Isham. A topos foundation for theories of physics: III. Quantum theory and the representation of physical quantities with arrows. *J. Math. Phys.*, 49, Issue 5:053517, 2008.
6. Andreas Döring and Christopher J. Isham. A topos foundation for theories of physics: IV. Categories of systems. *J. Math. Phys.*, 49, Issue 5:053518, 2008.
7. Andreas Döring and Christopher J. Isham. "What is a thing?": Topos theory and the foundations of physics. In Bob Coecke, editor, *New Structures for Physics*, volume 13 of *Lecture Notes in Physics*, pages 753–937. Springer Berlin Heidelberg, 2011.
8. C. Heunen, N.P. Landsman, and B. Spitters. A topos for algebraic quantum theory. *Communications in Mathematical Physics*, 291:63–110, 2009.
9. J. Hamilton, C. J. Isham, and J. Butterfield. A topos perspective on the Kochen-Specker theorem: III. Von Neumann algebras as the base category. *International Journal of Theoretical Physics*, 39(6):1413–1436, 2000.
10. Andreas Döring. Topos theory and 'neo-realist' quantum theory. In *Quantum Field Theory, Competitive Models*, pages 25–48. Birkhäuser, Basel, Boston, Berlin, 2009.
11. Andreas Döring. Topos quantum logic and mixed states. *Proceedings of the 6th International Workshop on Quantum Physics and Logic (QPL 2009), Oxford. Electronic Notes in Theoretical Computer Science*, 270(2), 2011.

12. Andreas Döring. Topos-based logic for quantum systems and bi-Heyting algebras. In *Logic and Algebraic Structures in Quantum Computing*. Cambridge University Press, Cambridge, 2016.
13. Andreas Döring. Generalised Gelfand spectra of nonabelian unital C^*-algebras I: Categorical aspects, automorphisms and Jordan structure. *ArXiv e-prints*, December 2012.
14. Andreas Döring. Generalised Gelfand spectra of nonabelian unital C^*-algebras I: Flows and time evolution of quantum systems. *ArXiv e-prints*, December 2012.
15. M.L. Dalla Chiara and R. Giuntini. Quantum logics. In *Handbook of Philosophical Logic*, pages 129–228. Kluwer, Dordrecht, 2002.
16. Pavel Pták and Sylvia Pulmannová. *Orthomodular Structures as Quantum Logics*. Kluwer Academic Publishers, London, 1991.
17. Garrett Birkhoff. *Lattice Theory*. American Mathematical Society, 3rd edition, 1967.
18. B.A. Davey and H.A. Priestly. *Introduction to Lattices and Order*. Cambridge University Press, Cambridge, second edition, 2002.
19. Robert Goldblatt. *Topoi: the Categorical Analysis of Logic*. Dover Publications, Inc., 2nd edition, 2006.
20. Peter T. Johnstone. *Sketches of an Elephant: A Topos Theory Compendium, Vol. 1*, volume 43 of *Oxford Logic Guides*. Oxford University Press, Oxford, 2002.
21. Peter T. Johnstone. *Sketches of an Elephant: A Topos Theory Compendium, Vol. 2*, volume 44 of *Oxford Logic Guides*. Oxford University Press, Oxford, 2003.
22. Saunders Mac Lane. *Categories for the working mathematician*, volume 5 of *Graduate Texts in Mathematics*. Springer, New York, second edition, 1978.
23. Saunders Mac Lane and Ieke Moerdijk. *Sheaves in Geometry and Logic, A First Introduction to Topos Theory*. Springer, New York, 1992.
24. Benjamin C. Pierce. *Basic category theory for computer scientists*. The MIT Press, London, England, 1991.
25. G. Kalmbach. *Orthomodular Lattices*. Academic Press, London, 1983.
26. Garrett Birkhoff and John von Neumann. The logic of quantum mechanics. In A.H. Taub, editor, *John von Neumann: collected works*, volume 4, pages 105–125. Pergamon Press, Oxford, 1962.
27. Stanley Burris and H.P. Sankappanavar. *A Course in Universal Algebra*. Springer-Verlag, 2012.
28. Steven Givant and Paul Halmos. *Introduction to Boolean Algebras*. Undergraduate Texts in Mathematics. Springer, New York, 2009.
29. Guram Bezhanishvili. Stone duality and Gleason covers through de Vries duality. *Topology and its Applications*, 157:1064–1080, 2010.
30. Benno van den Berg and Chris Heunen. No-go theorems for functorial localic spectra of noncommutative rings. In Proceedings 8th International Workshop on *Quantum Physics and Logic*, Nijmegen, Netherlands, October 27-29, 2011, volume 95 of *Electronic Proceedings in Theoretical Computer Science*, pages 21–25. Open Publishing Association, 2012.
31. John Harding and Mirko Navara. Subalgebras of orthomodular lattices. *Order - A Journal on the Theory of Ordered Sets and its Applications*, 28(3):549–563, 2011.
32. G.E. Reyes and H. Zolfaghari. Bi-Heyting algebras, toposes and modalities. *Journal of Philosophical Logic*, 25:25–43, 1996.
33. Andreas Döring. The physical interpretation of daseinisation. In Hans Halvorson, editor, *Deep Beauty*, pages 207–238. Cambridge University Press, Cambridge, 2011.
34. Benjamin Eva. Towards a paraconsistent quantum set theory. *Electronic Proceedings in Theoretical Computer Science*, 2015.

Bell's Local Causality is a d-Separation Criterion

Gábor Hofer-Szabó

Abstract This paper aims to motivate Bell's notion of local causality by means of Bayesian networks. In a locally causal theory any superluminal correlation should be screened off by atomic events localized in any so-called *shielder-off region* in the past of one of the correlating events. In a Bayesian network any correlation between non-descendant random variables are screened off by any so-called *d-separating set* of variables. We will argue that the shielder-off regions in the definition of local causality conform in a well defined sense to the d-separating sets in Bayesian networks.

Keywords Local causality · Shielder-off region · Bayesian network · d-separation

1 Introduction

John Bell's notion of local causality is one of the central notions in the foundations of relativistic quantum physics. Bell himself has returned to the notion of local causality from time to time providing a more and more refined formulation for it. The final formulation stems from Bell's posthumously published paper "La nouvelle cuisine." It reads as follows[1]:

> A theory will be said to be locally causal if the probabilities attached to values of local beables in a space-time region V_A are unaltered by specification of values of local beables in a space-like separated region V_B, when what happens in the backward light cone of V_A is already sufficiently specified, for example by a full specification of local beables in a space-time region V_C [1, 2, p. 239–240].

[1]For the sake of uniformity we slightly changed Bell's notation and figure.

G. Hofer-Szabó (✉)
Institute of Philosophy, Research Center for the Humanities, Országház u. 30, Budapest 1014, Hungary
e-mail: szabo.gabor@btk.mta.hu

M. Ozawa et al. (eds.), *Reality and Measurement in Algebraic Quantum Theory*,
Springer Proceedings in Mathematics & Statistics 261,
https://doi.org/10.1007/978-981-13-2487-1_2

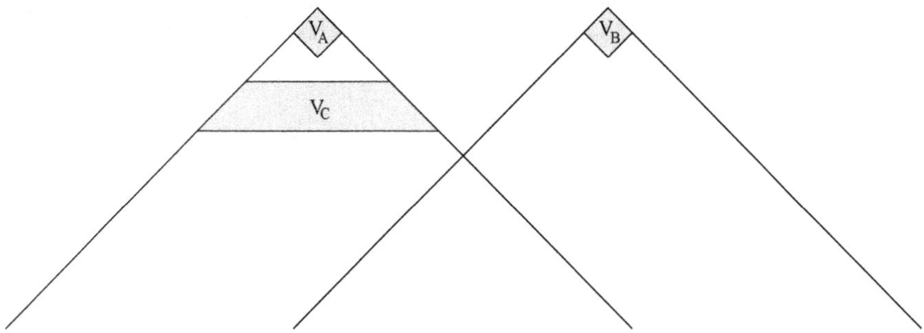

Fig. 1 Full specification of what happens in V_C makes events in V_B irrelevant for predictions about V_A in a locally causal theory

The figure Bell is attaching to his formulation of local causality is reproduced in Fig. 1 with Bell's original caption. In a rough translation, a theory is locally causal if any superluminal correlation can be screened-off by a "full specification of local beables in a space-time region" in the past of one of the correlating events.

The terms in quotation marks, however, need clarification. What are "local beables"? What is "full specification" and why is it important? Which are those regions in spacetime which, if fully specified, render superluminally correlating events probabilistically independent? The first two questions have attracted much interest among philosophers of science. As Bell puts it, "*be*ables of the theory are those entities in it which are, at least tentatively, to be taken seriously, as corresponding to something real" [1, 2, p. 234]. Furthermore, "it is important that events in V_C be specified completely. Otherwise the traces in region V_B of causes of events in V_A could well supplement whatever else was being used for calculating probabilities about V_A" [1, 2, p. 240].

The third question, however, concerning the localization of the screener-off regions has gained much less attention in the literature. How to characterize the regions which region V_C in Fig. 1 is an example of? Bell's answer is instructive but brief: "It is important that region V_C completely shields off from V_A the overlap of the backward light cones of V_A and V_B [1, 2, p. 240]." But why to shield off the common past of the correlating events? Why the region V_C cannot be in the remote past of V_A as for example in Fig. 2? Well, intuition dictates that in this latter case some event might occur *above* the shielder-off region but still *within* the common past establishing a correlation between events in V_A and V_B. This intuition is correct. The aim of this paper, however, is to provide a more precise explanation for the localization of the shielder-off regions in spacetime. This explanation will consists in drawing a parallel between local physical theories and Bayesian networks. It will turn out that *the*

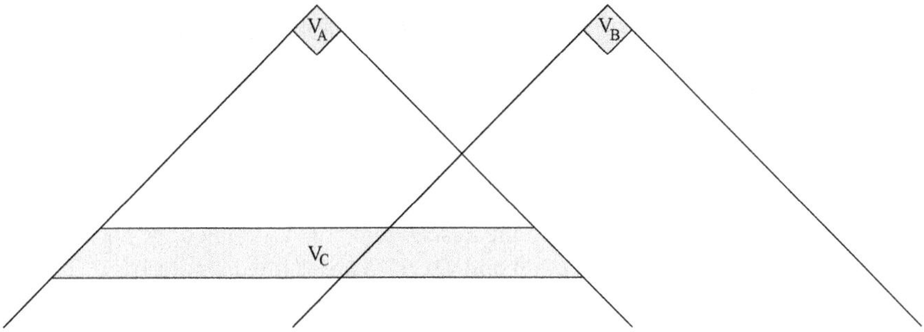

Fig. 2 A *not* completely shielding-off region V_C

shielder-off regions in the definition of local causality play an analogous role to the so-called d-separating sets of random variables in Bayesian networks.

There is a renewed interest in Bell's notion of local causality [22–24], its relation to separability [12]; the role of full specification in local causality [13, 32]; its role in relativistic causality [3, 4, 27]; its status as a local causality principle [10, 11, 28]. A similar closely related topic, the Common Cause Principle is also given much attention [15, 17, 26, 29]. On the other hand, there is also an intensive discussion on the applicability of the Causal Markov Condition in the EPR scenario [5, 21, 33, 34], Hausman and Woodward 1999. Despite the rich and growing literature on the topic I am unaware of any work relating Bayesian networks and especially d-separation directly to local causality. This paper intends to fill this gap. For a precursor of this paper investigating Causal Markov Condition in a specific local physical theory see [14]. For a comprehensive formally rigorous investigation of the relation of Bell's local causality to the Common Cause Principle and other relativistic locality concepts see [19]; for a more philosopher-friendly version see [20].

In the paper we will proceed as follows. In Sect. 2 we introduce the basics of the theory of Bayesian networks and the notion of d-separation and m-separation. In Sect. 3 we define the notion of a local physical theory and formulate Bell's notion of local causality within this framework. We prove our main claim in Sect. 4 and conclude in Sect. 5.

2 Bayesian Networks and d-Separation

A *Bayesian network* [6, 25] is a pair $(\mathscr{G}, \mathscr{V})$ where $\mathscr{G}$ is a directed acyclic graph and $\mathscr{V}$ is a set of random variables on a classical probability space (X, Σ, p) such that the elements $A, B \ldots$ of $\mathscr{V}$ are represented by the vertices of $\mathscr{G}$ and the arrows (directed edges) $A \to B$ on the graph represent that A is *causally relevant* for B. Two vertices are called *adjacent* if they are connected by an arrow. For a given $A \in \mathscr{V}$, the set of vertices that have directed edges in A is called the *parents* of A, denoted by $Par(A)$; the set of vertices from which a directed paths is leading to A is called the *ancestors* of A, denoted by $Anc(A)$; and finally the set of vertices that are endpoints of a directed paths from A is called the *descendants* of A, denoted by $Des(A)$. For a set $\mathscr{C}$ of vertices $Par(\mathscr{C})$, $Anc(\mathscr{C})$ and $Des(\mathscr{C})$ are defined similarly.

The set $\mathscr{V}$ is said to satisfy the *Causal Markov Condition* relative to the graph $\mathscr{G}$ if for any $A \in \mathscr{V}$ and any $B \notin Des(A)$ the following is true:

$$p(A \mid Par(A) \wedge B) = p(A \mid Par(A)) \tag{1}$$

or equivalently

$$p(A \wedge B \mid Par(A)) = p(A \mid Par(A))\, p(B \mid Par(A)) \tag{2}$$

That is conditioning on its parents any random variable will be probabilistically independent from any of its non-descendant. Non-descendants can be of two types: either ancestors or *collaterals* (non-descendants and non-ancestors). As we will see, being independent of collaterals is what relates the Causal Markov Condition to Bell's local causality.

Causal Markov Condition establishes a special conditional independence relation between some random variables of $\mathscr{V}$. But there are many other conditional independences. In a faithful Bayesian network these other conditional independences are all implied by the Causal Markov Condition by means of the so-called *d-separation* criterion. Let $\mathscr{P}$ be a *path* in $\mathscr{G}$, that is a sequence of adjacent vertices. A variable E on $\mathscr{P}$ is a *collider* if there are arrows to E from both its neighbors on $\mathscr{P}$ $(D \to E \leftarrow F)$. Now, let $\mathscr{C}$ be a set of vertices and let A and B two different vertices not in $\mathscr{C}$. The vertices A and B are said to be *d-connected* by $\mathscr{C}$ in $\mathscr{G}$ iff there exists a path $\mathscr{P}$ between A and B such that every non-collider on $\mathscr{P}$ is not in $\mathscr{C}$ and every collider is in $Anc(\mathscr{C}) \vee \mathscr{C}$. A and B are said to be *d-separated* by $\mathscr{C}$ in $\mathscr{G}$, iff they are not d-connected by $\mathscr{C}$ in $\mathscr{G}$.

The intuition behind d-separation is the following. A vertex E on a path (not at the endpoints) can be either a *collider* $(D \to E \leftarrow F)$, an *intermediary cause* $(D \to E \to F)$ or a *common cause* $(D \leftarrow E \to F)$. The idea here is that only intermediary and common causes (together called *non-colliders*) can transmit causal dependence and hence establish probabilistic dependence. This dependence can be *blocked* by conditioning on the non-collider. Colliders behave just the opposite way. They represent two events causing a common effect. These two causes are causally and probabilistically independent, but become dependent upon conditioning on their common effect. Moreover, they also become dependent upon conditioning on any of the descendants of the effect. Putting these together, the causal dependence on a path $\mathscr{P}$ connecting two vertices is blocked by a set $\mathscr{C}$ if either there is at least one non-collider on $\mathscr{P}$ which is in $\mathscr{C}$ or there is at least one collider E on $\mathscr{P}$ such that

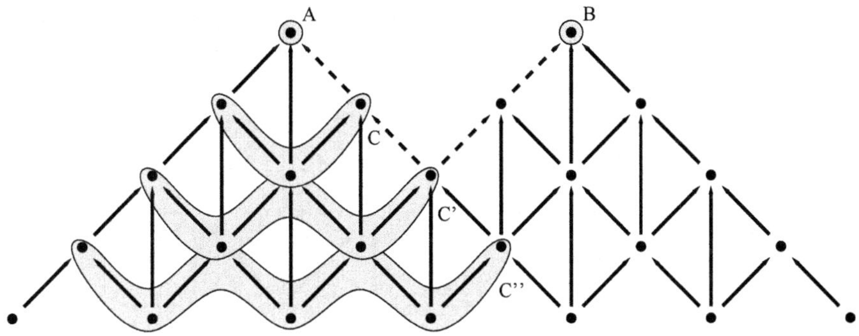

Fig. 3 A and B are d-separated by $\mathscr{C}$ and $\mathscr{C}'$ but d-connected by $\mathscr{C}''$

either E or a descendant of E is not in $\mathscr{C}$. The two vertices are d-separated by $\mathscr{C}$ if causal dependence is blocked on every path connecting them.

As an example for d-connection and d-separation consider the causal graph in Fig. 3. (The arrows are directed to up, left up and right up.) Let A be the left "peak" and B the right "peak" in the graph and let $\mathscr{C}$, $\mathscr{C}'$ and $\mathscr{C}''$ be the sets shown in the figure containing 3, 5 and 7 vertices, respectively. Then A and B are d-separated by $\mathscr{C}$ since the parents are always d-separating due to the Causal Markov Condition. A and B are d-separated also by $\mathscr{C}'$ since for every path connecting the peaks there is a non-collider in $\mathscr{C}'$. However, A and B are d-connected by $\mathscr{C}''$ since there is a path (denoted by a broken line in Fig. 3) connecting the peaks which contains only non-colliders outside $\mathscr{C}''$. Consequently, the following probabilistic relations hold:

$$p(A \wedge B \,|\, \mathscr{C}) = p(A \,|\, \mathscr{C})\, p(B \,|\, \mathscr{C}) \tag{3}$$

$$p(A \wedge B \,|\, \mathscr{C}') = p(A \,|\, \mathscr{C}')\, p(B \,|\, \mathscr{C}') \tag{4}$$

$$p(A \wedge B \,|\, \mathscr{C}'') \neq p(A \,|\, \mathscr{C}'')\, p(B \,|\, \mathscr{C}'') \tag{5}$$

Looking at in Fig. 3, what stands out immediately is that a set which is too far in the causal past of A cannot d-separate A from a collateral event since there might be paths connecting them "above" the set. As we will see, a similar moral will be valid in case of local causality: regions with are too far in the causal past of an event cannot screen it off from a spacelike separated event since there might be events "above" the region which can establish correlation between them.

In analyzing local causality sometimes we need to go beyond directed acyclic graphs. A graph which may contain both directed ($A \rightarrow B$) and bi-directed ($A \leftrightarrow B$) edges is called *mixed*. The d-separation criterion extended to mixed acyclic graphs

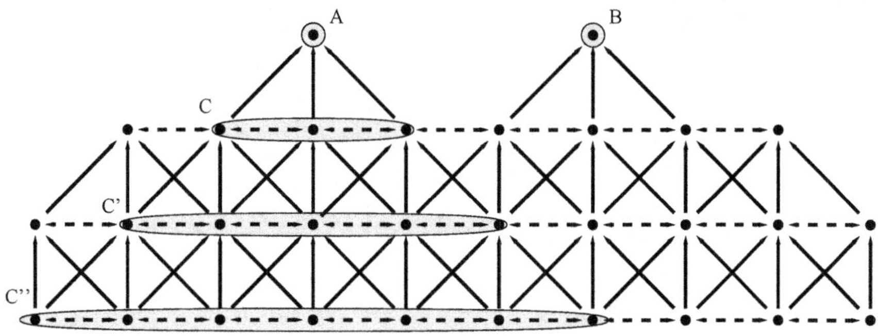

Fig. 4 A and B are m-separated by $\mathscr{C}$ but m-connected by both $\mathscr{C}'$ and $\mathscr{C}''$

is called *m-separation* [30, 31]. Two vertices A and B are said to be *m-connected* by $\mathscr{C}$ in a mixed acyclic graph $\mathscr{G}$ iff there exists a path $\mathscr{P}$ between A and B such that every non-collider on $\mathscr{P}$ is not in $\mathscr{C}$ and every collider is in $Anc(\mathscr{C}) \vee \mathscr{C}$. A and B are said to be *m-separated* by $\mathscr{C}$ in $\mathscr{G}$, iff they are not m-connected by $\mathscr{C}$ in $\mathscr{G}$. In a directed acyclic graph m-separation reduces to d-separation.

An example for a mixed acyclic graph is depicted in Fig. 4. Here the bi-directed edges are represented by dotted lines. Again, let A be the left "peak" and B the right "peak" in the graph and let $\mathscr{C}$, $\mathscr{C}'$ and $\mathscr{C}''$ be the sets shown in the figure containing 3, 5 and 7 vertices, respectively. Then A and B are m-separated by $\mathscr{C}$ but m-connected by both $\mathscr{C}'$ and $\mathscr{C}''$. The connecting path is the shortest path connecting A and B.

Now, let us connect the terminology of Bayesian networks to that of standard physics. Before doing that note that probability is commonly interpreted in Bayesianism subjectively as partial belief and in physics objectively as long-run relative frequency. This interpretative difference, however, does not undermine the analogy between local causality and d-separation, since Bayesian networks are well open to statistical interpretation and, conversely, there is a growing tendency to understand quantum physics in a subjectivist way.

Let us start with random variables. A random variable is a real-valued Borel-measurable function on X. Each random variable $A \in \mathscr{V}$ generates a sub-σ-algebra of Σ by the inverse image of the Borel sets:

$$\sigma(A) := \{A^{-1}(b) \mid b \in \mathscr{B}(\mathbb{R})\} \tag{6}$$

Similarly, each set $\mathscr{C}$ of n random variables generates a sub-σ-algebra of Σ by the inverse image of the n-dimensional Borel sets:

$$\sigma(\mathscr{C}) := \left\{(C_1, C_2 \ldots C_n)^{-1}(b) \mid C_i \in \mathscr{C},\ b \in \mathscr{B}(\mathbb{R}^n)\right\} \tag{7}$$

From this perspective d-separation tells us which sub-σ-algebras are probabilistically independent conditioned on which other sub-σ-algebras of Σ.

Now, instead of using σ-algebras it is more instructive to use a richer structure in physics, namely *von Neumann algebras*. Consider the characteristic functions on X projecting on the elements of Σ, called *events*. The set $\{\chi_S \mid S \in \Sigma\}$ of characteristic functions generates an abelian von Neumann algebra, namely $\mathscr{L}^\infty(X, \Sigma, p)$, the space of essentially bounded complex-valued functions on X. Starting from the characteristic functions of the sub-σ-algebra $\sigma(A)$, one arrives at a subalgebra of $\mathscr{L}^\infty(X, \Sigma, p)$. Denote this abelian von Neumann algebra determined by the random variable A by $\mathscr{N}_A$. Similarly, denote by $\mathscr{N}_\mathscr{C}$ the von Neumann algebra determined by a set $\mathscr{C}$ of random variables.

Instead of using a probability measure on Σ or on a sub-σ-algebra $\sigma(A)$, one can also use a state on the corresponding von Neumann algebra $\mathscr{N}_A$. A *state* ϕ is a positive linear functional of norm 1 on a von Neumann algebra. States on $\mathscr{N}_A$ and probability measures on $\sigma(A)$ mutually determine one another: a state restricted to the characteristic functions in $\mathscr{N}_A$ is a probability measure on $\sigma(A)$; and vice versa, integrating elements of $\mathscr{N}_A$ according to a probability measure on $\sigma(A)$ yields a state on $\mathscr{N}_A$.

Therefore, a conditional independence between random variables A and B given the set $\mathscr{C}$

$$p(A \wedge B \mid \mathscr{C}) = p(A \mid \mathscr{C})\, p(B \mid \mathscr{C}) \tag{8}$$

can be rewritten as follows: for any projection $A \in \mathscr{N}_A$, $B \in \mathscr{N}_B$ and $C \in \mathscr{N}_\mathscr{C}$:

$$\frac{\phi(A \wedge B \wedge C)}{\phi(C)} = \frac{\phi(A \wedge C)}{\phi(C)} \frac{\phi(B \wedge C)}{\phi(C)} \tag{9}$$

Although in this paper we stay at the classical level, the theory of von Neumann algebras is wide enough to incorporate also quantum physics. In this case the von Neumann algebras are nonabelian. The events, just like in the classical case, are represented by projections of the von Neumann algebras. In the quantum case conditional independence between the projection $A \in \mathscr{N}_A$ and $B \in \mathscr{N}_B$ given $C \in \mathscr{N}_\mathscr{C}$ reads as follows:

$$\frac{\phi(CABC)}{\phi(C)} = \frac{\phi(CAC)}{\phi(C)} \frac{\phi(CBC)}{\phi(C)} \tag{10}$$

which in the classical case reduces to (9).

The last point in converting the formalism of Bayesian networks into physics, is to swap the causal graph for spacetime. We can then replace the causal relations embodied in the causal graph by spatiotemporal relations of a given spacetime. Instead of saying that a random variable is the *ancestor* of another variable we will then say that an event is in the *past* of the other. But to do so first we need to *localize* events in spacetime that is we need to have an association of algebras of events to spacetime regions. Such a principled association is offered by the formalism of algebraic quantum field theory. Hence, in the next section we will introduce some elements of algebraic quantum field theory which is indispensable for our purpose which is to come up with a mathematically precise definition of Bell's notion of local causality.

3 Bell's Local Causality in a Local Physical Theory

Let $\mathcal{M}$ be a globally hyperbolic spacetime and let $\mathcal{K}$ be a covering collection of bounded, globally hyperbolic subspacetime regions of $\mathcal{M}$ such that $(\mathcal{K}, \subseteq)$ is a directed poset under inclusion $\subseteq$. A *local physical theory* is a net $\{\mathcal{A}(V), V \in \mathcal{K}\}$ associating algebras of events to spacetime regions which satisfies *isotony* and *microcausality* defined as follows [7, 8, 19, 20]:
Isotony. The net of local observables is given by the isotone map $\mathcal{K} \ni V \mapsto \mathcal{A}(V)$ to unital C^*-algebras, that is $V_1 \subseteq V_2$ implies that $\mathcal{A}(V_1)$ is a unital C^*-subalgebra of $\mathcal{A}(V_2)$. The *quasilocal algebra* $\mathcal{A}$ is defined to be the inductive limit C^*-algebra of the net $\{\mathcal{A}(V), V \in \mathcal{K}\}$ of local C^*-algebras.
Microcausality: $\mathcal{A}(V')' \cap \mathcal{A} \supseteq \mathcal{A}(V), V \in \mathcal{K}$, where primes denote spacelike complement and algebra commutant, respectively.

If the quasilocal algebra $\mathcal{A}$ of the local physical theory is commutative, we speak about a *local classical theory*; if $\mathcal{A}$ is noncommutative, we speak about a *local quantum theory*. For local classical theories microcausality fulfills trivially.

Given a state ϕ on the quasilocal algebra $\mathcal{A}$, the corresponding GNS representation $\pi_\phi : \mathcal{A} \to \mathcal{B}(\mathcal{H}_\phi)$ converts the net of C^*-algebras into a net of C^*-subalgebras of $B(\mathcal{H}_\phi)$. Closing these subalgebras in the weak topology one arrives at a net of local von Neumann observable algebras: $\mathcal{N}(V) := \pi_\phi(\mathcal{A}(V))''$, $V \in \mathcal{K}$. The *net* $\{\mathcal{N}(V), V \in \mathcal{K}\}$ *of local von Neumann algebras* also obeys isotony and microcausality, hence we can also refer to it as a local physical theory.

Given a local physical theory, we can turn now to the definition of Bell's notion of local causality. Recall that according to Bell a theory is locally causal if any superluminal correlation is screened-off by a "full specification of local beables in a space-time region V_C" as shown in Fig. 1. As indicated in the Introduction we need to address three questions. What are "local beables"? What is "full specification"? Which are the shielder-off regions? The brief answer to the first two questions is the following. In a local physical theory a "local beable" in a region V is an *element* of the local von Neumann algebra $\mathcal{N}(V)$. A "full specification" of local beables in region V is an *atomic element* of the local von Neumann algebra $\mathcal{N}(V)$. In this paper we do not comment on these two answers. For a more thoroughgoing discussion on why we think this to be the correct translation of Bell's intuition into our framework see [19, 20].

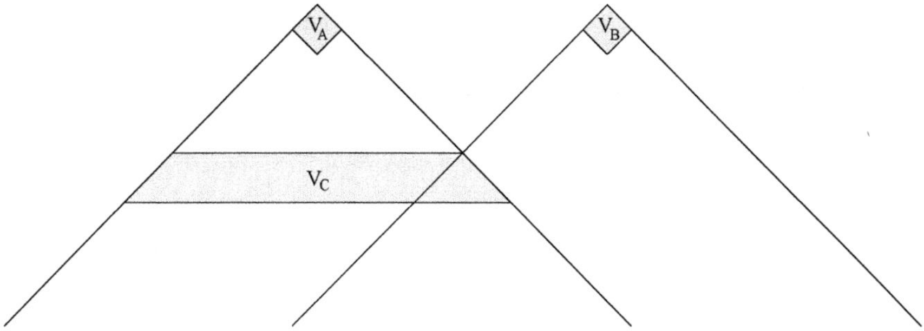

Fig. 5 A completely shielding-off region V_C intersecting with the common past of V_A and V_B

To the third question, which is the topic of our paper, the answer is this: a shielder-off region V_C is a region in the causal past of V_A which can block any causal influence on V_A arriving from the common past of V_A and V_B. But there is an ambiguity in this answer. Bell's Fig. 1 suggests that a shielder-off region should not intersect with the common past. Whereas the requirement of simply blocking causal influences from the past allows for also regions depicted in Fig. 5 intersecting with the common past. This means that one can define a *shielder-off region of* V_A *relative to* V_B either as a region V_C satisfying:

$\mathbf{L}_1 : V_C \subset J_-(V_A)$ (V_C is in the causal past of V_A),
$\mathbf{L}_2 : V_A \subset V_C''$ (V_C is wide enough such that its causal shadow contains V_A),
$\mathbf{L}_3^Q : V_C \subset V_B'$ (V_C is spacelike separated from V_B)

in tune with Bell's Fig. 1; or one can replace L_3^Q by the weaker requirement

$\mathbf{L}_3^C : J_-(V_C) \supset J_-(V_A) \cap J_-(V_B)$ (The causal past of V_C contains the common past of V_A and V_B)

allowing for regions such as in Fig. 2. It turns out that (with respect to the Bell inequalities, see [16, 18]) it is more appropriate to demand L_3^Q in case of a local *quantum* theory and L_3^C in case of a local *classical* theory (hence the superscripts). But note that as the covering regions become infinitely thin shrinking down to a Cauchy surface, requirement L_3^C coincides with requirement L_3^Q.

With all these considerations in mind Bell's notion of local causality in the framework of a local physical theory will be the following:

Definition 1 A local physical theory represented by a net $\{\mathcal{N}(V),\ V \in \mathcal{K}\}$ of von Neumann algebras is called *locally causal* (in Bell's sense), if

1. for any pair $A \in \mathcal{N}(V_A)$ and $B \in \mathcal{N}(V_B)$ of events represented by projections in spacelike separated regions $V_A, V_B \in \mathcal{K}$;
2. for every locally normal and faithful state ϕ establishing a correlation $\phi(AB) \neq \phi(A)\phi(B)$ between A and B;
3. for any spacetime shielder-off region V_C defined by requirements L_1, L_2 and L_3^Q/L_3^C;
4. for any event C in the set $\mathscr{C}$ of atomic events in $\mathscr{A}(V_C)$ the following screening-off condition holds:

$$\frac{\phi(CABC)}{\phi(C)} = \frac{\phi(CAC)}{\phi(C)} \frac{\phi(CBC)}{\phi(C)} \tag{11}$$

which for a local *classical* theory is equivalent to

$$p(A \wedge B \,|\, \mathscr{C}) = p(A \,|\, \mathscr{C})\, p(B \,|\, \mathscr{C}) \tag{12}$$

In short, a local physical theory is locally causal in Bell's sense if every superluminal correlation is screened off by all atomic events in all shielder-off region. (For many delicate questions such as what if the algebras are non-atomic, how this definition of local causality relates to the Common Cause Principle and the Bell inequalities see again [19, 20].)

The question left is, however: *why* shielder-off regions are characterized by requirements L_1, L_2 and L_3^Q/L_3^C? To this we turn in the next section.

4 Shielder-Off Regions are d-Separating

The point we are going to make in this Section is that shielder-off regions in the definition of local causality conform to d-separating sets in directed acyclic graphs and to m-separating sets in mixed acyclic graphs.

First we show how a local physical theory gives rise to a causal graph. Consider a local *classical* theory $\{\mathcal{N}(V), V \in \mathcal{K}\}$ where the covering collection is induced by a partition $\mathscr{T}$ of a spacetime $\mathscr{M}$. By *partition* we mean a countable set of disjoint, bounded spacetime regions such that their union is $\mathscr{M}$. The local classical theory $\{\mathcal{N}(V), V \in \mathcal{K}\}$ gives rise to a causal graph $\mathscr{G}$ as follows: Let the *vertices* of the $\mathscr{G}$ be the regions in the partition, $\{V \in \mathscr{T}\}$. For two vertices V_A and V_B, let there be an *edge* pointing from V_A and V_B, $V_A \to V_B$, iff there is a future directed causal curve from V_A to V_B such that the curve does not enter any region, except for V_A and V_B. It will turn out that the type of the graph we obtain is crucially depending on the partition $\mathscr{T}$ of the spacetime. Let us see some different cases.

If $\mathscr{M}$ is the 1+1 dimensional Minkowski spacetime, then it can be covered by double cones of equal size. (See Fig. 6.) The causal graph corresponding to this covering emerges simply by connecting those adjacent double cones which lie in the

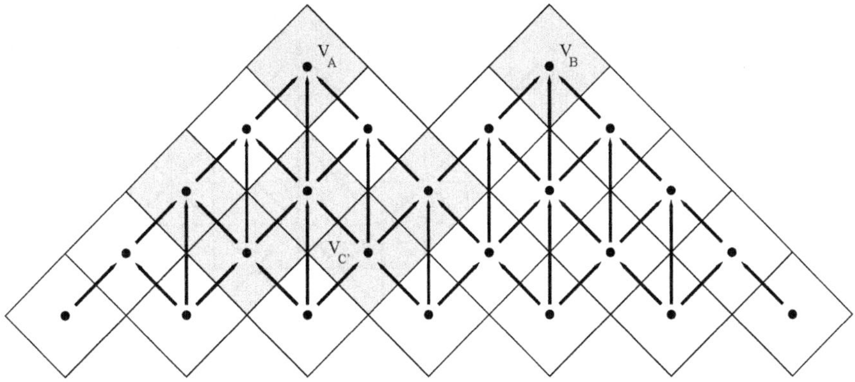

Fig. 6 The directed acyclic graph generated by double cones of equal size covering the 1+1 dimensional Minkowski spacetime

causal past of one another. What we get is just the directed acyclic graph depicted in Fig. 3 in Sect. 2.

Figure 6 is a kind of "superposition" of a spacetime diagram and a Bayesian network. Consider for example region $V_{\mathscr{C}'}$. Reading Fig. 6 as a spacetime diagram, one sees that $V_{\mathscr{C}'}$ is a *shielder-off region*. Reading Fig. 6 as a causal graph, one observes that the set $\mathscr{C}'$ corresponding to $V_{\mathscr{C}'}$ (depicted in Fig. 3) is a *d-separating set*. Similarly, one can check that the region associated to the d-separating set $\mathscr{C}$ in Fig. 3 is a shielder-off region and the region associated to the d-connecting set $\mathscr{C}''$ is not a shielder-off region.

A general spacetime $\mathscr{M}$ cannot be partitioned to globally hyperbolic regions, let alone to double cones. Still one can construct the causal graph corresponding to a partition $\mathscr{T}$. In Fig. 7 we illustrate such a construction where a 1+1 dimensional Minkowski spacetime is covered by boxes of equals size. (This example, in contrast to the previous one, can be generalized for a $3 + 1$-dimensional Minkowski spacetime covered by $3 + 1$-dimensional boxes of equals size.) The causal graph emerging from this construction is not a directed acyclic graph since it contains bi-directed edges: spacelike neighboring boxes will be *spouses*. What we get is a mixed acyclic graph depicted in Fig. 4. Again, confronting Figs. 4 and 7 one can see that the set $\mathscr{C}'$ is *not* an m-separating set and at the same time the corresponding region $V_{\mathscr{C}'}$ is *not* a shielder-off region of V_A relative to V_B.

The exact characterization of the graphs emerging from a different coverings of a given spacetime is a subtle question which we do not go into here. Instead we turn now to the construction of random variables. Let $\mathscr{N}(V)$ be the local von

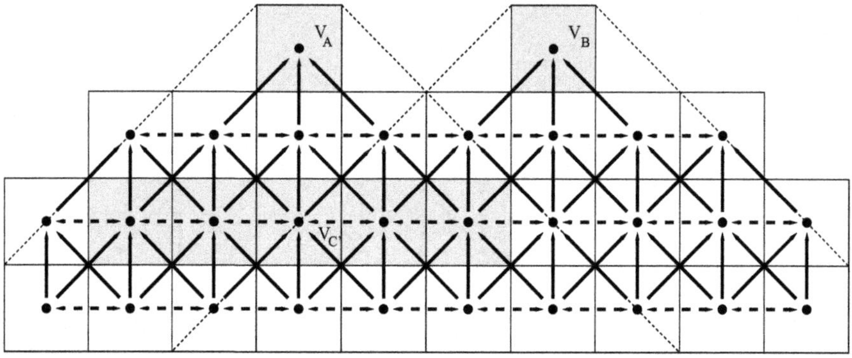

Fig. 7 The mixed acyclic graph generated by boxes of equals size covering of the 1+1 dimensional Minkowski spacetime

Neumann algebra associated to the spacetime region $V \in \mathcal{T}$. Denote by $\sigma(V)$ the sigma-algebra of the projections of $\mathcal{N}(V)$. Let the random variable associated to V be any Borel-measurable function from $\sigma(V)$ to $\mathcal{B}(\mathbb{R})$. Any state ϕ will then define a probability measure p on $\sigma(V)$ for any $V \in \mathcal{T}$ and, due to isotony of the net, also for any V which is a *finite* union of regions in $\mathcal{T}$. (Note that $\sigma(\mathcal{M})$ may not be a sigma-algebra since the quasilocal algebra $\mathcal{A}$ is not necessarily a von Neumann algebra, so it may not contain projections.)

In sum, any finite set of regions of a local classical theory $\{\mathcal{N}(V), V \in \mathcal{K}\}$ generated by a globally hyperbolic partition of $\mathcal{M}$ defines a pair $(\mathcal{G}, \mathcal{V})$. For certain specific coverings $\mathcal{G}$ will be a directed acyclic graph; in general, however, it will be a mixed graph.

Now, we state and prove the main claim of the paper.

Proposition 1 *Let G be a directed/mixed acyclic graph constructed from a local classical theory $\{\mathcal{N}(V), V \in \mathcal{K}\}$ where $\mathcal{K}$ is generated by a partition $\mathcal{T}$ of $\mathcal{M}$. Suppose that $\{\mathcal{N}(V), V \in \mathcal{K}\}$ is locally causal in the sense of Definition 1. For any V_A and V_B spacelike separated spacetime regions, call a set $\{V_i\} \subset \mathcal{K}$ a shielder-off set of regions for V_A if $\cup_i V_i$ is a shielder-off region for V_A characterized by the criteria L_1, L_2 and L_3^C. Then, any shielder-off set $\{V_i\}$ d-separates/m-separates V_A from V_B.*

Proof To prove Proposition 1, we have to show that $\{V_i\}$ blocks every path connecting V_A and V_B that is on every path there is at least one non-collider in $\{V_i\}$ or there is at least one collider V_E such that $V_E \notin Anc(\{V_i\}) \vee \{V_i\}$.

First consider those paths that contain no colliders. These paths need to pass through the set of common ancestors, $Anc(V_A) \wedge Anc(V_B)$. But due to L_3^C, the shielder-off set $\{V_i\}$ blocks every path connecting V_A and $Anc(V_A) \wedge Anc(V_B)$. Hence, $\{V_i\}$ blocks all the paths which contain no colliders.

So there remain only those paths to be blocked which contain at least one collider. There are two types of such paths: paths avoiding $\{V_i\}$ and path crossing $\{V_i\}$.

Consider first the paths avoiding $\{V_i\}$. Define the set

$$A^{cut} := (Anc(A) \vee A) \setminus (Anc(\{V_i\}) \vee \{V_i\})$$

Now, it is easy to see that no path which starts from V_A, avoids $\{V_i\}$ and contains only non-colliders can leave $Des(A^{cut})$. However, $V_B \notin Des(A^{cut})$, otherwise L_3^C would not hold. Hence, the path connecting V_A and V_B need to contain at least one collider $V_E \in Des(A^{cut})$. But $Des(A^{cut}) \wedge (Anc(\{V_i\}) \vee \{V_i\}) = \emptyset$, hence $V_E \notin Anc(\{V_i\}) \vee \{V_i\}$. Thus, the path is blocked by $\{V_i\}$.

Consider now the paths crossing $\{V_i\}$. Let $\mathscr{P} = (V_A, \ldots, V_D, V_E, \ldots, V_B)$ a path connecting V_A and V_B such that V_D is the last vertex before the path enters $\{V_i\}$ and V_E is the first vertex on the path which already is in $\{V_i\}$. We show that V_E cannot be a collider.

To see this, note that V_D has to be in A^{cut}, otherwise the subpath $\mathscr{P} = (V_A, \ldots, V_D)$ would contain at least one collider in $Des(A^{cut})$ and hence would be blocked. Now, suppose, contrary to our claim, that V_E is a collider. Then there is an arrow pointing from V_D to V_E. Hence, $V_D \in Anc(\{V_i\})$. But if V_D is both in A^{cut} and also in $Anc(\{V_i\})$, then $\{V_i\}$ cannot be a shielder-off set. Contradiction. Thus, V_E is a non-collider in $\{V_i\}$ and the path is blocked.

In sum, $\{V_i\}$ blocks every path connecting V_A and V_B, that is $\{V_i\}$ d-separates V_A from V_B. ■

The converse of Proposition 1 is not true: d-separating sets are not necessarily shielder-off sets. Reference [35] list algorithms to find the so-called *minimal d-separating sets* for two random variables A and B, that is sets that are d-separating but taking away any vertex from the set they will cease to be d-separating. It turns out that any minimal d-separating set is sitting in the *union* of the ancestors of A and B (including also A and B), $Anc(A) \vee Anc(B) \vee A \vee B$. However, a minimal d-separating set need not satisfy relations L_1, L_2 and L_3^C. For example the sets $\mathscr{D}$, $\mathscr{D}'$ and $\mathscr{D}''$ in Fig. 8 are all minimal d-separating sets but not shielder-off regions for A relative to B.

At any event, shielder-off regions are d-separating, and this was to be shown in this paper.

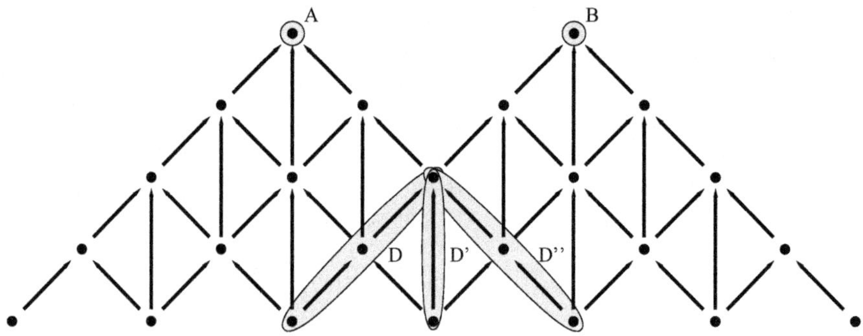

Fig. 8 Minimal d-separating but not shielder-off regions

5 Conclusions

The aim of the paper was to motivate Bell's definition of local causality by means of Bayesian networks. To this aim, first we constructed a causal graph from the covering collection of a spacetime. In certain cases the graph was a directed acyclic graph, in other cases only a mixed acyclic graph. Similarly, we have associated random variables to the local algebras of a local physical theory. By this move shielder-off regions turned out be specific d-separation (m-separating) sets on the causal graph. Hence, Bell's definition of local causality requiring that spacelike separated events should be screened-off by events in a shielder-off region turned out to be a d-separation criterion.

Acknowledgements I wish to thank Péter Vecsernyés for valuable discussions. This work has been supported by the Hungarian Scientific Research Fund, OTKA K-115593 and by the Bilateral Mobility Grant of the Hungarian and Polish Academies of Sciences, NM-104/2014.

References

1. J. S. Bell, "La nouvelle cuisine," in: J. Sarlemijn and P. Kroes (eds.), *Between Science and Technology*, Elsevier, (1990); reprinted in (Bell, 2004, 232-248).
2. J.S. Bell, *Speakable and Unspeakable in Quantum Mechanics*, (Cambridge: Cambridge University Press, 2004).
3. J. Butterfield, "Stochastic Einstein Locality Revisited," *Brit. J. Phil. Sci.*, **58**, 805-867, (2007).
4. J. Earman and G. Valente, "Relativistic causality in algebraic quantum field theory," *Int. Stud. Phil. Sci.*, **28 (1)**, 1-48 (2014).

5. C. Glymour, "Markov properties and quantum experiments," in W. Demopoulos and I. Pitowsky (eds.) *Physical Theory and its Interpretation*, (Springer, 117-126, 2006).
6. C. Glymour, R. Scheines and P. Spirtes, "Causation, Prediction, and Search," (Cambridge: The MIT Press, 2000).
7. R. Haag, *Local quantum physics*, (Heidelberg: Springer Verlag, 1992).
8. H. Halvorson, "Algebraic quantum field theory," in J. Butterfield, J. Earman (eds.), *Philosophy of Physics, Vol. I*, Elsevier, Amsterdam, 731-922 (2007).
9. D. M. Hausman and J. Woodward, "Independence, invariance and the causal Markov condition," *Brit. J. Phil Sci.*, **50** (4), 521–583 (1999).
10. J. Henson, "Comparing causality principles," *Stud. Hist. Phil. Mod. Phys.*, **36**, 519-543 (2005).
11. J. Henson, "Confounding causality principles: Comment on Rédei and San Pedro's "Distinguishing causality principles"," *Stud. Hist. Phil. Mod. Phys.*, **44**, 17-19 (2013a).
12. J. Henson, "Non-separability does not relieve the problem of Bell's theorem," *Found. Phys.*, **43**, 1008-1038 (2013b).
13. G. Hofer-Szabó, "Local causality and complete specification: a reply to Seevinck and Uffink," in U. Mäki, I. Votsis, S. Ruphy and G. Schurz (eds.) *Recent Developments in the Philosophy of Science: EPSA13 Helsinki*, Springer Verlag, 209-226 (2015a).
14. G. Hofer-Szabó, "Relating Bell's local causality to the Causal Markov Condition," *Found. Phys.*, **45(9)**, 1110-1136 (2015b).
15. G. Hofer-Szabó and P. Vecsernyés, "Reichenbach's Common Cause Principle in AQFT with locally finite degrees of freedom," *Found. Phys.*, **42**, 241-255 (2012a).
16. G. Hofer-Szabó and P. Vecsernyés, "Noncommuting local common causes for correlations violating the Clauser–Horne inequality," *J. Math. Phys.*, **53**, 12230 (2012b).
17. G. Hofer-Szabó and P. Vecsernyés, "Noncommutative Common Cause Principles in AQFT," *J. Math. Phys.*, **54**, 042301 (2013a).
18. G. Hofer-Szabó and P. Vecsernyés, "Bell inequality and common causal explanation in algebraic quantum field theory," *Stud. Hist. Phil. Mod. Phys.*, **44 (4)**, 404-416 (2013b).
19. G. Hofer-Szabó and P. Vecsernyés, "On the concept of Bell's local causality in local classical and quantum theory," *J. Math. Phys*, **56**, 032303 (2015).
20. G. Hofer-Szabó and P. Vecsernyés, "A generalized definition of Bell's local causality," *Synthese* **193(10)**, 3195-3207 (2016).
21. G. Hofer-Szabó, M. Rédei and L. E. Szabó, *The Principle of the Common Cause*, (Cambridge: Cambridge University Press, 2013).
22. T. Maudlin, "What Bell did," *J. Phys. A: Math. Theor.*, **47**, 424010 (2014).
23. T. Norsen, "Local causality and Completeness: Bell vs. Jarrett," *Found. Phys.*, **39**, 273 (2009).
24. T. Norsen, "J.S. Bell's concept of local causality," *Am. J. Phys*, **79**, 12, (2011).
25. J. Pearl, "Causality: Models, Reasoning, and Inference," Cambridge: (Cambridge University Press, 2000).
26. M. Rédei, "Reichenbach's Common Cause Principle and quantum field theory," *Found. Phys.*, **27**, 1309-1321 (1997).
27. M. Rédei, "A categorial approach to relativistic locality," *Stud. Hist. Phil. Mod. Phys.*, **48**, 137-146 (2014).
28. M. Rédei and I. San Pedro, "Distinguishing causality principles," *Stud. Hist. Phil. Mod. Phys.*, **43**, 84-89 (2012).
29. M. Rédei and J. S. Summers, "Local primitive causality and the Common Cause Principle in quantum field theory," *Found. Phys.*, **32**, 335-355 (2002).
30. T. S. Richardson and P. Spirtes, "Ancestral graph Markov models," *Ann. Statist.* **30**, 962-1030 (2002).
31. K. Sadeghi and S. Lauritzen, "Markov properties for mixed graphs," *Bernoulli.* **20/2**, 676-696 (2014).
32. M. P. Seevinck and J. Uffink, "Not throwing our the baby with the bathwater: Bell's condition of local causality mathematically 'sharp and clean'," in: Dieks, D.; Gonzalez, W.J.; Hartmann, S.; Uebel, Th.; Weber, M. (eds.) *Explanation, Prediction, and Confirmation The Philosophy of Science in a European Perspective*, Volume 2, 425-450 (2011).

33. M. Suárez, "Interventions and causality in quantum mechanics," *Erkenntnis*, **78**, 199-213 (2013).
34. M. Suárez and I. San Pedro "Causal Markov, robustness and the quantum correlations," in M. Suarez (ed.), *Probabilities, causes and propensities in physics*, 173-193. Synthese Library, 347, (Dordrecht: Springer, 2011)
35. J. Tian, A. Paz, and J. Pearl, "Finding minimal d-separating sets," *UCLA Cognitive Systems Laboratory, Technical Report* (R-254), (1998).

Local Operations and Completely Positive Maps in Algebraic Quantum Field Theory

Yuichiro Kitajima

Abstract Einstein introduced the locality principle which states that all physical effect in some finite space-time region does not influence its space-like separated finite region. Recently, in algebraic quantum field theory, Rédei captured the idea of the locality principle by the notion of operational separability. The operation in operational separability is performed in some finite space-time region, and leaves unchanged the state in its space-like separated finite space-time region. This operation is defined with a completely positive map. In the present paper, we justify using a completely positive map as a local operation in algebraic quantum field theory, and show that this local operation can be approximately written with Kraus operators under the funnel property.

Keywords Local operations · Completely positive maps
Algebraic quantum field theory

1 Introduction

Einstein [3] introduced the separability principle and the locality principle to show incompleteness of quantum mechanics. The separability principle says that 'any two spatially separated systems possess their own separate real states' [7, p. 173]. Einstein writes:

> [I]t is characteristic of these physical things that they are conceived of as being arranged in a space-time continuum. Further, it appears to be essential for this arrangement of the things introduced in physics that, at a specific time, these things claim an existence independent of one another, insofar as the these things 'lie in different parts of space'. ([3, p. 321]; Howard's translation [7, p. 187])

Y. Kitajima (✉)
Nihon University, 2-11-1 Shin-ei, Narashino, Chiba 275-8576, Japan
e-mail: kitajima.yuichirou@nihon-u.ac.jp

M. Ozawa et al. (eds.), *Reality and Measurement in Algebraic Quantum Theory*, Springer Proceedings in Mathematics & Statistics 261,
https://doi.org/10.1007/978-981-13-2487-1_3

Einstein introduced the locality principle in addition to the separability principle. Einstein writes:

> For the relative independence of spatially distant things (A and B), this idea is characteristic: an external influence on A has no immediate effect on B; this is known as the 'principle of local action', which is applied consistently only in field theory. The complete suspension of this basic principle would make impossible the idea of the existence of (quasi-) closed systems and, thereby, the establishment of empirically testable laws in the sense familiar to us. ([3, p. 322]; Howard's translation [7, p. 188])

This principle states that any physical effect in some finite space-time region does not influence its space-like separated finite region. Einstein [3] argued for the incompleteness of quantum mechanics under the locality principle and the separability principle.

According to Howard [7], the Bell inequality is a consequence of the separability and locality principle. Since the Bell inequality does not hold in algebraic quantum field theory and in quantum mechanics [5, 9, 11, 23–26], we must give up either separability or locality. Howard [7] argued that the separability principle must be abandoned, and that the locality principle holds in quantum theory. In the present paper we concentrate on the locality principle because it can be compatible with the violation of Bell inequalities.

Recently, in algebraic quantum field theory, Rédei [15, 17] captured the idea of the locality principle by the notion of operational separability (Definition 6), which had been introduced by Rédei and Valente [19]. The reason why he adopts the formalism of algebraic quantum field theory is that Einstein [3] says that physical things are conceived of as being arranged in a space-time continuum, and that observables in algebraic quantum field theory are 'explicitly regarded as localized in regions of the space-time continuum' [15, p. 1045].

The operation in operational separability is performed in some finite space-time region, and leaves unchanged the state in its space-like separated finite region. It is defined with a completely positive map. Valente [28] called such an operation a relatively local operation (Definition 7). On the other hand, there is another local operation. It is called an absolutely local operation, which is written with some operators in a local algebra which is associated with some open bounded region (Definition 7). This operation in some finite space-time region has no effects on the entire causal complement of this region. A difference between these two types of operations is that a relatively local operation is not necessarily written in terms of local operators while an absolutely local operation is given by local operators by definition. Valente [28] argued that the concept of absolutely local operation is too strong to express Einstein's locality principle because this principle simply demands that an operation performed in a system A leaves unchanged the state of another space-like separated system B.

There are two tasks here. One is to justify using a completely positive map as a local operation in algebraic quantum field theory. Another is to clarify the relation between these local operations. In the present paper, we show that a local operation in algebraic quantum field theory should be a completely positive map, and that a relatively local operation can be approximately written with some operators as well as an absolutely local operation.

The structure of the paper is as follows. We begin in Sect. 2 by reviewing the formalism of algebraic quantum field theory and notions of independence. In Sect. 3 we examine a definition of an operation. Usually a completely positive map is regarded as an operation. Although this assumption is natural in the case of nonrelativistic quantum mechanics, it is not transparent in the case of algebraic quantum field theory. We will justify using a completely positive map as a local operation in the case of algebraic quantum field theory (Theorem 1). We conclude, in Sect. 4, by examining a similarity between an absolutely local operation and a relatively local operation. An absolutely local operation is written with some operators. This representation is called the Kraus representation. On the other hand, a relatively local operation does not necessarily admit such a representation. By establishing a slightly generalized Kraus representation theorem (Theorem 4), it is shown that a relatively local operation can be approximately written with Kraus operators under the funnel property (Corollary 1).

2 Algebraic Quantum Field Theory

Algebraic quantum field theory exists in two versions: the Haag-Araki theory which uses von Neumann algebras on a Hilbert space, and the Haag-Kastler theory which uses abstract C*-algebras. Here we adopt the Haag-Araki theory. In this theory, each bounded open region $\mathcal{O}$ in the Minkowski space is associated with a von Neumann algebra $\mathfrak{N}(\mathcal{O})$ on a Hilbert space $\mathcal{H}$. Such a von Neumann algebra is called a local algebra.

In the present paper we use the following notation. For a subspace $\mathcal{K}$ of a Hilbert space $\mathcal{H}$, $\{\mathcal{K}\}^-$ stands for the closure of $\mathcal{K}$. $\mathbb{B}(\mathcal{H})$ is the set of all bounded operators on a Hilbert space $\mathcal{H}$. I stands for an identity operator on a Hilbert space. For a von Neumann algebra $\mathfrak{N}$ on a Hilbert space $\mathcal{H}$, $\mathfrak{N}'$ stands for the commutant of $\mathfrak{N}$ in $\mathbb{B}(\mathcal{H})$. For von Neumann algebras $\mathfrak{N}_1$ and $\mathfrak{N}_2$ on a Hilbert space $\mathcal{H}$, $\mathfrak{N}_1 \vee \mathfrak{N}_2$ stands for the von Neumann algebra generated by $\mathfrak{N}_1$ and $\mathfrak{N}_2$.

For an open bounded region $\mathcal{O}$ in the Minkowski space, $\mathcal{O}'$ stands for the causal complement of $\mathcal{O}$ and $\bar{\mathcal{O}}$ the closure of $\mathcal{O}$. A double cone in Minkowski space is the intersection of the causal future of a point x with the causal past of a point y to the future of x. Two double cones $\mathcal{O}_1$, $\mathcal{O}_2$ are said to be strictly space-like separated if there is a neighborhood $\mathcal{N}$ of zero such that $\mathcal{O}_1 + x$ is space-like separated from $\mathcal{O}_2$ for all $x \in \mathcal{N}$.

In the present paper, we assume the following axioms.

Definition 1 (Microcausality) [1, p. 10] Let $\mathcal{O}_1$ and $\mathcal{O}_2$ be bounded open regions in the Minkowski space. If $\mathcal{O}_1 \subseteq \mathcal{O}_2'$, then $\mathfrak{N}(\mathcal{O}_1) \subseteq \mathfrak{N}(\mathcal{O}_2)'$. This property is called microcausality.

Definition 2 (The funnel property) [21, Definition 6.14] For any pair $(\mathcal{O}, \tilde{\mathcal{O}})$ of double cones in the Minkowski space such that the closure of $\bar{\mathcal{O}} \subset \tilde{\mathcal{O}}$, there exists a type I factor $\mathfrak{N}$ such that $\mathfrak{N}(\mathcal{O}) \subset \mathfrak{N} \subset \mathfrak{N}(\tilde{\mathcal{O}})$. This property is called the funnel property.

The following property is derived from usual axioms of algebraic quantum field theory [1, Corollary 1.5.6].

Definition 3 Let $\mathcal{O}$ be a bounded open region in the Minkowski space. $\mathfrak{N}(\mathcal{O})$ is properly infinite.

Although there are some different notions of independence [6, 21], we use only two notions.

Definition 4 Let $\mathfrak{N}_1$ and $\mathfrak{N}_2$ be von Neumann algebras on a Hilbert space $\mathcal{H}$.

- $\mathfrak{N}_1$ and $\mathfrak{N}_2$ are called Schlieder independent if $A_1 A_2 \neq 0$ whenever $0 \neq A_1 \in \mathfrak{N}_1$ and $0 \neq A_2 \in \mathfrak{N}_2$.
- $\mathfrak{N}_1$ and $\mathfrak{N}_2$ are called split if there exists a type I factor $\mathfrak{N}$ such that $\mathfrak{N}_1 \subset \mathfrak{N} \subset \mathfrak{N}_2'$.

If two double cones $\mathcal{O}_1$ and $\mathcal{O}_2$ are strictly space-like separated, then $\mathfrak{N}(\mathcal{O}_1)$ and $\mathfrak{N}(\mathcal{O}_2)$ are split by Axioms Definitions 1 and 2. The following lemma shows that the split property is stronger than the Schlieder property.

Lemma 1 ([8, Theorem 5.5.4]) *Let $\mathfrak{N}$ be a factor on a Hilbert space $\mathcal{H}$. Then $AA' \neq 0$ for any nonzero operators $A \in \mathfrak{N}$ and $A' \in \mathfrak{N}'$.*

Lemma 1 shows that von Neumann algebras $\mathfrak{N}_1$ and $\mathfrak{N}_2$ are Schlieder independent if they are split. The following proposition is a characterization of Schlieder independence.

Proposition 1 ([4, Theorem 1 and Proposition 2] [6, Theorem 11.2.5 and Theorem 11.2.17]) *Let $\mathfrak{A}_1$ and $\mathfrak{A}_2$ be mutually commuting C*-subalgebras of a C*-algebra $\mathfrak{A}$. The following conditions are equivalent.*

1. *$\mathfrak{A}_1$ and $\mathfrak{A}_2$ are Schlieder independent.*
2. *$\|A_1 A_2\| = \|A_1\| \|A_2\|$ for any $A_1 \in \mathfrak{A}_1$ and $A_2 \in \mathfrak{A}_2$.*

3 Completely Positive Maps

In this section, we examine the reason why local operations are assumed to be completely positive in algebraic quantum field theory.

Definition 5 Let $\mathfrak{N}$ be a von Neumann algebra and let T be a linear map of $\mathfrak{N}$.

- T is called positive if $A \geq 0$ entails $T(A) \geq 0$.

- Let $[A_{jk}]$ be $n \times n$-matrix with entries A_{jk} in $\mathfrak{N}$. T is called completely positive if $[A_{jk}] \geq 0$ entails $[T(A_{jk})] \geq 0$ for any $n \in \mathbb{N}$.

It is natural to assume that an operation is a positive map because the probability after the process represented by the map T must be positive. Moreover, if we introduce an environmental system which is represented by a set $M_n(\mathbb{C})$ of all $n \times n$ matrices with complex entries, then $(T \otimes \mathrm{Id})(A)$ must be also positive for any positive operator A on $\mathbb{B}(\mathscr{H}) \otimes M_n(\mathbb{C})$, where Id denotes the identity map on $M_n(\mathbb{C})$. This is equivalent to the condition that T is completely positive. Therefore it is reasonable to assume that an operation is completely positive in the case of nonrelativistic quantum mechanics.

A completely positive map plays an important role in quantum measurements [13, 14]. It is also used as a local operation in algebraic quantum field theory [12, 15, 16, 18, 19, 28]. For example, a new concept of local states is defined in terms of a completely positive map [12]. But it is not transparent to use a completely positive map as an operation in algebraic quantum field theory because any local algebra which is associated with two space-like separated regions is not isomorphic to $\mathbb{B}(\mathscr{H}) \otimes M_n(\mathbb{C})$. Therefore, we examine how we can justify it in algebraic quantum field theory in Theorem 1.

We introduce a positive map T of $\mathfrak{N}_1$ such that it has an extension to $\mathfrak{N}_1 \vee \mathfrak{N}_2$ which is the identity map on $\mathfrak{N}_2$ to capture an idea that this operation is performed in the system $\mathfrak{N}_1$ and it does not influence the system $\mathfrak{N}_2$. To examine such an operation, we use the following lemma.

Lemma 2 ([29, Lemma] [21, Lemma 3.12]) *Let $\mathfrak{N}_1$ and $\mathfrak{N}_2$ be mutually commuting von Neumann algebras on a Hilbert space $\mathscr{H}$, and let T' be a positive map of $\mathfrak{N}_1 \vee \mathfrak{N}_2$ such that $T'(A_2) = A_2$ for all $A_2 \in \mathfrak{N}_2$. Then $T'(A_1A_2) = T'(A_1)A_2$ for any $A_1 \in \mathfrak{N}_1$ and $A_2 \in \mathfrak{N}_2$.*

By using this lemma, we can show the following fact.

Theorem 1 *Let $\mathfrak{N}_1$ and $\mathfrak{N}_2$ be mutually commuting von Neumann algebras which are Schlieder independent, let $\mathfrak{N}_2$ have either type II_1 direct summand or properly infinite one, and let T be a positive map of $\mathfrak{N}_1$. If there is a positive map T' of $\mathfrak{N}_1 \vee \mathfrak{N}_2$ such that*

$$T'(A_1) = T(A_1), \qquad T'(A_2) = A_2$$

for any $A_1 \in \mathfrak{N}_1$ and $A_2 \in \mathfrak{N}_2$, then T is completely positive.

Proof Since $\mathfrak{N}_2$ has have either type II_1 direct summand or properly infinite one, for any natural number $n \in \mathbb{N}$, there is a set $\{E_1, \ldots, E_n\}$ of mutually orthogonal and equivalent projections in $\mathfrak{N}_2$ [27, Proposition V.1.35 and Proposition V.1.36]. Thus there is a set $\{V_1, \ldots, V_n\}$ of partial isometries in $\mathfrak{N}_2$ such that $V_i^* V_i = E_1$ and $V_i V_i^* = E_i$ for any $i \in \{1, \ldots, n\}$.

Let $E_{jk} := V_j V_k^*$, let $M_n(\mathfrak{N}_1)$ be the set of all $n \times n$-matrices $[A_{jk}]$ with entries A_{jk} in $\mathfrak{N}_1$, and let

$$\mathfrak{C} := \left\{ \sum_{j,k=1}^{n} C_{jk}E_{jk} \middle| C_{jk} \in \mathfrak{N}_1, 1 \le j,k \le n \right\}.$$

$\mathfrak{C}$ is a linear subspace of $\mathfrak{N}_1 \vee \mathfrak{N}_2$, and is self-adjoint because $(C_{jk}E_{jk})^* = E^*_{jk}C^*_{jk} = C^*_{jk}E_{kj} \in \mathfrak{C}$ for any $C_{jk} \in \mathfrak{N}_1$. Furthermore, if $C_{jk}, C_{lm} \in \mathfrak{N}_1$, then $(C_{jk}E_{jk})(C_{lm}E_{lm}) = \delta_{kl}C_{jk}C_{lm}E_{jm} \in \mathfrak{C}$, where δ_{kl} equals 1 if $k = l$, and 0 if $k \ne l$. By linearity $\mathfrak{C}$ is closed under multiplication. Hence $\mathfrak{C}$ is a *-subalgebra of $\mathfrak{N}_1 \vee \mathfrak{N}_2$.

Let $M_n(\mathfrak{N}_1)$ be the set of all $n \times n$-matrices $[A_{ij}]$ with entries A_{ij} in $\mathfrak{N}_1$, and let α be a map of $M_n(\mathfrak{N}_1)$ to $\mathfrak{C}$ such that

$$\alpha\left([A_{jk}]\right) := \sum_{j,k=1}^{n} A_{jk}E_{jk} \tag{1}$$

for any $[A_{jk}] \in M_n(\mathfrak{N}_1)$. Clearly α is surjective. Given $S^{(s)}$ and $S^{(t)}$ in $\mathfrak{C}$, say

$$S^{(s)} = \sum_{j,k=1}^{n} A^{(s)}_{jk}E_{jk}, \qquad S^{(t)} = \sum_{j,k=1}^{n} A^{(t)}_{jk}E_{jk},$$

we have

$$\alpha([A^{(s)}_{jk}]^*) = \alpha([A^{(s)}_{jk}])^*, \tag{2}$$

$$\alpha([A^{(s)}_{jk}][A^{(t)}_{lm}]) = \alpha([A^{(s)}_{jk}])\alpha([A^{(t)}_{lm}]), \tag{3}$$

$$\begin{aligned} \|A^{(s)}_{jk} - A^{(t)}_{jk}\| &= \|A^{(s)}_{jk} - A^{(t)}_{jk}\| \|E_{jk}\| \\ &= \|(A^{(s)}_{jk} - A^{(t)}_{jk})E_{jk}\| \\ &= \|E_{jj}(S^{(s)} - S^{(t)})E_{kk}\| \\ &\le \|S^{(s)} - S^{(t)}\| \end{aligned} \tag{4}$$

because $\|E_{jk}\|^2 = \|E^*_{jk}E_{jk}\| = \|V_kV^*_jV_jV^*_k\| = \|E_k\| = 1$ and $\|A^{(s)}_{jk} - A^{(t)}_{jk}\| \|E_{jk}\| = \|(A^{(s)}_{jk} - A^{(t)}_{jk})E_{jk}\|$ by Proposition 1. Thus α is a faithful *-homomorphism of $M_n(\mathfrak{N}_1)$ to $\mathfrak{C}$, which entails that $\mathfrak{C}$ is a C*-algebra [27, p. 192].

Let T be a positive map of $\mathfrak{N}_1$, let T' be a positive map of $\mathfrak{N}_1 \vee \mathfrak{N}_2$ such that $T'(A_1) = T(A_1)$ and $T'(A_2) = A_2$ for any $A_1 \in \mathfrak{N}_1$ and $A_2 \in \mathfrak{N}_2$, and let $[A_{jk}]$ be a positive operator in $M_n(\mathfrak{N}_1)$. Then there is $[B_{jk}] \in M_n(\mathfrak{N}_1)$ such that $[A_{jk}] = [B_{jk}]^*[B_{jk}]$, so that $\sum_{j,k=1}^{n} A_{jk}E_{jk} = \alpha\left([A_{jk}]\right) = \alpha\left([B_{jk}]^*[B_{jk}]\right) = \alpha\left([B_{jk}]\right)^* \alpha\left([B_{jk}]\right) \ge 0$ by Equations (2) and (3). Since T' is positive on $\mathfrak{N}_1 \vee \mathfrak{N}_2$, $T'(\sum_{j,k=1}^{n} A_{jk}E_{jk}) \ge 0$. By Lemma 2,

$$\sum_{j,k=1}^{n} T(A_{jk})E_{jk} = \sum_{j,k=1}^{n} T'(A_{jk})E_{jk} = \sum_{j,k=1}^{n} T'(A_{jk}E_{jk}) = T'\left(\sum_{j,k=1}^{n} A_{jk}E_{jk}\right) \geq 0. \quad (5)$$

Since $\mathfrak{C}$ is a C*-algebra, there is an operator $D \in \mathfrak{C}$ such that $\sum_{j,k=1}^{n} T(A_{jk})E_{jk} = D^*D$ [8, Theorem 4.2.6]. Therefore

$$[T(A_{jk})] = \alpha^{-1}\left(\sum_{j,k=1}^{n} T(A_{jk})E_{jk}\right) = \alpha^{-1}(D^*D) = \alpha^{-1}(D)^*\alpha^{-1}(D) \geq 0. \quad (6)$$

Because n is an arbitrary natural number, T is completely positive on $\mathfrak{N}_1$. □

Let $\mathcal{O}_1$ and $\mathcal{O}_2$ be double cones such that $\mathcal{O}_1 \subset \mathcal{O}_2'$ and let T be a positive map of $\mathfrak{N}(\mathcal{O}_1)$. When T has an extension to $\mathfrak{N}(\mathcal{O}_1) \vee \mathfrak{N}(\mathcal{O}_2)$ which is the identity map on $\mathfrak{N}(\mathcal{O}_2)$, T can be regarded as an operation performed in $\mathcal{O}_1$ which does not influence a state in $\mathcal{O}_2$. Since $\mathfrak{N}(\mathcal{O}_1)$ and $\mathfrak{N}(\mathcal{O}_2)$ are split by Definitions 1 and 2, they are Schlieder independent by Lemma 1. By Definition 3, any local algebra is properly infinite. Thus, Theorem 1 entails that T is completely positive. Therefore it is reasonable to assume that a local operation performed in some region which does not influence its space-like separated region is completely positive in algebraic quantum field theory.

4 Relatively Local Operations

Rédei and Valente [19] introduced the notion of operational W*-separability to capture the idea that a causally well behaved operation exists.

Definition 6 (Operational W*-separability) [19, Definition 6] Let $\mathfrak{N}_1$ and $\mathfrak{N}_2$ be von Neumann subalgebras of a von Neumann algebra $\mathfrak{N}$. $\mathfrak{N}_1$ and $\mathfrak{N}_2$ are called operationally W*-separable in $\mathfrak{N}$ if the following two conditions are true:

1. If T is a normal completely positive map of $\mathfrak{N}$ such that $T(A_1) \in \mathfrak{N}_1$ for any $A_1 \in \mathfrak{N}_1$, there exists a normal completely positive map T' such that $T'(A_1) = T(A_1)$ and $T'(A_2) = A_2$ for any $A_1 \in \mathfrak{N}_1$ and $A_2 \in \mathfrak{N}_2$.
2. If T is a normal completely positive map of $\mathfrak{N}$ such that $T(A_2) \in \mathfrak{N}_2$ for any $A_2 \in \mathfrak{N}_2$, there exists a normal completely positive map T' such that $T'(A_2) = T(A_2)$ and $T'(A_1) = A_1$ for any $A_2 \in \mathfrak{N}_2$ and $A_1 \in \mathfrak{N}_1$.

The normal completely positive map T' in Definition 6 is performed in some finite space-time region, and leaves unchanged the state in its space-like separated finite region. Thus, this definition requires that there exists such a causally well behaved operation. The following proposition shows that operational W*-separability holds in algebraic quantum field theory.

Proposition 2 ([16, Proposition 2]; [18, Section 5]; [22, Theorem 5.2]) *Let assume microcausality (Definition 1) and the funnel property (Definition 2), let $\mathcal{O}_1$ and $\mathcal{O}_2$ be strictly space-like separated double cones. Then $\mathfrak{N}(\mathcal{O}_1)$ and $\mathfrak{N}(\mathcal{O}_2)$ are operationally W^*-separable in $\mathfrak{N}(\mathcal{O}_1) \vee \mathfrak{N}(\mathcal{O}_2)$.*

In this section, we examine the normal completely positive map T' in Definition 6. Valente [28] called it a relatively local operation. There is another local operation. It is called an absolutely local operation. Thus there are two types of local operations.

Definition 7 ([28, Section 3]) Let $\mathfrak{N}_1$ and $\mathfrak{N}_2$ be mutually commuting von Neumann algebras on a Hilbert space $\mathcal{H}$.

- A normal completely positive map T of $\mathbb{B}(\mathcal{H})$ is called an absolutely local operation in $\mathfrak{N}_1$ if there are operators K_i in $\mathfrak{N}_1$ such that

$$T(A) = \sum_{j \in J} K_j^* A K_j, \quad T(I) = I$$

for any $A \in \mathbb{B}(\mathcal{H})$.
- A normal completely positive map T of $\mathfrak{N}_1 \vee \mathfrak{N}_2$ is called a relatively local operation in $\mathfrak{N}_1$ with respect to $\mathfrak{N}_2$ if $T(A_1) \in \mathfrak{N}_1$ and $T(A_2) = A_2$ for any $A_1 \in \mathfrak{N}_1$ and $A_2 \in \mathfrak{N}_2$.

An absolutely local operation T in Definition 7 does not influence the system $\mathfrak{N}_1'$ which includes $\mathfrak{N}_2$ while a relatively local operation in Definition 7 does not influence only the system $\mathfrak{N}_2$. In the case of algebraic quantum field theory, an absolutely local operation in some region has no effect on the entire causal complement of this region. Although Clifton and Halvorson [2] discussed local disentanglement in terms of absolutely local operations, Valente [28] argued that an absolutely local operation is too strong because Einstein's locality principle simply demands that an operation performed in a system A leaves unchanged the state of another space-like separated system B.

There are two classical theorems characterizing a completely positive map. One is Stinespring representation theorem, and another Kraus representation theorem.

Theorem 2 (Stinespring representation theorem) *[20] Let $\mathfrak{A}$ be a unital C^*-algebra, let $\mathcal{H}$ be a Hilbert space, and let T be a completely positive map from $\mathfrak{A}$ to $\mathbb{B}(\mathcal{H})$. Then there exists a Hilbert space $\mathcal{K}$, a representation $\pi : \mathfrak{A} \to \mathbb{B}(\mathcal{K})$, and a bounded operator $W : \mathcal{H} \to \mathcal{K}$ such that*

$$T(A) = W^* \pi(A) W$$

for any $A \in \mathfrak{A}$.

Kraus representation theorem follows Stinespring representation theorem.

Theorem 3 (Kraus representation theorem) *[10] Let $\mathscr{H}$ be a Hilbert space and let T be a normal completely positive map of $\mathbb{B}(\mathscr{H})$ such that $0 < T(I) \leq I$. Then there are bounded operators K_j in $\mathbb{B}(\mathscr{H})$ such that*

$$T(A) = \sum_{j \in J} K_j^* A K_j, \quad 0 < \sum_{j \in J} K_j^* K_j \leq I$$

for any $A \in \mathbb{B}(\mathscr{H})$.

The operators K_i in Theorem 3 are called Kraus operators. If a normal completely positive map is defined on a proper subalgebra of $\mathbb{B}(\mathscr{H})$, it does not necessarily admit a decomposition with Kraus operators.

Here we examine a normal completely positive map T from a type I factor $\mathfrak{N}$ on a Hilbert space $\mathscr{H}$ to $\mathbb{B}(\mathscr{H})$. Note that if $T(A) \in \mathfrak{N}$ for any $A \in \mathfrak{N}$, we can apply Kraus representation theorem because there is a Hilbert space $\mathscr{K}$ such that $\mathfrak{N}$ is isomorphic to $\mathbb{B}(\mathscr{K})$. However, $T(\mathfrak{N})$ is not necessarily included in $\mathfrak{N}$, so we cannot use the original Kraus representation theorem. Yet, we show below (Theorem 4) that a representation theorem similar to Kraus representation theorem holds if the von Neumann algebra $\mathfrak{N}$ is a type I factor.

Theorem 4 *Let $\mathfrak{N}$ be a type I factor on a Hilbert space $\mathscr{H}$, and let T be a normal completely positive map of $\mathfrak{N}$ to $\mathbb{B}(\mathscr{H})$ such that $0 < T(I) \leq I$. Then there are bounded operators K_j in $\mathbb{B}(\mathscr{H})$ such that*

$$T(A) = \sum_{j \in J} K_j^* A K_j, \quad 0 < \sum_{j \in J} K_j^* K_j \leq I$$

for any $A \in \mathfrak{N}$.

Proof By Theorem 2, there is a representation π of $\mathfrak{N}$ on a Hilbert space $\mathscr{K}$ and a bounded operator $W : \mathscr{H} \to \mathscr{K}$ such that $T(A) = W^*\pi(A)W$ for any $A \in \mathfrak{N}$. Since T is normal, so is π. Since $\pi(I) > 0$ and $\mathfrak{N}$ is a type I factor, there exists a minimal projection $P_0 \in \mathfrak{N}$ such that $\pi(P_0) \neq 0$. Let $x_0 \in \mathscr{H}$ be a unit vector such that $P_0 x_0 = x_0$, let $y_0 \in \mathscr{K}$ be a unit vector such that $\pi(P_0)y_0 = y_0$, and let E_0 and Q_0 be projections whose ranges are $\{\pi(\mathfrak{N})y_0\}^-$ and $\{\mathfrak{N}x_0\}^-$, respectively. Then $E_0 \in \pi(\mathfrak{N})'$. For any $A \in \mathfrak{N}$, $P_0 A P_0 = \langle x_0, Ax_0 \rangle P_0$ since P_0 is a minimal projection. Thus $\langle y_0, \pi(A)y_0 \rangle = \langle y_0, \pi(P_0 A P_0)y_0 \rangle = \langle x_0, Ax_0 \rangle$ for any $A \in \mathfrak{N}$. Therefore there exists a unitary operator U_0 from $\{\pi(\mathfrak{N})y_0\}^-$ to $\{\mathfrak{N}x_0\}^-$ such that $\pi(A)E_0 = U_0^* A U_0$ for any $A \in \mathfrak{N}$ by [8, Proposition 4.5.3]. Let $V_0 := Q_0 U_0 E_0$. Then V_0 is an isometry from $\mathscr{K}$ to $\mathscr{H}$ such that $\pi(A)E_0 = V_0^* A V_0$ for any $A \in \mathfrak{N}$.

By Zorn's lemma, it can be shown that there are a maximal family $\{E_j \in \pi(\mathfrak{N})' | j \in J\}$ of mutually orthogonal projections in $\pi(\mathfrak{N})'$ and a family $\{V_j | j \in J\}$ of isometries from $\mathscr{K}$ to $\mathscr{H}$ such that the range of E_j is $\{\pi(\mathfrak{N})y_j\}^-$ for some unit vector $y_j \in \mathscr{K}$, and $\pi(A)E_j = V_j^* A V_j$ for any $A \in \mathfrak{N}$. Suppose that $\sum_{j \in J} E_j < I$. Let $F^0 := I - \sum_{j \in J} E_j$. Then there is a unit vector $y' \in F^0\mathscr{K}$. Since $\pi(I)y' = y' \neq 0$

and $\mathfrak{N}$ is a type I factor, there is a minimal projection $P^0 \in \mathfrak{N}$ such that $\pi(P^0)y' \neq 0$. Thus $\pi(P^0)F^0 \neq 0$. Let x^0 be a unit vector such that $P^0x^0 = x^0$, let y^0 be a unit vector such that $\pi(P^0)F^0y^0 = y^0$, and let E^0 be a projection whose range is $\{\pi(\mathfrak{N})y^0\}^-$. Then $E^0 \in \pi(\mathfrak{N})'$. Since $\pi(P^0)y^0 = y^0$ and $F^0y^0 = y^0$,

$$\langle \pi(A)y_j, \pi(B)y^0 \rangle = \langle \pi(B^*A)y_j, y^0 \rangle = \langle E_j\pi(B^*A)y_j, F^0y^0 \rangle = 0, \tag{7}$$

and

$$\langle y^0, \pi(A)y^0 \rangle = \langle y^0, \pi(P^0AP^0)y^0 \rangle = \langle x^0, Ax^0 \rangle \tag{8}$$

for any $j \in J$ and $A, B \in \mathfrak{N}$. Therefore $E_jE^0 = 0$ for any $j \in J$, and there exists an isometry V^0 from $\mathscr{K}$ to $\mathscr{H}$ such that $\pi(A)E^0 = V^{0*}AV^0$ for any $A \in \mathfrak{N}$. This contradicts the maximality of $\{E_j | j \in J\}$. Therefore, $\sum_{j \in J} E_j = I$.

Let $K_j := V_jW$ for any $j \in J$. Then

$$T(A) = W^*\pi(A)W = \sum_{j \in J} W^*\pi(A)E_jW = \sum_{j \in J} W^*V_j^*AV_jW = \sum_{j \in J} K_j^*AK_j \tag{9}$$

for any $A \in \mathfrak{N}$. Since $0 < T(I) \leq I$ and $T(I) = \sum_{j \in J} K_j^*K_j$, $0 < \sum_{j \in J} K_j^*K_j \leq I$. □

Under the funnel property (Definition 2), type I factors exist which are interpolated between local algebras of regions strictly contained in each other. By using Theorem 4, we show that a relatively local operation can be approximately written with Kraus operators in algebraic quantum field theory.

Corollary 1 *Let's assume microcausality (Definition 1) and the funnel property (Definition 2), let* $\tilde{\mathscr{O}}_1$ *and* $\tilde{\mathscr{O}}_2$ *be double cones such that* $\tilde{\mathscr{O}}_1 \subset \tilde{\mathscr{O}}_2'$, *and let* T *be a relatively local operation in* $\mathfrak{N}(\tilde{\mathscr{O}}_1)$ *with respect to* $\mathfrak{N}(\tilde{\mathscr{O}}_2)$. *For any double cones* $\mathscr{O}_1$ *and* $\mathscr{O}_2$ *such that* $\bar{\mathscr{O}}_1 \subset \tilde{\mathscr{O}}_1$ *and* $\bar{\mathscr{O}}_2 \subset \tilde{\mathscr{O}}_2$, *there are bounded operators* K_j *in* $\mathfrak{N}(\mathscr{O}_2)'$ *such that*

$$T(A) = \sum_{j \in J} K_j^*AK_j, \quad \sum_{j \in J} K_j^*K_j = I$$

for any $A \in \mathfrak{N}(\mathscr{O}_1) \vee \mathfrak{N}(\mathscr{O}_2)$.

Proof Let $\mathscr{O}_1$ and $\mathscr{O}_2$ be double cones such that $\bar{\mathscr{O}}_1 \subset \tilde{\mathscr{O}}_1$ and $\bar{\mathscr{O}}_2 \subset \tilde{\mathscr{O}}_2$. By Axiom 2, there are type I factors $\mathfrak{N}_1$ and $\mathfrak{N}_2$ such that $\mathfrak{N}(\mathscr{O}_1) \subset \mathfrak{N}_1 \subset \mathfrak{N}(\tilde{\mathscr{O}}_1)$ and $\mathfrak{N}(\mathscr{O}_2) \subset \mathfrak{N}_2 \subset \mathfrak{N}(\tilde{\mathscr{O}}_2)$. Then $\mathfrak{N}(\mathscr{O}_1) \vee \mathfrak{N}(\mathscr{O}_2) \subset \mathfrak{N}_1 \vee \mathfrak{N}_2 \subset \mathfrak{N}(\tilde{\mathscr{O}}_1) \vee \mathfrak{N}(\tilde{\mathscr{O}}_2)$, and $\mathfrak{N}_1 \vee \mathfrak{N}_2$ is a type I factor. By Theorem 4, there exists a set $\{K_j | j \in J\}$ of operators in $\mathbb{B}(\mathscr{H})$ such that

$$T(A) = \sum_{j \in J} K_j^*AK_j$$

for any $A \in \mathfrak{N}_1 \vee \mathfrak{N}_2$. $T(I) = I$ entails $\sum_{j \in J} K_j^*K_j = I$.

Since $T(A_2) = A_2$ for any $A_2 \in \mathfrak{N}(\mathcal{O}_2)$ and $T(I) = I$, $\sum_{j \in J}[K_j, A_2]^*[K_j, A_2] = 0$ [2, p. 13]. Thus $K_j \in \mathfrak{N}(\mathcal{O}_2)'$ for any $j \in J$. □

In Corollary 1, double cones $\mathcal{O}_1$ and $\mathcal{O}_2$ can approximate $\tilde{\mathcal{O}}_1$ and $\tilde{\mathcal{O}}_2$, respectively, as closely as possible. So we can say that T can be approximately written with operators in $\mathfrak{N}(\mathcal{O}_2)'$.

5 Conclusion

Einstein [3] introduced the locality principle which states that physical effects in some finite space-time region do not influence its space-like separated finite region. In algebraic quantum field theory, Rédei [15] captured the idea of the locality principle by the notion of operational W*-separability (Definition 6), which had been introduced by Rédei and Valente [19]. Valente [28] called such an operation a relatively local operation to distinguish it from an absolutely local operation which can be written with Kraus operators (Definition 7).

In the present paper, we examined two questions;

- Can we justify using a completely positive map as a local operation in algebraic quantum field theory?
- Can we write a relatively local operation with some operators?

Roughly speaking, complete positiveness of an operation T in a system A is equivalent to the condition that T performed in the system A does not influence a space-like separated system B which is represented by a set $M_n(\mathbb{C})$ of all $n \times n$ matrices with complex entries in the case of nonrelativistic quantum mechanics. But it is not obvious why a completely positive map is used as an operation in the case of algebraic quantum field theory because any local algebra which is associated with two space-like separated regions is not isomorphic to $\mathbb{B}(\mathcal{H}) \otimes M_n(\mathbb{C})$. In Theorem 1, we showed that an operation is completely positive in algebraic quantum field theory if it is performed in some region and does not influence its space-like separated region. Thus, it is reasonable to assume that a local operation is completely positive.

Valente [28] distinguished between absolutely local operations and relatively local operations. A difference between these operations is that a relatively local operation is not necessarily written with Kraus operators while an absolutely local operation is written with Kraus operators by definition (Definition 7). In the present paper, by generalizing slightly Kraus representation theorem (Theorem 4), it was shown that a relatively local operation can be approximately written with Kraus operators under the funnel property (Corollary 1).

Acknowledgements The author wishes to thank Masanao Ozawa for helpful comments on an earlier draft. The author is supported by the JSPS KAKENHI No.15K01123 and No.23701009.

References

1. Baumgärtel, H.: Operatoralgebraic Methods in Quantum Field Theory. Akademie Verlag, Berlin (1995)
2. Clifton, R., Halvorson, H.: Entanglement and open systems in algebraic quantum field theory. Studies in History and Philosophy of Modern Physics **32**(1), 1–31 (2001)
3. Einstein, A.: Quanten-Mechanik und Wirklichkeit. Dialectica **2**(3-4), 320–324 (1948)
4. Florig, M., Summers, S.J.: On the statistical independence of algebras of observables. Journal of Mathematical Physics **38**(3), 1318–1328 (1997)
5. Halvorson, H., Clifton, R.: Generic Bell correlation between arbitrary local algebras in quantum field theory. Journal of Mathematical Physics **41**(4), 1711–1717 (2000)
6. Hamhalter, J.: Quantum Measure Theory. Springer, Dordrecht (2013)
7. Howard, D.: Einstein on locality and separability. Studies in History and Philosophy of Science **16**(3), 171–201 (1985)
8. Kadison, R.V., Ringrose, J.R.: Fundamentals of the Theory of Operator Algebras: Elementary Theory, vol. 1. American Mathematical Society, Providence (1983)
9. Kitajima, Y.: EPR states and Bell correlated states in algebraic quantum field theory. Foundations of Physics **43**(10), 1182–1192 (2013)
10. Kraus, K.: States, effects, and operations fundamental notions of quantum theory. Lecture Notes in Physics **190** (1983)
11. Landau, L.J.: On the violation of Bell's inequality in quantum theory. Physics Letters A **120**(2), 54–56 (1987)
12. Ojima, I., Okamura, K., Saigo, H.: Local state and sector theory in local quantum physics. Letters in Mathematical Physics **106**(6), 741–763 (2016)
13. Okamura, K., Ozawa, M.: Measurement theory in local quantum physics. Journal of Mathematical Physics **57**(1), 015,209 (2016)
14. Ozawa, M.: Quantum measuring processes of continuous observables. Journal of Mathematical Physics **25**(1), 79–87 (1984)
15. Rédei, M.: Einstein's dissatisfaction with nonrelativistic quantum mechanics and relativistic quantum field theory. Philosophy of Science **77**(5), 1042–1057 (2010)
16. Rédei, M.: Operational independence and operational separability in algebraic quantum mechanics. Foundations of Physics **40**(9-10), 1439–1449 (2010)
17. Rédei, M.: Einstein meets von neumann: Locality and operational independence in algebraic quantum field theory. In: H. Halvorson (ed.) Deep Beauty: Understanding the Quantum World through Mathematical Innovation, pp. 343–361. Cambridge University Press, New York (2011)
18. Rédei, M., Summers, S.J.: When are quantum systems operationally independent? International Journal of Theoretical Physics **49**(12), 3250–3261 (2010)
19. Rédei, M., Valente, G.: How local are local operations in local quantum field theory? Studies in History and Philosophy of Modern Physics **41**(4), 346–353 (2010)
20. Stinespring, W.F.: Positive functions on C*-algebras. Proceedings of the American Mathematical Society **6**(2), 211–216 (1955)
21. Summers, S.J.: On the independence of local algebras in quantum field theory. Reviews in Mathematical Physics **2**(2), 201–247 (1990)
22. Summers, S.J.: Subsystems and independence in relativistic microscopic physics. Studies in History and Philosophy of Modern Physics **40**(2), 133–141 (2009)
23. Summers, S.J., Werner, R.: Bell's inequalities and quantum field theory. I. general setting. Journal of Mathematical Physics **28**(10), 2440–2447 (1987)
24. Summers, S.J., Werner, R.: Bell's inequalities and quantum field theory. II. Bell's inequalities are maximally violated in the vacuum. Journal of Mathematical Physics **28**(10), 2448–2456 (1987)
25. Summers, S.J., Werner, R.: Maximal violation of Bell's inequalities is generic in quantum field theory. Communications in Mathematical Physics **110**(2), 247–259 (1987)
26. Summers, S.J., Werner, R.: Maximal violation of Bell's inequalities for algebras of observables in tangent spacetime regions. Annales de l'IHP Physique théorique **49**(2), 215–243 (1988)

27. Takesaki, M.: Theory of Operator Algebras, vol. 1. Springer (2002)
28. Valente, G.: Local disentanglement in relativistic quantum field theory. Studies in History and Philosophy of Modern Physics **44**(4), 424–432 (2013)
29. Werner, R.: Local preparability of states and the split property in quantum field theory. Letters in Mathematical Physics **13**(4), 325–329 (1987)

Symmetries in Exact Bohrification

Klaas Landsman and Bert Lindenhovius

Abstract The 'Bohrification" program in the foundations of quantum mechanics implements Bohr's doctrine of classical concepts through an interplay between commutative and non-commutative operator algebras. Following a brief conceptual and mathematical review of this program, we focus on one half of it, called "exact" Bohrification, where a (typically noncommutative) unital C^*-algebra A is studied through its commutative unital C^*-subalgebras $C \subseteq A$, organized into a poset $\mathscr{C}(A)$. This poset turns out to be a rich invariant of A (Hamhalter in J Math Anal Appl 383:391–399, 2011, [19], Hamhalter in J Math Anal Appl 422:1103-1115, 2015, [20], Landsman in Bohrification: From classical concepts to commutative algebras. Chicago, Chicago University Press [34]). To set the stage, we first give a general review of symmetries in elementary quantum mechanics (i.e., on Hilbert space) as well as in algebraic quantum theory, incorporating $\mathscr{C}(A)$ as a new kid in town. We then give a detailed proof of a deep result due to Hamhalter (J Math Anal Appl 383:391–399, 2011, [19]), according to which $\mathscr{C}(A)$ determines A as a Jordan algebra (at least for a large class of C^*-algebras). As a corollary, we prove a new Wigner-type theorem to the effect that order isomorphisms of $\mathscr{C}(B(H))$ are (anti) unitarily implemented. We also show how $\mathscr{C}(A)$ is related to the orthomodular poset $\mathscr{P}(A)$ of projections in A. These results indicate that $\mathscr{C}(A)$ is a serious player in C^*-algebras and quantum theory.

Keywords Quantum physics · Symmetries · Wigner's Theorem · Operator algebras · Bohr's doctrine of classical concepts

---

K. Landsman (✉)
Institute for Mathematics, Astrophysics, and Particle Physics, Radboud University, Heyendaalseweg 135, 6525AJ Nijmegen, The Netherlands
e-mail: landsman@math.ru.nl

B. Lindenhovius
Department of Computer Science, Tulane University, 6823 St Charles Ave, New Orleans, LA 70118, USA
e-mail: alindenh@tulane.edu

M. Ozawa et al. (eds.), *Reality and Measurement in Algebraic Quantum Theory*, Springer Proceedings in Mathematics & Statistics 261,
https://doi.org/10.1007/978-981-13-2487-1_4

1 Bohrification

The Bohrification program is an attempt to relate the core of the Copenhagen Interpretation of quantum mechanics, viz. Bohr's *doctrine of classical concepts*, to the mathematical formalism of operator algebras created by von Neumann, as subsequently generalized into the theory of C^*-algebras by [17]. Other elements of the Copenhagen Interpretation, such as the rejection of the possibility to analyze what is going on during measurements, the closely related idea of the collapse of the wave-function (in the sense of a "second" time-evolution in quantum mechanics beside the primary unitary evolution governed by the Schrödinger equation), and the ensuing hybrid interpretation of quantum-mechanical states as mere catalogues of the probabilities attached to possible outcomes of experiments, are irrelevant for this paper (and in fact appear outdated to us). To introduce the doctrine, we quote the opening of the most (perhaps the only) systematic presentation of the Copenhagen Interpretation by one of its original authors (viz. Heisenberg):

> The Copenhagen interpretation of quantum theory starts from a paradox. Any experiment in physics, whether it refers to the phenomena of daily life or to atomic events, is to be described in the terms of classical physics. The concepts of classical physics form the language by which we describe the arrangement of our experiments and state the results. We cannot and should not replace these concepts by any others. Still the application of these concepts is limited by the relations of uncertainty. We must keep in mind this limited range of applicability of the classical concepts while using them, but we cannot and should not try to improve them [26, p. 44].

Despite their agreement about the central role of classical concepts in the study of quantum mechanics, there seems to have been an unresolved disagreement between Bohr and Heisenberg about their precise status [9], as follows:

- According to Bohr—haunted by his idea of Complementarity—only one classical concept (or sometimes one coherent family of classical concepts) applies to the experimental study of some quantum object at a time. But if it applies, it does so exactly, and has the same meaning as in classical physics (since Bohr held that any other meaning would simply be undefined). In a different experimental setup, some other classical concept may apply, which in classical physics would have been compatible with the previous one, but in quantum mechanics is not. Early examples of such "complementary" pairs, as presented e.g. in [5], are particle versus wave and space-time versus "causal" descriptions (by which Bohr means conservation laws). Later on, Bohr emphasized the complementarity of one "phenomenon" (i.e., an indivisible unit of a quantum object coupled to an experimental arrangement) against another (cf. [27]).
- Heisenberg, on the other hand, seems to have held a more relaxed attitude towards classical concepts, arguably inspired by his game-changing paper on the quantum-mechanical reinterpretation (*Umdeutung*) of mechanical and kinematical relations [24]. In this paper, he performed the act of what we now call quantization, in putting the observables of classical physics (i.e. functions on a phase space) on a new mathematical footing (i.e., they were turned into matrices), where they also

have new properties. In his second epoch-making paper [25] introducing the uncertainty relations, he then tried to find some operational meaning of these "reinterpreted" observables through measurement procedures. Since quantization applies to all classical observables at once, all classical concepts apply simultaneously, but approximately (ironically, [24] was inspired by Bohr's Correspondence Principle, but later on Bohr insisted on precise nature of classical concepts described above).

This ideological split between Bohr and Heisenberg is still with us, as it leads to a similar break-up of the Bohrification program into two parts. The overall idea of Bohrification is to interpret classical concepts as commutative C^*-algebras, and hence the two parts in question are mathematically distinguished by the specific relationship between a given noncommutative C^*-algebra A and the commutative C^*-algebras that give physical "access" to A. Bohr's view on the precise nature of classical concepts comes back mathematically in *exact* Bohrification, which studies (unital) commutative C^*-subalgebras C of a given (unital) noncommutative C^*-algebra A. Heisenberg's interpretation of the doctrine of classical concepts, on the other hand, resurfaces in *asymptotic* Bohrification, which involves asymptotic inclusions (i.e. deformations) of commutative C^*-algebras into noncommutative ones. The precise relationship between Bohr's and Heisenberg's views, and hence also between exact and asymptotic Bohrification, remains to be clarified; their joint existence is unproblematic, however, since the two programs complement each other.

As reviewed in [34] and explained in detail in [35], asymptotic Bohrification provides a mathematical setting for the measurement problem, spontaneous symmetry breaking, the classical limit of quantum mechanics, the thermodynamic limit of quantum statistical mechanics, and the Born rule for probabilities construed as long-run frequencies, whereas exact Bohrification turns out to be an appropriate framework for Gleason's Theorem, the Kadison–Singer conjecture, the Born rule (for single case probabilities), and, initially via the topos-theoretic approach to quantum mechanics, intuitionistic quantum logic. In the context of the present paper it should be mentioned that the poset $\mathscr{C}(A)$ we will be concerned with has its origins in the reinterpretation of the Kochen–Specker Theorem in the language of topos theory by Isham and Butterfield [31]. In the setting of von Neumann algebras this led [21] to a poset similar to $\mathscr{C}(A)$ (though crucially with the opposite ordering), which was studied in great detail by [12–15]. The poset $\mathscr{C}(A)$ as we use it was introduced by [30], again in the context of topos theory.

In this paper we discuss the virtues of exact Bohrification in providing a new invariant $\mathscr{C}(A)$ for unital C^*-algebras A, defined as the poset of all unital commutative C^*-subalgebras of a unital C^*-algebra A (that share the unit of A), ordered by inclusion. We start with a general discussion of symmetries in elementary quantum mechanics on Hilbert space in Sect. 2, which culminates in our Wigner Theorem for $\mathscr{C}(B(H))$. Moving to general (unital) C^*-algebras A in Sect. 3, we discuss the place of $\mathscr{C}(A)$ amidst some comparable constructions A gives rise to, viz. its (pure) state space, its Jordan algebra structure, its effect algebra, and its (orthocomplemented) poset of projections. In Sect. 4 we give a complete and independent proof of Hamhalter's [19] great theorem to the effect that for a large class of C^*-algebras A, order isomorphisms of $\mathscr{C}(A)$ are induced by Jordan automorphisms of A (this theorem was predated by an analogous result by [11] for von Neumann algebras, which may have been the first of its kind). Hamhalter's theorem is also the key lemma in our Wigner Theorem for $\mathscr{C}(B(H))$. We close our paper in Sect. 5 with a study of the relationship between $\mathscr{C}(A)$ and the poset $\mathscr{P}(A)$ of projections in A.

2 Symmetries in Quantum Theory on Hilbert Space

Even in elementary quantum mechanics, where $A = B(H)$, i.e., the C^*-algebra of all bounded operators on some Hilbert space H, the concept of a symmetry is already diverse, as least apparently, since a non-commutative C^*-algebra like $B(H)$ gives rise to numerous "quantum structures". The main examples are:

1. The ***normal pure state space*** $\mathscr{P}_1(H)$, i.e., the set of one-dimensional projections on H, with a "transition probability" $\tau : \mathscr{P}_1(H) \times \mathscr{P}_1(H) \to [0, 1]$ defined by
$$\tau(e, f) = \mathrm{Tr}\,(ef). \tag{1}$$
2. The ***normal state space*** $\mathscr{D}(H)$, which is the convex set of all density operators ρ on H (i.e., $\rho \geq 0$ and $\mathrm{Tr}\,(\rho) = 1$).
3. The ***self-adjoint operators*** $B(H)_{\mathrm{sa}}$ on H, seen as a Jordan algebra.
4. The ***effects*** $\mathscr{E}(H) = [0, 1]_{B(H)}$ on H, i.e., the set of all $a \in B(H)_{\mathrm{sa}}$ for which $0 \leq a \leq 1_H$, seen as a convex poset.
5. The ***projections*** $\mathscr{P}(H)$ on H, seen as an orthocomplemented lattice.
6. The ***unital commutative*** C^****-subalgebras*** $\mathscr{C}(B(H))$ ***of*** $B(H)$, seen as a poset.

Each structure comes with its own notion of a symmetry (whose name has been chosen for historical reasons and—except for the first and third—is not standard):

Definition 1 Let H be a Hilbert space (not necessarily finite-dimensional).

1. A ***Wigner symmetry*** is a bijection $\mathsf{W} : \mathscr{P}_1(H) \to \mathscr{P}_1(H)$ that satisfies
$$\mathrm{Tr}\,(\mathsf{W}(e)\mathsf{W}(f)) = \mathrm{Tr}\,(ef), \quad e, f \in \mathscr{P}_1(H). \tag{2}$$
2. A ***Kadison symmetry*** is an affine bijection $\mathsf{K} : \mathscr{D}(H) \to \mathscr{D}(H)$.
3. A ***Jordan symmetry*** is an invertible ***Jordan map*** $\mathsf{J} : B(H)_{\mathrm{sa}} \to B(H)_{\mathrm{sa}}$, where the latter is an $\mathbb{R}$-linear map that satisfies either one of the equivalent conditions
$$\mathsf{J}(a \circ b) = \mathsf{J}(a) \circ \mathsf{J}(b); \tag{3}$$
$$\mathsf{J}(a^2) = \mathsf{J}(a)^2, \tag{4}$$
where the ***Jordan product*** $\circ$ is defined by $a \circ b = \frac{1}{2}(ab + ba)$, so that $a^2 = a \circ a$. Equivalently, a Jordan symmetry is a Jordan automorphism of $B(H)$, see below.

4. A ***Ludwig symmetry*** is an affine order isomorphism $\mathsf{L} : \mathscr{E}(H) \to \mathscr{E}(H)$.
5. A ***von Neumann symmetry*** is an order isomorphism $\mathsf{N} : \mathscr{P}(H) \to \mathscr{P}(H)$ that preserves the orthocomplementation, i.e. $\mathsf{N}(1_H - e) = 1_H - \mathsf{N}(e)$, $e \in \mathscr{P}(H)$.
6. A ***Bohr symmetry*** is an order isomorphism $\mathsf{B} : \mathscr{C}(B(H)) \to \mathscr{C}(B(H))$.

In no. 2 (and 4) being ***affine*** means that K (and similarly L) preserves convex sums, i.e., for $t \in (0, 1)$ and $\rho_1, \rho_2 \in \mathscr{D}(H)$, $\mathsf{K}(t\rho_1 + (1-t)\rho_2) = t\mathsf{K}\rho_1 + (1-t)\mathsf{K}\rho_2$. In nos. 4–6, an ***order isomorphism*** O of the given poset is a bijection such that $x \leq y$ if and only if $\mathsf{O}(x) \leq \mathsf{O}(y)$. In no. 3 one may complexify J to a $\mathbb{C}$-linear map

$$\mathsf{J}_{\mathbb{C}} : B(H) \to B(H) \tag{5}$$

by writing $a \in B(H)$ as $a = b + ic$, with $b = \frac{1}{2}(a + a^*)$ and $c = -\frac{1}{2}i(a - a^*)$, so that $b^* = b$ and $c^* = c$, and putting

$$\mathsf{J}_{\mathbb{C}}(a) = \mathsf{J}(b) + i\mathsf{J}(c). \tag{6}$$

If J satisfies (3)–(4) for each $a, b \in B(H)_{\mathrm{sa}}$, then $\mathsf{J}_{\mathbb{C}}$ satisfies (3)–(4) for each $a, b \in B(H)$ (with $\mathsf{J} \rightsquigarrow \mathsf{J}_{\mathbb{C}}$) as well as

$$\mathsf{J}_{\mathbb{C}}(a^*) = \mathsf{J}_{\mathbb{C}}(a)^*. \tag{7}$$

Conversely, one may restrict such a $\mathsf{J}_{\mathbb{C}}$ to the self-adjoint part $B(H)_{\mathrm{sa}}$ of $B(H)$, so that Jordan symmetries are essentially the same thing as Jordan automorphisms, i.e., $\mathbb{C}$-linear maps (5) that satisfy (7) and (3)–(4) with $\mathsf{J} \rightsquigarrow \mathsf{J}_{\mathbb{C}}$.

It is well known that the first four notions of symmetry are equivalent (see for example [1, 6, 10, 39]). If $\dim(H > 2)$, as a corollary to Gleason's Theorem the fifth notion is also equivalent to all of these [18], and, under the same assumption, so is the sixth [19]. We now sketch these equivalences; combine the above references or see [35] for complete proofs.

1. There is a bijective correspondence between:
   - Wigner symmetries $\mathsf{W} : \mathscr{P}_1(H) \to \mathscr{P}_1(H)$;
   - Kadison symmetries $\mathsf{K} : \mathscr{D}(H) \to \mathscr{D}(H)$, viz.

$$\mathsf{W} = \mathsf{K}_{|\mathscr{P}_1(H)}; \tag{8}$$

$$\mathsf{K}\left(\sum_i \lambda_i e_{\upsilon_i}\right) = \sum_i \lambda_i \mathsf{W}(\upsilon_{\upsilon_i}), \tag{9}$$

   where $\rho = \sum_i \lambda_i e_{\upsilon_i}$ is a spectral expansion of $\rho \in \mathscr{D}(H)$ in terms of a basis of eigenvector υ_i of ρ with eigenvalues λ_i, where $\lambda_i \geq 0$ and $\sum_i \lambda_i = 1$.
   It is a nontrivial fact that (9) is well defined (despite non-uniqueness of the spectral expansion in case that ρ has degenerate spectrum). Furthermore, K maps

$\mathscr{P}_1(H) \subset \mathscr{D}(H)$ into itself because $\mathscr{P}_1(H) = \partial_e\mathscr{D}(H)$ and affine bijections of convex sets restrict to bijections of their extreme boundaries. Finally, (8) preserves transition probabilities because an affine bijection $\mathsf{K} : \mathscr{D}(H) \to \mathscr{D}(H)$ extends to an isometric isomorphism $\mathsf{K}_1 : B_1(H)_{\rm sa} \to B_1(H)_{\rm sa}$ with respect to the trace-norm $\|\cdot\|_1$, and for any $e, f \in \mathscr{P}_1(H)$ we have

$$\|e - f\|_1 = 2\sqrt{1 - \mathrm{Tr}\,(ef)}. \tag{10}$$

2. There is a bijective correspondence between:
   - Kadison symmetries $\mathsf{K} : \mathscr{D}(H) \to \mathscr{D}(H)$;
   - Jordan symmetries $\mathsf{J} : B(H)_{\rm sa} \to B(H)_{\rm sa}$,

   such that for any $a \in B(H)_{\rm sa}$ one has

$$\mathrm{Tr}\,(\mathsf{K}(\rho)a) = \mathrm{Tr}\,(\rho\mathsf{J}(a)). \tag{11}$$

   To see this, we identify $\mathscr{D}(H)$ with the set $S_n(B(H))$ of normal states ω on $B(H)$ through $\omega(a) = \mathrm{Tr}\,(\rho a)$, so that with slight abuse of notation Eq. (11) reads

$$(\mathsf{K}\omega)(a) = \omega(\mathsf{J}(a)). \tag{12}$$

   This defines K in terms of J. Conversely, we identify $B(H)_{\rm sa}$ with the set $A_b(S_n(B(H)))$ of all real-valued bounded affine functions on the convex set $S_n(B(H))$ through the Gelfand-like transform $a \leftrightarrow \hat{a}$, where $\hat{a}(\omega) = \omega(a)$; here the nontrivial analytic facts are that the functions $\hat{a}$ exhaust $A_b(S_n(B(H)))$ and that $\|a\| = \|\hat{a}\|_\infty$. We now define a map

$$\hat{\mathsf{J}} : A_b(S_n(B(H))) \to A_b(S_n(B(H)))$$

   in terms of K in the obvious way, i.e., by $(\hat{\mathsf{J}}\hat{a})(\omega) = \hat{a}(\mathsf{K}(\omega))$. This, in turn, defines J in terms of K, which again yields (12). The map $\hat{\mathsf{J}}$ trivially preserves the (pointwise) order as well as the unit (function) in $A_b(S_n(B(H)))$, so that the corresponding map J preserves the usual partial order on $\leq B(H)_{\rm sa}$ (i.e. $a \leq b$ iff $b - a = c^2$ for some $c \in B(H)_{\rm sa}$) as well as the unit (operator) 1_H in $B(H)_{\rm sa}$. Finally, for invertible linear maps these properties are equivalent to the fact that J is a Jordan symmetry.
3. There is a bijective correspondence between:
   - Jordan symmetries $\mathsf{J} : B(H)_{\rm sa} \to B(H)_{\rm sa}$;
   - Ludwig symmetries $\mathsf{L} : \mathscr{E}(H) \to \mathscr{E}(H)$.

   Since $\mathscr{E}(H) \subset B(H)_{\rm sa}$, we may simply restrict J to $\mathscr{E}(H)$ so as to obtain L. Since a Jordan automorphism preserves order as well as the unit, the inequality $0 \leq a \leq 1_H$ characterizing $a \in \mathscr{E}(H)$ is preserved, i.e., $0 \leq \mathsf{J}(a) \leq 1_H$. In other words, J preserves $\mathscr{E}(H)$, whose order it preserves, too. Convexity is obvious, since $\mathsf{L} = \mathsf{J}_{|\mathscr{E}(H)}$ comes from a linear map. Conversely, since L is an order isomorphism, it must satisfy $\mathsf{L}(0) = 0$ (as well as $\mathsf{L}(1_H) = 1_H$), since 0 is the bottom element of $\mathscr{E}(H)$ as an ordered set (and 1_H is its the top element). One can show that this property plus convexity yields a linear extension J of L from $\mathscr{E}(H)$ to $B(H)_{\text{sa}}$, which is unital as well order-preserving, and hence is a Jordan symmetry.

4. If $\dim(H) > 2$, then there is a bijective correspondence between:

   - Jordan symmetries $\mathsf{J} : B(H)_{\text{sa}} \to B(H)_{\text{sa}}$;
   - von Neumann symmetries $\mathsf{N} : \mathscr{P}(H) \to \mathscr{P}(H)$.

   Jordan symmetries restrict to order isomorphisms of $\mathscr{P}(H) \subset B(H)_{\text{sa}}$; the only nontrivial point is that the order in $\mathscr{P}(H)$ (i.e., $e \leq f$ iff $ef = e$, which is the case iff $eH \subseteq fH$) coincides with the order inherited from $B(H)_{\text{sa}}$. Conversely, one may attempt to extend some map $\mathsf{N} : \mathscr{P}(H) \to \mathscr{P}(H)$ to $B(H)_{\text{sa}}$ by first supposing that $a \in B(H)_{\text{sa}}$ has a finite spectral decomposition $a = \sum_j \lambda_j f_j$, where (f_j) is a family of mutually orthogonal projections and $\lambda_j \in \mathbb{R}$, and putting

$$\mathsf{J}(a) = \sum_j \lambda_j \mathsf{N}(f_j). \tag{13}$$

   For general a, one then hopes to be able to use the spectral theorem in order to extend J to all of $B(H)_{\text{sa}}$ by continuity. It is far from trivial that this construction works and yields an $\mathbb{R}$-linear map, but it does. The proof relies on Gleason's Theorem (whence the assumption $\dim(H) > 2$), which in turn can be invoked because von Neumann symmetries preserve all suprema in $\mathscr{P}(H)$. The extension J thus obtained is positive and unital, and hence is a Jordan symmetry.

5. If $\dim(H) > 2$, then there is a bijective correspondence between:

   - Jordan symmetries $\mathsf{J} : B(H)_{\text{sa}} \to B(H)_{\text{sa}}$;
   - Bohr symmetries $\mathsf{B} : \mathscr{C}(B(H)) \to \mathscr{C}(B(H))$.

   Given J, as explained above we first complexify it so as to obtain a Jordan automorphism $\mathsf{J}_{\mathbb{C}} : B(H) \to B(H)$. It is a standard result that such maps are isometric. If $C \subset B(H)$ is commutative, then so is its image $\mathsf{J}_{\mathbb{C}}(C)$, since commutativity of C is equivalent to associativity of the Jordan product within C, and hence is preserved under Jordan maps. Furthermore, since $\mathsf{J}_{\mathbb{C}}$ is an isometry on C, its image is (norm) closed, and by (7) it is also self-adjoint. Finally, Jordan automorphisms preserve the unit 1_H, so that if C is a unital commutative C^*-subalgebra of $B(H)$, then so is $\mathsf{J}_{\mathbb{C}}(C)$. Thus J induces a map B by $\mathsf{B}(C) = \mathsf{J}_{\mathbb{C}}(C)$.
   Trivially, if $C \subseteq D$ in $B(H)$, so that $C \leq D$ in $\mathscr{C}(B(H))$, then $\mathsf{J}_{\mathbb{C}}(C) \subseteq \mathsf{J}_{\mathbb{C}}(D)$ in $B(H)$, so that $\mathsf{J}(C) \leq \mathsf{J}(D)$ in $\mathscr{C}(B)$. It follows that B is an order isomorphism. The converse, i.e., the fact that any Bohr symmetry is induced by a Jordan symmetry in the said way, is very deep [19]; see Theorem 4 below.

In view of these equivalences and Wigner's Theorem, we may conclude:

Theorem 1 *Let H be a Hilbert space, with* $\dim(H) > 2$ *in nos. 5 and 6.*

1. *Each Wigner symmetry takes the form* $\mathsf{W}(e) = ueu^*$ *($e \in \mathcal{P}_1(H)$);*
2. *Each Kadison symmetry takes the form* $\mathsf{K}(\rho) = u\rho u^*$ *($\rho \in \mathcal{D}(H)$);*
3. *Each Jordan symmetry takes the form* $\mathsf{J}(a) = uau^*$*; ($a \in B(H)_{\mathrm{sa}}$);*
4. *Each Ludwig symmetry takes the form* $\mathsf{L}(a) = uau^*$ *($a \in \mathcal{E}(H)$);*
5. *Each von Neumann symmetry takes the form* $\mathsf{N}(e) = ueu^*$ *($e \in \mathcal{P}(H)$);*
6. *Each Bohr symmetry takes the form* $\mathsf{B}(C) = uCu^*$ *($C \in \mathcal{C}(B(H))$),*

where in all cases the operator u is either unitary or anti-unitary, and is uniquely determined by the symmetry in question "up to a phase" (that is, u and u' implement the same symmetry by conjugation iff $u' = zu$, where $z \in \mathbb{T}$).

Of these six results, only the first and the third seem to have a direct proof; see e.g. Simon [6, 41], respectively. Neither of these proofs is particularly elegant, so that especially a direct proof of no. 6 would be welcome.

3 Symmetries in Algebraic Quantum Theory

In this section we generalize the above analysis from $A = B(H)$ to arbitrary C^*-algebras A, which for simplicity we assume to have a unit 1_A.

1. The ***pure state space*** $P(A) = \partial_e S(A)$ of A is the extreme boundary of the state space $S(A)$, seen as a uniform space equipped with a transition probability

$$\tau(\omega, \omega') = \inf\{\omega(a) \mid a \in A, 0 \leq a \leq 1_A, \omega'(a) = 1\}. \tag{14}$$

   If $A = B(H)$ and ω, ω' lie in the normal pure state space $P_n(B(H))$ of $B(H)$, a simple computation [33] shows that the above expression reproduces the standard quantum-mechanical transition probabilities (1), but compared to this special case one novel aspect of $P(A)$ is that all pure states are now taken into account (as opposed to merely the *normal* ones, which notion is undefined for general C^*-algebras anyway). Another is that in order to obtain the desired equivalence with other structures, the set $P(A)$ should carry a uniform structure, namely the w^*-uniformity inherited from A^*. Thus a ***Wigner symmetry*** of A is a uniformly continuous bijection $\mathsf{W} : P(A) \to P(A)$ with uniformly continuous inverse that preserves transition probabilities, i.e., that satisfies

$$\tau(\mathsf{W}(\omega)\mathsf{W}(\omega')) = \tau(\omega, \omega'), \quad \omega, \omega' \in P(A). \tag{15}$$

2. The ***state space*** $S(A)$ is the set of all states on A, seen as a compact convex set in the w^*-topology inherited from the embedding $S(A) \subset A^*$. Hence a ***Kadison symmetry*** of A is an affine homeomorphism $\mathsf{K} : S(A) \to S(A)$. Compared to the

case $A = B(H)$, firstly *all* states are now taken into account (instead of all *normal* states), and secondly we have added a continuity condition on K.

3. Any C^*-algebra A defines an associated ***Jordan algebra***–more precisely, a JB-algebra if the norm is taken into account, cf. [22]–namely $A_{\rm sa}$ equipped with the commutative product $a \circ b = \frac{1}{2}(ab + ba)$. A ***Jordan symmetry*** J of A is a Jordan isomorphism of $(A_{\rm sa}, \circ)$ (equivalently, a unital linear order isomorphism of $(A_{\rm sa}, \leq)$, cf. [1], Prop. 4.19).
4. The ***effects*** in A comprise the order unit interval $\mathscr{E}(A) = [0, 1_A]$, i.e., the set of all $a \in A_{\rm sa}$ such that $0 \leq a \leq 1_A$, seen as a convex poset as for $B(H)$. Hence a ***Ludwig symmetry*** of A is an affine order isomorphism $\mathsf{L} : \mathscr{E}(A) \to \mathscr{E}(A)$.
5. The projections $\mathscr{P}(A)$ in A form an orthocomplemented poset with $e \leq f$ iff $ef = e$ and $e^{\perp} = 1_A - e$; if A is a von Neumann algebra or more generally an AW^*-algebra or a Rickart C*-algebra, $\mathscr{P}(A)$ is even an orthocomplemented lattice. A ***von Neumann symmetry*** of A is an invertible map $\mathsf{N} : \mathscr{P}(A) \to \mathscr{P}(A)$ that preserves 0 and $\perp$ (and hence preserves 1) and satisfies $\varphi(x \vee y) = \varphi(x) \vee \varphi(y)$ if $x \leq y^{\perp}$ (in which case $x \vee y$ is defined, as is always the case if $\mathscr{P}(A)$ is a lattice).
6. The poset $\mathscr{C}(A)$ lying at the heart of exact Bohrification consists of all *commutative* C^*-subalgebras of A that contain the unit 1_A, partially ordered by inclusion. A ***Bohr symmetry*** of A, then, is an order isomorphism $\mathsf{B} : \mathscr{C}(A) \to \mathscr{C}(A)$.

The structures 1, 2, 3, and 4 are equivalent, as follows.

Theorem 2 *Let A and B be unital C^*-algebras. The relation $f = \varphi^*$ (that is, $f(\omega)(a) = \omega(\varphi(a))$, where $a \in A_{\rm sa}$ and $\omega \in P(A)$ or $\omega \in S(A)$, and φ and f are specified below) gives a bijective correspondence between:*

1. *Jordan isomorphisms $\varphi : A_{\rm sa} \to B_{\rm sa}$;*
2. *Bijections $f : P(B) \to P(A)$ that preserve transition probabilities and are w^*-uniformly continuous along with their inverse;*
3. *Affine homeomorphisms $f : S(B) \to S(A)$.*

See [40] for $1 \leftrightarrow 2$ in this theorem and see [1], Corollary 4.20, for $1 \leftrightarrow 3$. The first of these equivalences also follows from the reconstruction of $(A_{\rm sa}, \circ)$ from $P(A)$ in [33], whereas the second follows from the reconstruction of $(A_{\rm sa}, \circ)$ from $S(A)$ in [2]. The equivalence $3 \leftrightarrow 4$ on the above list 1–6 is proved in the same way as for $A = B(H)$.

The case of projections is more complicated, since many C^*-algebras have few projections (think of $A = C([0, 1])$). Therefore, the poset $\mathscr{P}(A)$ of all projections in A can only be as informative as the four invariants just discussed under special assumptions on A. In the absence of a general result, we single out the class of AW^*-algebras as a particularly nice one in so far as abundance of projections is concerned. Recall that a C^*-algebra A is an AW^*-algebra if for each nonempty subset $S \subseteq A$ there is a projection $e \in \mathscr{P}(A)$ so that $R(S) = eA$, where the *right-annihilator* $R(S)$ of $S \subseteq A$ is defined as $R(S) = \{a \in A \mid ba = 0 \forall b \in S\}$, and $R(a) \equiv R(\{a\})$. It follows that if it exists, e is uniquely determined by S. For example, all von Neumann algebras are AW^*-algebras, so this class is vast. See [3].

The key result of interest to our theme is then provided by Hamhalter's generalization of Dye's Theorem to AW^*-algebras [20]:

Theorem 3 *Let A and B be AW^*-algebras and let $\mathsf{N} : \mathscr{P}(A) \to \mathscr{P}(B)$ be an isomorphism of the corresponding orthocomplemented projection lattices that in addition preserves arbitrary suprema. If A has no summand isomorphic to either $\mathbb{C}^2$ or $M_2(\mathbb{C})$, then there is a unique Jordan isomorphism $\mathsf{J} : A_{\rm sa} \to B_{\rm sa}$ that extends N (and hence Jordan isomorphisms are characterized by their values on projections).*

We omit the proof, as it is not related to our main topic of interest $\mathscr{C}(A)$; the proof follows from a theorem of [29] on projections in AW^*-algebras and Hamhalter's own Gleason's Theorem for homogeneous AW^*-algebras.

We now move to the posets $\mathscr{C}(A)$. Since the structures 1–4 are equivalent, we may pick the one that is most convenient for a comparison with $\mathscr{C}(A)$, which turns out to be the Jordan algebra structure of A. Henceforth A and B are unital C^*-algebras, and we define a ***weak Jordan isomorphism***, also called a *quasi Jordan isomorphism*, of A and B as an invertible map $\mathsf{J} : A_{\rm sa} \to B_{\rm sa}$ whose restriction to each subspace $C_{\rm sa}$ of $A_{\rm sa}$, where $C \in \mathscr{C}(A)$, is linear and preserves the Jordan product $\circ$ (so that a Jordan symmetry of A alone is a weak Jordan automorphism of of A). Such a map complexifies to a map $\mathsf{J}_{\mathbb{C}} : A \to B$ in the same way as for $A = B = B(H)$. If no confusion arises, we write J for $\mathsf{J}_{\mathbb{C}}$.

Proposition 1 *Given a weak Jordan isomorphism $\mathsf{J} : A_{\rm sa} \to B_{\rm sa}$, the ensuing map $\mathsf{B} : \mathscr{C}(A) \to \mathscr{C}(B)$ defined by $\mathsf{B}(C) = \mathsf{J}_{\mathbb{C}}(C)$ is an order isomorphism.*

Note that as an argument of B the symbol C is a point in the poset $\mathscr{C}(A)$, whereas as an argument of $\mathsf{J}_{\mathbb{C}}$ it is a subset of A, so that $\mathsf{J}_{\mathbb{C}}(C)$ stands for $\{\mathsf{J}_{\mathbb{C}}(c) \mid c \in C\}$. The proof is elementary and is practically the same as for the special case $A = B = B(H)$; see also [19], Proposition 1.1.

4 Hamhalter's Theorem

The converse, however, is a deep result, due to [19], Theorem 3.4.

Theorem 4 *Let A and B be unital C^*-algebras and let $\mathsf{B} : \mathscr{C}(A) \to \mathscr{C}(B)$ be an order isomorphism. Then there is a weak Jordan isomorphism $\mathsf{J} : A_{\rm sa} \to B_{\rm sa}$ such that $\mathsf{B}(C) = \mathsf{J}_{\mathbb{C}}(C)$ for each $C \in \mathscr{C}(A)$. Moreover, if A is isomorphic to neither $\mathbb{C}^2$ nor $M_2(\mathbb{C})$, then J is uniquely determined by B, so in that case there is a bijective correspondence $\mathsf{J} \leftrightarrow \mathsf{B}$ between weak Jordan symmetries J of A and Bohr symmetries B of A.*

The question whether the weak Jordan isomorphism in question is a Jordan isomorphism will be postponed to Theorem 5 below.

Before proving Theorem 4, let us explain why $\mathbb{C}^2$ and $M_2(\mathbb{C})$ are exceptional.

- The only order isomorphism of the poset $\mathscr{C}(\mathbb{C}^2) \cong \{0, 1\}$ (with $0 \equiv \mathbb{C} \cdot 1_2$ and $1 \equiv \mathbb{C}^2$) is the identity map, which is induced by both the map $(a, b) \mapsto (b, a)$ and by the identity map on $\mathbb{C}^2$ (each of which is a weak Jordan automorphism).
- The poset $\mathscr{C}(M_2(\mathbb{C}))$ has a bottom element $0 \equiv \mathbb{C} \cdot 1_2$, as before, but no top element; each element $C \neq \mathbb{C} \cdot 1_2$ of $\mathscr{C}(M_2(\mathbb{C}))$ is a unitary conjugate of the diagonal subalgebra $D_2(\mathbb{C})$, with $0 \leq C$ but no other orderings. Furthermore, $C \cap \tilde{C} = \mathbb{C} \cdot 1_2$ whenever $C \neq \tilde{C}$. Hence any order isomorphism of $\mathscr{C}(M_2(\mathbb{C}))$ maps $\mathbb{C} \cdot 1_2$ to itself and permutes the C's. Thus each map $\mathrm{J} : M_2(\mathbb{C})_{\mathrm{sa}} \to M_2(\mathbb{C})_{\mathrm{sa}}$ whose complexification $\mathrm{J}_{\mathbb{C}} : M_2(\mathbb{C}) \to M_2(\mathbb{C})$ shuffles the C's isomorphically (as C^*-algebras) gives a weak Jordan automorphism. For example, take $(a, b) \mapsto (b, a)$ on $D_2(\mathbb{C})$ and the identity on each $C \neq D_2(\mathbb{C})$; this induces the identity map on $\mathscr{C}(M_2(\mathbb{C}))$. It follows that there are vastly more weak Jordan automorphisms of $M_2(\mathbb{C})$ than there are order isomorphisms of $\mathscr{C}(M_2(\mathbb{C}))$.

The proof of Theorem 4 deserves a section on its own; we roughly follow [19], but add various details and also take some different turns. The main differences with the original proof by Hamhalter are the following. Firstly, we give an order-theoretic characterization of u.s.c. decompositions of the form π_K (and hence of algebras in $\mathscr{C}(C(X))$ that are the unitization of some ideal) by three axioms as stated in Lemma 3.1.1 in [16], whereas Hamhalter uses Proposition 7 in [38], which gives a different characterization of unitizations of ideals. Furthermore, Hamhalter only treats Lemma 1 in full generality (cf. Theorem 2.3 in [19]), whereas in our opinion it is very instructive to take the case of finite sets first, where many of the key ideas already appear in a setting where they are not overshadowed by topological complications. Finally, our proof of Lemma 2 differs from Hamhalter's proof (cf. Preposition 3.1 in [19]).

Proof The key to the proof lies in the commutative case, which can be reduced to topology. If $A = C(X)$, any $C \in \mathscr{C}(A)$ induces an equivalence relation $\sim_C$ on X by

$$x \sim_C y \text{ iff } f(x) = f(y) \,\forall f \in C. \tag{16}$$

This, in turn, defines a partition $X = \bigsqcup_\lambda K_\lambda$ of X (henceforth called π), whose blocks $K_\lambda \subseteq X$ are the equivalence classes of $\sim_C$. To study a possible inverse of this procedure, for any closed subset $K \subset X$ we define the ideal

$$I_K = C(X; K) = \{f \in C(X) \mid f(x) = 0 \,\forall x \in K\}, \tag{17}$$

in $C(X)$, and its unitization $\dot{I}_K = I_K \oplus \mathbb{C} \cdot 1_X$, which evidently consists of all continuous functions on X that are constant on K. If X is finite (and discrete), each partition π of X defines some unital C^*-algebra $C \subseteq C(X)$ through

$$C = \bigcap_{K_\lambda \in \pi} \dot{I}_{K_\lambda}, \tag{18}$$

which consists of all $f \in C(X)$ that are constant on each block K_λ of the given partition π. In that case, the correspondence $C \leftrightarrow \pi$, where π is defined by the equivalence relation $\sim_C$ in (16), gives a bijection between $\mathscr{C}(C(X))$ and the set $\mathfrak{P}(X)$ of all partitions of X. For example, the subalgebra $C = \dot{I}_K$ corresponds to the partition consisting of K and all singletons not lying in K. Given the already defined partial order on $\mathscr{C}(C(X))$ (i.e., $C \leq D$ iff $C \subseteq D$), we may promote this bijection to an order isomorphism of posets if we define a partial order on $\mathfrak{P}(X)$ to be the *opposite* of the usual one in which $\pi \leq \pi'$ (where π and π' consist of blocks $\{K_\lambda\}$ and $\{K'_{\lambda'}\}$, respectively) iff each K_λ is contained in some $K'_{\lambda'}$ (i.e., π is finer than π'). This partial ordering makes $\mathfrak{P}(X)$ a complete lattice, whose bottom element consists of all singletons on X and whose top element just consists of X itself: the former corresponds to $C(X)$, which is the top element of $\mathscr{C}(C(X))$, whilst the latter corresponds to $\mathbb{C} \cdot 1_X$, which is the bottom element of $\mathscr{C}(C(X))$.

For general compact Hausdorff spaces X, since $C(X)$ is sensitive to the topology of X the equivalence relation (16) does not induce arbitrary partitions of X. It turns out that each $C \in \mathscr{C}(C(X))$ induces an ***upper semicontinuous partition*** (abbreviated by ***u.s.c. decomposition***) of X, i.e.,

- Each block K_λ of the partition π is closed;
- For each block K_λ of π, if $K_\lambda \subseteq U$ for some open $U \in \mathscr{O}(X)$, then there is $V \in \mathscr{O}(X)$ such that $K_\lambda \subseteq V \subseteq U$ and V is a union of blocks of π (in other words, if K is such a block, then $V \cap K = \emptyset$ implies $K = \emptyset$).

This can be seen as follows. Firstly, if we equip π with the quotient topology with respect to the the natural map $q : X \to \pi, x \mapsto K_\lambda$ if $x \in K_\lambda$, then π is compact, for X is compact. Moreover, π is Hausdorff: let K_λ and K_μ be two distinct *points* in π. Recall that $x, y \in K_\lambda$ if and only if $f(x) = f(y)$ for each $f \in C$. Since $K_\lambda \neq K_\mu$, there is some $x \in K_\lambda$, some $y \in K_\mu$ and some $f \in C$ such that $f(x) \neq f(y)$, whence there are open disjoint $U, V \subseteq \mathbb{C}$ such that $f(x) \in U$ and $f(y) \in V$.

Define $\hat{f} : \pi \to \mathbb{C}$ by $\hat{f}(K_\lambda) = f(x)$ for some $x \in K_\lambda$. By definition of K_λ this is independent of the choice of $x \in K_\lambda$, hence $\hat{f}$ is well defined. Again by definition, we have $f = \hat{f} \circ q$, hence $q^{-1}(\hat{f}^{-1})[U] = f^{-1}[U]$, which is open in X since f is continuous. Since π is equipped with the quotient topology, it follows that $\hat{f}^{-1}[U]$ is open in π, and similarly $\hat{f}^{-1}[V]$ is open. Moreover, we have $\hat{f}(K_\lambda) = f(x)$ and $f(x) \in U$, hence $K_\lambda \in \hat{f}^{-1}[U]$, and similarly, $K_\mu \in \hat{f}^{-1}[V]$. We conclude that π is also Hausdorff. Since q is a continuous map between compact Hausdorff spaces, it follows that q is closed. It now follows from Theorem 9.9 in [42]—which also gives further background on decompositions—that π is a u.s.c. decomposition.

Consequently, by the same maps (16) and (18), the poset $\mathscr{C}(C(X))$ is anti-isomorphic to the poset $\mathfrak{F}(X)$ of all u.s.c. decompositions of X (this proves that $\mathfrak{F}(X)$ is a complete lattice, since $\mathscr{C}(C(X))$ is). This is still a complicated poset; assuming X to be larger than a singleton, the next step is to identify the simpler poset $\mathscr{F}_2(X)$ of all closed subsets of X containing at least two elements within $\mathfrak{F}(X)$, where (as above) we identify a closed $K \subseteq X$ with the (u.s.c.) partition π_K of X whose blocks are K and all singletons not lying in K (note that the poset $\mathscr{F}(X)$ of all closed subsets of X is less useful, since any singleton in $\mathscr{F}(X)$ gives rise to the bottom element of $\mathfrak{F}(X)$). To do so, we first recall that β is said to ***cover*** α in some poset if $\alpha < \beta$, and $\alpha \leq \gamma < \beta$ implies $\alpha = \gamma$. If the poset has a bottom element, then its covers are called *atoms*. Furthermore, note that since the bottom element 0 of $\mathfrak{F}(X)$ consists of singletons, the atoms in $\mathfrak{F}(X)$ are the partitions of the form $\pi_{\{x_1,x_2\}}$ (where $x_1 \neq x_2$). It follows that some partition $\pi \in \mathfrak{F}(X)$ lies in $\mathscr{F}_2(X) \subset \mathfrak{F}(X)$ iff exactly one of the following conditions holds:

- π is an atom in $\mathfrak{F}(X)$, i.e., $\pi = \pi_{\{x_1,x_2\}}$ for some $x_1, x_2 \in X$, $x_1 \neq x_2$;
- π covers three (distinct) atoms in $\mathfrak{F}(X)$, in which case $\pi = \pi_{\{x_1,x_2,x_3\}}$ where all x_i are different, which covers the atoms $\pi_{\{x_1,x_2\}}$, $\pi_{\{x_1,x_3\}}$, and $\pi_{\{x_2,x_3\}}$;
- If $\alpha \neq \beta$ are atoms in $\mathfrak{F}(X)$ such that $\alpha \leq \pi$ and $\beta \leq \pi$, there is an atom $\gamma \leq \pi$ such that there are three (distinct) atoms covered by $\alpha \vee \gamma$ and three (distinct) atoms covered by $\beta \vee \gamma$. In that case, $\pi = \pi_K$ where K has more than three elements: if $\alpha = \pi_{\{x_1,x_2\}}$ and $\beta = \pi_{\{x_3,x_4\}}$, then due to the assumption $\alpha \neq \beta$, the set $\{x_1, x_2, x_3, x_4\}$ (which lies in K) has at least three distinct elements, say $\{x_1, x_2, x_3\}$. Hence we may take $\gamma = \pi_{\{x_2,x_3\}}$, in which case $\alpha \vee \gamma = \pi_{\{x_1,x_2,x_3\}}$, which covers the atoms α, γ, and $\pi_{\{x_1,x_3\}}$. Likewise, we have $\beta \vee \gamma = \pi_{\{x_2,x_3,x_4\}}$, which covers three atoms β, γ, and $\pi_{\{x_2,x_4\}}$.

In order to see that π satisfying the third condition must be of the form π_K, assume the converse. So π contains two blocks K_λ and K_μ consisting of two or more elements. Say $\{x_1, x_2\} \subseteq K_\lambda$ and $\{x_3, x_4\} \subseteq K_\mu$. Then $\alpha = \pi_{\{x_1,x_2\}}$ and $\beta_{\{x_3,x_4\}}$ are atoms such that $\alpha, \beta < \pi$, and there is an atom $\gamma = \pi_{\{x_5,x_6\}} \leq \pi$ such that there are three atoms covered by $\alpha \vee \gamma$, and there are three atoms covered by $\beta \vee \gamma$. It follows from the second condition that $\alpha \vee \gamma = \pi_L$ with L a three-point set. This implies that $\{x_1, x_2\} \cap \{x_5, x_6\}$ is not empty, from which it follows that $\alpha \vee \gamma = \pi_{\{x_1,x_2,x_5,x_6\}}$. Similarly, we find $\beta \vee \gamma = \pi_{\{x_3,x_4,x_5,x_6\}}$. Since $\{x_1, x_2, x_5, x_6\}$ and $\{x_3, x_4, x_5, x_6\}$ overlap, we obtain $\alpha \vee \beta \vee \gamma = \pi_{\{x_1,x_2,x_3,x_4,x_5,x_6\}}$. Moreover, $\alpha, \beta, \gamma \leq \pi$, so $\alpha \vee \beta \vee \gamma \leq \pi$. However, since $x_1, x_2 \in K_\lambda$, we must have $\{x_1, x_2, x_3, x_4, x_5, x_6\} \subseteq K_\lambda$ by definition of the order on $\mathfrak{F}(X)$. But since $x_3, x_4 \in K_\mu$, we must also have $\{x_1, x_2, x_3, x_4, x_5, x_6\} \subseteq K_\mu$, which is not possible, since K_λ and K_μ are distinct blocks, hence disjoint. We conclude that π can have only one block K of two or more elements, hence $\pi = \pi_K$.

Thus $\mathscr{F}_2(X) \subset \mathfrak{F}(X)$ has been characterized order-theoretically. Moreover,

$$\pi = \vee_{x \in X} \pi_{K(x)}, \tag{19}$$

where $K(x)$ is the unique block of X that contains x. Hence $\mathscr{F}_2(X)$ determines $\mathfrak{F}(X)$.

Let X and Y be compact Hausdorff spaces of cardinality at least two (so that the empty set and singletons are excluded). By the previous analysis, an order isomorphism $\mathsf{B} : \mathscr{C}(C(X)) \to \mathscr{C}(C(Y))$ is equivalent to an order isomorphism $\mathfrak{F}(X) \to \mathfrak{F}(Y)$, which in turn restricts to an order isomorphism $\mathscr{F}_2(X) \to \mathscr{F}_2(Y)$.

Lemma 1 *If X and Y are compact Hausdorff spaces of cardinality at least two, then any order isomorphism $\mathsf{F} : \mathscr{F}_2(X) \to \mathscr{F}_2(Y)$ is induced by a homeomorphism $\varphi : X \to Y$ via $\mathsf{F}(F) = \varphi(F)$, i.e., $\mathsf{F}(F) = \cup_{x \in F}\{\varphi(x)\}$. Moreover, if X and Y have cardinality at least three, then φ is uniquely determined by F.*

We first prove this for finite X, where $\mathscr{F}_2(X)$ simply consists of all subsets of X having at least two elements, etc. It is easy to see that X and Y must have the same cardinality $|X| = |Y| = n$. If $n = 2$, then $\mathscr{F}_2(X) = X$ etc., so there is only one map F, which is induced by each of the two possible maps $\varphi : X \to Y$, so that φ exists but fails to be unique. If $n > 2$, then F must map each subset of X with $n - 1$ elements to some subset of Y with $n - 1$ elements, so that taking complements we obtain a unique bijection $\varphi : X \to Y$. To show that φ induces F, note that the meet $\wedge$ in $\mathscr{F}_2(X)$ is simply intersection $\cap$, and also that for any $F \in \mathscr{F}_2(X)$,

$$F = \cup_{x\in F}\{x\} = \cap_{x\notin F}\{x\}^c = (\cup_{x\notin F}\{x\})^c,$$

where $A^c = X\backslash A$. Since F is an order isomorphism it preserves $\wedge = \cap$, so that

$$\mathsf{F}(F) = \cap_{x\notin F}\mathsf{F}(\{x\}^c) = \cap_{x\notin F}X\backslash\{\varphi(x)\} = (\cup_{x\notin F}\{\varphi(x)\})^c = \cup_{x\in F}\{\varphi(x)\}.$$

Now assume that X is infinite. Let $x \in X$. If x is not isolated, we define $\varphi(x)$ as follows. Let $\mathscr{O}(x)$ denote the set of all open neighborhoods of x. Since x is not isolated, each $O \in \mathscr{O}(x)$ contains at least another element, so $\overline{O} \in \mathscr{F}_2(X)$. Moreover, finite intersections of elements of $\{\overline{O} : O \in \mathscr{O}(x)\}$ are still in $\mathscr{F}_2(X)$. Indeed, if $O_1, \ldots, O_n \in \mathscr{O}(x)$, then $O_1 \cap \cdots \cap O_n$ is an open set containing x, and since $\overline{O_1 \cap \cdots \cap O_n} \subseteq \overline{O_1} \cap \cdots \cap \overline{O_n}$, it follows that $\overline{O_1} \cap \cdots \cap \overline{O_n} \in \mathscr{F}_2(X)$. Since F is an order isomorphism, we find that finite intersections of $\{\mathsf{F}(\overline{O}) : O \in \mathscr{O}(x)\}$ are contained in $\mathscr{F}_2(Y)$. This implies that $\{\mathsf{F}(\overline{O}) : O \in \mathscr{O}(x)\}$ satisfies the finite intersection property. As Y is compact, it follows that $I_x = \bigcap_{O\in\mathscr{O}(x)} \mathsf{F}(\overline{O})$ is non-empty. We can say more: it turns out that I_x contains exactly one element. Indeed, assume that there are two different points $y_1, y_2 \in I_x$. Then $\{y_1, y_2\} \in \mathscr{F}_2(Y)$, so $\mathsf{F}^{-1}(\{y_1, y_2\}) \in \mathscr{F}_2(X)$. Since $\{y_1, y_2\} \in \mathsf{F}(\overline{O})$ for each $O \in \mathscr{O}(x)$, we also find that $\mathsf{F}^{-1}(\{y_1, y_2\}) \subseteq \overline{O}$ for each $O \in \mathscr{O}(x)$. This implies that

$$\mathsf{F}^{-1}(\{y_1, y_2\}) \subseteq \bigcap_{O\in\mathscr{O}(x)} \overline{O} = \{x\},$$

where the last equality holds by normality of X. But this is a contradiction with $\mathsf{F} : \mathscr{F}_2(X) \to \mathscr{F}_2(Y)$ being a bijection. So I_x contains exactly one point. We define $\varphi(x)$ such that $\{\varphi(x)\} = I_x$. Notice that $\varphi(x)$ cannot be isolated in Y, since if we assume otherwise, then $Y \setminus \{\varphi(x)\}$ must be a co-atom in $\mathscr{F}_2(Y)$, whence $\mathsf{F}^{-1}(Y \setminus \{\varphi(x)\})$ is a co-atom in $\mathscr{F}_2(X)$, which must be of the form $X \setminus \{z\}$ for some isolated $z \in X$. Since x is not isolated, we cannot have $x = z$, so $X \setminus \{z\}$ is an open neighborhood of x, which is even clopen since z is isolated. By definition of $\varphi(x)$, we must have $\varphi(x) \in \mathsf{F}(X \setminus \{z\})$, but $\mathsf{F}(X \setminus \{z\}) = Y \setminus \{\varphi(x)\}$. We found a contradiction, hence $\varphi(x)$ cannot be isolated. Now assume that x is an isolated point. Then $X \setminus \{x\}$ is a co-atom in $\mathscr{F}_2(X)$, so $\mathsf{F}(X \setminus \{x\})$ is a co-atom in $\mathscr{F}_2(Y)$, too. Clearly this implies that $\mathsf{F}(X \setminus \{x\}) = Y \setminus \{y\}$ for some unique $y \in Y$, which must be isolated, since $Y \setminus \{y\}$ is closed. We define $\varphi(x) = y$.

In an analogous way, F^{-1} induces a map $\psi : Y \to X$. We shall show that φ and ψ are each other's inverses. Let $x \in X$ be isolated. We have seen that $\varphi(x)$ must be isolated as well, and that $\varphi(x)$ is defined by the equation $\mathsf{F}(X \setminus \{x\}) = Y \setminus \{\varphi(x)\}$. Since F is an order isomorphism, we have $X \setminus \{x\} = \mathsf{F}^{-1}(Y \setminus \{\varphi(x)\})$. Since $\varphi(x)$ is isolated, we find by definition of ψ that $\psi(\varphi(x)) = x$. In a similar way we find that $\varphi(\psi(y)) = y$ for each isolated $y \in Y$. Now assume that x is not isolated and let $F \in \mathscr{F}_2(X)$ such that $x \in F$. Then

$$\begin{aligned}\{\varphi(x)\} &= \bigcap_{O \in \mathscr{O}(x)} \mathsf{F}(\overline{O}) \subseteq \bigcap \{\mathsf{F}(\overline{O}) : O \text{ open}, F \subseteq O\} \\ &= \mathsf{F}\left(\bigcap \{\overline{O} : O \text{ open}, F \subseteq O\}\right) = \mathsf{F}(F),\end{aligned}$$

where the last equality follows by completely regularity of X. The penultimate equality follows from the following facts. Firstly, the set $\bigcap\{\overline{O} : O \text{ open}, F \subseteq O\}$ is closed since it is the intersection of closed sets. Moreover, the intersection contains more than one point, since F contains two or more points and $F \subseteq \overline{O}$ for each O. Hence $\bigcap\{\overline{O} : O \text{ open}, F \subseteq O\} \in \mathscr{F}_2(X)$, and since F is an order isomorphism, it preserves infima, which justifies the penultimate equality. Hence $\varphi(x) \in \mathsf{F}(F)$ for each $F \in \mathscr{F}_2(X)$ containing x. Since x is not isolated, $\varphi(x)$ is not isolated either. Hence in a similar way, we find that $\psi(\varphi(x)) \in \mathsf{F}^{-1}(G)$ for each $G \in \mathscr{F}_2(Y)$ containing $\varphi(x)$. Let $z = \psi(\varphi(x)$. Combining both statements, we find that $z \in F$ for each $F \in \mathscr{F}_2(X)$ such that $x \in F$. In other words, $z \in \bigcap\{F \in \mathscr{F}_2(X) : x \in F\}$. Since x is not isolated, we each $O \in \mathscr{O}(x)$ contains at least two points. Hence

$$\bigcap\{F \in \mathscr{F}_2(X) : x \in F\} \subseteq \bigcap\{\overline{O} : O \in \mathscr{O}(x)\} = \{x\},$$

where we used complete regularity of X in the last equality. We conclude that $z = x$, so $\psi(\varphi(x)) = x$. In a similar way, we find that $\varphi(\psi(y)) = y$ for each non-isolated $y \in Y$. We conclude that φ is a bijection with $\varphi^{-1} = \psi$.

We have to show that if $F \in \mathscr{F}_2(X)$, then $\varphi[F] = \mathsf{F}(F)$. Let $x \in F$. When we proved that φ is a bijection, we already noticed that $\varphi(x) \in \mathsf{F}(F)$ if x is not isolated. If x is isolated in X, then we first assume that F has at least three points. Since $\{x\}$ is open, $G = F \setminus \{x\}$ is closed. Since F contains at least three points, $G \in \mathscr{F}_2(X)$. So G is covered by F in $\mathscr{F}_2(X)$, so $\mathsf{F}(F)$ covers $\mathsf{F}(G)$. It follows that there must be an element $y_G \in Y \setminus \mathsf{F}(G)$ such that

$$\mathsf{F}(F) = \mathsf{F}(G \cup \{x\}) = \mathsf{F}(G) \cup \{y_G\}.$$

We have $G \cup \{x\}, X \setminus \{x\} \in \mathscr{F}_2(X)$, so

$$\begin{aligned}\mathsf{F}(G) &= \mathsf{F}(G \cup \{x\} \cap X \setminus \{x\}) = \mathsf{F}(G \cup \{x\}) \cap \mathsf{F}(X \setminus \{x\}) \\ &= (\mathsf{F}(G) \cup \{y_G\}) \cap (Y \setminus \{\varphi(x)\}),\end{aligned}$$

where $\mathsf{F}(X \setminus \{x\}) = Y \setminus \{\varphi(x)\}$ by definition of values of φ at isolated points. Since $x \notin G$ and F preserves inclusions, this latter equation also implies $\mathsf{F}(G) \subseteq Y \setminus \{\varphi(x)\}$. Hence we find

$$\mathsf{F}(G) = (\mathsf{F}(G) \cup \{y_G\}) \cap (Y \setminus \{\varphi(x)\}) = \mathsf{F}(G) \cup (\{y_G\} \cap Y \setminus \{\varphi(x)\}).$$

Thus we obtain $\{y_G\} \cap Y \setminus \{\varphi(x)\} \subseteq \mathsf{F}(G)$, but since $y_G \notin \mathsf{F}(G)$, we must have $\varphi(x) = y_G$. As a consequence, we obtain $\mathsf{F}(F) = \mathsf{F}(G) \cup \{\varphi(x)\}$, so $\varphi(x) \in \mathsf{F}(F)$.

Summarizing, if F has at least three points, then $\varphi(x) \in \mathsf{F}(F)$ for $x \in F$, regardless whether x is isolated or not. So $\varphi[F] \subseteq \mathsf{F}(F)$ for each $F \in \mathscr{F}_2(X)$ such that F has at least three points. Let $F \in \mathscr{F}_2(X)$ have exactly two points. Then there are $F_1, F_2 \in \mathscr{F}_2(X)$ with exactly three points such that $F = F_1 \cap F_2$. Then since φ is a bijection and F as an order isomorphism both preserve intersections in $\mathscr{F}_2(X)$, we find

$$\varphi[F] = \varphi[F_1 \cap F_2] = \varphi[F_1] \cap \varphi[F_2] \subseteq \mathsf{F}(F_1) \cap \mathsf{F}(F_2) = \mathsf{F}(F_1 \cap F_2) = \mathsf{F}(F).$$

So $\varphi[F] \subseteq \mathsf{F}(F)$ for each $F \in \mathscr{F}_2(X)$. In a similar way, we find $\varphi^{-1}[G] \subseteq \mathsf{F}^{-1}[G]$ for each $G \in \mathscr{F}_2(Y)$. So if we substitute $G = \mathsf{F}(F)$, we obtain $\varphi^{-1}[\mathsf{F}(F)] \subseteq F$. Since φ is a bijection, it follows that $\mathsf{F}(F) = \varphi[F]$ for each $F \in \mathscr{F}_2(X)$. As a consequence, φ induces a one-one correspondence between closed subsets of X and closed subsets of Y. Hence φ is a homeomorphism.

This proves Lemma 1. The special case of Theorem 4 where A and B are commutative now follows if we combine all steps so far:

1. The Gelfand isomorphism allows us to assume $A = C(X)$ and $B = C(Y)$, as above;
2. The order isomorphism $\mathsf{B} : \mathscr{C}(A) \to \mathscr{C}(B)$ determines and is determined by an order isomorphism $F : \mathfrak{F}(X) \to \mathfrak{F}(Y)$ of the underlying lattices of u.s.c. decompositions;
3. Because of (19), the order isomorphism F in turn determines and is determined by an order isomorphism $\mathsf{F} : \mathscr{F}_2(X) \to \mathscr{F}_2(Y)$;
4. Lemma 1 yields a homeomorphism $\varphi : X \to Y$ inducing $\mathsf{F} : \mathscr{F}_2(X) \to \mathscr{F}_2(Y)$;
5. The inverse pullback $(\varphi^{-1})^* : C(X) \to C(Y)$ is an isomorphism of C^*-algebras, which (running backwards) reproduces the initial map $\mathsf{B} : \mathscr{C}(C(X)) \to \mathscr{C}(C(Y))$.

Therefore, in the commutative case we apparently obtain rather more than a weak Jordan isomorphism $\mathsf{J} : A_{\rm sa} \to B_{\rm sa}$; we even found an isomorphism $\mathsf{J} : A \to B$ of C^*-algebras. However, if A and B are commutative, the condition of linearity on each commutative C^*-subalgebra C of A includes $C = A$, so that (after complexification) weak Jordan isomorphisms are the same as isomorphisms of C^*-algebras.

We now turn to the general case, in which A and B are both noncommutative (the case where one, say A, is commutative but the other is not cannot occur, since $\mathscr{C}(A)$ would be a complete lattice but $\mathscr{C}(B)$ would not). Let D and E be maximal abelian C^*-subalgebras of A, so that the corresponding elements of $\mathscr{C}(A)$ are maximal in the order-theoretic sense. Given an order isomorphism $\mathsf{B} : \mathscr{C}(A) \to \mathscr{C}(B)$, we restrict the map B to the down-set $\downarrow D = \mathscr{C}(D)$ in $\mathscr{C}(A)$ so as to obtain an order homomorphism $\mathsf{B}_{|D} : \mathscr{C}(D) \to \mathscr{C}(B)$. The image of $\mathscr{C}(D)$ under B must have a maximal element (since B is an order isomorphism), and so there is a maximal commutative C^*-subalgebra $\tilde{D}$ of B such that $\mathsf{B}_{|D} : \mathscr{C}(D) \to \mathscr{C}(\tilde{D})$ is an order isomorphism. Applying the previous result, we obtain an isomorphism $\mathsf{J}_D : D \to \tilde{D}$ of commutative C^*-algebras that induces $\mathsf{B}_{|D}$. The same applies to E, so we also have an isomorphism $\mathsf{J}_E : E \to \tilde{E}$ of commutative C^*-algebras that induces $\mathsf{B}_{|E}$. Let $C = D \cap E$, which lies in $\mathscr{C}(A)$. We now show that J_D and J_E coincide on C. There are three cases.

1. $\dim(C) = 1$. In that case $C = \mathbb{C} \cdot 1_A$ is the bottom element of $\mathscr{C}(A)$, so it must be sent to the bottom element $\tilde{C} = \mathbb{C} \cdot 1_B$ of $\mathscr{C}(B)$, whence the claim.
2. $\dim(C) = 2$. This the hard case dealt with below.
3. $\dim(C) > 2$. This case is settled by the uniqueness claim in Lemma 1.

So assume $\dim(C) = 2$. In that case, $C = C^*(e)$ for some proper projection $e \in \mathscr{P}(A)$, which is equivalent to C being an atom in $\mathscr{C}(A)$. Recall that all our C^*-algebras are unital, and that by assumption C^*-subalgebras share the unit of the ambient C^*-algebra, hence $C^*(e)$ contains the unit of A. Hence $\tilde{C} \equiv \mathsf{B}(C) = \mathsf{B}_{|D}(C) = \mathsf{B}_{|E}(C)$ is an atom in $\mathscr{C}(B)$, which implies that $\tilde{C} = C^*(\tilde{e})$ for some projection $\tilde{e} \in \mathscr{P}(B)$. If $\mathsf{J}_D(e) = \mathsf{J}_E(e)$ we are done, so we must exclude the case $\mathsf{J}_D(e) = \tilde{e}$, $\mathsf{J}_E(e) = 1_B - \tilde{e}$. This analysis again requires a case distinction:

$$\dim(eAe) = \dim(e^{\perp}Ae^{\perp}) = 1; \tag{20}$$

$$\dim(eAe) = 1, \;\; \dim(e^{\perp}Ae^{\perp}) > 1; \tag{21}$$

$$\dim(eAe) > 1, \;\; \dim(e^{\perp}Ae^{\perp}) > 1, \tag{22}$$

where $e^{\perp} = 1_A - e$. Each of these cases is nontrivial, and we need another lemma.

Lemma 2 *Let $C \in \mathscr{C}(A)$ be maximal (i.e., $C \subset A$ is maximal abelian).*

1. *For each projection $e \in \mathscr{P}(C)$ we have $\dim(eCe) = 1$ iff $\dim(eAe) = 1$.*
2. *We have $\dim(C) = 2$ iff either $A \cong \mathbb{C}^2$ or $A \cong M_2(\mathbb{C})$.*

Proof For the first claim $\dim(eAe) = 1$ clearly implies $\dim(eCe) = 1$. For the converse implication, assume *ad absurdum* that $\dim(eAe) > 1$, so that there is an $a \in A$ for which $eae \neq \lambda \cdot e$ for any $\lambda \in \mathbb{C}$. If also $\dim(eCe) = 1$, then any $c \in C$ takes the form $c = \mu \cdot e + e^{\perp}ce^{\perp}$ for some $\mu \in \mathbb{C}$. Indeed, since c, e, $e^{\perp}$ commute within C,

$$c = ce + ce^{\perp} = ce^2 + c(e^{\perp})^2 = ece + e^{\perp}ce^{\perp} = \mu e + e^{\perp}ce^{\perp},$$

where the last equality follows since $ece \in eCe$, which is spanned by e. This implies that $eae \in C'$ (where C' is the commutant of C within A), and since C is maximal abelian, we have $C = C'$, whence $eae \in C$. Now $eae = e(eae)e$, hence $eae \in eCe$, whence $eae = \lambda \cdot e$ for some $\lambda \in \mathbb{C}$. Contradiction.

According to Exercise 4.6.12 in Kadison and Ringrose [32], showing that a C^*-algebra is finite-dimensional if it has finite-dimensional maximal abelian $*$-subalgebra, the assumption $\dim(C) = 2$ implies that A is finite-dimensional. The well-known theorem stating that every finite-dimensional C^*-algebra is isomorphic to a direct sum of matrix algebras then easily yields the second claim.

Having proved Lemma 2, we move on the analyze the cases (20)–(22).

- Equation (20) implies that C is maximal, as follows. Any element $a \in A$ is a sum of eae, $e^\perp ae^\perp$, $eae^\perp$, and $e^\perp ae$; nonzero elements of $C' = \{e\}'$ can only be of the first two types. If (20) holds, then $\dim(C') = 2$, but since C is abelian we have $C \subseteq C'$ and since $\dim(C) = 2$ we obtain $C' = C$. Lemma 2 then implies that either $A \cong \mathbb{C}^2$ or $A \cong M_2(\mathbb{C})$. These C^*-algebras have been analyzed after the statement of Theorem 4, and since those two A's conversely imply (20), we may exclude them in dealing with (21)–(22). By Lemma 2 (applied to D and E instead of C), in what follows we may assume that $\dim(D) > 2$ and $\dim(E) > 2$ (as D and E are maximal).
- Equation (21) implies $\dim(eD) = 1$. Assuming $\mathsf{J}_D(e) = \tilde{e}$, this implies $\dim(\tilde{e}\tilde{D}) = 1$ (since J_D is an isomorphism). Applying Lemma 1 to B gives $\dim(\tilde{e}B\tilde{e}) = 1$ (since $\tilde{D}$ is maximal). If also $\dim((1_B - \tilde{e})B(1_B - \tilde{e})) = 1$, then $\dim(\tilde{D}) = 2$, whence $\dim(D) = 2$, which we excluded. Hence $\dim((1_B - \tilde{e})B(1_B - \tilde{e})) > 1$. Applied to J_E this gives $\mathsf{J}_E(e) = \tilde{e}$, and hence J_D and J_E coincide on $C = C^*(e)$.
- Equation (22) implies that $\dim(eDe) > 1$ as well as $\dim(e^\perp Ee^\perp) > 1$ (apply Lemma 1 to D and E, respectively). Since $\dim(eDe) > 1$, there is some $a \in D$ such that e and $a' = eae \in D$ are linearly independent, and similarly there is some $b \in E$ such that $b' = e^\perp be^\perp$ is linearly independent from $e^\perp$. Then a', b', e commute (in fact, $a'b' = b'a' = 0$), so that we may form the abelian C^*-algebras $C_1 = C^*(e, a') \subseteq D$ and $C_2 = C^*(e, b') \subseteq E$, which (also containing the unit 1_A) both have dimension at least three. We also form $C_3 = C^*(e, a', b')$, which contains C_1 and C_2 and hence is at least three-dimensional, too. Because D and E are maximal abelian, C_3 must lie in both D and E. Applying the abelian case of the theorem already proved to D and E, as before, but replacing C used so far by C_3, we find that J_D and J_E coincide on C_3 (as its dimension is >2). In particular, $\mathsf{J}_D(e) = \mathsf{J}_E(e)$.

To finish the proof, we first note that Theorem 4 holds for $A = B = \mathbb{C}$ by inspection, whereas the cases $A \cong B \cong \mathbb{C}^2$ or $\cong M_2(\mathbb{C})$ have already been discussed.

In all other cases we define $\mathsf{J} : A_{\mathrm{sa}} \to B_{\mathrm{sa}}$ by putting $\mathsf{J}(a) = \mathsf{J}_D(a)$ for any maximal abelian unital C^*-subalgebra D containing $C = C^*(a)$ and hence a; as we just saw, this is independent of the choice of D. Since each J_D is an isomorphism of commutative C^*-algebras, J is a weak Jordan isomorphism. Finally, uniqueness of J (under the stated restriction on A) follows from Lemma 1. □

Theorem 4 begs the question if we can strengthen weak Jordan isomorphisms to Jordan isomorphism. This hinges on the extendibility of weak Jordan isomorphisms to linear maps (which are automatically Jordan isomorphisms). The problem of whether a quasi-linear map (i.e., a map that is linear on commutative subalgebras) is linear has been studied by [8] for the case of C*-algebras without a quotient isomorphic to $M_2(\mathbb{C})$. More suitable for our setting is a generalization of Gleason's Theorem, proven by [7], and thoroughly discussed in [18]. More generally, one can rely on Dye's Theorem for AW*-algebras. For the exact statement and its proof we refer to [20], who used it to prove the following result:

Theorem 5 *Let A and B be unital AW*-algebras, where A contains neither* $\mathbb{C}^2$ *nor* $M_2(\mathbb{C})$ *as a summand. Then there is a bijective correspondence between order isomorphisms* $\mathsf{B} : \mathscr{C}(A) \to \mathscr{C}(B)$ *and Jordan isomorphisms* $\mathsf{J} : A_{\mathrm{sa}} \to B_{\mathrm{sa}}$.

We note that a version of this theorem in the setting of von Neumann algebras is proven in [11]. If $A = B = B(H)$, then the ordinary Gleason Theorem suffices to yield the crucial lemma for Wigner's Theorem for Bohr symmetries (i.e. Theorem 1.6).

5 Projections

Given some C^*-algebra A, the orthocomplemented poset $\mathscr{P}(A)$ of projections in A satisfies the following two conditions:

1. if $p \leq q^{\perp}$, then $p \vee q$ exists (and is equal to $p + q$ in A);
2. if $p \leq q$, then $q = p \vee (p^{\perp} \wedge q)$.

We say that $\mathscr{P}(A)$ is an *orthomodular* poset. We note that $\mathscr{P}(A)$ is Boolean if A is commutative, but the converse implication does not hold. Indeed, [4] showed the existence of a non-commutative C^*-algebra whose only projections are trivial (and hence form a Boolean algebra). Notwithstanding such cases, it might be interesting to investigate in which ways $\mathscr{C}(A)$ and $\mathscr{P}(A)$ determine each other. In one direction we have the following result (cf. Theorem 6.4.4 in [37]):

Theorem 6 *Let A and B be* C^**-algebras. Then any order isomorphism* $\mathscr{C}(A) \to \mathscr{C}(B)$ *induces an isomorphism of orthomodular posets* $\mathscr{P}(A) \to \mathscr{P}(B)$.

Its proof is based on the following observations. Firstly, a C^*-algebra A is called ***approximately finite-dimensional***, abbreviated by ***AF***, if there is a directed collection $\mathscr{D}$ of finite-dimensional C^*-subalgebras of A such that $A = \overline{\bigcup \mathscr{D}}$. A commutative C^*-algebra A is AF if and only if it is generated by its projections, which form a Boolean algebra, since A is commutative. One can show that the Gelfand spectrum of A is homeomorphic to the Stone spectrum of $\mathscr{P}(A)$, and that there is an equivalence between the category of Boolean algebras with Boolean morphisms and the category of commutative AF-algebras with *-homomorphisms.

Given a C^*-algebra A, we let $\mathscr{C}_{\mathrm{AF}}(A)$ be the subposet of $\mathscr{C}(A)$ consisting of all commutative C^*-subalgebras of A that are AF. There are several order-theoretic criteria whether or not an element $C \in \mathscr{C}(A)$ belongs to $\mathscr{C}_{\mathrm{AF}}(A)$, which is important since any order-theoretic criterion assures that an order isomorphism $\mathscr{C}(A) \to \mathscr{C}(B)$ restricts to an order isomorphism $\mathscr{C}_{\mathrm{AF}}(A) \to \mathscr{C}_{\mathrm{AF}}(B)$. Firstly, $C \in \mathscr{C}_{\mathrm{AF}}(A)$ if and only if C is the supremum of a directed subset of the ***compact*** elements of $\mathscr{C}(A)$, where $B \in \mathscr{C}(A)$ is called compact if for each directed subset $\mathscr{D}$ of $\mathscr{C}(A)$ the inclusion $B \subseteq \bigvee \mathscr{D}$ implies that $B \subseteq D$ for some $D \in \mathscr{D}$; note that the supremum $\bigvee \mathscr{D}$ of any directed subset $\mathscr{D}$ exists, and is given by $\overline{\bigcup \mathscr{D}}$. The compact elements of $\mathscr{C}(A)$ turn out to be the finite-dimensional elements of $\mathscr{C}(A)$, so clearly C is AF if and only if it is the directed supremum of compact elements. Another criterion for $C \in \mathscr{C}_{\mathrm{AF}}(A)$ is that C is the supremum of some collection of atoms in $\mathscr{C}(A)$. The intuition behind this criterion is that $C \in \mathscr{C}_{\mathrm{AF}}(A)$ if and only if it is generated by its projections, and any atom of $\mathscr{C}(A)$ is a C^*-subalgebra of A that is generated by single proper projection in A. For details, we refer to [28].

Secondly, given some orthomodular poset P, we say that some subset $D \subseteq P$ is a ***Boolean subalgebra*** of P if it is a Boolean algebra in its relative order for which the meet, join, and orthocomplementation agree with the meet, join, and orthocomplementation, respectively on P. We denote the poset of Boolean subalgebras of P ordered by inclusion by $\mathscr{B}(P)$. The equivalence between the categories of Boolean algebras and of commutative AF-algebras yields the following proposition:

Proposition 2 *Let A be a C^*-algebra. The map*

$$\mathscr{C}_{\mathrm{AF}}(A) \to \mathscr{B}(\mathscr{P}(A)); \tag{23}$$
$$C \mapsto \mathscr{P}(C), \tag{24}$$

is an order isomorphism with inverse $B \mapsto C^*(B)$, where $C^*(B)$ denotes the C^*-subalgebra of A generated by B.

A modification of the Harding–Navara Theorem (cf. Remark 4.4 in [23]) states that if P and Q are orthomodular posets, then an order isomorphism $\Phi : \mathscr{B}(P) \to \mathscr{B}(Q)$ is induced by an orthomodular isomorphism $\varphi : P \to Q$ via $B \mapsto \varphi[B]$. Moreover, this orthomodular isomorphism is unique if P does not have ***blocks***, i.e., maximal Boolean subalgebras of four elements. In combination with the previous proposition, this proves the following (cf. Theorem 6.4.4 in [37]):

Theorem 7 *Let A and B be C^*-algebras. Then for each order isomorphism*

$$\Phi : \mathscr{C}_{\mathrm{AF}}(A) \to \mathscr{C}_{\mathrm{AF}}(B)$$

there is an orthomodular isomorphism $\varphi : \mathscr{P}(A) \to \mathscr{P}(B)$ such that

$$\Phi(C) = C^*(\varphi[\mathscr{P}(C)]),$$

for each $C \in \mathscr{C}_{\mathrm{AF}}(A)$, which is unique if $\mathscr{P}(A)$ does not have blocks of four elements.

Theorem 6 now is an easy consequence of Theorem 7.

Acknowledgements The first author has been supported by Radboud University and Trinity College (Cambridge). The second author was supported by the Netherlands Organisation for Scientific Research (NWO) under TOP-GO grant no. 613.001.013.

References

1. Alfsen, E.M., Shultz, F.W. (2001). *State Spaces of Operator Algebras*. Basel: Birkhäuser.
2. Alfsen, E.M., Shultz, F.W. (2003). *Geometry of State Spaces of Operator Algebras*. Basel: Birkhäuser.
3. Berberian, S.K. (1972). *Baer *-rings*. Springer-Verlag.
4. Blackadar, B. (1981). A simple unital projectionless C^*-algebra. *Journal of Operator Theory* 5, 63–71.
5. Bohr, N. (1928). The quantum postulate and recent developments of atomic theory (Como lecture). *Nature suppl.* April 14, 1928, pp. 580–590.
6. Bratteli, O., Robinson, D.W. (1987). *Operator Algebras and Quantum Statistical Mechanics. Vol. I: C^*- and W^*-Algebras, Symmetry Groups, Decomposition of States*. 2nd Ed. Berlin: Springer.
7. Bunce, L.J., Wright, J.D.M. (1992). *The Mackey-Gleason Problem*. Bulletin of the American Mathematical Society Vol. 26, No. 2, 288–293.
8. Bunce, L.J., Wright, J.D.M. (1996). *The quasi-linearity problem for C^*-algebras*. Pacific Journal of Mathematics Vol. 172, No. 1, 41–47.
9. Camilleri, K. (2009). *Heisenberg and the Interpretation of Quantum Mechanics: The Physicist as Philosopher*. Cambridge: Cambridge University Press.
10. Cassinelli, G., De Vito, E., Lahti, P.J., Levrero, A. (2004). *The Theory of Symmetry Actions in Quantum Mechanics*. *Lecture Notes in Physics* 654. Berlin: Springer-Verlag.
11. Döring, A., Harding, J. (2010). Abelian subalgebras and the Jordan structure of a von Neumann algebra. arXiv:1009.4945.
12. Döring, A., Isham, C.J. (2008a). *A topos foundation for theories of physics: I. Formal languages for physics*, Journal of Mathematical Physics 49, Issue 5, 053515.
13. Döring, A., Isham, C.J. (2008b). *A topos foundation for theories of physics: II. Daseinisation and the liberation of quantum theory*, Journal of Mathematical Physics 49, Issue 5, 053516.
14. Döring, A., Isham, C.J. (2008c). *A topos foundation for theories of physics: III. Quantum theory and the representation of physical quantities with arrows* $\breve{\delta}(\hat{A}) : \underline{\Sigma} \to \underline{\mathbb{R}}^{\leftrightarrow}$, Journal of Mathematical Physics 49, Issue 5, 053517.
15. Döring, A., Isham, C.J. (2008d). *A topos foundation for theories of physics: IV. Categories of systems*, Journal of Mathematical Physics 49, Issue 5, 053518.
16. Firby, P.A. (1973). Lattices and compactifications, II. *Proceedings of the London Mathematical Society* 27, 51–60.
17. Gelfand, I.M., Naimark, M.A. (1943). On the imbedding of normed rings into the ring of operators in Hilbert space. *Sbornik: Mathematics* 12, 197–213.
18. Hamhalter, J. (2004). *Quantum Measure Theory*. Dordrecht: Kluwer Academic Publishers.
19. Hamhalter, J. (2011). Isomorphisms of ordered structures of abelian C^*-subalgebras of C^*-algebras, *Journal of Mathematical Analysis and Applications* 383, 391–399.
20. Hamhalter, J. (2015). Dye's Theorem and Gleason's Theorem for AW^*-algebras. *Journal of Mathematical Analysis and Applications* 422, 1103–1115.
21. Hamilton, J., Isham, C.J., Butterfield, J. (2000). Topos perspective on the Kochen–Specker Theorem: III. Von Neumann Algebras as the base category. *International Journal of Theoretical Physics* 39, 1413–1436.
22. Hanche-Olsen, H., Størmer, E. (1984). *Jordan Operator Algebras*. Boston: Pitman.
23. Harding, J. , Navara, M. (2011). Subalgebras of Orthomodular Lattices. *Order* 28, 549–563.
24. Heisenberg, W. (1925). Über quantentheoretische Umdeutung kinematischer und mechanischer Beziehungen. *Zeitschrift für Physik* 33, 879–893.
25. Heisenberg, W. (1927). Über den anschaulichen Inhalt der quantentheoretischen Kinematik und Mechanik. *Zeitschrift für Physik* 43, 172–198.
26. Heisenberg, W. (1958). *Physics and Philosophy: The Revolution in Modern Science*. London: Allen & Unwin.
27. Held, C. (1994). The Meaning of Complementarity. *Studies in History and Philosophy of Science* 25, 871–893.

28. Heunen, C., Lindenhovius, A.J. (2015). Domains of C^*-subalgebras. *Proceedings of the 30th annual ACM/IEEE symposium on Logic in Computer Science* pp. 450–461.
29. Heunen, C., Reyes, M.L. (2014). Active lattices determine AW^*-algebras. *Journal of Mathematical Analysis and Applications* 416, 289–313.
30. Heunen, C., Landsman, N.P., Spitters, B. (2009). A topos for algebraic quantum theory. *Communications in Mathematical Physics* 291, 63–110.
31. Isham, C.J., Butterfield, J. (1998). Topos perspective on the Kochen–Specker theorem. I. Quantum states as generalized valuations. *International Journal of Theoretical Physics* 37, 2669–2733.
32. Kadison, R.V., Ringrose, J.R. (1983). Fundamentals of the Theory of Operator Algebras. Vol 1: Elementary Theory. New York: Academic Press.
33. Landsman, N.P. 1998. *Mathematical Topics Between Classical and Quantum Mechanics*. New York: Springer-Verlag.
34. Landsman, N.P. (2016). Bohrification: From classical concepts to commutative algebras. To appear in *Niels Bohr in the 21st Century*, eds. J. Faye, J. Folse. Chicago: Chicago University Press. arXiv:1601.02794.
35. Landsman, N.P. (2017). *Bohrification: From Classical Concepts to Commutative Operator Algebras*. In preparation.
36. Lindenhovius, A.J. (2015). Classifying finite-dimensional C^*-algebras by posets of their commutative C^*-subalgebras. arXiv:1501.03030.
37. Lindenhovius, A.J. (2016). $\mathscr{C}(A)$. PhD Thesis, Radboud University Nijmegen.
38. Mendivil, F. (1999). Function algebras and the lattices of compactifications. *Proceedings of the American Mathematical Society* 127, 1863–1871.
39. Moretti, V. (2013). *Spectral Theory and Quantum Mechanics*. Mailand: Springer-Verlag.
40. Shultz, F.W. (1982). Pure states as dual objects for C^*-algebras. *Communications in Mathematical Physics* 82, 497–509.
41. Simon, B. (1976). Quantum dynamics: from automorphism to Hamiltonian. *Studies in Mathematical Physics: Essays in Honor of Valentine Bargmann*, pp. 327–349. Lieb, E., Simon, B., Wightman, A.S., eds. Princeton: Princeton University Press.
42. Willard, S. (1970). *General Topology*. Reading: Addison-Wesley Publishing Company.

Categorial Local Quantum Physics

Miklós Rédei

Abstract Categorial local quantum field theory was suggested as a new paradigm for quantum field theory by Brunetti, Fredenhagen and Verch in 2003 (Commun Math Phys 237:31–68, [7]). In this paradigm quantum field theory is defined to be a covariant functor from the category of certain spacetimes (with isometric embeddings as morphisms) into the category of C^*-algebras (with injective C^*-algebra homomorphisms as morphisms). Further properties of the functor are stipulated axiomatically on the basis of physical considerations. The present paper suggests an additional axiom on the functor that expresses independence of systems as morphism co-possibility. It is argued that this axiom is very natural because it has a direct physical interpretation. The relation of the axiom system containing the morphism co-possibility axiom to other axiom systems is investigated. It will be seen that this axiom system is strictly stronger than the axiom system originally formulated in Brunetti, Fredenhagen, Verch (Commun Math Phys 237:31–68, 2003, [7]), and it is conjectured that it is strictly weaker than the ones formulated in subsequent development of categorial quantum field theory in which the functor is required to be extensible to a tensor functor. Determining the precise status of the axiom system based on morphism co-possibility as independence needs further analysis.

Keywords Quantum field theory · Category theory · Operator algebra theory

1 The Main Idea of the Categorial Paradigm for Quantum Field Theory

In their seminal paper, [7], Brunetti, Fredenhagen and Verch initiated a new approach to quantum field theory. The new approach is based on category theory. The theory

---

M. Rédei
Department of Philosophy, Logic and Scientific Method,
London School of Economics and Political Science, Houghton Street,
London WC2A 2AE, UK
e-mail: m.redei@lse.ac.uk

M. Ozawa et al. (eds.), *Reality and Measurement in Algebraic Quantum Theory*,
Springer Proceedings in Mathematics & Statistics 261,
https://doi.org/10.1007/978-981-13-2487-1_5

was further developed in a series of papers [5, 6, 14, 18] (the recent papers [12, 13] give an overview of the framework).

This new, categorial approach generalizes substantially the Haag–Kastler algebraic axiomatization of quantum field theory (for monographic presentation of the Haag–Kastler axiomatization see [2, 15, 16]). The main motivation for the Brunetti-Fredenhagen-Verch generalization is the desire to develop quantum field theory in a general (curved) spacetime. To do this one needs a formalism that is flexible enough to accommodate any (physically reasonable) spacetime as background to quantum field theory. The Haag–Kastler approach is unsatisfactory from this perspective because it relies on certain axioms (e.g. covariance with respect to the group of symmetries of the spacetime, spectrum condition, existence of vacuum state) which are framed in terms of a preferred representation of the Poincaré group. In a curved spacetime, however, there are no non-trivial global symmetries; hence none of the standard axioms that rely on the existence of a global symmetry make sense in a quantum field theory on a general, curved spacetime. Brunetti, Fredenhagen and Verch summarize these motivations this way:

> Quantum field theory incorporates two main principles into quantum physics, locality and covariance. Locality expresses the idea that quantum processes can be localized in space and time (and, at the level observable quantities, that causally separated processes are exempt from any uncertainty relations restricting their commensurability). The principle of covariance within *special* relativity states that there are no preferred Lorentzian coordinates for the description of physical processes, and thereby the concept of an absolute space as an arena for physical phenomena is abandoned. Yet it is meaningful to speak of events in terms of spacetime points as entities of a given, fixed spacetime background, in the setting of special relativistic physics.
>
> In general relativity, however, spacetime points loose this a priori meaning. The principle of general covariance forces one to regard spacetime points simultaneously as members of several, locally diffeomorphic spacetimes. It is rather the relations between distinguished events that have physical interpretation.
>
> This principle should also be observed when quantum field theory in presence of gravitational fields is discussed.
>
> Quantum field theory ... is a *covariant functor* ... in the ... fundamental and physical sense of implementing the principles of locality and general covariance... [7] [p. 61–78]

The covariant functor Brunetti, Fredenhagen and Verch refer to in the above quotation is between two concrete categories (see Sect. 2 for the properties of these two categories):

- $(\mathfrak{Man}, hom_{\mathfrak{Man}})$
  The category of spacetimes with isometric embeddings of spacetimes as morphisms.
- $(\mathfrak{Alg}, hom_{\mathfrak{Alg}})$
  The category of C^*-algebras with injective C^*-algebra homomorphisms as morphisms.

The properties of the functor are fixed axiomatically: one requires the functor to have certain features that express "locality" alluded to in the quotation above. It is a priori more or less clear that this can be done in more than one ways. It will be seen in

Sect. 2 that different axioms have indeed been formulated in the papers [5–7, 13, 14, 18]. The different axiomatizations differ in how they express independence of physical systems pertaining to causally disjoint spacetime regions.

The present paper suggests an axiomatization in which the axiom expressing independence of systems differs from the ones in the aforementioned papers. The independence axiom proposed here is the categorial morphism-co-possibility, introduced first in [20]. I will argue that the independence axiom suggested is very natural because it has a direct physical interpretation. Having different axiomatizations, the question of their relation emerges as a non-trivial problem. Some results will be recalled in Sect. 4 that clarify some of the relations but there remain open questions in this regard. It will be seen that the axiom system proposed in this paper is strictly stronger than the axiom system originally formulated in [7], and it is conjectured that it is strictly weaker than the ones formulated in subsequent developments of categorial quantum field theory in which the functor is required to be extensible to a tensor functor. Determining the status of the axiom system based on morphism–co-possibility as independence needs further analysis.

2 The Covariant Functor of Categorial Local Quantum Physics

The category $(\mathfrak{Man}, hom_{\mathfrak{Man}})$ is specified by the following stipulations (see [7] for more details):

- The objects in $Obj(\mathfrak{Man})$ are 4 dimensional C^∞ spacetimes (M, g) with a Lorentzian metric g and such that (M, g) is Hausdorff, connected, time oriented and globally hyperbolic.
- The morphisms in $hom_{\mathfrak{Man}}$:

$$\psi\colon (M_1, g_1) \to (M_2, g_2)$$

  are isometric smooth embeddings such that

  - ψ preserves the time orientation;
  - ψ is causal in the following sense: if the endpoints $\gamma(a), \gamma(b)$ of a timelike curve $\gamma\colon [a, b] \to M_2$ are in the image $\psi(M_1)$, then the whole curve is in the image: $\gamma(t) \in \psi(M_1)$ for all $t \in [a, b]$.
  - The composition of morphisms is the usual composition of maps.

The category $(\mathfrak{Alg}, hom_{\mathfrak{Alg}})$ is defined as:

- The objects in $Obj(\mathfrak{Alg})$ are unital C^*-algebras.
- The morphisms are injective, unit preserving C^*-algebra homomorphisms

$$\alpha\colon \mathscr{A}_1 \to \mathscr{A}_2$$

The composition of morphisms is the usual composition of C^*-algebra homomorphisms.

Categorial local quantum field theory is then defined as a functor:

Definition 1 A locally covariant quantum field theory is a covariant functor $\mathscr{F}$ between the categories $(\mathfrak{Man}, hom_{\mathfrak{Man}})$ and $(\mathfrak{Alg}, hom_{\mathfrak{Alg}})$:

- For any object (M, g) in $\mathfrak{Man}$ the $\mathscr{F}(M, g)$ is a C^*-algebra in $\mathfrak{Alg}$.
- For any homomorphism ψ in $hom_{\mathfrak{Man}}$ the $\mathscr{F}(\psi)$ is a C^*-algebra homomorphism in $hom_{\mathfrak{Alg}}$

such that the following hold

$$\mathscr{F}(\psi_1 \circ \psi_2) = \mathscr{F}(\psi_1) \circ \mathscr{F}(\psi_2)$$
$$\mathscr{F}(id_{\mathfrak{Man}}) = id_{\mathfrak{Alg}}$$

The physical interpretation of the elements in the definition is along the lines of local quantum physics as this is understood in the Haag–Kastler version of algebraic quantum field theory: The functor $\mathscr{F}$ assigns to a spacetime manifold (M, g) an operator algebra $\mathscr{F}(M, g)$ of observables measurable in M. This explicit association of the observables with a specific spacetime embodies an elementary but fundamental aspect of locality: the idea that any measurement, observation, and interaction can only take place at a particular location in spacetime. Following the terminology introduced in [20], I call this kind of locality "spatio-temporal locality".

3 Causal Locality Conditions on the Covariant Functor

The interpretation of $\mathscr{F}(M, g)$ as the algebra of observables measurable in M motivates imposing further conditions on the functor $\mathscr{F}$. The further conditions express "locality", understood as conditions ensuring harmony of the assignment $(M, g) \mapsto \mathscr{F}(M, g)$ with the causal structure of the spacetimes. Following the terminology introduced in [20], I call this kind of locality "causal locality" to distinguish it from "spatio-temporal locality", which does not involve causal content explicitly: Spatio-temporal locality expresses the fact that $\mathscr{F}$ explicitly specifies the spatio-temporal location of observables in such a way that spatio-temporal locality of observables is in harmony with the subsystem relation. That is to say, the content of spatio-temporal locality is that a physical system's set of observables are a subset of the set of observables of a system if the latter system's spatio-temporal locality region contains that of the former (this is expressed by the covariance property of the functor). While extremely natural, spatio-temporal locality is crucially important: it is a conceptual pre-condition without which causal locality cannot be formulated at all [20]; furthermore, as emphasized by [15], all the physical information is contained in the association of observables with spacetime regions. This is reflected by the fact that in the Haag–Kastler version of algebraic quantum field theory it holds that the local algebras pertaining to typical spacetime regions are all isomorphic [8], [15] [p. 225]; thus the physical content of the theory is contained in the way the isomorphic algebras are related to each other via the isotony relation constrained by the causal locality condition.

Causal locality so interpreted cannot be expressed as a single condition: A spacetime has a causal structure that specifies both causally *independent* and causally *dependent* spacetime regions. Accordingly, causal locality conditions to be imposed on the functor $\mathscr{F}$ should regulate the behavior of $\mathscr{F}$ from the perspective of both causally independent and dependent spacetime regions. The most basic stipulations were formulated in [7]: Einstein Locality and Time Slice axiom, they are recalled in the next subsection.

3.1 The BASIC Axioms: Einstein Locality and Time Slice

Definition 2 The covariant functor of categorial quantum field theory

$$\mathscr{F} : (\mathfrak{Man}, hom_{\mathfrak{Man}}) \to (\mathfrak{Alg}, hom_{\mathfrak{Alg}})$$

should satisfy

- Causal Locality – Independence: Einstein Causality:

$$\Big[\mathscr{F}(\psi_1)\Big(\mathscr{F}(M_1, g_1)\Big), \mathscr{F}(\psi_2)\Big(\mathscr{F}(M_2, g_2)\Big)\Big]_-^{\mathscr{F}(M,g)} = \{0\} \tag{1}$$

whenever

$$\psi_1 : (M_1, g_1) \to (M, g) \tag{2}$$
$$\psi_2 : (M_2, g_2) \to (M, g) \tag{3}$$

and $\psi_1(M_1)$ and $\psi_2(M_2)$ are spacelike in M, where $[\ ,\ ]_-^{\mathscr{F}(M,g)}$ in (1) denotes the commutator in the C^*-algebra $\mathscr{F}(M, g)$.

- Causal Locality – Dependence: Time slice axiom: If (M, g) and (M', g') and the embedding

$$\psi : (M, g) \to (M', g')$$

are such that $\psi(M, g)$ contains a Cauchy surface for (M', g') then

$$\mathscr{F}(\psi)\mathscr{F}(M, g) = \mathscr{F}(M', g')$$

I call the axiom system specified by Definition 2 BASIC.

In what sense is Einstein Causality a causal independence condition? The standard answer is that Einstein Causality entails no superluminal signaling with respect to non-selective operations represented by local Kraus operations: A completely positive, unite preserving map (operation) T on local algebra $\mathscr{F}(M, g)$ of the form

$$T(X) = \sum_i W_i^* X W_i \tag{4}$$

is called a local Kraus operation represented by the local Kraus operators W_i, if all W_i are in $\mathscr{F}(M, g)$ and sum up to the identity:

$$\sum_i W_i^* W_i = I \tag{5}$$

(See [3, 17] for the definition and elementary facts about operations, including operations that are not Kraus representable.) Given spacetimes (M_1, g_1), (M_2, g_2) and (M, g) with embeddings ψ_1, ψ_2 (2)–(3) such that $\psi_1(M_1)$ and $\psi_2(M_2)$ are spacelike in M, any Kraus operation T_1 on $\mathscr{F}(M_1, g_1)$ can be extended to the local algebra $\mathscr{F}(M, g)$ to a Kraus operation T by

$$T(A) \doteq \sum_i \mathscr{F}(\psi_1)(W_i^*) A \mathscr{F}(\psi_1)(W_i) \qquad A \in \mathscr{F}(M, g) \tag{6}$$

Einstein Locality together with (5) entails then that the restriction of T to the algebra $\mathscr{F}(\psi_2)(\mathscr{F}(M_2, g_2))$ is the identity map. Thus the state of system localized in spacetime (M_2, g_2) and viewed as a subsystem of $\mathscr{F}(M, g)$ remains unaffected by performing the operation T_1 on system in spacetime (M_1, g_1) (viewed as a subsystem of $\mathscr{F}(M, g)$). This is the content of no-signaling. A particular case of no-signaling is when the Kraus operators W_i in (4) are one dimensional projections, giving no-signaling with respect to the projection postulate.

Einstein Causality does not entail however no superluminal signaling with respect to *general* spatio-temporally local operations; i.e. with respect to operations T on $\mathscr{F}(M_1, g_1)$ that are not Kraus representable. An example of such an operation is the Accardi-Cecchini state-preserving conditional expectation [1] in the context of the Haag–Kastler quantum field theory (see [23] for details).

More generally: Einstein Causality does not, in and of itself, entail what is called operational subsystem independence: That any two (non-selective) operations performed on spacelike separated subsystems S_1, S_2 of system S are jointly implementable as a single operation on S [22]. Given the significance of the concept of subsystem independence in quantum field theory [24, 25], one should ensure that the axioms of the categorial approach to quantum field theory express subsystem independence. One way to do this is to formulate a categorial version of subsystem independence and postulate it axiomatically as a required feature of the functor $\mathscr{F}$. The natural independence notion in a concrete category is morphism co-possibility. This notion was introduced in [20] in the specific context of the category $(\mathfrak{Alg}, hom_{\mathfrak{Alg}})$ but I formulate it in an arbitrary concrete category in the next subsection.

3.2 Amending the Basic Axioms by Adding Morphism Co-possibility as Subsystem Independence

Let $(\mathfrak{C}, hom_{\mathfrak{C}})$ be a concrete category and $Mor_{\mathfrak{C}}$ be a class of morphisms between objects of $\mathfrak{C}$. The morphism class $Mor_{\mathfrak{C}}$ can be the same as $hom_{\mathfrak{C}}$, but this is not required: $Mor_{\mathfrak{C}}$ can be larger than $hom_{\mathfrak{C}}$. The class of morphisms $Mor_{\mathfrak{C}}$ should be viewed as a variable in the independence notion, specified by the following definition. Different choices of $Mor_{\mathfrak{C}}$ yield qualitatively different independence concepts.

Definition 3 Given objects C_1, C_2 and C in $\mathfrak{C}$ and homomorphisms $h_1: C_1 \to C$ and $h_2: C_2 \to C$ in $hom_{\mathfrak{C}}$, the objects $h_1(C_1)$ and $h_2(C_2)$ are said to be $Mor_{\mathfrak{C}}$-independent in C, if for any two morphisms $m_1: h_1(C_1) \to h_1(C_1)$ and $m_2: h_2(C_2) \to h_2(C_2)$ in $Mor_{\mathfrak{C}}$, there exists a morphism $m: C \to C$ in $Mor_{\mathfrak{C}}$ that coincides with m_1 on $h_1(C_1)$ and coincides with m_2 on $h_2(C_2)$.

It is intuitively clear why $Mor_{\mathfrak{C}}$-independence of objects $h_1(C_1)$ and $h_2(C_2)$ is an independence condition: fixing morphism m_1 on object $h_1(C_1)$ does not interfere with fixing any morphism m_2 on object $h_1(C_1)$ and vice versa. That is to say, morphisms can be independently chosen on these objects seen as parts of object C. This independence notion is a natural categorial generalization of the concept known as subsystem independence [24, 25]. One can recover all the major subsystem independence concepts that occur in algebraic quantum (field) theory by taking the category $(\mathfrak{Alg}, hom_{\mathfrak{Alg}})$ and choosing, as morphism class $Mor_{\mathfrak{Alg}}$, special subclasses of the class of all non-selective operations (unit preserving completely positive, linear maps) $Op_{\mathfrak{Alg}}$ [19].

Given the concept of $Op_{\mathfrak{Alg}}$-independence, it is natural to impose it on the covariant functor $\mathcal{F}$ representing quantum field theory in order to express causal locality in terms of it:

Definition 4 The covariant functor of categorial quantum field theory

$$\mathcal{F}: (\mathfrak{Man}, hom_{\mathfrak{Man}}) \to (\mathfrak{Alg}, hom_{\mathfrak{Alg}})$$

should satisfy

- Causal Locality – Independence: $Op_{\mathfrak{Alg}}$-independence: whenever

$$\psi_1 : (M_1, g_1) \to (M, g) \tag{7}$$

$$\psi_2 : (M_2, g_2) \to (M, g) \tag{8}$$

and $\psi_1(M_1)$ and $\psi_2(M_2)$ are spacelike in M, the objects $\mathscr{F}(M_1, g_1)$ and $\mathscr{F}(M_2, g_2)$ are $Op_{\mathfrak{Alg}}$-independent in the sense of Definition 3, taking $Op_{\mathfrak{Alg}}$ as $Mor_{\mathfrak{Alg}}$.

The axiom system that requires quantum field theory to be a covariant functor having the features of Einstein Locality, Time Slice axiom and $Op_{\mathfrak{Alg}}$-independence, is called **OPIND**.

One can strengthen $Op_{\mathfrak{Alg}}$-independence into $Op_{\mathfrak{Alg}}$-independence *in the product sense* by requiring the morphism m in Definition 3 that extends m_1 and m_2 to factorize across $h_1(C_1)$ and $h_2(C_2)$:

$$m(AB) = m(A)m(B) = m_1(A) = m_2(B) \qquad A \in h_1(C_1) \quad B \in h_2(C_2) \tag{9}$$

This leads to a natural strengthening of the axiom system **OPIND**: by requiring that the extension T in (6) factorizes across the algebras $\mathscr{F}(\psi_1)(\mathscr{F}(M_1, g_1))$ and $\mathscr{F}(\psi_2)(\mathscr{F}(\psi_2))$:

$$T(AB) = T(A)T(B) \qquad A \in \mathscr{F}(\psi_1)(\mathscr{F}(M_1, g_1)), \quad B \in \mathscr{F}(\psi_2)(\mathscr{F}(M_2, g_2)) \tag{10}$$

I call the strengthened axiom system **OPIND**$^{\times}$.

3.3 *Amending the Basic Axioms by Adding the Categorial Split Property*

Categorial split property was introduced in [6] (also see [11]). The categorial split property is the categorial version of what is known the funnel property of the Haag–Kastler net of local algebras:

Definition 5 The functor $\mathscr{F}$ has the categorial split property if the following two conditions hold:

1. For spacetimes $(M, g_M), (N, g_N)$ in $\mathfrak{Man}$ and morphism $\psi\colon (M, g_M) \to (N, g_N)$ such that the closure of $\psi(M, g_M)$ is compact, connected and in the interior of M, there exists a type I von Neumann factor $\mathscr{R}$ such that

$$\mathscr{F}(\psi)(\mathscr{F}(M, g_M)) \subset \mathscr{R} \subset \mathscr{F}(N, g_N) \tag{11}$$

2. σ-continuity of the $\mathscr{F}(\psi')$ with respect to the inclusion $\mathscr{R} \subset \mathscr{R}'$, where $\psi'\colon (M, g_M) \to (L, g_L)$ and

$$(\mathscr{F}(\psi') \circ \mathscr{F}(\psi))(\mathscr{F}(M, g_M)) \subset \mathscr{F}(\psi')(\mathscr{R}) \tag{12}$$

$$\subset \mathscr{F}(\psi')(\mathscr{F}(N, g_N)) \subset \mathscr{R}' \subset \mathscr{F}(L, g_L) \tag{13}$$

For later purposes I recall the notion of weak additivity of the functor $\mathscr{F}$:

Definition 6 The functor $\mathscr{F}$ satisfies weak additivity if for any spacetime (M, g) and any family of spacetimes (M_i, g_i) with morphisms $\psi_i : (M_i, g_i) \to (M, g)$ such that

$$M \subseteq \cup_i \psi_i(M_i) \tag{14}$$

we have

$$\mathscr{F}(M, g) = \overline{\cup_i \mathscr{F}(\psi_i)(\mathscr{F}(M_i, g_i)))}^{norm} \tag{15}$$

I call **BASIC+SPLIT** the axiom system that requires of the covariant functor $\mathscr{F}$ to have weak additivity and the categorial split property, in addition to Einstein Locality and Time Slice axiom.

3.4 The Tensor Axiom

The axiom system BASIC was modified by Brunetti and Fredenhagen by replacing the Einstein Causality condition by an axiom that requires a tensorial property of $\mathscr{F}$ (Axiom 4 in [5]; also see [14]). To formulate this axiom one has to extend $(\mathfrak{Man}, hom_{\mathfrak{Man}})$ and $(\mathfrak{Alg}, hom_{\mathfrak{Alg}})$ to tensor categories.

The category $(\mathfrak{Man}^{\otimes}, hom^{\otimes}_{\mathfrak{Man}})$ has, by definition, as its objects *finite* disjoint unions of objects from $\mathfrak{Man}$, and the empty set as unit object. (Thus the objects in $\mathfrak{Man}^{\otimes}$ are no longer connected spacetimes.) By definition, the morphisms $\psi^{\otimes}$ in $hom^{\otimes}_{\mathfrak{Man}}$ are maps of the form

$$\psi^{\otimes} : M_1 \sqcup M_2 \sqcup \ldots \sqcup M_n \to M \tag{16}$$

$\sqcup$ denoting the disjoint union) such that

- the restriction of $\psi^{\otimes}$ to any M_i are morphisms in the category $(\mathfrak{Man}, hom_{\mathfrak{Man}})$;
- the images $\psi^{\otimes}(M_i)$ and $\psi^{\otimes}(M_j)$ of the spacetimes M_i are spacelike in M for $i \neq j$.

One can take $(\mathfrak{Alg}^{\otimes}, hom^{\otimes}_{\mathfrak{Alg}})$ to be the tensor category of C^*-algebras with respect to the minimal tensor product of C^*-algebras, with the set of complex numbers as unit object and with the homomorphisms $hom^{\otimes}_{\mathfrak{Alg}}$ being identical to $hom_{\mathfrak{Alg}}$: the class of injective C^*-algebra homomorphisms.

To define the tensorial features of the functor, we need some notation first. Let $\psi_i : M_i \to N_i$ be embeddings of disjoint spacetimes M_i $(i = 1, 2)$ such that the images $\psi_1(M_1)$ and $\psi_2(M_2)$ are causally disjoint in $N_1 \cup N_2$. Then $\psi_1 \otimes \psi_2$ denotes the map

$$(\psi_1 \otimes \psi_2) : M_1 \sqcup M_2 \to N_1 \cup N_2 \tag{17}$$

$$(\psi_1 \otimes \psi_2)(x) \doteq \begin{cases} \psi_1(x) & \text{if } x \in M_1 \\ \psi_2(x) & \text{if } x \in M_2 \end{cases} \tag{18}$$

Clearly, the map $(\psi_1 \otimes \psi_2)$ is a morphism in the category $(\mathfrak{Man}^{\otimes}, hom^{\otimes}_{\mathfrak{Man}})$. The tensor product $\alpha_1 \otimes \alpha_2$ of two injective C^*-algebra homomorphisms α_1 and α_2 on the tensor product $\mathscr{A}_1 \otimes \mathscr{A}_2$ of C^*-algebras $\mathscr{A}_1$ and $\mathscr{A}_2$ is defined in the usual way as the extension to $\mathscr{A}_1 \otimes \mathscr{A}_2$ of the map

$$(\mathscr{A}_1 \otimes \mathscr{A}_2) \ni A_1 \otimes A_2 \mapsto \alpha_1(A_1) \otimes \alpha_2(A_2) \tag{19}$$

Let $\iota_1 \colon M_1 \to M_1 \otimes M_2$ denote the trivial embedding of spacetime M_1 into the disjoint union $M_1 \otimes M_2$. One can then require that the covariant functor $\mathscr{F}$ be a tensor functor in the sense of the following definition:

Definition 7 The covariant functor

$$\mathscr{F}^{\otimes} \colon (\mathfrak{Man}^{\otimes}, hom^{\otimes}_{\mathfrak{Man}}) \to (\mathfrak{Alg}^{\otimes}, hom^{\otimes}_{\mathfrak{Alg}}) \tag{20}$$

is called a *tensor* functor if for any two spacetimes $M_1, M_2 \in \mathfrak{Man}$ with $M_1 \cap M_2 = \emptyset$ and embeddings $\psi_1 \colon M_1 \to N$ and $\psi_2 \colon M_2 \to N$ with causally disjoint images in N we have

$$\mathscr{F}^{\otimes}(\emptyset) = \mathbb{C} \tag{21}$$

$$\mathscr{F}^{\otimes}(\iota_1)(A_1) = A_1 \otimes I \qquad A_1 \in \mathscr{F}^{\otimes}(M_1) \tag{22}$$

$$\mathscr{F}^{\otimes}(\iota_2)(A_2) = I \otimes A_2 \qquad A_2 \in \mathscr{F}^{\otimes}(M_2) \tag{23}$$

$$\mathscr{F}^{\otimes}(M_1 \otimes M_2) = \mathscr{F}^{\otimes}(M_1) \otimes \mathscr{F}^{\otimes}(M_2) \tag{24}$$

$$\mathscr{F}^{\otimes}(\psi_1 \otimes \psi_2) = \mathscr{F}^{\otimes}(\psi_1) \otimes \mathscr{F}^{\otimes}(\psi_2) \tag{25}$$

I call **TENSOR** the axiom system that requires the functor $\mathscr{F}$ to be extendible to a tensor functor $\mathscr{F}^{\otimes}$ between the tensor categories $(\mathfrak{Man}^{\otimes}, hom^{\otimes}_{\mathfrak{Man}})$ and $(\mathfrak{Alg}^{\otimes}, hom^{\otimes}_{\mathfrak{Alg}})$, in addition to the Time Slice axiom.

4 Relation of Axiom Systems

Given the axiom systems **BASIC**, **OPIND**, **OPIND**$^{\times}$, **BASIC+SPLIT** and **TENSOR**, two questions arise:

- What is their logical relation?
- Assuming that they are not equivalent, which one is the most suitable one?

The next few propositions summarize some relations among the axiom systems. In the rest of this section (M_1, g_1), (M_2, g_2) and (M, g) are objects from $\mathfrak{Man}$, ψ_1 and

ψ_2 are morphisms from $hom_{\mathfrak{Man}}$ such that

$$\psi_1 : (M_1, g_1) \to (M, g) \tag{26}$$
$$\psi_2 : (M_2, g_2) \to (M, g) \tag{27}$$

and $\psi_1(M_1)$ and $\psi_2(M_2)$ are spacelike in M.

Proposition 1

$$\textit{BASIC+SPLIT} \Leftrightarrow \textit{TENSOR}$$

Proposition 1 is the combined content of Theorem 1 in [5] and Theorem 2.5 in [6]. The role of the split property and weak additivity in the equivalence claim is to pick out the minimal (also called: spatial) tensor product in the category of C^*-algebras from the other possible tensor products as the suitable one to define the extension of the functor $\mathscr{F}$ to the tensor functor $\mathscr{F}^{\otimes}$.

Proposition 10 in [22] entails that $Op_{\mathfrak{Alg}}$-independence in the product sense of algebras $\mathscr{F}(\psi_1)(\mathscr{F}(M_1, g_1))$ and $\mathscr{F}(\psi_2)(\mathscr{F}(M_1, g_1))$ is equivalent to the condition that the algebra in $\mathscr{F}(M, g)$ generated by algebras $\mathscr{F}(\psi_1)(\mathscr{F}(M_1, g_1))$ and $\mathscr{F}(\psi_2)(\mathscr{F}(M_1, g_1))$ is isomorphic to the tensor product $\mathscr{F}(\psi_1)(\mathscr{F}(M_1, g_1)) \otimes \mathscr{F}(\psi_2)(\mathscr{F}(M_1, g_1))$. So we have

Proposition 2

$$\textit{BASIC+SPLIT} \Leftrightarrow \textit{TENSOR} \Leftarrow \textit{OPIND}^{\times}$$

Does **TENSOR** entail **OPIND**$^{\times}$? Since the tensor product of two operations is again an operation [4] [p. 190], (see also Proposition 9 in [22]), it follows that if $\mathscr{F}$ can be extended to a tensor functor $\mathscr{F}^{\otimes}$, then the algebras $\mathscr{F}(\psi_1)(\mathscr{F}(M_1, g_1))$ and $\mathscr{F}(\psi_2)(\mathscr{F}(M_1, g_1))$ are $Op_{\mathfrak{Alg}}$-independent in the product sense *in the tensor product* $\mathscr{F}(\psi_1)(\mathscr{F}(M_1, g_1)) \otimes \mathscr{F}(\psi_2)(\mathscr{F}(M_1, g_1))$. Although this tensor product algebra is a C^*-subalgebra of $\mathscr{F}(M, g)$, since operations on subalgebras need not be extendible to operations to superalgebras, [3], the $Op_{\mathfrak{Alg}}$-independence of algebras $\mathscr{F}(\psi_1)(\mathscr{F}(M_1, g_1))$ and $\mathscr{F}(\psi_2)(\mathscr{F}(M_1, g_1))$ in the tensor product algebra $\mathscr{F}(\psi_1)(\mathscr{F}(M_1, g_1)) \otimes \mathscr{F}(\psi_2)(\mathscr{F}(M_1, g_1))$ does *not* entail, without further assumptions, the $Op_{\mathfrak{Alg}}$-independence of $\mathscr{F}(\psi_1)(\mathscr{F}(M_1, g_1))$ and $\mathscr{F}(\psi_2)(\mathscr{F}(M_1, g_1))$ in $\mathscr{F}(M, g)$. A C^*-algebra $\mathscr{A}$ is called injective just in case the following holds: Given any C^*-algebra $\mathscr{B}$ and a C^*-subalgebra $\mathscr{B}_0 \subseteq \mathscr{B}$, every completely positive map $T_0 \colon \mathscr{B}_0 \to \mathscr{A}$ has an extension to $\mathscr{B}$ to a completely positive map $T \colon \mathscr{B} \to \mathscr{A}$. Thus **TENSOR** entails **OPIND**$^{\times}$ holds if the local algebras $\mathscr{F}(M, g)$ are injective. In the Haag–Kastler quantum field theory the local algebras associated with double cones can be proved to be hyperfinite hence injective [8], [15] [p. 225], [9], [10] [Theorem 6]. But it is not clear to me whether injectivity of the algebras $\mathscr{F}(M, g)$ can be proved to be a consequence of **BASIC+SPLIT**. Thus the following problem seems to be open:

Problem 1

$$\textbf{BASIC+SPLIT} \Leftrightarrow \textbf{TENSOR} \overset{?}{\not\Rightarrow} \textbf{OPIND}^{\times}$$

OPIND$^{\times}$ obviously entails **OPIND**. I conjecture that the converse is not true. To prove this rigorously one would have to display a model of the axioms such that **OPIND** holds but **OPIND**$^{\times}$ does not. I am not aware of such models; however, this conjecture is supported by the following two facts: (i) C^*-independence is strictly weaker than C^*-independence in the product sense ([24], also see Proposition 1 in [22]); (ii) operational C^*-independence in the product sense is equivalent to C^*-independence in the product sense when the C^*-subalgebras commute (Proposition 10 in [22]).

We have seen in Sect. 3.1 that **BASIC** does not entail **OPIND**. Since **OPIND** entails **BASIC** by definition, the logical relationship of the different axioms can be summarized in the following diagram (question mark ? next to the arrows indicating open questions):

BASIC+SPLIT $\Leftrightarrow$ **TENSOR**

$\Uparrow$ $\Downarrow$?

OPIND$^{\times}$

$\Uparrow$? $\Downarrow$

OPIND

$\Downarrow\not\Uparrow$

BASIC

Assessing the suitability of the different axiom systems, one has to ask which of the axioms has a direct physical interpretation. From this perspective I regard **OPIND** as the most suitable one: $Op_{\mathfrak{Alg}}$-independence has a clear operational physical content. This content is not captured fully by **BASIC**, and both **OPIND**$^{\times}$ and the (possibly equivalent) **TENSOR** seem to require too much: a particular (product) form of independence that does not seem to be justifiable by some specific physical considerations.

What matters ultimately, however, is which of the axiom systems allows a sufficient number of models that describe physically relevant quantum fields. Axiom systems in physics, in quantum field theory in particular, should possess the right balance of two features that are pulling in different directions: restricting their models and, at the same time, being sufficiently non-categorical, allowing a large number of models describing physical systems [21].

Acknowledgements Research supported in part by the Hungarian Scientific Research Found (OTKA). Contract numbers: K-115593 and K100715.

References

1. L. Accardi and C. Cecchini. Conditional expectations in von Neumann algebras and a theorem of Takesaki. *Journal of Functional Analysis*, 45(245-273), 1982.
2. H. Araki. *Mathematical Theory of Quantum Fields*, volume 101 of *International Series of Monograps in Physics*. Oxford University Press, Oxford, 1999. Originally published in Japanese by Iwanami Shoten Publishers, Tokyo, 1993.
3. W. Arveson. Subalgebras of C^*-algebras. *Acta Mathematica*, 123:141–224, 1969.
4. B. Blackadar. *Operator Algebras: Theory of C*-Algebras and von Neumann Algebras*. Encyclopaedia of Mathematical Sciences. Springer, 1. edition, 2005.
5. R. Brunetti and K. Fredenhagen. Quantum field theory on curved backgrounds. In C. Bär and K. Fredenhagen, editors, *Quantum Field Theory on Curved Spacetimes*, volume 786 of *Lecture Notes in Physics*, chapter 5, pages 129–155. Springer, Dordrecht, Heidelberg, London, New York, 2009.
6. R. Brunetti, K. Fredenhagen, I. Paniz, and K. Rejzner. The locality axiom in quantum field theory and tensor products of C^*-algebras. *Reviews of Mathematical Physics*, 26:1450010, 2014. arXiv:1206.5484 [math-ph].
7. R. Brunetti, K. Fredenhagen, and R. Verch. The generally covariant locality principle – a new paradigm for local quantum field theory. *Communications in Mathematical Physics*, 237:31–68, 2003. arXiv:math-ph/0112041.
8. D. Buchholz, C. D'Antoni, and K. Fredenhagen. The universal structure of local algebras. *Commununications in Mathematical Physics*, 111:123–135, 1987.
9. A. Connes. Classification of injective factors. Cases II_1, II_∞, III_λ, $\lambda \neq 1$. *The Annals of Mathematics*, 104:73–115, 1976.
10. A. Connes. *Non-commutative Geometry*. Academic Press, San Diego and New York, 1994.
11. C.J. Fewster. The split property for quantum field theories in flat and curved spacetimes. *Abhandlungen aus dem Mathematischen Seminar der Universität Hamburg*, 86:153–175, 2016. arXiv:1601.06936 [math-ph].
12. C.J. Fewster and R. Verch. Algebraic quantum field theory in curved spacetimes. In R. Brunetti, C. Dappiaggi, K. Fredenhagen, and J. Yngvason, editors, *Advances in Algebraic Quantum Field Theory*, Mathematical Physics Studies, pages 125–189. Springer International Publishing, Switzerland, 2015.
13. K. Fredenhagen and K. Reijzner. Quantum field theory on curved spacetimes: Axiomatic framework and examples. *Journal of Mathematical Physics*, 57:031101, 2016.
14. K. Fredenhagen and K. Rejzner. Local covariance and background independence. In F. Finster, O. Müller, M. Nardmann, J. Tolksdorf, and E. Zeidler, editors, *Quantum Field Theory and Gravity. Conceptual and Mathematical Advances in the Search for a Unified Framework*, pages 15–24. Birkhäuser Springer Basel, Basel, 2012. arXiv:1102.2376 [math-ph].
15. R. Haag. *Local Quantum Physics: Fields, Particles, Algebras*. Springer Verlag, Berlin and New York, 1992.
16. S.S. Horuzhy. *Introduction to Algebraic Quantum Field Theory*. Kluwer Academic Publishers, Dordrecht, 1990.
17. K. Kraus. *States, Effects and Operations*, volume 190 of *Lecture Notes in Physics*. Springer, New York, 1983.
18. I. Paniz. Tensor pruducts of C^*-algebras and independent systems. Master's thesis, Institut für Theoretische Physik, Universität Hamburg, Hamburg, 2012.
19. M. Rédei. Operational independence and operational separability in algebraic quantum mechanics. *Foundations of Physics*, 40:1439–1449, 2010.
20. M. Rédei. A categorial approach to relativistic locality. *Studies in History and Philosophy of Modern Physics*, 48:137–146, 2014.
21. M. Rédei. Hilbert's 6th problem and axiomatic quantum field theory. *Perspectives on Science*, 22:80–97, 2014.
22. M. Rédei and S.J. Summers. When are quantum systems operationally independent? *International Journal of Theoretical Physics*, 49:3250–3261, 2010.

23. M. Rédei and G. Valente. How local are local operations in local quantum field theory? *Studies in History and Philosophy of Modern Physics*, 41:346–353, 2010.
24. S.J. Summers. On the independence of local algebras in quantum field theory. *Reviews in Mathematical Physics*, 2:201–247, 1990.
25. S.J. Summers. Subsystems and independence in relativistic microphysics. *Studies in History and Philosophy of Modern Physics*, 40:133–141, 2009. arXiv:0812.1517 [quant-ph].

Part II
Quantum Information

Reverse Data-Processing Theorems and Computational Second Laws

Francesco Buscemi

Abstract Drawing on an analogy with the second law of thermodynamics for adiabatically isolated systems, Cover argued that data-processing inequalities may be seen as second laws for "computationally isolated systems," namely, systems evolving without an external memory. Here we develop Cover's idea in two ways: on the one hand, we clarify its meaning and formulate it in a general framework able to describe both classical and quantum systems. On the other hand, we prove that also the reverse holds: the validity of data-processing inequalities is not only necessary, but also sufficient to conclude that a system is computationally isolated. This constitutes an information-theoretic analogue of Lieb's and Yngvason's entropy principle. We finally speculate about the possibility of employing Maxwell's demon to show that adiabaticity and memorylessness are in fact connected in a deeper way than what the formal analogy proposed here *prima facie* seems to suggest.

Keywords Second law · Statistical comparison theory · Blackwell theorem Degradability ordering

1 Introduction

Cover, in the attempt to set the second law of thermodynamics in a computational framework, concludes his work with the following suggestive observations [1]:

> The second law of thermodynamics says that uncertainty increases in closed physical systems and that the availability of useful energy decreases. If one can make the concept of "physical information" meaningful, it should be possible to *augment the statement of the second law of thermodynamics with the statement, "useful information becomes less available."* Thus the ability of a physical system to act as a computer should slowly degenerate as the system becomes more amorphous and closer to equilibrium. A perpetual computer should be impossible [emphasis added].

F. Buscemi (✉)
Nagoya University, Nagoya 464-8601, Japan
e-mail: buscemi@is.nagoya-u.ac.jp

M. Ozawa et al. (eds.), *Reality and Measurement in Algebraic Quantum Theory*,
Springer Proceedings in Mathematics & Statistics 261,
https://doi.org/10.1007/978-981-13-2487-1_6

Cover's analysis can be summarized as follows. He first argues, more or less implicitly, that the computational analogue of an adiabatically isolated system should be taken to be a system evolving—i.e., computing—without an external memory. (For this reason, in what follows we use the term "computationally isolated" as a synonym for"memoryless.") This observation leads him to consider stochastic memoryless processes, in particular discrete-time Markov chains. Cover then shows that, while entropy can increase or decrease in this setting, thus violating the thermodynamical second law, *relative entropy* instead never increases. We refer to this statement as Cover's "computational second law.[1]" On a technical side, what Cover proves in [1] is an expression of the monotonicity of the relative entropy under the action of a noisy channel. Thus Cover's second law is in fact a particular *data-processing inequality* [2, 3], and we can imagine that there are as many computational second laws as there are data-processing inequalities, all formalizing the idea that the information content of a system cannot increase without the presence of an external memory.[2]

Cover hence shows that the condition of being memoryless is sufficient for a system to obey data-processing inequalities, i.e., computational second laws. The question we address in this paper concerns the other direction: is it possible to show that the memoryless condition is also necessary for the validity of all data-processing inequalities? Equivalently stated: is it true that a system, if it *is not* computationally isolated, will necessarily violate some data-processing inequality? It is important to address these questions, if we want to understand how far the analogy between memorylessness and adiabaticity can be pushed. Here, in particular, we have in mind Lieb's and Yngvason's formulation of the second law of thermodynamics [7], according to which a non-decreasing entropy is not only necessary but also sufficient for the existence of an adiabatic process connecting two thermodynamical states.[3]

The aim of this paper is to provide a comprehensive framework that is able to answer the above questions. More specifically, we prove here a family of *reverse data-processing theorems*, showing that as soon as a system is not computationally isolated, it must necessarily violate a data-processing inequality. The framework we construct is quite general and it can be applied to classical, quantum, and hybrid classical/quantum systems. In fact, it may even be extended in principle to generalized operational theories as it involves only basic notions like states, effects, and operations; this development is however beyond the scope of the present work.

Thus we are able to strengthen Cover's computational second law in two ways: on the one hand, we give it a converse, in a way that is analogous to what Lieb and Yngvason did the second law of thermodynamics. On the other hand, we include in the analysis the possibility of dealing with quantum systems and quantum memories.

---

[1]Since the entropy of a distribution p is the *negative* of the relative entropy of p with respect to the uniform distribution, it is clear that Cover's computational second law formally constitutes a relaxation of the second law of thermodynamics. Indeed, the former is satisfied in situations violating the latter. We will say more about the relation between thermodynamical and computational second laws in Sect. 6.

[2]Relations between data-processing inequalities, the second law of thermodynamics, and statistical mechanics have been studied also by Merhav in [4–6].

[3]More on this point can be found in Sect. 6.

The paper is organized as follows. We being in Sect. 2 with reviewing the data-processing inequality for a classical Markov chain. This is the encoding–channel–decoding model considered by Shannon to describe the simplest communication scenario. In this scenario we prove our first *reverse data-processing theorem*. We also show how this relates with the theory of comparison of noisy channels, as introduced by Shannon [8] and later developed by Körner and Marton [9]. In Sect. 3 we state and prove a lemma that allows us to extend our considerations to the quantum case, and discuss the notion of *quantum statistical morphisms*. In Sect. 4 we study the case of a system, processing quantum information but outputting only classical data, and prove the corresponding reverse data-processing theorem. Section 5 presents the general case of a fully quantum computer, i.e., a process with quantum inputs and quantum outputs. Finally, in Sect. 6, we briefly discuss about analogies and differences between thermodynamical and computational second laws. In particular, we speculate about the possibility that Maxwell's paradox (his "demon") may enable a deeper relation between adiabatic processes and memoryless processes, going beyond the formal analogy considered in this work. At the end of the paper, three appendices are available: the first, reviewing conventions, notations, and terminology used in this work; the second, containing a version of the minimax theorem; and the third, presenting (just for the sake of completeness) an elementary proof of the separation theorem for convex sets.

This work contains ideas that were presented during the Sixth Nagoya Winter Workshop (NWW2015) held in Nagoya on 9–13 March 2015. Part of the technical results presented here were first introduced in previous papers by the author [10–15], building upon works of Shmaya [16] and Chefles [17].

2 A Reverse-Data Processing Theorem for Classical Channels

A *data-processing inequality* is a mathematical statement formalizing the fact that the information content of a signal cannot be increased by post-processing. As there are many ways to quantify information, so there are many corresponding data-processing inequalities. Such inequalities, however, despite formalizing the same intuitive concept, are not all *logically equivalent*: some may be stronger than (i.e., imply) others, some may be easier to prove, some may be better suited for a particular problem at hand. Data-processing inequalities usually find application in information theory when proving that a given approach (coding strategy) is optimal: if a better coding were possible, that would result in the violation of one or more data-processing inequalities, thus leading to an absurd. In this sense, data-processing inequalities provide a sort of "sanity check" of the result.

One of the simplest scenarios in which a data-processing inequality can be formulated is the following [2, 3]. Given are two noisy channels $w_1 : \mathscr{X} \to \mathscr{Y}$ and $w_2 : \mathscr{Y} \to \mathscr{Z}$. Then, for any set $\mathscr{U}$ and any initial joint distribution $p(x, u)$, the

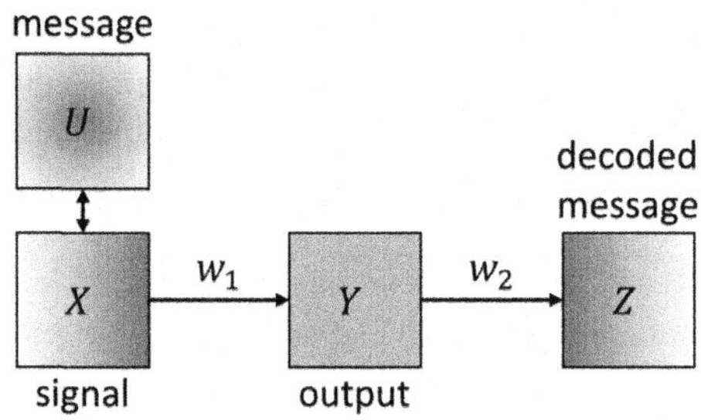

Fig. 1 Shannon's basic communication scheme: a message U is encoded on the signal X (i.e., a joint distribution (U, X) is given), which is transmitted to the receiver via the communication channel w_1. The receiver obtains the output Y and processes it according to the decoding function (another channel w_2) to obtain the recovered message Z

joint distribution $\sum_x w_2(z|y)w_1(y|x)p(x,u)$ satisfies the following inequalities:

$$I(U;Y) \geq I(U;Z) .$$

[Notations and definitions used here and in what follows are given for completeness in Appendix 1.] Referring to the situation depicted in Fig. 1 and interpreting U as the message, X as the signal, w_1 as the communication channel, Y as the output signal, w_2 as the decoding, and Z as the recovered message, the above inequality formalizes the fact that the information content carried by the signal about the message cannot be increased by any decoding performed locally at the receiver. Of course, this does not mean that decoding should be avoided (actually, in most cases a decoding is necessary to make the signal readable to the receiver), but that no decoding is able to add *a posteriori* more information to what is already carried by the signal.

Data-processing inequalities hence provide necessary conditions for the "locality" of the information-processing device. Namely, data-processing inequalities must be obeyed whenever the physical process carrying the message from the sender to the receiver is composed by computationally isolated parts (encoding, transmission, decoding, etc.). Any information that is communicated must be transmitted via a physical signal: as such, in the absence of an external memory, information can only decrease, never increase, along the transmission. Hence,"locality" in this sense can be understood as the condition that the process $U \to X \to Y \to Z$ forms a Markov chain. For this reason, we refer to such locality as "Markov locality," in order to avoid confusion with other connotations of the word.[4]

In this paper we aim to derive statements that provide *sufficient* conditions for Markov locality, in the form of a set of information-theoretic inequalities. We refer to such statements as *reverse data-processing theorems*. For example, a first attempt in this direction would be to prove the following:

[4]In this work, memoryless process, Markov local process, and computationally isolated process are all synonyms. We prefer however to maintain all three terms because they in fact highlight different aspects of the same information-theoretic concept.

> Given are two noisy channels $w : \mathscr{X} \to \mathscr{Y}$ and $w' : \mathscr{X} \to \mathscr{Z}$. Suppose that, for any set $\mathscr{U}$ and for any initial joint distribution $p(x, u)$, the resulting distributions $\sum_x w(y|x)p(x, u)$ and $\sum_x w'(z|x)p(x, u)$ always satisfy the inequality $I(U; Y) \geq I(U; Z)$. Then there exists a noisy channel $\varphi : \mathscr{Y} \to \mathscr{Z}$ such that $w'(z|x) = \sum_y \varphi(z|y)w(y|x)$.

Notice that, in the above statement, the two given channels w and w' are assumed to have the same input alphabet: this is a consequence of the fact that we are now formulating a *reverse* data-processing theorem, so that the existence of a Markov-local decoding (the channel φ) is something to be proved, rather than being a datum. Interpreting the four random variables (U, X, Y, Z) as before, if the reverse data-processing theorem holds, then we can conclude that *any violation of Markov locality* is detectable, in the precise sense that the data-processing inequality has to be violated at some point along the communication process.

2.1 Comparison of Noisy Channels

A reverse data-processing theorem can be understood as a statement about the comparison of two noisy channels. Hence we want to introduce ordering relations between noisy channels, capturing the idea that one channel is able to transmit "more information" than another. This problem, first considered by Shannon [8], is intimately related to the theory of statistical comparisons [18–21], even though this connection was not made until recently [22]. The theory of comparison of noisy channels received a thorough treatment by Körner and Marton, who in Ref. [9] introduce the following definitions (the notation used here follows [23]):

Definition 1 Given are two noisy channels, $w : \mathscr{X} \to \mathscr{Y}$ and $w' : \mathscr{X} \to \mathscr{Z}$.

(i) the channel w is said to be *less noisy* than w' if and only if, for any set $\mathscr{U}$ and any joint distribution $p(x, u)$, the resulting distributions $\sum_x w(y|x)p(x, u)$ and $\sum_x w'(z|x)p(x, u)$ always satisfy the inequality

$$H(U|Y) \leq H(U|Z) \; ; \tag{1}$$

(ii) the channel w is said to be *degradable* into w' if and only if there exists another channel $\varphi : \mathscr{Y} \to \mathscr{Z}$ such that

$$w'(z|x) = \sum_y \varphi(z|y)w(y|x) \; . \tag{2}$$

□

Since $I(U; Y) \geq I(U; Z)$ if and only if $H(U|Y) \leq H(U|Z)$, we immediately notice that the reverse data-processing theorem, as tentatively formulated above, is equivalent to the implication (i) $\Longrightarrow$ (ii): indeed, the reverse implication, (ii) $\Longrightarrow$ (i), is the usual data-processing inequality. Körner and Marton provide an explicit counterexample showing that

$$\text{degradable} \underset{\not\Leftarrow}{\Longrightarrow} \text{less noisy} . \qquad (3)$$

This means that, if a reverse data-processing theorem holds, it must be formulated differently.

2.2 *Replacing H with* $H_{\min}$

Even though we know that "less noisy" does not imply"degradable," in what follows we show that just a slight formal modification in the definition of "less noisy, " Eq. (1), is enough to obtain the sought-after reverse data-processing theorem. Such a slight modification consists in replacing, in point (i) of Definition 1, the Shannon conditional entropy $H(\cdot|\cdot)$ with the conditional min-entropy $H_{\min}(\cdot|\cdot)$.

Theorem 1 *Given are two noisy channels* $w : \mathscr{X} \to \mathscr{Y}$ *and* $w' : \mathscr{X} \to \mathscr{Z}$. *The following are equivalent:*

(i) for any set $\mathscr{U}$ *and for any initial joint distribution* $p(x, u)$*, the resulting distributions* $\sum_x w(y|x)p(x, u)$ *and* $\sum_x w'(z|x)p(x, u)$ *always satisfy the inequality*

$$H_{\min}(U|Y) \le H_{\min}(U|Z) ; \qquad (4)$$

(ii) w *is degradable into* w'*, namely, there exists another channel* $\varphi : \mathscr{Y} \to \mathscr{Z}$ *such that* $w'(z|x) = \sum_y \varphi(z|y)w(y|x)$.

Proof (ii) $\Longrightarrow$ (i) is a direct consequence of the data-processing inequality for $H_{\min}$. Suppose that there exists another conditional probability distribution $\varphi(z|y)$ such that $w'(z|x) = \sum_y \varphi(z|y)w(y|x)$. This means that the random variable Z is obtained locally from Y, i.e., the four random variables (U, X, Y, Z) form a Markov chain $U \to X \to Y \to Z$. This implies that (4) holds.

In order to prove (i) $\Longrightarrow$ (ii), let us assume that the inequality in (4) holds for any initial joint distribution $p(x, u)$. Exponentiating both sides, and using Eq. (59), this is equivalent to

$$P_{\text{guess}}(U|Y) \ge P_{\text{guess}}(U|Z) , \qquad (5)$$

namely,

$$\max_{\varphi} \sum_{u,y,x} \varphi(u|y)w(y|x)p(x, u) \ge \max_{\varphi'} \sum_{u,z,x} \varphi'(u|z)w'(z|x)p(x, u) , \qquad (6)$$

for all choices of $p(x, u)$. In the above equation, the noisy channels φ and φ' represent the decision functions that the statistician designs in order to optimally guess the value of U.

Let us choose U such that its support coincides with that of Z, i.e., $\mathscr{U} \equiv \mathscr{Z}$. We can therefore denote its states by z'. Let us also fix the guessing strategy on the right-hand side of (6) to be $\varphi'(z'|z) \equiv \delta_{z',z}$, i.e., 1 if $z' = z$ and 0 otherwise. Then, we know that there exists a decision function $\varphi(z'|y)$ such that

$$0 \geq \sum_{z',z,x} \delta_{z',z} w'(z|x) p(x, z') - \sum_{z',y,x} \varphi(z'|y) w(y|x) p(x, z') \tag{7}$$

$$= \sum_{z',x} w'(z'|x) p(x, z') - \sum_{z',y,x} \varphi(z'|y) w(y|x) p(x, z') \tag{8}$$

$$= \sum_{z',x} \left[w'(z'|x) p(x, z') - \sum_{y} \varphi(z'|y) w(y|x) p(x, z') \right] \tag{9}$$

$$= \sum_{z',x} \left[w'(z'|x) - \sum_{y} \varphi(z'|y) w(y|x) \right] p(x, z') \, . \tag{10}$$

In other words, for any $p(x, z')$, there exists a $\varphi(z'|y)$ such that the above inequality holds. This is equivalent to say that

$$\max_{p} \min_{\varphi} \sum_{z',x} \left[w'(z'|x) - \sum_{y} \varphi(z'|y) w(y|x) \right] p(x, z') \leq 0 \, . \tag{11}$$

We now invoke the minimax theorem (in the form reported in Appendix 2, Theorem 4) and exchange the order of the two optimizations:

$$\min_{\varphi} \max_{p} \sum_{z',x} \left[w'(z'|x) - \sum_{y} \varphi(z'|y) w(y|x) \right] p(x, z') \leq 0 \, . \tag{12}$$

Let us now introduce the quantity

$$\Delta_{\varphi}(z', x) \triangleq w'(z'|x) - \sum_{y} \varphi(z'|y) w(y|x) \, . \tag{13}$$

First of all, we notice that the maximum in Eq. (12) is reached when the distribution $p(x, z')$ is entirely concentrated on an entry where $\Delta_{\varphi}(z', x)$ is maximum, that is,

$$0 \geq \min_{\varphi} \max_{p} \sum_{z',x} \left[w'(z'|x) - \sum_{y} \varphi(z'|y) w(y|x) \right] p(x, z') \tag{14}$$

$$= \min_{\varphi} \max_{z',x} \Delta_{\varphi}(z'x) \,. \tag{15}$$

In general, $\Delta_{\varphi}(z', x)$ does not have a definite sign, however, since $\sum_{z',x} \Delta_{\varphi}(z', x) = 0$ (as a consequence of the normalization of probabilities), it must be that $\max_{z',x} \Delta_{\varphi}(z', x) \geq 0$ (otherwise, of course, one would have $\sum_{z',x} \Delta_{\varphi}(z', x) < 0$). The above inequality hence implies that $\min_{\varphi} \max_{z',x} \Delta_{\varphi}(z', x) = 0$. In turns this implies, again because $\sum_{z',x} \Delta_{\varphi}(z', x) = 0$, that $\Delta_{\varphi}(z', x) = 0$ for all z' and x. In other words, we showed that there exists a $\varphi(z'|y)$ such that

$$w'(z'|x) = \sum_{y} \varphi(z'|y) w(y|x), \tag{16}$$

for all z', x, which coincides with the definition of degradability. □

Remark 1 From the proof we see that in point (ii) of Theorem 1 it is possible to restrict, without loss of generality, the random variable U to be supported on the set $\mathscr{Z}$, i.e., the same supporting the output of w'. □

3 The Fundamental Lemma for Quantum Channels

The following lemma plays a crucial role in the derivation of reverse data-processing theorems valid in the quantum case.

Lemma 1 *Let $\Phi_A : \mathsf{L}(\mathscr{H}_A) \to \mathsf{L}(\mathscr{H}_B)$ and $\Phi'_A : \mathsf{L}(\mathscr{H}_A) \to \mathsf{L}(\mathscr{H}_{B'})$ be two quantum channels. For any set $\mathscr{U} = \{u\}$, the following are equivalent:*

(i) for all ensembles $\{p(u); \omega_A^u\}$,

$$P_{\text{guess}}\{p(u); \Phi_A(\omega_A^u)\} \geq P_{\text{guess}}\{p(u); \Phi'_A(\omega_A^u)\} \,; \tag{17}$$

(ii) for any POVM $\{Q_{B'}^u\}$, there exists a POVM $\{P_B^u\}$ such that

$$\mathrm{Tr}\big[\Phi'_A(\omega_A)\, Q_{B'}^u\big] = \mathrm{Tr}\big[\Phi_A(\omega_A)\, P_B^u\big] \,, \tag{18}$$

for all $u \in \mathscr{U}$ and all $\omega_A \in \mathsf{D}(\mathscr{H}_A)$.

Proof The fact that (ii) implies (i) follows by definition of guessing probability. We therefore prove the converse, namely, that (i) implies (ii).

Let us rewrite condition (17) explicitly as follows: for all ensembles $\{p(u); \omega_A^u\}$,

$$\max_P \sum_u p(u)\mathrm{Tr}\left[\Phi_A(\omega_A^u)\ P_B^u\right] \geq \max_Q \sum_u p(u)\mathrm{Tr}\left[\Phi'_A(\omega_A^u)\ Q_{B'}^u\right] , \tag{19}$$

where the maxima are taken over all possible POVMs. Introduce now an auxiliary Hilbert space $\mathscr{H}_R \cong \mathscr{H}_A$, and denote by ϕ_{RA}^+ a fixed maximally entangled in $\mathsf{D}(\mathscr{H}_R \otimes \mathscr{H}_A)$. Construct then the Choi operators corresponding to channels Φ and Φ', namely,

$$\chi_{RB} \triangleq (\mathrm{id}_R \otimes \Phi_A)\phi_{RA}^+ \qquad \text{and} \qquad \chi'_{RB'} \triangleq (\mathrm{id}_R \otimes \Phi'_A)\phi_{RA}^+ . \tag{20}$$

Noticing that, for any ensemble $\{p(u); \omega_A^u\}$ with $\sum_u p(u)\omega_A^u = I_A/d_A$, there exists a POVM $\{E_R^u\}$ such that $p(u)\omega_A^u = \mathrm{Tr}_R\left[\phi_{RA}^+\ (E_R^u \otimes I_A)\right]$, we immediately see that, if condition (19) above holds, then, for any POVM $\{E_R^u\}$,

$$\max_P \sum_u \mathrm{Tr}\left[\chi_{RB}\ (E_R^u \otimes P_B^u)\right] \geq \max_Q \sum_u \mathrm{Tr}\left[\chi'_{RB'}\ (E_R^u \otimes Q_{B'}^u)\right] . \tag{21}$$

We now prove that condition (21) above in turns implies that, for any collection of Hermitian operators $\{O_R^u\}$,

$$\max_P \sum_u \mathrm{Tr}\left[\chi_{RB}\ (O_R^u \otimes P_B^u)\right] \geq \max_Q \sum_u \mathrm{Tr}\left[\chi'_{RB'}\ (O_R^u \otimes Q_{B'}^u)\right] . \tag{22}$$

The crucial observation here is that, given a collection of Hermitian operators $\{O_R^u\}$, we can always derive from it a POVM $\{E_R^u\}$ given by

$$E_R^u \triangleq \frac{1}{\alpha|\mathscr{U}|}\left\{O_R^u + \alpha I_R - \frac{1}{|\mathscr{U}|}\Sigma_R\right\} , \tag{23}$$

with $\Sigma_R \triangleq \sum_u O_R^u$ and $\alpha > 0$ sufficiently large so that $O_R^u + \alpha I_R - |\mathscr{U}|^{-1}\Sigma_R$ is nonnegative for all u. Therefore, assuming that inequality (21) holds for any POVM $\{E_R^u\}$, we have that

$$\begin{aligned}
&\max_P \sum_u \mathrm{Tr}\left[\chi_{RB}\ (O_R^u \otimes P_B^u)\right] \\
&= \alpha|\mathscr{U}| \max_P \sum_u \mathrm{Tr}\left[\chi_{RB}\ (E_R^u \otimes P_B^u)\right] - \alpha \mathrm{Tr}[\chi_{RB}] + \frac{1}{|\mathscr{U}|}\mathrm{Tr}[\chi_{RB}\ (\Sigma_R \otimes I_B)] \\
&= \alpha|\mathscr{U}| \max_P \sum_u \mathrm{Tr}\left[\chi_{RB}\ (E_R^u \otimes P_B^u)\right] - \alpha + \frac{1}{|\mathscr{U}|}\mathrm{Tr}[\mathrm{Tr}_B[\chi_{RB}]\ \Sigma_R] \\
&\geq \alpha|\mathscr{U}| \max_Q \sum_u \mathrm{Tr}\left[\chi'_{RB'}\ (E_R^u \otimes Q_{B'}^u)\right] - \alpha + \frac{1}{|\mathscr{U}|}\mathrm{Tr}\left[\mathrm{Tr}_{B'}\left[\chi'_{RB'}\right]\ \Sigma_R\right] \qquad (24) \\
&= \max_Q \sum_u \mathrm{Tr}\left[\chi'_{RB'}\ (O_R^u \otimes Q_{B'}^u)\right],
\end{aligned}$$

for any collection of Hermitian operators $\{O_R^u\}$. Inequality (24) above is a consequence of condition (21) together with the identity $\mathrm{Tr}_B[\chi_{RB}] = \mathrm{Tr}_{B'}\big[\chi'_{RB'}\big] = I_R/d_R$. Hence we showed that condition (22) holds if condition (21) holds, even though the former looks at first sight more general than the latter. The vice versa is true simply because any POVM is, in particular, a family of Hermitian operators.

Let us now denote by $\mathscr{L}(\mathscr{U})$ the set of operator tuples

$$\mathbf{a} \equiv \left(a^u : u \in \mathscr{U}\right) , \qquad a^u \in \mathsf{L}_H(\mathscr{H}_R) , \tag{25}$$

with inner product

$$\mathbf{a} \cdot \mathbf{b} \triangleq \sum_u \mathrm{Tr}\left[a^u b^u\right] . \tag{26}$$

We then define $\mathscr{C}(\chi; \mathscr{U})$ as the convex subset of $\mathscr{L}(\mathscr{U})$ containing tuples $\mathbf{b}$ such that $b^u \triangleq \mathrm{Tr}_B\big[\chi_{RB}\,(I_R \otimes P_B^u)\big]$, for varying POVM $\{P_B^u\}$. [The fact that $\mathscr{C}(\chi; \mathscr{U})$ is convex is a direct consequence of the fact that the set of POVMs supported on $\mathscr{U}$ is convex.] In the same way, we also define $\mathscr{C}'(\chi'; \mathscr{U})$. For the sake of simplicity of notation, when no confusion arises, we simply denote $\mathscr{C}(\chi; \mathscr{U})$ as $\mathscr{C}$ and $\mathscr{C}'(\chi'; \mathscr{U})$ as $\mathscr{C}'$. Using this notation, condition (22) becomes

$$\max_{\mathbf{b}\in\mathscr{C}} \mathbf{a} \cdot \mathbf{b} \geq \max_{\mathbf{b}'\in\mathscr{C}'} \mathbf{a} \cdot \mathbf{b}' , \tag{27}$$

for all $\mathbf{a} \in \mathscr{L}(\mathscr{U})$. [Here $a^u = O_R^u$.]

Hence, we turned the initial conditions involving guessing probabilities into a family of linear constraints on two convex sets, $\mathscr{C}$ and $\mathscr{C}'$. Then, a direct application of the separation theorem for convex sets (see Corollary 2 in Appendix 3), leads us to conclude that

$$\mathscr{C}(\chi; \mathscr{U}) \supseteq \mathscr{C}'(\chi'; \mathscr{U}) . \tag{28}$$

In other words, condition (17) in the statement of the lemma implies that, for any POVM $\{Q_{B'}^u\}$, there exists a POVM $\{P_B^u\}$ such that

$$\mathrm{Tr}_B\big[\chi_{RB}\,(I_R \otimes P_B^u)\big] = \mathrm{Tr}_{B'}\big[\chi'_{RB'}\,(I_R \otimes Q_{B'}^u)\big] , \tag{29}$$

for all $u \in \mathscr{U}$.

The final step consists in noticing that any state ω_A can be written as $\mathrm{Tr}_R\big[\phi^+_{RA}\,(E_R \otimes I_A)\big]$ for some $E_R \in \mathsf{L}_+(\mathscr{H}_R)$. Therefore, multiplying both sides of (29) by E_R and taking the trace, we obtain

$$\mathrm{Tr}\big[\Phi_A(\omega_A)\,P_B^u\big] = \mathrm{Tr}\big[\chi_{RB}\,(E_R \otimes P_B^u)\big] \tag{30}$$

$$= \mathrm{Tr}\big[\chi'_{RB'}\,(E_R \otimes Q_{B'}^u)\big] \tag{31}$$

$$= \mathrm{Tr}\big[\Phi'_A(\omega_A)\,Q_{B'}^u\big] , \tag{32}$$

which of course holds for any choice of E_R, that is, ω_A, as claimed. □

Remark 2 As explained in the paragraph following Eq. (20), the above proof shows that, in particular, the ensembles $\{p(u); \omega_A^u\}$ in point (i) can be restricted, without loss of generality, to ensembles with maximally mixed average, i.e., $\sum_u p(u)\omega_A^u \propto I_A$. □

Remark 3 We notice that point (ii) can be alternatively formulated as follows: for any POVM $\{Q_{B'}^u\}$, there exists a POVM $\{P_B^u\}$ such that

$$\left(\Phi'\right)^\dagger \left(Q_{B'}^u\right) = \Phi^\dagger \left(P_B^u\right) , \tag{33}$$

for all $u \in \mathscr{U}$, where $\Phi^\dagger$ denotes the trace-dual defined in Eq. (55). □

3.1 Quantum Statistical Morphisms

Let us now choose the set $\mathscr{U}$ in Lemma 1 so that its size $|\mathscr{U}|$ is equal to $(\dim \mathscr{H}_{B'})^2$. Assuming that channels Φ and Φ' actually satisfy either (17) or (18), let us set the POVM $\{Q_{B'}^u\}$ to be informationally complete, that is, $\mathrm{span}\{Q_{B'}^y\} = \mathsf{L}(\mathscr{H}_{B'})$. Then, if $\{P_B^u\}$ is any POVM satisfying the equality (33) in Remark 3, the relation

$$Q_{B'}^y \longmapsto P_B^y , \qquad y \in \mathscr{Y} , \tag{34}$$

can be used to define a linear map $\Gamma : \mathsf{L}(\mathscr{H}_B) \to \mathsf{L}(\mathscr{H}_{B'})$ with the following properties:

1. let $\{\Xi_{B'}^y\}$ be the unique dual of $\{Q_{B'}^y\}$, in the sense that $X_{B'} = \sum_y \mathrm{Tr}\left[Q_{B'}^y\, X_{B'}\right] \Xi_{B'}^y$, for all $X_{B'} \in \mathsf{L}(\mathscr{H}_{B'})$; then the action of Γ is given by $\Gamma(\cdot) = \sum_y \mathrm{Tr}\left[P_B^y \cdot\right]\, \Xi_{B'}^y$;
2. Γ is Hermiticity-preserving, i.e., $X = X^\dagger$ implies that $\Gamma(X) = [\Gamma(X)]^\dagger$;
3. Γ is trace-preserving;
4. $\Phi' = \Gamma \circ \Phi$.

In particular, the map Γ, as defined above, is positive and trace-preserving on the output (meant as the whole linear range) of Φ. In order to prove this, let $X_A \in \mathsf{L}(\mathscr{H}_A)$ be any operator such that $\Phi_A(X_A) \geq 0$. (Notice that X_A need not be positive itself.) Then $\Gamma_B(\Phi_A(X_A)) \geq 0$. This is because $\Gamma \circ \Phi = \Phi'$ and we know, from Eq. (18), that for any positive operator $Q_{B'}$ there exists a positive operator P_B such that $\mathrm{Tr}[Q_{B'}\, \Gamma_B(\Phi_A(X_A))] = \mathrm{Tr}\left[Q_{B'}\, \Phi'_A(X_A)\right] = \mathrm{Tr}[P_B\, \Phi_A(X_A)]$. Hence, we know that for any positive operator $Q_{B'}$, $\mathrm{Tr}[Q_{B'}\, \Gamma_B(\Phi_A(X_A))] \geq 0$ whenever $\Phi_A(X_A) \geq 0$, which is the definition of positivity.

Following the terminology of [24, 25], the following definition was introduced in [10]:

Definition 2 Given a channel $\Phi : \mathsf{L}(\mathscr{H}_A) \to \mathsf{L}(\mathscr{H}_B)$, a linear map $\Gamma : \mathsf{L}(\mathscr{H}_B) \to \mathsf{L}(\mathscr{H}_C)$ is said to be a *quantum statistical morphism* of Φ if and only if, for any state ω_A and any POVM $\{Q_C^y\}$, there exists a POVM $\{P_B^y\}$ such that

$$\mathrm{Tr}\big[(\Gamma_B \circ \Phi_A)(\omega_A)\ Q_C^y\big] = \mathrm{Tr}\big[\Phi_A(\omega_A)\ P_B^y\big]\ , \tag{35}$$

for all y. □

It is easy to verify that an everywhere positive trace-preserving linear map is always a statistical morphisms for *any channel*, as long as the composition between the two is well defined. Then, the natural question is whether a linear map defined as Γ above can always be extended to become positive and trace-preserving *everywhere*, not only on the range of Φ. The question was answered in the negative by Matsumoto, who gave an explicit counterexample in Ref. [26].

Vice versa, one may ask whether any linear map that is positive and trace-preserving on the range of Φ is a well-defined statistical morphism of Φ or not. Also in this case, the answer is in the negative: the fact that condition (35) must hold *for any* POVM (in particular, for any number of outcomes) is strictly stronger than just positivity, for which is enough if condition (35) holds *only for* two-outcome POVMs.

Statistical morphisms hence lie somewhere in between linear maps that are positive and trace-preserving (PTP) everywhere, and those that are so only on the range of Φ:

$$\text{PTP everywhere} \underset{\not\Leftarrow}{\Longrightarrow} \text{stat. morph. of } \Phi \underset{\not\Leftarrow}{\Longrightarrow} \text{PTP on range}(\Phi)\ . \tag{36}$$

We summarize the contents of this section in one definition and one corollary.

Definition 3 Given are two quantum channels $\Phi : \mathsf{L}(\mathscr{H}_A) \to \mathsf{L}(\mathscr{H}_B)$ and $\Phi' : \mathsf{L}(\mathscr{H}_A) \to \mathsf{L}(\mathscr{H}_{B'})$. For a given set $\mathscr{U}$, we say that Φ is *$\mathscr{U}$-sufficient* for Φ', in formula,

$$\Phi \succeq_{\mathscr{U}} \Phi'\ , \tag{37}$$

if and only if either of the conditions in Lemma 1 hold. □

Corollary 1 *Given are two quantum channels $\Phi : \mathsf{L}(\mathscr{H}_A) \to \mathsf{L}(\mathscr{H}_B)$ and $\Phi' : \mathsf{L}(\mathscr{H}_A) \to \mathsf{L}(\mathscr{H}_{B'})$. The following are equivalent:*

(i) $\Phi \succeq_{\mathscr{U}} \Phi'$, *for any set* $\mathscr{U}$;
(ii) *there exists a quantum statistical morphism* $\Gamma : \mathsf{L}(\mathscr{H}_B) \to \mathsf{L}(\mathscr{H}_{B'})$ *of* Φ *such that* $\Phi' = \Gamma \circ \Phi$.

Remark 4 Using the correspondence between ensembles and bipartite states, together with the relation between guessing probability and conditional min-entropy, given in Appendix 1 in Eqs. (54) and (60), we notice that the condition $\Phi \succeq_{\mathscr{U}} \Phi'$ can be equivalently written as

$$H_{\min}(U|B) \leq H_{\min}(U|B'), \tag{38}$$

where the entropies are computed with respect to states $(\mathrm{id}_U \otimes \Phi_A)(\omega_{UA})$ and $(\mathrm{id}_U \otimes \Phi'_A)(\omega_{UA})$, respectively. This is equivalent to the formulation used in Theorem 1. □

4 A Semiclassical (Semiquantum) Reverse-Data Processing Theorem

We consider in this section the case in which the output of a quantum channel is classical, in the precise sense that the range is supported on a commutative subalgebra.

Theorem 2 *Given are two quantum channels $\Phi : \mathsf{L}(\mathscr{H}_A) \to \mathsf{L}(\mathscr{H}_B)$ and $\Phi' : \mathsf{L}(\mathscr{H}_A) \to \mathsf{L}(\mathscr{H}_{B'})$. Assuming that the output of Φ' is classical, i.e.,*

$$[\Phi'(X), \Phi'(Y)] = 0, \qquad \forall X, Y \in \mathsf{L}(\mathscr{H}_A), \tag{39}$$

the following are equivalent:

(i) $\Phi \succeq_{\mathscr{U}} \Phi'$, for any set $\mathscr{U}$;
(ii) $\Phi \succeq_{\mathscr{U}} \Phi'$, for a set $\mathscr{U}$ such that $|\mathscr{U}| = \dim \mathscr{H}_{B'}$;
(iii) there exists a quantum channel $\Psi : \mathsf{L}(\mathscr{H}_B) \to \mathsf{L}(\mathscr{H}_{B'})$ such that $\Phi' = \Psi \circ \Phi$.

Proof Since the implications (iii) $\Longrightarrow$ (i) $\Longrightarrow$ (ii) are either trivial of direct consequence of the data-processing inequality for the guessing probability, we only prove the implication (ii) $\Longrightarrow$ (iii).

Since $|\mathscr{U}| = \dim \mathscr{H}_{B'}$, we can use the elements $u \in \mathscr{U}$ to label an orthonormal basis $\{|u\rangle : u \in \mathscr{U}\}$ of $\mathscr{H}_{B'}$. Assuming (ii), we know from Lemma 1 that (18) holds, so, in particular, we know that there exists a POVM $\{P_B^u\}$ such that

$$\mathrm{Tr}\big[\Phi'_A(\omega_A)\ |u\rangle\langle u|_{B'}\big] = \mathrm{Tr}\big[\Phi(\omega_A)\ P_B^u\big]\ , \tag{40}$$

for all u and all $\omega_A \in \mathsf{D}(\mathscr{H}_A)$.

We now use the fact that the output of Φ' is classical and assume that any operator in the range of Φ' can be diagonalized on the basis $\{|u\rangle\}$. This means that

$$\Phi'_A(\cdot) = \sum_{u \in \mathscr{U}} \mathrm{Tr}\big[\Phi'_A(\cdot)\ |u\rangle\langle u|_{B'}\big]\, |u\rangle\langle u|_{B'} \ . \tag{41}$$

Using Eq. (40), and defining a measure-and-prepare channel $\Psi : \mathsf{L}(\mathscr{H}_B) \to \mathsf{L}(\mathscr{H}_{B'})$ by the relation

$$\Psi(\cdot) \triangleq \sum_u \mathrm{Tr}\big[\cdot\ P_B^u\big]\, |u\rangle\langle u|_{B'} \ , \tag{42}$$

we finally have that $\Phi' = \Psi \circ \Phi$. □

Remark 5 In order to highlight the perfect analogy with Theorem 1, we recall that the relation between guessing probability and conditional min-entropy (see Appendix 1) allows us to rewrite points (i) and (ii) of Theorem 2 as:

$$H_{\min}(U|B) \le H_{\min}(U|B') \ . \tag{43}$$

See also Remark 4 above. □

Remark 6 It is possible to show that Theorem 1 becomes a corollary of Theorem 2. Consider in fact the situation in which both Φ and Φ' are *classical-quantum channels*, namely, $\Phi : \mathscr{X} \to \mathsf{L}(\mathscr{H}_B)$ and $\Phi' : \mathscr{X} \to \mathsf{L}(\mathscr{H}_{B'})$, with $\Phi(x) \triangleq \rho_B^x \in \mathsf{D}(\mathscr{H}_B)$ and $\Phi'(x) \triangleq \sigma_{B'}^x \in \mathsf{D}(\mathscr{H}_{B'})$. Assume moreover that $[\rho^x, \rho^{x'}] = 0$ and $[\sigma^x, \sigma^{x'}] = 0$, for all $x, x' \in \mathscr{X}$. We are hence in a scenario much more restricted than that of Theorem 2: in fact, by identifying commuting states with the probability distributions of their eigenvalues, we recover the classical framework and the statement of Theorem 1. □

Remark 7 Theorem 1, the classical reverse data-processing inequality, has thus two different proofs: one using the minimax theorem and another using the separation theorem for convex sets. Despite the fact that minimax theorem and separation theorem are ultimately equivalent [27], the minimax theorem allows for an easier treatment of the approximate case, which is a very relevant point but goes beyond the scope of the present contribution. The interested reader may refer to Refs. [15, 28]. □

5 A Fully Quantum Reverse Data-Processing Theorem

We consider in this section the case of two completely general quantum channels, with the only restriction that the input space is the same for both.

Theorem 3 *Given are two quantum channels, $\Phi : \mathsf{L}(\mathscr{H}_A) \to \mathsf{L}(\mathscr{H}_B)$ and $\Phi' : \mathsf{L}(\mathscr{H}_A) \to \mathsf{L}(\mathscr{H}_{B'})$, and an auxiliary Hilbert space $\mathscr{H}_{B''} \cong \mathscr{H}_{B'}$. The following are equivalent:*

(i) $\mathrm{id}_{B''} \otimes \Phi_A \succeq_{\mathscr{U}} \mathrm{id}_{B''} \otimes \Phi'_A$, *for any set* $\mathscr{U}$;
(ii) $\mathrm{id}_{B''} \otimes \Phi_A \succeq_{\mathscr{U}} \mathrm{id}_{B''} \otimes \Phi'_A$, *for a set* $\mathscr{U}$ *such that* $|\mathscr{U}| = \dim(\mathscr{H}_{B''} \otimes \mathscr{H}_{B'}) = (\dim \mathscr{H}_{B'})^2$;
(iii) *there exists a quantum channel* $\Psi : \mathsf{L}(\mathscr{H}_B) \to \mathsf{L}(\mathscr{H}_{B'})$ *such that* $\Phi' = \Psi \circ \Phi$.

Remark 8 In terms of the conditional min-entropy, points (i) and (ii) above can be written as

$$H_{\min}(U|B''B) \leq H_{\min}(U|B''B') , \tag{44}$$

with obvious meaning of symbols. See also Remarks 4 and 5 above. □

Proof Since the implication (iii) $\Longrightarrow$ (i) $\Longrightarrow$ (ii) is straightforward, we prove here only that (ii) $\Longrightarrow$ (iii).

Let $\mathscr{H}_{B'''}$ be a further auxiliary Hilbert space such that $\mathscr{H}_{B'''} \cong \mathscr{H}_{B''} \cong \mathscr{H}_{B'}$. We begin by showing that, if $\mathrm{id}_{B''} \otimes \Phi_A \succeq_{\mathscr{U}} \mathrm{id}_{B''} \otimes \Phi'_A$, then, for any POVM $\{Q^u_{B''B'}\}$, there exists a POVM $\{P^u_{B''B}\}$ such that

$$\begin{aligned} &\mathrm{Tr}_{B''B'}\left[(\phi^+_{B'''B''} \otimes \Phi'_A(\cdot))\,(I_{B'''} \otimes Q^u_{B''B'})\right] \\ &= \mathrm{Tr}_{B''B}\left[(\phi^+_{B'''B''} \otimes \Phi_A(\cdot))\,(I_{B'''} \otimes P^u_{B''B})\right] , \end{aligned} \tag{45}$$

where $\phi^+_{B'''B''}$ is a maximally entangled state in $\mathscr{H}_{B'''} \otimes \mathscr{H}_{B''}$. In fact, Lemma 1 states that, for any POVM $\{Q^u_{B''B'}\}$, there exists a POVM $\{P^u_{B''B}\}$ such that

$$\mathrm{Tr}\left[(\mathrm{id}_{B''} \otimes \Phi'_A)(\cdot_{B''A})\; Q^u_{B''B'}\right] = \mathrm{Tr}\left[(\mathrm{id}_{B''} \otimes \Phi_A)(\cdot_{B''A})\; P^u_{B''B}\right] , \tag{46}$$

for all $u \in \mathscr{U}$. In particular, for any family of states $\{\xi^x_{B''}\}_x$ on $\mathscr{H}_{B''}$, we have

$$\mathrm{Tr}\left[(\mathrm{id}_{B''} \otimes \Phi'_A)(\xi^x_{B''} \otimes \cdot_A)\; Q^u_{B''B'}\right] = \mathrm{Tr}\left[(\mathrm{id}_{B''} \otimes \Phi_A)(\xi^x_{B''} \otimes \cdot_A)\; P^u_{B''B}\right] , \tag{47}$$

for all u and all x. Let us choose $\xi^x_{B''} = \mathrm{Tr}_{B'''}\left[\phi^+_{B'''B''}\,(\Xi^x_{B'''} \otimes I_{B''})\right]$ for some complete set of positive operators $\{\Xi^x_{B'''}\}_x$. Hence Eq. (47) becomes

$$\mathrm{Tr}\left[(\mathrm{id}_{B'''} \otimes \mathrm{id}_{B''} \otimes \Phi'_A)(\phi^+_{B'''B''} \otimes \cdot_A)\;(\Xi^x_{B'''} \otimes Q^u_{B''B'})\right] \tag{48}$$

$$= \mathrm{Tr}\left[(\mathrm{id}_{B'''} \otimes \mathrm{id}_{B''} \otimes \Phi_A)(\phi^+_{B'''B''} \otimes \cdot_A)\;(\Xi^x_{B'''} \otimes P^u_{B''B})\right] , \tag{49}$$

for all u and all x. But since the family $\{\Xi^x_{B'''}\}_x$ has been chosen to be complete, the above equality implies the equality of the operators in Eq. (45).

Now, we can use generalized teleportation and show that

$$\Phi'_A(\cdot) = \sum_u W^u_{B'''}\left\{\mathrm{Tr}_{B''B'}\left[(\phi^+_{B'''B''} \otimes \Phi'_A(\cdot))\,(I_{B'''} \otimes \beta^u_{B''B'})\right]\right\}(W^u_{B'''})^\dagger , \tag{50}$$

where $\{\beta^u_{B''B'''} : u \in \mathscr{U}\}$ are the $(\dim \mathscr{H}_{B'})^2$ projectors onto the Bell states, and $\{W^u_{B'''} : u \in \mathscr{U}\}$ are suitable isometries from $\mathscr{H}_{B'''}$ to $\mathscr{H}_{B'}$. But then, using Eq. (45) with $Q^u_{B''B'} = \beta^u_{B''B'}$, we obtain

$$\Phi'_A(\cdot) = \sum_u W^u_{B'''} \left\{ \mathrm{Tr}_{B''B}\left[(\phi^+_{B'''B''} \otimes \Phi_A(\cdot)) \, (I_{B'''} \otimes P^u_{B''B}) \right] \right\} (W^u_{B'''})^\dagger \, . \tag{51}$$

Hence, defining a quantum channel $\Psi : \mathsf{L}(\mathscr{H}_B) \to \mathsf{L}(\mathscr{H}_{B'})$ as

$$\Psi(\cdot) \triangleq \sum_u W^u_{B'''} \left\{ \mathrm{Tr}_{B''B}\left[(\phi^+_{B'''B''} \otimes \cdot) \, (I_{B'''} \otimes P^u_{B''B}) \right] \right\} (W^u_{B'''})^\dagger \, , \tag{52}$$

we finally have that $\Phi' = \Psi \circ \Phi$, as claimed. □

Remark 9 Theorem 3 holds also if the identity channel $\mathrm{id}_{B''}$ is replaced by a *complete* channel, namely, a channel $\Upsilon : \mathsf{L}(\mathscr{H}_{B''}) \to \mathsf{L}(\mathscr{H}_{B''})$ that is bijective (in the sense of the linear map): linearly independent inputs are transformed into linearly independent outputs. This is so because linearly independent states $\xi^x_{B''}$ in Eq. (47) remain linearly independent after the action of Υ. In this way, the proof can continue along the same lines.

We notice, in particular, that a channel can be complete despite being entanglement breaking or measure-and-prepare. This implies that the ensembles used to probe channels $\mathrm{id}_{B''} \otimes \Phi_A$ and $\mathrm{id}_{B''} \otimes \Phi'_A$ can always be chosen, without loss of generality, to comprise separable states only. □

6 The Computational Second Law: An Analogy

The aim of this section is to construct an analogy, clarifying and somehow strengthening that given by Cover [1], between data-processing theorems and the second law of thermodynamics. In what follows we abandon a formally rigorous language, preferring instead a generic language better suited to highlight the similarities and differences between thermodynamics and information theory.

Theorems 1, 2, and 3, a part from the formal complications necessary to describe classical and quantum systems together, have all the same simple interpretation that we summarize in two statements (A) and (B):

Direct statement: the information that the signal carries about the message (*any* message) cannot increase along a Markov local process; (A)

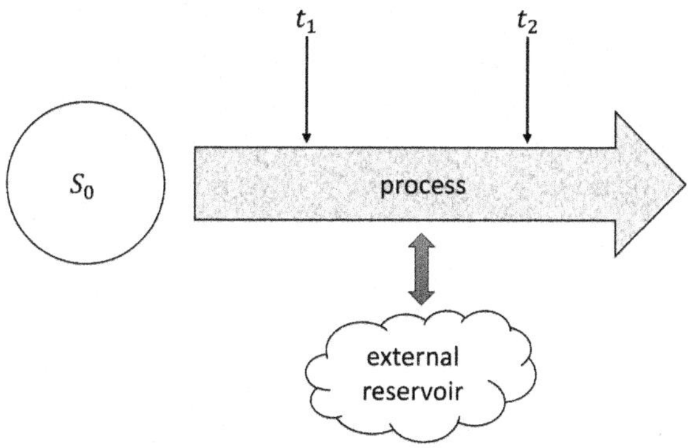

Fig. 2 Suppose that a system, prepared at time t_0, undergoes a process, and that we observe it at two later times $t_1 \geq t_0$ and $t_2 \geq t_1$. **Thermodynamical case**: Clausius' principle and Lieb's and Yngvason's entropy principle state that $\Delta H = H(S_2) - H(S_1) \geq 0$ if and only if the process bringing the system from t_1 to t_2 can be realized adiabatically (i.e., exchanging only work and no heat). This is equivalent to say that: (i) a decrease in entropy can only be achieved by exchanging heat with an external reservoir; (ii) if the process cannot be realized adiabatically then there is some initial configuration S_0 for which a decrease in entropy occurs. **Information-theoretic case**: the data-processing inequality and the reverse data-processing theorems state that $\Delta H_{\min} = H_{\min}(U|S_2) - H_{\min}(U|S_1) \geq 0$ for all U, if and only if the process bringing the system from t_1 to t_2 is Markov local (i.e., there exists a memoryless channel Ψ such that $S_2 = \Psi(S_1)$). This is equivalent to say that: (i) a decrease in the conditional min-entropy can only be achieved in the presence of an external memory storing information about the message and feeding it back into the system at later times; (ii) if the process is not Markov local, then there exists some initial message-signal joint distribution for which a decrease of $H_{\min}$ occurs

and

> **Reverse statement**: if the information that the signal carries *never* increases along a given process, then such a process admits a Markov local realization. (B)

The direct statement corresponds to Cover's law, as formulated in [1] (see the quotation at the beginning of this paper). Here "useful information" is precisely the information that the signal carries about the message, and it is measured by the conditional min-entropy, which is directly related to the guessing probability. The reverse statement, which is consequence of the reverse data-processing theorems that we proved, corresponds to Lieb's and Yngvason's entropy principle [7].

In order to make our discussion more concrete, let us consider a thermodynamical system prepared at time t_0 and evolving through successive times $t_1 \geq t_0$ and $t_2 \geq t_1$, as depicted in Fig. 2. The second law of thermodynamics, in the formulation usually attributed to Clausius, states that the following inequality is necessarily obeyed:

$$\Delta H \geq \frac{\Delta Q}{T}, \tag{53}$$

where $\Delta H = H(S_2) - H(S_1)$ is the change in thermodynamical entropy of the system and ΔQ is the heat *absorbed by* the system.[5] The above equation basically says that the only way to decrease the entropy of a system is to extract heat from the system. This implies that, if a system is adiabatically isolated (i.e., no heat is exchanged, only mechanical work), then its entropy cannot decrease. Equivalently stated: a decrease in entropy represents a definite witness of the fact that the system *is not* adiabatically isolated and is dumping heat in the environment.

This part of the second law can be seen as the analogue of statement (A) above, that is, the usual data-processing inequality. Suppose now that the system S is an information signal. As before, it is prepared at time t_0 and then it undergoes a process that is information-theoretic, rather than thermodynamical. If we observe the signal at two times $t_1 \geq t_0$ and $t_2 \geq t_1$, then we know that if the process is Markov local, then the data-processing inequality holds, namely, the information carried by the signal cannot increase going from t_1 to t_2. Therefore, any increase in the information carried by the signal is a definite witness of the fact that the process *is not* Markov local, namely, that *an external memory was used as a side resource at some point along the process.*

We now come to the reverse statement (B), arguing that it is the analogue of Lieb's and Yngvason's entropy principle. The latter states that, assuming the validity of a set of axioms about simple thermodynamical systems,[6] a non-decreasing entropy between t_1 and t_2 is not only necessary (Clausius' principle) but also *sufficient* for the existence of an adiabatic process between the two times. It is clear that the analogy works in this case too: *the reverse data-processing theorems we proved constitute the information-theoretic analogue of Lieb's and Yngvason's entropy principle.* An overview of the analogy is summarized in the table below.

Despite the tantalizing analogies, there are however two points (at least) that we should keep in mind before jumping to rash conclusions. The first one is that, while in thermodynamics a process is usually given by an initial state and a final state, in information theory a process is a channel, which acts upon *any input* it receives.

The second point is that the relation presented here between adiabaticity and Markov locality (or memorylessness) has been discussed only on a formal level, but no claim has been made about any *quantitative* relation between the two concepts. However, we would like to conclude this paper speculating about the possibility to envisage a deeper relation between adiabaticity and Markov locality, going beyond the formal analogy presented above. An adiabatically isolated system cannot exchange heat, but can interact with a mechanical device and exchange work with it. Since it is possible to imagine a purely mechanical memory (at least of finite size), it seems that the presence of a memory, in itself, should not violate adiabaticity. But then, a scenario similar to that of Maxwell's demon immediately comes to mind. Indeed, Maxwell's demon violates the second law using nothing but its memory:

---

[5] In the precise sense that ΔQ is positive if heat is injected into the system and negative if heat is extracted from the system; see, e.g., Ref. [29].

[6] The most important and debated of which is the *comparability hypothesis*: the interested reader may refer to Uffink [30].

Table 1 Summary of the analogies between the second law of thermodynamics and its computational analogue discussed here

Thermodynamical setting	Information-theoretic setting
Thermodynamical system S	Message U encoded on signal S
Entropy $H(S)$	Conditional min-entropy $H_{\min}(U\|S)$
Clausius' principle	Data-processing inequality
Lieb–Yngvason entropy principle	Reverse data-processing theorem
Adiabatically isolated system	Computationally isolated system
Adiabatic process	Markov local (memoryless) process
Heat sink/reservoir	External memory

its actions, included the measurements it performs, are assumed to be otherwise perfectly adiabatic.[7] Hence, it seems that adiabaticity does not play well with the presence of an external memory, even if this is taken to be perfectly mechanical. This fact suggests that adiabaticity and Markov locality may be even closer than what the analogies in Table 1 *prima facie* seem to suggest. This and other questions are left open for future investigations.

Acknowledgements It is a pleasure to thank, in alphabetical order, Ettore Bernardi, Jeremy Butterfield, Weien Chen, Giulio Chiribella, Giacomo Mauro D'Ariano, Nilanjana Datta, Gábor Hofer-Szabó, Koji Maruyama, Keiji Matsumoto, Milán Mosonyi, Masanao Ozawa, Veiko Palge, Paolo Perinotti, David Reeb, and Mark Wilde, whose comments helped in shaping the present ideas at various stages during the past few years. This work was supported in part by JSPS KAKENHI, grants no. 26247016 and no. 17K17796. Support from the program for FRIAS-Nagoya IAR Joint Project Group is also acknowledged.

Appendix 1: Definitions and Notations

Here we review some basic notions and clarify the notation that is used in the paper. The reader familiar with the standard toolbox used in quantum information theory (see, e.g., Ref. [31]) can safely skip to the next section.

All set and spaces considered here are finite or finite dimensional. We denote sets as $\mathscr{X}, \mathscr{Y}, \mathscr{Z}, \mathscr{U}, \ldots$ and their elements as $x, y, z, u, \ldots$. Sets support probability distributions, for example, $p(x)$. When we speak of a random variable, for example, X, we mean that it is supported by the set $\mathscr{X}$, in the sense that its states are labeled by $x \in \mathscr{X}$, and that each state can occur with probability $p(x) = \Pr\{X = x\}$. When a pair (or a triple etc) of random variables are considered, we write (X, Y) to mean a bipartite random variable supported on the cartesian product $\mathscr{X} \times \mathscr{Y} = \{(x, y) : x \in \mathscr{X}, y \in \mathscr{Y}\}$ and distributed with joint probability $p(x, y)$. Classical noisy channels are represented by conditional input–output probability distributions $w(y|x)$: in this

[7] Thus the demon can be imagined as a "perfect clockwork."

case we understand that the channel w has input alphabet $\mathscr{X}$ and output alphabet $\mathscr{Y}$ and write $w : \mathscr{X} \to \mathscr{Y}$.

Quantum systems are labeled by $A, B, C, \ldots$ and their corresponding finite dimensional Hilbert spaces are denoted as $\mathscr{H}_A, \mathscr{H}_B, \mathscr{H}_C, \ldots$. The set of linear operators on a Hilbert space $\mathscr{H}$ is denoted as $\mathsf{L}(\mathscr{H})$, the set of Hermitian operators as $\mathsf{L}_H(\mathscr{H})$, the set of positive semidefinite operators as $\mathsf{L}_+(\mathscr{H})$, and the set of density operators (or states), i.e., positive semidefinite with unit trace, as $\mathsf{D}(\mathscr{H})$. Vectors are denoted as kets $|\phi\rangle$, while if we write ϕ we mean the corresponding state, that is, the projector $|\phi\rangle\langle\phi|$. Given an orthonormal basis $\{|x\rangle : x \in \mathscr{X}\}$ for a Hilbert space, we sometime call the set of orthogonal projectors $|x\rangle\langle x|$ "flags," since these can be used to model a classical random variable with distinguishable states. For example, given a random variable X with states $x \in \mathscr{X}$ and distribution $p(x)$, we will often think of it as"embedded" in a Hilbert space $\mathscr{H}_X$, with $\dim \mathscr{H}_X = |\mathscr{X}|$, and described by the state $\sum_x p(x)\,|x\rangle\langle x|$. This is a convention commonly used in quantum information theory as it significantly simplifies the analysis of hybrid classical-quantum scenarios.

A family $\{P_A^x : x \in \mathscr{X}\}$ of operators $P_A^x \in \mathsf{L}_+(\mathscr{H}_A)$ such that $\sum_x P_A^x = I_A$ is called a POVM on $\mathscr{H}_A$. An ensemble is given by giving a set $\mathscr{X}$, a probability distribution $p(x)$ and a family of states $\rho_A^x \in \mathsf{L}(\mathscr{H}_A)$: we denote it for brevity as $\{p(x); \rho_A^x\}$, where the set $\mathscr{X}$ is usually understood from the context. Extending the idea mentioned in the preceding paragraph of embedding classical random variables in orthogonal states of a suitable Hilbert space, it is also common to interpret an ensemble as a bipartite state as follows:

$$\{p(x); \rho_A^x\}_{x\in\mathscr{X}} \iff \rho_{XA} \triangleq \sum_{x\in\mathscr{X}} p(x)\,|x\rangle\langle x|_X \otimes \rho_A^x\ . \tag{54}$$

A linear map $\Phi : \mathsf{L}(\mathscr{H}_A) \to \mathsf{L}(\mathscr{H}_B)$ is said to be a *quantum channel* if and only if it is completely positive and trace-preserving. Given a linear map $\Phi : \mathsf{L}(\mathscr{H}_A) \to \mathsf{L}(\mathscr{H}_B)$, its *trace-dual* $\Phi^\dagger : \mathsf{L}(\mathscr{H}_B) \to \mathsf{L}(\mathscr{H}_A)$ is the linear map defined by the relation

$$\mathrm{Tr}\big[X\,\Phi^\dagger(Y)\big] \triangleq \mathrm{Tr}[\Phi(X)\,Y]\ , \tag{55}$$

for all $X \in \mathsf{L}(\mathscr{H}_A)$ and all $Y \in \mathsf{L}(\mathscr{H}_B)$. Φ is a channel if and only if $\Phi^\dagger$ is completely positive and unit-preserving, i.e., $\Phi_B^\dagger(I_B) = I_A$.

Given a pair of random variables (X, U), the guessing probability of U given X is

$$P_{\mathrm{guess}}(U|X) \triangleq \max_{\varphi} \sum_u \varphi(u|x)p(x,u) \tag{56}$$

$$= \sum_x \max_u p(u,x)\ , \tag{57}$$

where the optimization is done over all channels (decoding strategies) $\varphi : \mathscr{X} \to \mathscr{U}$. In other words, it is the probability of correctly guessing U using the ideal observer decoding strategy on X.

The quantum analogue of this is the problem of correctly guessing U given an ensemble of quantum states $\{p(u); \rho_A^u\}$. In this case, the role of datum X is played by the quantum system A and the guessing probability is

$$P_{\text{guess}}(U|A) \triangleq \max_P \sum_u \text{Tr}\left[P_A^u \, \rho_A^u\right] , \tag{58}$$

where the optimization is done over all POVMs $\{P_A^u : u \in \mathscr{U}\}$. Notice that in this paper we only consider the case of guessing a classical random variable given a quantum system, so in the expression $P_{\text{guess}}(U|A)$ the roles of U (random variable) and A (quantum system) should always be clearly understandable from the context.

Entropies

The letter H is used to denote the entropy. More precisely, in the case of classical random variables $H(X) \triangleq -\sum_x p(x) \log_2 p(x)$; in the case of a quantum state ρ_A, $H(A) \triangleq -\sum_i \lambda_i \log_2 \lambda_i$, where the λ's are the eigenvalues of ρ_A. Following common terminology, the entropy of a classical variable is called the Shannon entropy, while the entropy of a state is called the von Neumann entropy.

Given a pair of random variables (X, Y), the conditional entropy is $H(X|Y) = H(XY) - H(Y)$ and the mutual information is $I(X; Y) = H(X) + H(Y) - H(XY) = H(X) - H(X|Y)$. Given a bipartite state $\rho_{AB} \in \mathsf{D}(\mathscr{H}_A \otimes \mathscr{H}_B)$, all the definitions are extended by analogy, for example, $H(A|B) = H(AB) - H(B)$, where $H(AB)$ is the von Neumann entropy of ρ_{AB} and $H(B)$ is the von Neumann entropy of the reduced state $\rho_B = \text{Tr}_A[\rho_{AB}]$.

von Neumann and Shannon entropies are not the only entropies that are relevant in information theory. Lately, in particular, alternative entropies have been found to play a central role in various information-theoretic scenarios. Such entropies, whose classification is beyond the scope of this work, include for example Rényi entropies and, in particular, min- and max-entropies, see, e.g., Ref. [32]. The one that is relevant for this work is the so-called *conditional min-entropy* which is given by

$$H_{\min}(U|X) = -\log_2 P_{\text{guess}}(U|X) \tag{59}$$

in the case of two classical random variables, and

$$H_{\min}(U|A) = -\log_2 P_{\text{guess}}(U|A) \tag{60}$$

in the case of an ensembles of quantum states. In fact, $H_{\min}(U|A)$ is the conditional min-entropy of the classical-quantum state ρ_{XA} defined in Eq. (54).

Appendix 2: The Minimax Theorem

Here we state a form of the Minimax Theorem as needed in the proof of Theorem 1, see, e.g., Lemma 4.13 in Ref. [21]:

Theorem 4 *Let $\mathscr{S} \subset \mathbb{R}^s$ be a closed convex set and $\mathscr{L} \subset \mathbb{R}^d$ be a polytope. If $f : \mathscr{S} \times \mathscr{L} \to \mathbb{R}$ is continuous and satisfies*

$$f\left[\alpha y_1 + (1-\alpha)y_2, z\right] = \alpha f(y_1, z) + (1-\alpha) f(y_2, z) \tag{61}$$

$$f\left[y, \alpha z_1 + (1-\alpha)z_2\right] = \alpha f(y, z_1) + (1-\alpha) f(y, z_2)\,, \tag{62}$$

for all $\alpha \in [0, 1]$, $y, y_1, y_2 \in \mathscr{S}$, and $z, z_1, z_2 \in \mathscr{L}$, then

$$\max_{z\in\mathscr{L}} \min_{y\in\mathscr{S}} f(y, z) = \min_{y\in\mathscr{S}} \max_{z\in\mathscr{L}} f(y, z)\,. \tag{63}$$

In proving Theorem 1 we specialize the above statement to the case in which $\mathscr{S}$ is the set of classical channels $\varphi : Y \to Z$ (indeed convex and closed) and $\mathscr{L}$ is the set of joint probability distributions on $\mathscr{X} \times \mathscr{Z}$ (indeed a polytope). Last thing to check is that conditions (61) and (62) hold: this is a consequence of the fact that the function in the case considered is actually linear in both its variables.

Appendix 3: The Separation Theorem

Here we give an elementary geometrical proof of the Hahn-Banach separation theorem in its simplest case, i.e. where the sets considered are closed and bounded. For a more general treatment the interested reader may refer to, e.g., Ref. [33].

Theorem 5 *Let $C \in \mathbb{R}^n$ be a closed and bounded convex set, and let $y \in \mathbb{R}^n$ be a vector that does not belong to C, i.e. $y \notin C$. Then, there exists a vector $k \in \mathbb{R}^n$ and a constant $\alpha \in \mathbb{R}$ such that $k \cdot x < \alpha < k \cdot y$, for all $x \in C$. We say that the hyperplane $\mathscr{L} := \{z \in \mathbb{R}^n : z \cdot k = \alpha\}$ separates C and y strictly.*

Proof Let $x_0 \in C$ be a point such that

$$\|x_0 - y\| = \min_{x\in C} \|x - y\| > 0. \tag{64}$$

Its existence is guaranteed by the Weierstrass' extreme value theorem. The strict inequality comes from the fact that $y \notin C$, by assumption.

Let us now define

$$k := y - x_0 \tag{65}$$

and

$$\alpha := \frac{1}{2}(k \cdot x_0 + k \cdot y) = \frac{1}{2}(y \cdot y - x_0 \cdot x_0). \tag{66}$$

We note now that

$$\begin{aligned} k \cdot y &= (y - x_0) \cdot y \\ &= \frac{1}{2}\{(y - x_0) \cdot y + (y - x_0) \cdot y\} \\ &= \frac{1}{2}\{(y - x_0) \cdot (y - x_0 + x_0) + (y - x_0) \cdot y\} \\ &= \frac{1}{2}\{(y - x_0)x_0 + (y - x_0) \cdot y\} + \frac{1}{2}(y - x_0) \cdot (y - x_0) \\ &> \alpha, \end{aligned} \tag{67}$$

and that

$$\begin{aligned} k \cdot x_0 &= (y - x_0) \cdot x_0 \\ &= \frac{1}{2}\{(y - x_0) \cdot x_0 + (y - x_0) \cdot x_0\} \\ &= \frac{1}{2}\{(y - x_0) \cdot (x_0 + y - y) + (y - x_0) \cdot x_0\} \\ &= \frac{1}{2}\{(y - x_0)x_0 + (y - x_0) \cdot y\} + \frac{1}{2}(y - x_0) \cdot (x_0 - y) \\ &= \frac{1}{2}\{(y - x_0)x_0 + (y - x_0) \cdot y\} - \frac{1}{2}(y - x_0) \cdot (y - x_0) \\ &< \alpha, \end{aligned} \tag{68}$$

Now, let us consider any $x \in C$. By convexity, $(1 - p)x_0 + px \in C$, for any $p \in [0, 1]$. Then, we have that

$$\begin{aligned} \|x_0 - y\|^2 &= \min_{x \in C} \|x - y\|^2 \\ &\leq \|(1 - p)x_0 + px - y\|^2 \\ &= \|(1 - p)(x_0 - y) + p(x - y)\|^2 \\ &= (1 - p)^2 \|x_0 - y\|^2 + 2p(1 - p)(x_0 - y) \cdot (x - y) + p^2 \|x - y\|^2, \end{aligned} \tag{69}$$

where we used the formula $\|w_0 + w_1\|^2 = \|w_0\|^2 + \|w_1\|^2 + 2w_0 \cdot w_1$, valid for all $w_0, w_1 \in \mathbb{R}^n$. Therefore,

$$0 \leq p(p - 2) \|x_0 - y\|^2 + 2p(1 - p)(x_0 - y) \cdot (x - y) + p^2 \|x - y\|^2. \tag{70}$$

Let us now consider the case $p \neq 0$. Then,

$$0 \leq (p - 2) \|x_0 - y\|^2 + 2(1 - p)(x_0 - y) \cdot (x - y) + p \|x - y\|^2, \tag{71}$$

and, taking the limit for $p \to 0$, we finally obtain

$$\begin{aligned} 0 &\le -2\,\|x_0 - y\|^2 + 2(x_0 - y)\cdot(x - y) \\ &\le 2\,\{(x_0 - y)\cdot(x - y) - (x_0 - y)\cdot(x_0 - y)\} \\ &= 2\,\{(x_0 - y)\cdot x - (x_0 - y)\cdot x_0\} \\ &= 2\,\{-k\cdot x + k\cdot x_0\}\,, \end{aligned} \tag{72}$$

which implies that $k\cdot x \le k\cdot x_0 < \alpha < k\cdot y$, for any $x \in C$, as claimed. □

For our purpose the following reformulation of Theorem 5 is particularly useful:

Corollary 2 *Let C_1 and C_2 be two closed and bounded convex sets in $\mathbb{R}^n$. Then, $C_1 \supseteq C_2$ if and only, for every vector $k \in \mathbb{R}^n$,*

$$\max_{x\in C_1} k\cdot x \ge \max_{y\in C_2} k\cdot y. \tag{73}$$

References

1. Cover, T. M.: Which processes satisfy the second law. In *Physical Origins of Time Asymmetry* (eds. Halliwell, J. J., Pérez-Mercader, J. & Zurek, W. H.) 98–107 (Cambridge University Press, 1996).
2. Csiszár, I., Körner, J.: *Information Theory*. (Cambridge University Press, Second Edition, 2011).
3. Cover, T. M., Thomas, J. A.: *Elements of Information Theory*. (John Wiley & Sons, Second Edition, 2006)
4. Merhav, N.: *Physics of the Shannon limits*. IEEE Trans. Inform. Theory **56**(9), pages 4274–4285 (2010).
5. Merhav, N.: *Data processing theorems and the second law of thermodynamics*. IEEE Trans. on Inform. Theory **57**(8), pages 4926–4939 (2011).
6. Merhav, N.: *Statistical physics and information theory*. Foundations and Trends in Communications and Information Theory **6**(1-2), pages 1–212 (2009).
7. Lieb, E. H., Yngvason, J.: *The physics and mathematics of the second law of thermodynamics*. Physics Reports **310**, 1–96 (1999).
8. Shannon, C. E.: A note on a partial ordering for communication channels. Information and Control **1**, 390–397 (1958).
9. Körner, J., and Marton, K.: Comparison of two noisy channels. *Topics in information theory*, (16):411–423, 1977.
10. Buscemi, F.: *Comparison of Quantum Statistical Models: Equivalent Conditions for Sufficiency*. Commun. Math. Phys. **310**, 625–647 (2012).
11. Buscemi, F.: *All Entangled Quantum States Are Nonlocal*. Phys. Rev. Lett. **108**, 200401 (2012).
12. Buscemi, F., Datta, N., Strelchuk, S.: *Game-theoretic characterization of antidegradable channels*. Journal of Mathematical Physics **55**, 92202 (2014).
13. Buscemi, F.: *Complete Positivity, Markovianity, and the Quantum Data-Processing Inequality, in the Presence of Initial System-Environment Correlations*. Phys. Rev. Lett. **113**, 140502 (2014).
14. Buscemi, F., Datta, N.: *Equivalence between divisibility and monotonic decrease of information in classical and quantum stochastic processes*. Phys. Rev. A **93**, 12101 (2016).

15. Buscemi, F.: *Degradable channels, less noisy channels, and quantum statistical morphisms: an equivalence relation*. Probl. Inf. Trans. to appear. arXiv:1511.08893 [quant-ph].
16. Shmaya, E.: *Comparison of information structures and completely positive maps*. J. Phys. A: Math. Gen. **38**, 9717 (2005).
17. Chefles, A.: *The Quantum Blackwell Theorem and Minimum Error State Discrimination*. arXiv:0907.0866 [quant-ph] (2009).
18. Blackwell, D.: Equivalent Comparisons of Experiments. *The Annals of Mathematical Statistics*, 24(2):265–272, 1953.
19. Torgersen, E.: *Comparison of Statistical Experiments*. Encyclopedia of Mathematics and its Applications. (Cambridge University Press, 1991).
20. Cohen, J.E., Kemperman, J.H.B., and Zbăganu, G.: *Comparisons of Stochastic Matrices, with Applications in Information Theory, Statistics, Economics, and Population Sciences*. (Birkhäuser, 1998).
21. Liese, F., Miescke, K.-J.: *Statistical Decision Theory*. (Springer New York, 2008).
22. Raginsky,M.: Shannon meets Blackwell and Le Cam: Channels, codes, and statistical experiments. *2011 IEEE International Symposium on Information Theory Proceedings*, pages 1220–1224, July 2011.
23. El Gamal, A. A.: Broadcast Channels With And Without Feedback. *Circuits, Systems and Computers, 1977. Conference Record. 1977 11th Asilomar Conference on*, pages 180–183, 1977.
24. Morse, N., and Sacksteder, R.: *Statistical Isomorphism*. Ann. Math. Stat. **37**, 203-214 (1966).
25. Čencov, N. N.: *Statistical Decision Rules and Optimal Inference*. (American Mathematical Society, 1982).
26. Matsumoto, K.: An example of a quantum statistical model which cannot be mapped to a less informative one by any trace preserving positive map. arXiv:1409.5658 [quant-ph, stat] (2014).
27. Frenk, J. B. G., Kassay, G., and Kolumbán, J.: *On equivalent results in minimax theory*. European Journal of Operational Research **157**, 46–58 (2004).
28. Jenčová, A.: *Comparison of quantum channels and statistical experiments*. ISIT 2016. arXiv:1512.07016 [quant-ph].
29. Borgnakke, C., Sonntag, R. E.:*Fundamentals of Thermodynamics*. (John Wiley & Sons, 2009).
30. Uffink, J.: *Bluff Your Way in the Second Law of Thermodynamics*. Studies in History and Philosophy of Science Part B: Studies in History and Philosophy of Modern Physics **32**, 305–394 (2001).
31. Wilde, M. M.:*Quantum Information Theory*. (Cambridge University Press, 2013.) Also available at arXiv:1106.1445 [quant-ph].
32. Tomamichel, M.: *Quantum Information Processing with Finite Resources*. SpringerBriefs in Mathematical Physics **5**, 2016.
33. Rockafellar, R.T.: *Convex Analysis*. (Princeton University Press, 1970).

Trust-Free Verification of Steering: Why You Can't Cheat a Quantum Referee

Michael J. W. Hall

Abstract It was believed until recently that the verification of quantum entanglement and quantum steering, between two parties, required trust in at least one of the parties and their devices, in contrast to the verification of Bell nonseparability. It has since been shown that this is not the case: the need for trust, in verifying two parties share a given quantum correlation resource, can be replaced by quantum refereeing, in which the referee sends quantum signals rather than classical signals to untrusted parties. The existence of such quantum-refereed games is discussed, with particular emphasis on how they make it impossible for the parties to cheat. The example of a particular quantum-refereed steering game is used to show explicitly how measurement-device independence is achieved via 'quantum programming' of untrusted measurement devices; how cheating is prevented by the steered party being unable to distinguish sufficiently well between two sets of nonorthogonal signal states; and that cheating remains impossible when one-way communication is allowed from the steered party to the steering party. This game has been recently implemented experimentally, and is of particular interest both in accounting for any imperfections in the referee's preparation of signal states, and in suggesting the future possibility of secure two-sided quantum key distribution with Bell-local states.

Keywords Quantum refereeing · Quantum steering · Quantum games

1 Introduction

Pure quantum states shared between two parties are rather simple with regard to possible types of correlations: they are either factorisable or entangled. If they are factorisable, then all correlations are trivial. If they are entangled then they are also

M. J. W. Hall (✉)
Centre for Quantum Dynamics, Griffith University, Brisbane, QLD 4111, Australia
e-mail: michael.hall@griffith.edu.au

M. Ozawa et al. (eds.), *Reality and Measurement in Algebraic Quantum Theory*,
Springer Proceedings in Mathematics & Statistics 261,
https://doi.org/10.1007/978-981-13-2487-1_7

Bell nonseparable, steerable, have quantum discord, etc. Thus, the latter properties only become distinguishable for mixed quantum states. This was first clearly pointed out by Werner, who showed there are mixed quantum states that are both entangled and Bell separable [1]. Later it was shown that there are also mixed quantum states that are both unentangled and discordant [2], and states that are both Bell separable and steerable [3]. In this way a hierarchical structure of quantum correlations has emerged, as reviewed in Sect. 2.

This hierarchy is of interest not just for foundational reasons: different types of quantum correlation reflect the resources needed to accomplish various tasks of physical interest. Thus, for example, entanglement is necessary to optimally distinguish any two quantum channels [4]; steerability is necessary for subchannel discrimination [5] and allows one-sided secure key distribution [6]; and Bell nonseparability allows two-sided secure key distribution and randomness generation [7]. It is therefore important to be able to verify or witness the level of correlation of a claimed resource. This is typically done via testing for the violation of suitable inequalities, as is discussed in Sect. 3.

In Sect. 4, it is recalled how verifying a given degree of correlation can be recast as a quantum correlation game, in which a referee sends signals to the parties, receives corresponding outputs from them, and calculates the value of a suitable payoff function from the correlations between the inputs and outputs. Until recently, such games for verifying steering and entanglement were thought to require trust by the referee in at least one of the parties and their devices: a hierarchy of trust mirrored the hierarchy of correlations [8]. However, based on pioneering work by Buscemi, it is now known that trust can be replaced by 'quantum refereeing', in which the referee sends quantum signals rather than classical signals to the parties [9–14].

Section 5 explores in depth how cheating by the parties is prevented in quantum-refereed games, using a steering game as an example. A formal proof is given for why the parties can only win this game if they share a steerable resource, before considering the physical reasons behind this. It is shown explicitly how the impossibility of discriminating between sets of nonorthogonal signals from the referee prevents the success of possible cheating strategies. It is also shown that the quantum signals from the referee effectively 'program' the measurement device of the receiving party, where the corresponding programs cannot be distinguished from one another, preventing any 'hacking' by the parties. Finally, it is shown that cheating remains impossible when one-way communication from the steered party to the steering party is permitted during the game.

Section 6 considers experimental implementations of quantum-refereed correlation games, including the need for modification of payoffs due to imperfect preparation of signal states by the referee, and the robustness of these games when the signal states are transmitted through a noisy channel. Results from a recent experiment for a quantum-refereed steering game are briefly recalled [13], that demonstrates trust-free verification of steering is in principle possible using a Bell-local resource.

Finally, a brief discussion is given in Sect. 7.

2 Hierarchy of Quantum Correlations

We consider the hierarchical structure of quantum correlations in more detail in this section, within a standard two-party scenario. In particular, we assume that two distant parties, Alice and Bob, can generate a set of statistical correlations in the following way. On each run, Alice makes some measurement labelled by x, and obtains a result labelled by a. Similarly, on each run Bob makes some measurement labelled by y, and obtains a result labelled by b. Over many runs, therefore, they are able to estimate the set of joint probabilities $\{p(a, b|x, y)\}$. An aim of physics is to explain these joint probabilities and the statistical correlations that they generate. We will consider three types of physical explanation in particular.

2.1 *Entanglement and Separability*

First, we can search for a separable quantum model of the correlations, where Alice's and Bob's measurement statistics are generated by a set of local quantum states on two Hilbert spaces H_A and H_B respectively. In particular, we say there is a separable quantum state model of the correlations on $H_A \otimes H_B$ if and only if they have the form

$$p(a, b|x, y) = \sum_\lambda p(\lambda)\, p_Q(a|x, \rho_\lambda^A)\, p_Q(b|y, \rho_\lambda^B). \tag{1}$$

Here λ denotes a classical random variable with probability density $p(\lambda)$, ρ_λ^A and ρ_λ^B denote density operators on H_A and H_B, and $p_Q(m|M, \rho)$ denotes a quantum probability distribution for state ρ and measurement M, i.e,

$$p_Q(m|M, \rho) = \mathrm{Tr}\left[E_m^M \rho\right] \tag{2}$$

for some positive operator valued measure (POVM) $\{E_m^M\}$ (thus, $E_m^M \geq 0$ and $\sum_m E_m^M = \hat{1}$).

It follows that correlations with a separable quantum state model on $H_A \otimes H_B$ are equivalently described by the separable quantum state

$$\rho^{AB} := \sum_\lambda p(\lambda)\, \rho_\lambda^A \otimes \rho_\lambda^B \tag{3}$$

on $H_A \otimes H_B$. Conversely, correlations with no such separable quantum state model are defined to be *entangled* with respect to $H_A \otimes H_B$.

2.2 Steering and Spooky Action at a Distance

The concepts of entanglement and steering were introduced by Schrödinger [15], in his response to the famous Einstein–Podolsky–Rosen (EPR) paper of 1935 [16]. In particular, he used 'steering' to denote the property that, for a shared quantum state, Alice can typically control, via her choice of measurement, the corresponding set of local quantum states that Bob's system is described by. This steering of Bob's local state by a remote measurement is the 'spooky action at a distance' that Einstein so disliked about quantum mechanics [17].

Clearly, there is no steering of the above type in the case that the statistical correlations between Alice and Bob can be explained via some fixed set of local quantum states for Bob: in this case Alice's measurements have no effect. A simple example is a factorisable state, $\rho^A \otimes \rho^B$, where Bob's local state is always described by ρ^B independently of Alice's actions. This consideration led Wiseman et al. to formally define EPR-steering in terms of the existence or otherwise of a local hidden state (LHS) model for one of the parties [3].

In particular, a given set of joint probabilities $\{p(a, b|x, y)\}$ is defined to have a local hidden state model for Bob, on Hilbert space H_B, if and only if

$$p(a, b|x, y) = \sum_\lambda p(\lambda)\, p(a|x, \lambda)\, p_Q(b|y, \rho_\lambda^B). \tag{4}$$

Here $p_Q(m|M, \rho)$ is a quantum probability distribution as per Eq. (2), and $p(a|x, \lambda)$ can be an arbitrary probability distribution. Thus, all correlations are explained via some pre-existing set of local quantum states for Bob on H_B. Conversely, correlations that do not admit such an LHS model are defined to be *EPR steerable* from Alice to Bob, with respect to H_B [3].

EPR steerability from Bob to Alice is similarly defined with respect to an LHS model for Alice, relative to some Hilbert space H_A. Thus, the concept of steering is inherently asymmetric.

Comparison of Eqs. (1) and (4) show that, unlike separable state models of correlations, LHS models do not require that the steering party (Alice in this case), is described by the laws of quantum mechanics. They only require that there is some local statistical description for her outcomes, $p(a|x, \lambda)$. Thus, for example, in any such model Bob's local statistics are subject to the Heisenberg uncertainty principle, but Alice's need not be. This underlies tests for the existence of such models via steering inequalities [20], as will be seen below.

2.3 Bell Nonseparability and Local Hidden Variables

The Einstein–Podolsky–Rosen paper of 1935 further inspired the consideration of an even more general class of correlation models: local hidden variable (LHV) models.

In particular, a given set of joint probabilities $\{p(a, b|x, y)\}$ is defined to have a local hidden variable model if and only if

$$p(a, b|x, y) = \sum_{\lambda} p(\lambda)\, p(a|x, \lambda)\, p(b|y, \lambda), \tag{5}$$

where both $p(a|x, \lambda)$ and $p(b|y, \lambda)$ can be arbitrary probability distributions. Conversely, if there is no such model, the correlations are said to be Bell nonseparable (or Bell nonlocal).

Such models were introduced by Bell [18], who famously showed how the nonexistence of such models for a given set of correlations can be tested experimentally, via what are now called Bell inequalities. Note that LHV models do not make any assumptions about how the local statistics are generated—in particular, unlike quantum separability and local quantum state models, there is no assumption that any particular theory, such as quantum mechanics, is valid.

3 Witnessing the Hierarchy

It is a logical consequence of the above definitions that joint quantum states on a given Hilbert space $H_A \otimes H_B$ have the hierarchical ordering

$$\text{Bell nonseparability} \implies \text{EPR steering} \implies \text{entanglement}, \tag{6}$$

according to the type of correlations they can generate via suitable measurements. This hierarchy is strict: there are steerable states that are not Bell nonseparable, and entangled states that are not steerable [3]. A nice example is provided by the Werner states of two qubits, defined by

$$\rho_W := W\,|\Psi_-\rangle\langle\Psi_-| + (1-W)\,\frac{1}{4}\hat{1}\otimes\hat{1} = \frac{1}{4}\left(\hat{1}\otimes\hat{1} - W\sum_j \sigma_j\otimes\sigma_j\right), \tag{7}$$

where $|\Psi_-\rangle$ denotes the singlet state, $\sigma_1, \sigma_2, \sigma_3$ are the Pauli spin operators, and $-1/3 \le W \le 1$. Werner states are Bell nonseparable for $W > 1/\sqrt{2}$, EPR steerable for $W > 1/2$, and entangled for $W > 1/3$ [1, 3].

Membership of each class in the hierarchy can be witnessed via suitable correlation inequalities. To see this, we consider the simplest case where Alice and Bob can each make two possible measurements, x_1 and x_2 for Alice and y_1 and y_2 for Bob, with each measurement having two possible outcomes ± 1. We will denote the respective outcomes by a_1, a_2, b_1 and b_2.

First, if an LHV model as per Eq. (5) can predict the outcomes of each possible measurement (i.e., it is deterministic), it is easy to check that that these predetermined outcomes must satisfy

$$a_1b_1 + a_1b_2 + a_2b_1 - a_2b_2 = a_1(b_1 + b_2) + a_2(b_1 - b_2) = \pm 2$$

for each run. Hence, the correlations satisfy the Bell inequality [19]

$$|\langle a_1b_1\rangle + \langle a_1b_2\rangle + \langle a_2b_1\rangle - \langle a_2b_2\rangle| \leq 2. \tag{8}$$

This inequality may similarly be shown to hold for nondeterministic LHV models of the correlations [19], and is well known to be violated by some two-qubit states (in particular, it is violated by two-qubit Werner states with $W > 1/\sqrt{2}$).

Second, if an LHS model for Bob on a qubit space, as per Eq. (4), can predict the outcomes of Alice's possible measurements, and Bob's measurements y_1, y_2 correspond to measurements of σ_1 and σ_2 on his qubit, then for each local state ρ^B_λ one has

$$\langle a_1b_1\rangle_\lambda + \langle a_2b_2\rangle_\lambda = \mathrm{Tr}\left[\rho^B_\lambda(a_1\sigma_1 + a_2\sigma_2)\right].$$

Now, the eigenvalues of the qubit operator $\pm\sigma_1 + \pm\sigma_2$ are $\pm\sqrt{2}$ for any choice of the signs. Hence, averaging over λ, one obtains the EPR steering inequality [20]

$$|\langle a_1\sigma_1\rangle + \langle a_2\sigma_2\rangle| \leq \sqrt{2}. \tag{9}$$

This inequality easily generalises to the case that Alice's outcomes are not predetermined [20]. A simple calculation shows that it is violated, for example, by two-qubit Werner states with $W > 1/\sqrt{2}$ (a stronger steering inequality, violated for $W > 1/\sqrt{3}$, will be given further below).

Third and finally, a quantum separable model for the correlations on a two-qubit space, in the case that both Alice and Bob measure σ_1 and σ_2 (i.e., $x_j = y_j = \sigma_j$), implies via Eq. (1) that

$$\langle a_1b_1\rangle_\lambda + \langle a_2b_2\rangle_\lambda = \sum_{j=1}^{2} \mathrm{Tr}\left[\rho^A_\lambda\sigma_j\right]\,\mathrm{Tr}\left[\rho^B_\lambda\sigma_j\right] = \sum_{j=1}^{2} m_j(\lambda)\,n_j(\lambda),$$

where $m(\lambda)$ and $n(\lambda)$ denote the Bloch vectors of $\rho^A(\lambda)$ and $\rho^B(\lambda)$ respectively. Hence, since these Bloch vectors are at most of unit length, one obtains the entanglement witness inequality [21]

$$|\langle\sigma_1 \otimes \sigma_1\rangle + \langle\sigma_2 \otimes \sigma_2\rangle| \leq \sum_\lambda p(\lambda)|m(\lambda)|\,|n(\lambda)| \leq 1, \tag{10}$$

where the triangle and Schwarz inequalities have been used. This inequality is violated by, for example, two-qubit Werner states with $W > 1/2$.

4 Quantum Correlation Games

It is possible to recast Bell, steering and entanglement inequalities, such as those in Eqs. (8)–(10), into the form of games, played by Alice and Bob to convince a referee that they share a resource that is entangled, steerable, or Bell nonseparable. Alice and Bob are not allowed to communicate with each other during the game, although they can agree on a prearranged strategy beforehand. On each run the referee, Charlie say, sends a measurement label x to Alice, and receives a corresponding measurement outcome a from Alice. Similarly, Charlie sends a measurement label y to Bob, and receives a corresponding measurement outcome b. From many runs, Charlie can estimate the probabilities in the set $\{p(a, b|x, y)\}$, and determine whether they violate the inequality being tested.

This inequality may also be used to calculate a suitable payoff, $\wp(a, b, x, y)$, to Alice and Bob on each run of the game. For example, consider the entanglement inequality in Eq. (10). In this case, a suitable payoff function is $\wp(a, b, x, y) = ab\,\delta_{xy}/p(x, y) - 1$, where $p(x, y)$ denotes the joint probability that the referee sends x to Alice and y to Bob in any run. The corresponding average payoff is therefore

$$\bar{\wp} := \sum_{a,b,x,y} \wp(a, b, x, y)\, p(a, b|x, y)\, p(x, y) = \langle\sigma_1 \otimes \sigma_1\rangle + \langle\sigma_2 \otimes \sigma_2\rangle - 1. \quad (11)$$

Thus, Alice and Bob can only win the game, i.e., score a positive average payoff, if they can violate the entanglement inequality in Eq. (10).

4.1 *Cheating in Classically-Refereed Games: A Hierarchy of Trust*

For the case of Bell games, corresponding to the referee testing whether Alice and Bob share a Bell nonseparable resource, Charlie does not have to trust Alice or Bob, nor their measurement devices. He may regard them as 'black boxes', into which values of x and y are input, and from which values of a and b are output. As long as the correlations between these inputs and outputs violate a Bell inequality, there is no LHV model that can explain them. Bell games are said to be device independent.

However, for steering games and entanglement games the situation is different. Suppose, for example, that Alice and Bob claim that they share a two-qubit entangled state that violates the entanglement inequality in Eq. (10), whereas in fact they share no quantum state at all. Can the referee be confident that they cannot win the corresponding correlation game?

The answer is no: Alice and Bob (or their devices) can cheat. They can, for example, share a predetermined list of $+1$s and -1s, such as $\{1, -1, -1, 1, -1, \ldots\}$, and on the nth run each return the nth member of the list as their output. In this way they will maximally violate the entanglement inequality in Eq. (10), with a value of

2 for the left hand side, and will obtain the maximal possible average payoff of +1 in Eq. (11). The same cheating strategy will clearly also allow them to violate the steering inequality in Eq. (9).

For some time it was thought, therefore, that trust was required to verify entanglement and steering. In particular, entanglement inequalities are with respect to particular POVMs for each of Alice and Bob, such as in Eq. (10) while steering inequalities are with respect to particular POVMs for the steered party, such as in Eq. (9). However, the referee has no mechanism for ensuring that these POVMs are actually measured to generate the reported outcomes, and so must simply trust this is the case.

Thus, until recently, the standard picture was that tests of entanglement require trust in both Alice and Bob and their devices; tests of EPR steering require trust in the steered party and their device; and tests of Bell nonseparability require no trust at all [8]. This picture places limitations on applications of quantum correlations. For example, it implies that EPR-steering can only be used for one-sided secure key distribution, due to the need to trust the steered party [6].

4.2 Preventing Cheating: Quantum-Refereed Games

Surprisingly, however, it turns out that the above hierarchy of trust can be dispensed with! The idea, first proposed by Buscemi for the case of entanglement [9], and elaborated on and generalised to steering by Cavalcanti et al. [10], is for the referee to replace the need for trust by quantum channels.

For example, in the case of verifying EPR steering, instead of sending a label y to Bob via a classical communication channel, and trusting him and his devices to implement the corresponding measurement, Charlie sends a quantum state ω_y via a *quantum* channel. By choosing a suitable set of such states and a corresponding payoff function, this makes it impossible for Alice and Bob to demonstrate EPR steering unless they genuinely share a steerable state. Thus, they can outwit a classical referee, but not a quantum referee.

The underlying physical mechanism for overcoming cheating is that the quantum states sent by the referee are nonorthogonal. Such states cannot be unambiguously distinguished, preventing Alice and Bob from knowing which correlation is being tested on a given run [10, 12, 14] . Together with a suitable payoff function, this completely undermines any cheating strategy for simulating quantum correlations that they do not actually share. As will be seen in the next section, quantum refereeing may also be regarded as a means of 'quantum programming' measurement devices: no one other than the referee knows precisely what instructions the devices have been given.

5 Example: A Quantum-Refereed Steering Game

Buscemi proved that a suitable quantum-refereed game exists for verifying the entanglement of any given entangled state [9].This was generalised by Cavalcanti et al. to prove the existence of a suitable quantum-refereed game for verifying the steerability of any given EPR-steerable state [10]. However, these existence proofs gave no explicit method for constructing such games. This was remedied for entanglement games by Branciard et al. who showed how to construct a quantum-refereed game for each possible entangled state [11]. Remarkably, Rosset et al. showed that one can even construct quantum-refereed entanglement games in which the requirement of no communication between Alice and Bob can be removed: such communication does not enable them to cheat [12].

Kocsis et al. similarly showed how to construct a suitable quantum-refereed steering game that tests for violation of a given EPR steering inequality [13]. An example of such a game is discussed in this section, with an emphasis on understanding why the game requires no trust in either of Alice and Bob (even when Bob is permitted one-way communication with Alice).

5.1 Rules of the Game

We now consider the following example of a quantum-refereed steering game. On each run of the game, Charlie sends Alice a classical signal $j \in 1, 2, 3$, and Bob a qubit signal corresponding to an eigenstate of σ_j, i.e., a density operator $\omega^C_{j,s} := (1/2)(\hat{1} + s\sigma_j)$ with $s = \pm 1$. Alice is required to return a value $a = \pm 1$, while Bob is required to return a value $b = 0$ or 1. The average payoff function is defined to be

$$\bar{\wp} := 2 \sum_{j,s} \left(s\langle ab\rangle_{j,s} - \frac{1}{\sqrt{3}} \langle b\rangle_{j,s} \right), \tag{12}$$

where $\langle\cdot\rangle_{j,s}$ denotes the average over those runs with a given value of j and s. The game is won if Alice and Bob can achieve an average payoff $\bar{\wp} > 0$.

As per the general proof of Kocsis et al. and as shown directly for this particular game below, Alice and Bob can win only if they genuinely share a steerable state [13]. In fact, as will be shown below, they can win only if they can violate the known EPR steering inequality [20]

$$\langle a_1\sigma_1\rangle + \langle a_2\sigma_2\rangle + \langle a_3\sigma_3\rangle \leq \sqrt{3}, \tag{13}$$

where a_j denotes Alice's outcome for input signal j. Indeed, we will see that the average payoff function is equal to the amount of violation of this inequality that they can achieve with a two-qubit shared state. Note that this steering inequality is a simple generalisation of the one in Eq. (9), and is proved the same way, noting that the eigenvalues of the qubit operator $\pm\sigma_1 \pm \sigma_2 \pm \sigma_3$ are $\pm\sqrt{3}$ for any choice of signs.

5.2 *Why Cheating is Impossible*

Suppose that Alice and Bob do not share a steerable resource. Hence, by definition, there must be an LHS model for Bob on some Hilbert space H_B (not necessarily a qubit space), i.e., all correlations between Alice and Bob are described by a model as per Eq. (4) for some ensemble of hidden states ρ_λ^B on some Hilbert space H_B. Now, since Bob receives an unknown state from Charlie, the most general action he can take to return a value $b = 0$ or 1 is to measure some POVM $\{E_0^{BC}, E_1^{BC}\}$ on the combination of his local hidden state and the received state, and return the outcome. Note this includes, for example, strategies such as first making a measurement on the unknown state to try and determine the value of j and s and then making a corresponding measurement on his local state.

It follows that the average value of ab, when Charlie sends Alice j and Bob the state ω_{js}^C, is given by

$$\begin{aligned}\langle ab\rangle_{j,s} &= \sum_\lambda p(\lambda)\,\langle a\rangle_{j,\lambda}\,\langle b\rangle_{j,s,\lambda} \\ &= \sum_\lambda p(\lambda)\,\langle a\rangle_{j,\lambda}\,\mathrm{Tr}_{BC}[E_1^{BC}\rho_\lambda^B \otimes \omega_{js}^C] \\ &= \sum_\lambda p(\lambda)\,\langle a_j\rangle_\lambda\,\mathrm{Tr}[X_\lambda^C \omega_{js}^C],\end{aligned}$$

where we rewrite $\langle a\rangle_{j,\lambda}$ as $\langle a_j\rangle_\lambda$ (i.e., $a_j = \pm 1$ denotes Alice's outcome for input j), and define the positive operator X_λ^C on H_C by $X_\lambda^C := \mathrm{Tr}_B[E_1^{BC}\rho_\lambda^B \otimes \hat{1}_C]$. Defining the density operator τ_λ^C, probability density $q(\lambda)$ and positive constant N by

$$\tau_\lambda^C := X_\lambda^C/\mathrm{Tr}\left[X_\lambda^C\right], \qquad q(\lambda) := p(\lambda)\mathrm{Tr}\left[X_\lambda^C\right]/N, \qquad N := \sum_\lambda p(\lambda)\mathrm{Tr}\left[X_\lambda^C\right],$$

then yields

$$\langle ab\rangle_{j,s} = N\sum_\lambda q(\lambda)\,\langle a_j\rangle_\lambda\,\mathrm{Tr}\left[\omega_{js}^C\tau_\lambda^C\right]. \tag{14}$$

One similarly finds

$$\langle b\rangle_{j,s} = N\sum_\lambda q(\lambda)\,\mathrm{Tr}\left[\omega_{js}^C\tau_\lambda^C\right]. \tag{15}$$

Hence, noting from the definition of the states ω_{js}^{C} in Sect. 5.1 that $\sum_s s\,\omega_{js}^{C} = \sigma_j$ and $\sum_s \omega_{js}^{C} = \hat{1}$, the value of the average payoff in Eq. (12) is given by

$$\begin{aligned}\bar{\wp} &= 2N \sum_{\lambda, j} q(\lambda) \left(\langle a_j \rangle_\lambda \, \mathrm{Tr}\left[\sigma_j \tau_\lambda^C\right] - \frac{1}{\sqrt{3}} \right) \\ &= 2N \left[\sum_j \langle a_j \sigma_j \rangle_{\mathrm{LHS}} - \sqrt{3} \right], \qquad (16)\end{aligned}$$

where the average is with respect to the LHS model defined by the probability density $q(\lambda)$ and the corresponding hidden states τ_λ^C on Charlie's qubit Hilbert space. Hence, using the steering inequality in Eq. (13), $\bar{\wp} \le 0$, i.e., Alice and Bob cannot with the game with a nonsteerable resource, as claimed.

The above proof that Alice and Bob cannot cheat is, necessarily, somewhat formal in nature. The focus in the remainder of this section is on giving some physical insight into why cheating is impossible, and also showing how Alice and Bob can win the game if they do share a suitable steering resource.

5.3 Connection to Unambiguous State Discrimination

Some insight is gained by considering a possible cheating strategy that Alice and Bob could employ if they share no quantum state. It is clear from Eq. (12) that the average payoff is maximised if Bob returns the outcome $b = 0$ whenever $sa = -1$. Now, Alice has no access to the value of s (she is only sent the value of j), but Bob can in principle try to estimate s from the state ω_{js}^{C} sent to him by the referee. Hence, an obvious cheating strategy is for Alice to always return the result $a = 1$, and for Bob to return the value $b = 1$ if his estimated value of s is 1 and $b = 0$ otherwise. Note that this strategy can also be easily varied to the case where Alice returns $a = \pm 1$ according to a preagreed list, while Bob returns $b = 1$ if and only if his estimated value of s equals this value. This variation allows Alice to return seemingly random outputs, while yielding the same average payoff.

If Bob can precisely determine the value of s, the above strategy results in the maximum possible average payoff,

$$\bar{\wp} = 2 \sum_j \left(1 - \frac{1}{\sqrt{3}} \right) = 2(3 - \sqrt{3}). \qquad (17)$$

More generally, if $p(+|s, j)$ denotes the probability that Bob estimates $s = +1$ when the state ω_{js}^{C} is sent by the referee, then the strategy yields

$$\sum_s s\langle b\rangle_{j,s} = p(+|+,j) - p(+|-,j), \qquad \sum_s \langle b\rangle_{j,s} = p(+|+,j) + p(+|-,j),$$

and hence the average payoff is given by

$$\begin{aligned}\bar{\wp} &= 2\sum_j \left[\left(1-\frac{1}{\sqrt{3}}\right)p(+|+,j) - \left(1+\frac{1}{\sqrt{3}}\right)p(+|-,j)\right] \\ &= 6\left[\left(1-\frac{1}{\sqrt{3}}\right)\bar{p}(+|+) - \left(1+\frac{1}{\sqrt{3}}\right)\bar{p}(+|-)\right]. \end{aligned} \tag{18}$$

Here, $\bar{p}(+|+) := (1/3)\sum_j p(+|+,j)$ is the average probability that Bob correctly identifies $s = 1$, while $\bar{p}(+|-) := (1/3)\sum_j p(+|-,j)$ is the average probability that Bob wrongly identifies $s = 1$, i.e., a false positive.

It follows immediately that the condition for this cheating strategy to be successful is that the ratio of true positives to false positives satisfies

$$\frac{\bar{p}(+|+)}{\bar{p}(+|-)} > \frac{\sqrt{3}+1}{\sqrt{3}-1} \approx 3.732. \tag{19}$$

This might not appear too much to ask. But, in fact, it is impossible. Bob's aim is to successfully distinguish the set of states $\{\omega^C_{j+}\}$ from the set of states $\{\omega^C_{j-}\}$, where these two sets are clearly not mutually orthogonal. Unfortunately for Bob, quantum mechanics places strong constraints on the success with which such sets can be unambiguously distinguished.

In particular, for Bob to estimate the value of s, he will have to measure some POVM $\{M_\pm\}$ on the states sent to him by Charlie. It follows, recalling that $\sigma^C_{js} = \frac{1}{2}(1 + s\sigma_j)$, that

$$\bar{p}(+|s) = (1/3)\sum_j \mathrm{Tr}\left[M_+\omega^C_{js}\right] = (1/6)\sum_j \mathrm{Tr}\left[M_+(1+s\sigma_j)\right].$$

Further, the requirement $M_+ > 0$ implies that $M_\pm = \mu(1 + m\cdot\sigma)$ for some $\mu > 0$ and 3-vector m of length no greater than unity. Substitution then gives

$$\frac{\bar{p}(+|+)}{\bar{p}(+|-)} = \frac{3+\sum_j m_j}{3-\sum_j m_j} \le \frac{\sqrt{3}+1}{\sqrt{3}-1}, \tag{20}$$

where the inequality is easily obtained by maximising $\sum_j m_j$ subject to the constraint $m\cdot m \le 1$ (equality corresponds to $m_j \equiv 1/\sqrt{3}$).

Comparison of Eqs. (19) and (20) immediately shows that the cheating strategy fails: Bob cannot make any measurement that distinguishes the input states sufficiently for him to estimate s to the required degree of accuracy.

5.4 *Quantum Refereeing as Quantum Programming of Measurement Devices*

It also of interest to give some insight as to how Alice and Bob can win this quantum-refereed steering game, when they do share a suitable steerable quantum state ρ. In particular, the states sent by the referee can be regarded as 'programming' Bob's devices to make corresponding measurements, where neither Bob nor his devices are able to cheat by 'reading' the program (as this would again correspond to distinguishing between sets of nonorthogonal states).

In the general case, suppose that Alice measures the POVM $\{E_a^x\}$ on receipt of classical input x from the referee, and Bob measures the POVM $\{E_b^{BC}\}$ on receipt of quantum input ω_y^C from the referee. The joint outcome probability distribution corresponding to x and y then follows as

$$p(a,b|x,y) = \mathrm{Tr}\left[(E_a^x \otimes E_b^{BC})(\rho \otimes \omega_y^C)\right] = \mathrm{Tr}\left[(E_a^x \otimes M_b^y)\rho\right], \tag{21}$$

where $M^y := \{M_b^y\}$ is an 'induced' or 'programmed' POVM on Bob's Hilbert space, defined by

$$M_b^y := \mathrm{Tr}_C[E_b^{BC}\omega_y^C]. \tag{22}$$

Thus, unlike a classically-refereed game, in which the referee sends a classical signal y and Bob chooses a corresponding measurement to make on his system, a quantum referee sends a quantum signal ω_y^C that determines Bob's corresponding measurement on his shared system: Bob's measurement is 'quantum programmed' by the referee (although Bob retains some freedom via his choice of $\{E_b^{BC}\}$).

For the particular quantum-refereed steering game defined in Sect. 5.1, consider the case where E_1^{BC} corresponds to the projection onto a singlet state, i.e.,

$$E_1^{BC} = |\psi_-\rangle\langle\psi_-| = \frac{1}{4}\left(\hat{1} \otimes \hat{1} - \sum_j \sigma_j \otimes \sigma_j\right). \tag{23}$$

Thus, Bob makes a partial Bell-state measurement on the combination of the state sent by Charlie and his component of the state ρ that he shares with Alice. Note that Bell-state measurements are natural in the context of quantum-refereed correlation games, as they are an integral part of the existence proofs by Buscemi [9] and Cavalcanti et al. [10]. The adequacy of *partial* Bell-state measurements was discovered by Branciard et al. for entanglement games [11], and generalised to steering games by Kocsis et al. [13]. It can be shown more generally that it is always best for Bob to make an entangling (i.e., non-factorisable) measurement.

Equations (22) and (23) yield the corresponding programmed POVM element

$$M_1^{js} = \frac{1}{8}\mathrm{Tr}_C\left[\left(\hat{1}\otimes\hat{1} - \sum_k \sigma_k\otimes\sigma_k\right)(\hat{1}+s\sigma_j)\right] = \frac{1}{4}(\hat{1}-s\sigma_j) = \tfrac{1}{2}\omega^C_{j,-s}. \tag{24}$$

Thus, when an eigenstate of σ_j is sent by the referee, the programmed POVM element is proportional to the projection onto the orthogonal eigenstate. In this way the referee effectively receives information about measurements of σ_j on Bob's component of the shared state. Bob cannot cheat, however, because he does not know which σ_j measurement is actually programmed in any run.

If Alice further measures $-\sigma_j$ on receipt of signal j from the referee, and they share a Werner state as in Eq. (7), then substitution into Eq. (12) yields the average payoff [13]

$$\bar{\wp} = -\sum_j \mathrm{Tr}\left[\sigma_j\otimes\sigma_j\rho_W\right] - \sqrt{3} = 3W - \sqrt{3}. \tag{25}$$

Thus, Alice and Bob can win the game for any value $W > 1/\sqrt{3}$.

More generally, it is not difficult to show that whatever measurement Alice makes, the average payoff for the game under a partial Bell-state measurement by Bob corresponds precisely to the degree by which the corresponding steering inequality in Eq. (13) is violated.

5.5 Relaxing Communication Restrictions?

As mentioned previously, Rosset et al. have demonstrated the existence of quantum-refereed entanglement games for which the requirement that Alice and Bob do not communicate during the game can be relaxed [12]. Here the extent to which a similar relaxation is possible for quantum-refereed steering games is investigated, for the example in Sect. 5.1. It turns that that while allowing communication from Alice to Bob permits cheating, allowing communication from Bob to Alice does not.

In particular, for the steering game in Sect. 5.1, suppose first that Alice can send classical signals to Bob. They can then cheat as follows: Alice passes the input j she receives from the referee on to Bob. Bob uses this information to measure σ_j on the corresponding state ω^C_{js} he has received from the referee, thus determining s. This in turn allows a perfect implementation of the cheating strategy in Sect. 5.3, yielding a positive average payoff of $2(3-\sqrt{3})$ as per Eq. (17). Hence, one-way communication from Alice to Bob cannot be permitted in this game.

Conversely, however, suppose that Bob can send classical signals to Alice, but not vice versa. From the form of the average payoff in Eq. (12), the only way to cheat is for Alice to ensure that her output satisfies $a = sb$ as often as possible, so that a positive sign dominates in the first term. Since Bob can send her his value of b, she therefore only further needs a reliable estimate of s from him. But, as we have already seen in Sect. 5.3, Bob's best possible estimate of s is just not good enough to be of any help to her. Hence, one-way communication from Bob to Alice, i.e., from the steered party to the steering party, can be permitted without compromising the game.

6 Experiment: Trust-Free Verification of EPR Steering

A number of experimental implementations of quantum-refereed entanglement games have now been performed [22–24], as well as an experimental implementation of a quantum-refereed steering game [13]. The latter is of particular interest for two reasons: (i) it provides a proof of principle for trust-free verification of EPR steering, raising the possibility of two-sided secure quantum key distribution without the need for a Bell nonseparable state; and (ii) it explicitly accounts for imperfections in the referee's preparation of the states he sends to Bob.

In particular, in any experimental implementation of a quantum-refereed game, the referee cannot perfectly prepare the states intended to be sent to the untrusted party or parties. The accuracy of preparation must therefore be taken into account to prevent cheating by Alice and Bob. As an extreme example, suppose for the steering game in Sect. 5.1 that the referee prepares the states $\tilde{\omega}^C_{js} := \frac{1}{2}(1 + s\sigma_1)$, independently of j, rather than eigenstates of σ_j. Then Bob can unambiguously determine s by measuring σ_1, and then implement the cheating strategy of Sect. 5.3 to achieve a positive average payoff of $2(3 - \sqrt{3})$.

The way to account for imperfect preparations is to modify the payoff function for the game, based on tomography of the prepared states. For example, in Ref. [13] the modified average payoff

$$\bar{\wp}(r) := 2\sum_{j,s}\left(s\langle ab\rangle_{j,s} - \frac{r}{\sqrt{3}}\langle b\rangle_{j,s}\right) \tag{26}$$

was used, where $r \geq 1$ is a tomographically-determined measure of the imperfect preparation, with $r = 1$ corresponding to perfect preparation. Recalling $b \geq 0$, one has $\bar{\wp}(r) < \bar{\wp}(1)$ for $r > 1$, and so a positive average payoff is harder to achieve for imperfect state preparation—indeed so much harder that Alice and Bob are prevented form cheating, as follows from an argument analogous to that in Sect. 5.2 [13].

It should further be noted that any noise, in the quantum channel used to send states to the untrusted parties, does not compromise quantum-refereed games [13]. For example, for the steering game in Sect. 5.1, suppose that Bob receives the states $\phi(\omega^C_{js})$, where ϕ is a completely positive trace preserving (CPTP) map describing the quantum channel used by the referee. To show this does not allow any cheating by Alice and Bob, note first that

$$\mathrm{Tr}_{BC}[E^{BC}_b \rho^B_\lambda \otimes \phi(\omega^C_{js})] = \mathrm{Tr}_{BC}[\tilde{E}^{BC}_b \rho^B_\lambda \otimes \omega^C_{js}], \tag{27}$$

for any joint POVM $\{E_b^{BC}\}$ on $H_B \otimes H_C$, where the modified POVM $\{\tilde{E}_b^{BC}\}$ is defined by $\tilde{E}_b^{BC} := (I_B \otimes \phi^*)(E_b^{BC})$. Here I_B denotes the identity map on H_B, and ϕ^* denotes the dual map defined by $\mathrm{Tr}\,[X\phi(Y)] = \mathrm{Tr}\,[\phi^*(X)Y]$. It is easily checked from this definition that $\{\tilde{E}_b^{BC}\}$ is indeed a POVM. Hence, the proof in Sect. 5.2 that Alice and Bob cannot cheat goes through just as before, using the modified POVM $\{\tilde{E}_b^{BC}\}$ in place of $\{E_b^{BC}\}$.

It is worth noting that this robustness of quantum-refereed games under noise may reduce the degree to which the average payoff needs to be modified to account for imperfect preparation. For example, if the experimentally prepared states $\tilde{\omega}_y^C$ can be written in terms of the intended states as $\tilde{\omega}_y^C = \phi(\omega_y^C)$, for some CPTP map ϕ, then no modification at all is necessary. More generally, however, some tradeoff between finding a suitable ϕ and modifying the payoff will be necessary.

The experiment reported by Kocsis et al. implemented the quantum-refereed steering game in Sect. 5.1 using optical polarisation qubits and partial Bell-state measurements, where imperfect state preparation by the referee required a modified average payoff $\bar{\wp}(r)$ as per Eq. (26), with $r = 1.081$ [13]. Positive average payoffs of 1.09 ± 0.03 and 0.05 ± 0.04 were obtained, for shared Werner states with $W = 0.98$ and $W = 0.698$ respectively, thus confirming a shared steering resource without any trust in Alice and Bob. The latter case is of particular interest, as this Werner state does not violate any known Bell inequality, including the Bell inequality in Eq. (8) (which requires $W \geq 1/\sqrt{2} \approx 0.707$).

7 Discussion

Quantum-refereed games remove the need for trust in parties and their devices, when verifying the correlation strength of a shared resource, by replacing trust with quantum signal states. The physical mechanism by which cheating is prevented, including when communication restrictions are relaxed, has been studied in detail in Sect. 5, using a particular quantum-refereed steering game as an example.

It should be noted that while quantum refereeing regains measurement device independence, in the verification of any level of the hierarchy of quantum correlations, this is at some cost. First, while there is no need to trust Alice and Bob or their devices, the referee must be able to trust his own characterisation of the quantum states he sends. Second, for Alice and Bob to win a quantum-refereed game, the parties that are sent a quantum state by the referee must be able to perform a joint measurement on that state and their local state. Hence, the technical demands are higher than for tests of Bell nonseparability, in which only classical signals need be sent. Finally, as is seen in the proof given in Sect. 5.2, the referee must trust that Alice and Bob's devices are subject to the laws of quantum mechanics (although no particular quantum model of the devices need be assumed).

Future work includes exploring whether allowing one-way communication from the steered party to the steering party, as per the example in Sect. 5.5, can be generalised to all quantum-refereed steering games; and finding an explicit protocol for secure two-sided quantum key distribution based on a steerable but Bell-nonlocal resource. It would also be of interest to investigate whether proposed measures of steerability of quantum states in the literature [5, 25, 26] respect the ordering induced by quantum-refereed steering games defined in Ref. [10].

Acknowledgements This work was supported by the Australian Research Council, Project No. DP 140100648. I thank Cyril Branciard, Francesco Buscemi, Yeong-Cherng Liang, Geoff Pryde, Dylan Saunders, Ernest Tan and Howard Wiseman for useful discussions.

References

1. Werner, R.F.: Quantum states with EPR correlations admitting a hidden variable model. Phys. Rev. A **40**, 4277–4281 (1989)
2. Ollivier, H., Zurek, W.H.: Quantum discord: a measure of the quantumness of correlations. Phys. Rev. Lett. **88**, 017901 (2001)
3. Wiseman, H.M., Jones, S.J., Doherty, A.C.: Steering, entanglement, nonlocality, and the EPR paradox. Phys. Rev. Lett. **98**, 140402 (2007)
4. Piani, M., Watrous, J.: Phys. All entangled states are useful for channel discrimination. Phys. Rev. Lett. **102**, 250501 (2009).
5. Piani, M., Watrous, J.: Necessary and sufficient quantum information characterization of Einstein-Podolsky-Rosen steering. Phys. Rev. Lett. **114**, 060404 (2015)
6. Branciard, C., Cavalcanti, E.G., Walborn, S.P., Scarani, V., Wiseman, H.M.: One-sided device-independent quantum key distribution: security, feasibility, and the connection with steering. Phys. Rev. A **85**, 010301(R) (2012)
7. Brunner, N., Cavalcanti, D., Pironio, S., Scarani, V, Wehner, S.: Bell nonlocality. Rev. Mod.Phys. **86**, 419–478 (2014)
8. Jones, S.J., Wiseman, H.M., Doherty, A.C.: Entanglement, Einstein-Podolsky-Rosen correlations, Bell nonlocality, and steering. Phys. Rev. A **76**, 052116 (2007)
9. Buscemi, F.: All entangled states are nonlocal. Phys. Rev. Lett. **108**, 200401 (2012)
10. Cavalcanti, E.C.G., Hall, M.J.W., Wiseman, H.M.: Entanglement verification and steering when Alice and Bob cannot be trusted. Phys. Rev. A **87**, 032306 (2013)
11. Branciard, C., Rosset, D., Liang, Y.-C., Gisin, N.: Measurement-device independent entanglement witness for all entangled quantum states. Phys. Rev. Lett. **110**, 060405 (2013)
12. Rosset, D., Branciard, C., Gisin, N., Liang, Y.-C.: Entangled states cannot be classically simulated in generalized Bell experiments with quantum inputs. NJP **15**, 05302 (2013)
13. Kocsis, S., Hall, M.J.W., Bennet, A.J., Saunders, D.J., Pryde, G.J.: Experimental measurement-device-independent verification of quantum steering. Nat. Commun. **6**, 5886 (2015)
14. Lim, C.C.W.: Optimality of semiquantum nonlocality in the presence of high inconclusive rates. Phys. Rev. A **93**, 020101(R) (2016)
15. Schrödinger, E.: Discussion of probability relations between separated systems. Proc. Camb. Phil. Soc. **31**, 555–563 (1935)
16. Einstein, A., Podolsky, B., Rosen, N. Can quantum-mechanical description of physical reality be considered complete? Phys. Rev. **47**, 777–780 (1935)
17. Einstein, A.: Autobiographical notes. In: Albert Einstein: Philosopher-Scientist. Ed. by P.A. Schilpp, Library of the Living Philosophers, Evanston (1949), pp. 1–94
18. Bell, J.S.: On the Einstein Podolsky Rosen paradox. Physics **1**, 195–200 (1964)
19. Clauser, M.F., Horne, M.A., Shimony, A., Holt, R.A.: Proposed experiment to test local hidden-variable theories. Phys. Rev. Lett. **23**, 880-884 (1969)

20. Cavalcanti, E.G., Jones, S.J., Wiseman, H.M., Reid, M.D.: Experimental criteria for steering and the EPR paradox. Phys. Rev. A **80**, 032112 (2009)
21. Terhal, B.M.: Bell inequalities and the separability criterion. Phys. Lett A **271**, 319 (2000)
22. Xu, P. *et al.*: Implementation of a measurement-device-independent entanglement witness. Phys. Rev. Lett. **112**, 140506 (2014)
23. Nawareg, M., Muhammad, S., Amselem, E., Bourennane, M.: Sci. Rep. **5**, 8048 (2015)
24. Verbanis, E., Martin, A., Rosset, D., Lim, C.C.W., Thew, R.T., Zbinden, H.: Resource-efficient measurement device independent entanglement witness. Phys. Rev. Lett. **116**, 190501 (2016)
25. Skrzypczyk, P., Navascués, M., Cavalcanti, D.: Quantifying Einstein-Podolsky-Rosen Steering, Phys. Rev. Lett. **112**, 180404 (2014)
26. Costa, A.C.S., Angelo, R.M.: Quantification of Einstein-Podolski-Rosen steering for two-qubit states. Phys. Rev. A **93**, 020103(R) (2016)

Dynamics and Statistics in the Operator Algebra of Quantum Mechanics

Holger F. Hofmann

Abstract Physics explains the laws of motion that govern the time evolution of observable properties and the dynamical response of systems to various interactions. However, quantum theory separates the observable part of physics from the unobservable time evolution by introducing mathematical objects that are only loosely connected to the actual physics by statistical concepts and cannot be explained by any conventional sets of events. Here, I examine the relation between statistics and dynamics in quantum theory and point out that the Hilbert space formalism can be understood as a theory of ergodic randomization, where the deterministic laws of motion define probabilities according to a randomization of the dynamics that occurs in the processes of state preparation and measurement.

Keywords Weak values · Unitary transformations · Action
Ozawa uncertainties · Planck's constant

1 Introduction

Quantum theory is unique in the history of science. No other theory of natural phenomena has caused as much confusion about the relation between logical concepts and experimental observations. It may therefore be necessary to take a step back and examine the reasons for the confusion without hastily committing to one of the many ideological camps that have sprung up in the course of the scientific discussion. To do so, we should remind ourselves that the scientific method is to resolve controversies by a direct appeal to shared experience in the form of experimental observations. If quantum theory is really a scientific theory, all controversies can be decided by focusing the discussion on the experimental evidence. Specifically, we need to take

H. F. Hofmann (✉)
Graduate School of Advanced Sciences of Matter,
Hiroshima University, Kagamiyama 1-3-1, Higashi Hiroshima 739-8530, Japan
e-mail: hofmann@hiroshima-u.ac.jp

M. Ozawa et al. (eds.), *Reality and Measurement in Algebraic Quantum Theory*,
Springer Proceedings in Mathematics & Statistics 261,
https://doi.org/10.1007/978-981-13-2487-1_8

care that all our statements, no matter how sophisticated or abstract, can be fully explained in terms of their relevance for possible experimental observations.

The historical problem of quantum theory is that it was developed with a minimum of experimental input, using extrapolations that were motivated mostly by the beauty of the mathematical formalism [16]. However, technology has advanced to the point where we can finally control and measure individual quantum systems. Interestingly, the effects we can now observe are correctly described and predicted by the original formalism, and yet we have not been able to resolve the paradoxes associated with quantum theory. Indeed, the number of paradoxes has only increased [1–3, 6, 15, 30, 31], and many of these paradoxes have been confirmed experimentally without providing any hint of an underlying physical reality [9, 13, 27, 33, 38, 42, 43, 47, 49]. At the heart of all of these perplexing paradoxes lies the fact that we do not understand the quantum processes used to measure and control the physical properties of quantum systems. It is here where a proper revision of quantum mechanics should start: how does the established formalism deal with the problem of measurement and control?

An interesting contribution to this important question has been provided by Ozawa, who showed that the uncertainties of quantum measurements can be much lower than textbook formulations of the uncertainty principle suggest [37]. Most importantly, this result was derived entirely from the algebraic structure of the Hilbert space formalism, without any speculations about the underlying realities. Experimental studies are possible and have been realized [4, 12, 39, 40, 46], but these methods rely on indirect evaluations of the uncertainties, illustrating the fundamental problem that it is impossible to obtain the uncertainty free value of the target observable in conjunction with the uncertain outcome of an individual measurement. The dilemma of quantum measurements is that one cannot go back in time and obtain the value of a different observable for the same system. In Ozawa's theory, the problem is solved by using the operator formalism to define the value of a physical property mathematically, but critics of this approach tend to insist on definitions of uncertainties that are based only on the experimentally observable statistics of measurement outcomes - a notion that is extremely restrictive in the context of quantum mechanics [7, 8, 11, 45].

It seems to me that the present discussions are missing the actual point. Clearly, Ozawa's theory is valid within the stage set by the formalism. The confusion arises because the self-adjoint operators used to describe physical properties cannot be identified with the measurement outcomes through which we experience the physical property. To solve this problem, we need to review why quantum theory seems to introduce physical properties in two different and essentially incompatible ways - both as qualitative measurement outcomes with possible statistical errors and as quantitative shifts of pointer position averages associated with the external measurement setup (the "meter system"). In the formalism, this dualism between quality and quantity is represented by operators, with the measurement operators of positive valued operator measures (POVMs) describing the qualitative outcome and the self-adjoint operators associated with observable properties of the system describing the quantity that is responsible for the pointer shift of the meter [36]. As I will show in the following, the problem can be addressed by considering the peculiar role of unitary dynamics in the formalism, which leads to a new understanding of the action in quantum statistics [25]. From the mathematical side, the close relation between unitary transformations and self-adjoint operators established by their corresponding eigenstates indicates that the eigenstate projectors represent time-averaged orbits of the dynamics, and not just the selection of a specific subset from a set of pre-determined realities. In the theoretical description of a quantum measurement, the dephasing processes associated with the observation of a precise outcome correspond to a dynamical randomization along the complete orbit. Importantly, it is not possible to separate this ergodic orbit into individual phase space points. Both the experimental evidence and the theoretical description therefore suggest that each measurement samples the complete dynamics generated by the target observable.

In this paper, I will explain the relation between the elements of Hilbert space algebra and the experimental processes used in the laboratory. It is then possible to see that the algebra describes the fundamental laws of physics that govern physical interactions at the absolute limit of control set by the fundamental constant $\hbar$. In particular, I will address the origin of probabilities and the reason why quantum statistics is different from classical phase space statistics. The central result is that our understanding of experiments and experimental evidence cannot be based on preconceived notions of reality, but should instead emerge from the laws of causality that relate phenomena to physical objects. Quantum mechanics only appears strange and confusing because we fail to include the role of the dynamics in these causality relations. At the order of magnitude defined by the constant $\hbar$, Hilbert space is needed to express the dynamical structure of physical processes, which is more fundamental than the cruder notion of static realities commonly used in classical physics.

2 The Physics of Hilbert Space

Many introductions to quantum mechanics start from the assumption that physical systems are described by a "state". The problem with such an introduction is twofold. Firstly, real systems are usually in motion, and secondly, the word "state" has no meaning until we explain how the "state" can describe a specific situation found in the real world. Interestingly, the closest practical analogy to the use of the term "state" in quantum theory is found in statistical physics, where thermal states are described by ergodic averages of their motion, with each orbit obtaining a statistical weight according to the energy of the orbit. In fact, we can see that the analogy works perfectly in quantum mechanics, where the thermal state is given by the density operator

$$\hat{\rho} = \sum_n \frac{1}{Z} \exp\left(-\frac{E_n}{k_B T}\right) \mid \psi_n\rangle\langle\psi_n \mid . \tag{1}$$

The canonical partition function Z is defined as in classical physics and the projectors on the energy eigenstates $| \psi_n\rangle$ take over the role of the orbits of energy E_n.

Thermal states are time independent by definition. In quantum theory, this is particularly easy to see, since the energy eigenstates are also eigenstates of the unitary transformation $\hat{U}(t)$ that describes the time evolution of states. In fact, the similarity between the theoretical representation of time evolution and the representation of time independent ergodic states is a non-trivial feature of quantum theory that should not be underestimated. I hope that the arguments I am presenting here will draw more attention to this fact, and to the necessary consequences for our understanding of physics. Specifically, the time evolution is represented by an operator of the form

$$\hat{U}(t) = \sum_n \exp\left(-i\frac{S_n}{\hbar}\right) | \psi_n\rangle\langle\psi_n |, \tag{2}$$

where the action S_n is given by the product of energy and time, $E_n t$. Two observations are important here. Firstly, no such operator exists in classical physics, and this makes it extremely difficult to identify the actual relations between classical concepts and the Hilbert space algebra. Secondly, the action S_n is the quantity that defines the amount of change induced by $\hat{U}$, and it is here that the fundamental constant $\hbar$ obtains a physical meaning.

Experimentally, we can control systems by manipulating their interactions using the available forces, very often in the form of rather strong electromagnetic fields. Unfortunately, most systems are also experiencing a wide range of completely uncontrolled interactions, and this often limits the quality of control to a level where quantum effects cannot be seen. Note that the presence of these uncontrolled interactions means that the mathematical structure of classical physics is not confirmed by any experimental results, since the correspondence between experimental result and classical theory is merely an approximate fit valid at very limited resolution. Differential equations are only successful in describing real world physics because their solutions roughly approximate those patterns in our experience of nature that are robust to the extra noise of real life physics. To investigate the actual laws of physics, we need to remove these extra noise sources, and that is quite difficult. In many cases, it involves vacuum chambers and highly specialized methods of cooling.

To "prepare" a quantum state, we usually start by isolating and cooling a physical system, which results in an isolated ground state - the $T \to 0$ limit of Eq. (1). We can then obtain the desired state by applying fields, the effects of which are described by unitary transformations defined by an action S_n as shown in Eq. (2). A quantum state provides a mathematical summary of these processes, which should allow us to understand the observable effects of our "preparation" in interactions with other objects in the laboratory. This is the point where quantum theory causes the most misunderstandings. Firstly, the mathematical description is so abstract that we usually fail to see the relation with the actual physics of quantum state preparation. Secondly, the description of the measurement process is also given in abstract terms, making it impossible to identify the outcomes of measurements with "elements of reality".

The latter problem is well known and has led to the controversies about different interpretations of quantum mechanics. What we can say for sure is that the quantum state is not a conventional description of physical reality, since it does not describe the system by assigning precise values to the observable properties of the object. Likewise, quantum measurement theory does not provide us with a conventional description of causality, where the measurement outcome is simply an effect caused by a well-defined property of the object.

The standard textbook solution to the measurement problem is to assume that a precise measurement of a physical property $\hat{A}$ will result in an outcome given by an eigenvalue A_a of the self-adjoint operator $\hat{A}$, where the probability of the outcome is given by the projection on the eigenstate $\mid a\rangle$ of the operator. The problem with this approach is that it only applies to a very narrow range of measurements, and these kinds of measurements are not really representative of physics in general. Thus, the measurement postulate fails to connect the description of physical properties by self-adjoint operators with the experimental reality of physics in the laboratory. A proper understanding of both state preparation and measurement requires a closer look at the physics that is being summarized by the mathematical expressions. The question is whether the Hilbert space formalism itself already gives us some clues about the relations between the physics of state preparation and measurement on the one side, and the mathematics of state vectors and projectors on the other. Based on recent research, I would say that the essential insight is contained in the representation of dynamics by unitary transformations, as represented by the relation between Eqs. (1) and (2).

3 Quantum Ergodicity

Let us start with the problem of state preparation. The starting point is a cooling process which involves random interactions that have no specific time dependence. As a result, the system is left in a completely random phase of its motion, which is why thermal statistics can be derived using the ergodic hypothesis that identifies ensemble averages with time averages. In quantum mechanics, state preparation most often starts from an energetic ground state. However, the Hilbert space formalism makes no fundamental distinction between ground states and excited states. Motion is described by Eq. (2), and in that equation, energy eigenstates are stationary because they represent ergodic averages over the motion described by $\hat{U}(t)$. This fact can be confirmed by considering an alternative method of state preparation, where an arbitrary physical property $\hat{A}$ is determined by a precise measurement. This requires an interaction that conserves $\hat{A}$ while changing all other properties according to a random force ϕ that represents the back-action of the meter on the system. The effect on an arbitrary initial state $\rho(\text{in})$ is given by

$$\rho(\text{out}) = \lim_{L\to\infty} \frac{1}{L} \int_0^L \exp\left(-i\frac{\phi}{\hbar}\hat{A}\right) \quad \hat{\rho}(\text{in}) \quad \exp\left(i\tfrac{\phi}{\hbar}\hat{A}\right)$$
$$= \quad \sum_a \mid a\rangle\langle a \mid \quad \hat{\rho}(\text{in}) \quad \mid a\rangle\langle a \mid . \tag{3}$$

Thus the pre-condition of a preparation by measurement is a randomization of the dynamics along the trajectory represented by $\mid a\rangle\langle a \mid$. The loss of coherence between different eigenstates finds its physical meaning in the randomization of the dynamics along a. We should therefore not think of quantum states as representations of the physical quantity A_a given by the eigenvalue of $\hat{A}$, but as complete orbits of the dynamics generated by the physical property $\hat{A}$. This is precisely why the action plays such a fundamental role in quantum physics.

The important message here is that state preparation is not just "knowledge of the property A_a." The quantum state also contains a memory of the dynamics by which A_a was determined. That is the fundamental reason why we cannot just add information about a different physical property $\hat{B}$ to an initial state $\mid a\rangle\langle a \mid$. The orbit $\mid b\rangle\langle b \mid$ is fundamentally different from the orbit $\mid a\rangle\langle a \mid$, and there is no"joint orbit" of a and b. Nevertheless, there is a kind of intersection between the two orbits, and this intersection obtains its physical meaning when a precise measurement of $\hat{B}$ is performed after the preparation of $\mid a\rangle\langle a \mid$. Specifically, the measurement is just the time reverse of a quantum state preparation, and the reason why the outcome B_b should not be mistaken for a measurement independent "element of reality" is that it can only be obtained after the system was driven through the complete orbit described by $\mid b\rangle\langle b \mid$. Note that this observation is closely related to the role of the eigenvalues in the dynamics generated by an operator. The original motivation for the formulation of Eq. (2) was that the frequencies of dipole oscillations in atomic transitions correspond to differences between the energy levels. In other words, the differences between energy eigenvalues $E_n - E_m$ correspond to periodicities T in the dynamics of the system,

$$E_n - E_m = \frac{2\pi\hbar}{T_{nm}}. \tag{4}$$

Importantly, T_{nm} is a property of the complete orbit generated by the operator of energy $\hat{H}$. Therefore, the energy eigenvalues E_n cannot just represent the energy of a single point along the orbit, but need to be associated with the dynamics of the entire orbit. Experimental observation of quantized values necessarily require interactions that sample the complete orbit generated by the observable. Physical effects that do not involve a complete orbit cannot resolve a specific eigenvalue. The emergence of quantized eigenvalues is therefore a feature of the dynamics, and not a static reality of the non-interacting system.

We can now get a better physical understanding of the textbook version of a quantum measurement by considering the relation between the initial randomization of the dynamics represented by $\mid a\rangle\langle a \mid$ and the final randomization represented by $\mid b\rangle\langle b \mid$. The measurement outcome b is obtained from a because the two orbits intersect, and the statistical weight of the intersection between the orbits, which corresponds to the dwell time of a in b (or equivalently, of b in a), is given by the well known formula

$$P(b|a) = \mathrm{Tr}\,(|\, b\rangle\langle b \mid a\rangle\langle a\,|) \,. \tag{5}$$

This standard rule of quantum statistics therefore represents a relation between the dynamics along a and the dynamics along b, which corresponds to the classical phase space geometry of ergodic orbits.

In general, a quantum system will also evolve in time between the initial preparation and the final measurement, so it may be useful to take a closer look at the way that the unitary transformation in Eq. (2) connects state preparation and measurement when the operators $\hat{A}$, $\hat{B}$ and $\hat{H}$ do not commute. In that case, the eigenstates $\mid a\rangle$ and $\mid b\rangle$ can be represented by superpositions of the eigenstates of energy, and the time dependent probability of finding a after a time t is

$$\begin{aligned} P(b|a;t) &= \langle b \mid \hat{U}(t) \mid a\rangle\langle a \mid \hat{U}^{\dagger}(t) \mid b\rangle \\ &= \left| \sum_n \exp\left(-i\frac{S_n}{\hbar}\right) \langle b \mid \psi_n\rangle\langle\psi_n \mid a\rangle \right|^2 . \end{aligned} \tag{6}$$

In this context, it is interesting to consider how much time it will take to get from a to b. In quantum mechanics, this is a somewhat ambiguous question, since we can only determine the probability of b at a time t. A meaningful answer is only obtained if the superposition of energy eigenstates in a results in a highly localized peak in the time dependence of this probability. It is therefore more useful to ask at what time the probability of b is maximal for an initial state $\mid \phi(a)\rangle$ centered around a and an average energy of E. For such a localized state, the maximal probability of b is reached when the time evolution of the phases in Eq. (6) cancels out the phase differences that exist at $t = 0$. We can therefore conclude that the transformation distance between a and b is given by the energy dependent quantum phases, which can be evaluated in terms of the energy dependent action

$$S_n(\text{max.}) = \hbar\, \mathrm{Arg}(\langle b \mid \psi_n\rangle\langle\psi_n \mid a\rangle). \tag{7}$$

This relation shows that the phases in the eigenstate decompositions of $\mid a\rangle$ and $\mid b\rangle$ have a clear physical meaning: they describe the transformation distance between a and b along the orbits ψ_n. It is possible to connect this to the classical notion of a transformation distance as the time t needed to get from a to b along an orbit of specific energy E. In Eq. (6), the action of the time evolution is given by $S_n = E_n t$. Phases in the vicinity of E_n are equal when the energy gradient of $S_n(\text{max.})$ is compensated by the gradient of $S_n = E_n t$, which is given by the time t. For that purpose, we can approximate the action $S_n(\text{max.})$ by a continuous function of energy $S(E)$, where the continuous energy E represents the expectation value of a minimum uncertainty state centered around a and E. We can then use the energy dependence of the quantum mechanical action phase $S_n(\text{max.})$ to determine the classical limit of the time of propagation between a and b at energy E,

$$t(b, a, E) = \frac{\partial}{\partial E} S(b, a, E). \tag{8}$$

As discussed in more detail in [25], we can use this relation to derive quantum mechanical phases directly from the classical description of the dynamics. In fact, the notion of transformation distance allows us to derive quantum interference effects from the classical laws of dynamics, which provides a physical explanation of the main differences between quantum statistics and classical phase space statistics. Specifically, it is possible to derive a phase space analog of quantum statistics that incorporates the transformation distance in the form of complex phases for the joint and conditional probabilities that relate non-commuting physical properties to each other.

4 Phase Space Analogs and Their Limitations

The identification of projection operators with orbits raises an important question about the physics of Hilbert space. Why is it that the orbits cannot be expressed as a sequence of points that correspond to the changing values of physical properties along the orbit? Why is it that the intersection between two orbits does not identify a phase space point defined by the pair of eigenvalues that characterize the two orbits? We can actually use the concept of transformation distance to address this question.

Classical phase space points provide a compact description of all physical properties. For example, the intersection of the orbits a and b would provide a well defined value for the energy E, and this value would be found where the transformation distance between a and b along E was zero,

$$\frac{\partial}{\partial E} S(b, a, E) = 0. \tag{9}$$

In quantum mechanics, this relation can be no more than an approximation. If we look at the definition of transformation distance in Eq. (7), we can see that this approximation relates to a stationary phase in Hilbert space. As shown in [18], this phase also appears in weak measurements of the probability of finding E_n conditioned by an initial state a and a final state b,

$$S(b, a, E_n) = \hbar \text{Arg}\left(\frac{\langle b \mid \psi_n\rangle\langle\psi_n \mid a\rangle}{\langle b \mid a\rangle}\right). \tag{10}$$

It is interesting to note that coarse graining this complex weak value over an energy interval ΔE will eliminate contributions with action derivatives much greater than $\hbar/\Delta E$, leaving only results in the vicinity of the classical solution $E(b, a)$ [18, 20, 21]. This means that weak values establish a physically meaningful link between Hilbert space and classical phase space. What is even more astonishing is that the mathematics of this phase space analogy was already discovered in the early days of quantum mechanics, when it was constructed from the operator algebra as an alternative to the Wigner distribution [10, 29, 35]. Unfortunately, these mathematical insights were mostly forgotten by the time that Aharonov, Albert and Vaidman introduced weak measurements and their result, the weak values [2]. It was therefore not immediately recognized that the oddities of weak values merely describe the differences between classical phase space concepts and their more accurate description in the Hilbert space formalism. However, recent experimental demonstrations have shown that weak measurements can be used to directly measure quantum states as weak joint probabilities of two non-commuting observables [5, 32, 34, 41, 48]. These results show that weak values represent the quantum mechanical analog of classical phase space statistics, including the non-classical correlations between physical properties that cannot be measured jointly. It is also worth noting that weak values can also be observed at finite measurement strengths, indicating that the algebra of weak values provides an experimentally relevant description of non-classical correlations [17, 19, 22, 26, 28, 42, 44, 50]. In fact, many of the recent experimental investigations of quantum paradoxes have used weak measurements to show that the paradoxical features can be understood as a direct consequence of the negative weak conditional probabilities associated with action phases of $\hbar\pi$ in Eq. (10) [9, 13, 24, 27, 33, 38, 42, 43, 47, 49].

With this large number of results from both experiment and theory, it is rather surprising that so little attention has been paid to the role that the operator algebra plays in determining the non-classical statistics that are observed by weak measurements and related methods. As shown in [23], it is actually possible to argue that the Hilbert space algebra itself defines the ordered product of the projection operators as the only reasonable representation of joint probabilities for the possible measurement outcomes of two non-commuting observables. We can now understand this result in terms of the identification between projectors and orbits discussed above. The joint statistical weights of two orbits in a quantum state $\hat{\rho}$ are then given by

$$\begin{aligned}\rho(a, b) &= \langle \mid b\rangle\langle b \mid a\rangle\langle a \mid\rangle \\ &= \langle b \mid a\rangle\langle a \mid \hat{\rho} \mid b\rangle.\end{aligned} \tag{11}$$

It is fairly easy to see that this is a complete description of the state $\hat{\rho}$ for any two basis systems with non-zero mutual overlaps $\langle b \mid a\rangle$. In fact, this expression was already introduced as a phase space analog by Dirac in 1945, and is therefore often referred to as the Dirac distribution [10]. In the same work, Dirac also introduced weak values as a mathematical description of operators. In terms of the operator algebra, we can see that the weak values for all combinations of a and b give a complete description of the operator $\hat{M}$ as

$$\hat{M} = \sum_{a,b} \frac{\langle b \mid \hat{M} \mid a\rangle}{\langle b \mid a\rangle} \mid b\rangle\langle b \mid a\rangle\langle a \mid . \tag{12}$$

Thus weak values are closely associated with the idea that the product of projection operators represents the intersection of two orbits and therefore corresponds to the closest analogy to a phase space point that can be defined in quantum physics.

A complete description of the operator algebra associated with complex joint probabilities has been given in [20]. For the present purpose, it is sufficient to note that the strangeness of the statistics associated with weak values and complex probabilities arises from the dynamical relations between the physical properties. It is therefore not possible to assign an eigenstate $\mid m\rangle$ of the operator $\hat{M}$ to the combination of orbits (a, b). Instead, the contribution of the orbit m to the intersection of the orbits a and b is given by a complex conditional probability,

$$P(m|a, b) = \frac{\langle b \mid m\rangle\langle m \mid a\rangle}{\langle b \mid a\rangle}, \tag{13}$$

where the probability $P(m)$ of finding m for a specific Dirac distribution $\rho(a, b)$ is given by the standard form for conditional probabilities,

$$P(m) = \sum_{a,b} P(m|a, b)\rho(a, b). \tag{14}$$

As shown in Eqs. (7) and (10), the complex conditional probability $P(m|a, b)$ describes transformation distances between the different orbits rather than joint realities [18, 21]. It is therefore necessary to distinguish the reality of a precise measurement outcome from the dynamical relations between physical properties.

5 The Relation Between Mathematics and Physical Reality

We now turn to the central question that has caused so much confusion in quantum physics. How do the physical properties of a system appear in the outcomes of an actual experiment? As mentioned at the end of Sect. 2, the concept of measurement given by most textbooks of quantum mechanics is actually too narrow to accommodate all of the possible interactions of a physical system. In a more general description of measurements, the outcome m is represented by an operator $\hat{E}_m$, so that the probability of obtaining m for a quantum state $\mid \psi\rangle$ is given by

$$P(m|\psi) = \langle \psi \mid \hat{E}_m \mid \psi\rangle. \tag{15}$$

Effectively, the operator $\hat{E}_m$ describes the conditional probability of obtaining m for arbitrary initial conditions $\mid \psi\rangle$. But what is the relation between the measurement outcome m and the physical properties of the system? The specific realization of the measurement should give a non-trivial answer, and that answer must somehow enter into the Hilbert space description as well.

As mentioned in the introduction above, an interesting solution to the problem was presented by Ozawa [37] and has recently been investigated in a number of experiments [4, 12, 39, 40, 46]. In this approach, the operator $\hat{A}$ is used to represent the target observable, and a quantitative estimate $\tilde{A}_m$ is associated with each measurement outcome m. The measurement error is then given by the difference between the operator $\hat{A}$ and the value $\tilde{A}_m$. Since this expression is itself an operator, it needs to be evaluated using the operator algebra. The expression for the total measurement error derived by Ozawa is

$$\varepsilon^2(A) = \sum_m \langle \psi \mid (\tilde{A}_m - \hat{A})\hat{E}_m(\tilde{A}_m - \hat{A}) \mid \psi \rangle. \tag{16}$$

As we discuss in a recent paper [36], this relation makes a non-trivial statement about the relation between quantitative properties and measurement outcomes. Specifically, we show that the only possible definition of joint statistical weights for the eigenstate outcomes a and the actual outcomes m that is consistent with the quantitative definition of the error in Eq. (16) is given by

$$P(m, a|\psi) = \mathrm{Re}(\langle \psi \mid \hat{E}_m \mid a \rangle \langle a \mid \psi \rangle). \tag{17}$$

Since the algebra of Hilbert space corresponds directly to the algebra of classical probabilities, the optimal estimate is then given by the real part of the weak value of $\hat{A}$, as already pointed out by Hall in [14], soon after the initial concept had been introduced by Ozawa.

In the present context, it is important to realize that the weak values are optimal estimates because they accurately summarize the causality relations between non-commuting physical properties in the Hilbert space formalism. To understand the relation between measurement outcomes and causality better, one should keep in mind that the initial state $\mid \psi \rangle$ represents an orbit generated by a specific physical property $\hat{B}$, so that $\mid \psi \rangle$ is an eigenstate of $\hat{B}$ with an eigenvalue of B_ψ. We can now add the quantity $\hat{A}$ to the value of $\hat{B}$ to obtain a new quantity $\hat{M}$, and this quantity defines a new set of orbits $\mid m \rangle$. A quantitative measurement of $\hat{M}$ identifies the eigenvalue M_m of the final orbit. Since the quantity $\hat{A}$ is defined as the difference between $\hat{M}$ and $\hat{B}$, it is clear that its value should be

$$A(\psi, m) = M_m - B_\psi. \tag{18}$$

Here, the quantity $A(\psi, m)$ does not refer to an orbit generated by $\hat{A}$. Instead, it related the orbits expressed by $\mid \psi \rangle$ and $\mid m \rangle$ to each other. Specifically, M_m and B_ψ are eigenstates of $\mid m \rangle$ and $\mid \psi \rangle$ for operators $\hat{M}$ and $\hat{B}$, defined in such a way that $\hat{A}$ can be expressed as the operator sum $\hat{M} + \hat{B}$ as shown in [36],

$$A(\psi, m) = \frac{\langle m \mid (\hat{M} + \hat{B}) \mid \psi \rangle}{\langle m \mid \psi \rangle} \qquad (19)$$
$$= \frac{\langle m \mid \hat{A} \mid \psi \rangle}{\langle m \mid \psi \rangle}.$$

Note that in the general case the values of $A(\psi, m)$ are complex, requiring a non-hermitian operator $\hat{M}$ for the assignment of complex values m_m to the measurement outcomes m. However, such an assignment is not necessarily meaningless since the purpose of the present analysis is to identify a precise relation between the value $A(\psi, m)$ of $\hat{A}$ and the eigenvalues M_m and B_ψ, where the statistical errors in the quantitative relation are zero. Ozawa's error relation confirms this expectation by defining the contribution of m to the error $\varepsilon^2(A)$ as

$$\varepsilon(A, m) = \langle m \mid (\hat{A} - \tilde{A}_m) \mid \psi \rangle$$
$$= \left(A(\psi, m) - \tilde{A}_m \right) \langle m \mid \psi \rangle. \qquad (20)$$

This contibution is zero whenever the estimate $\tilde{A}_m$ is equal to the complex weak value conditioned by ψ and m. In the example above, $\tilde{A}_m = A(\psi, m)$ is only possible when the weak value $A(\psi, m)$ is real, so that a measurement error of $\varepsilon^2(A) = 0$ is only possible when all of the weak values associated with different measurement outcomes m are real. However, there is no logically binding reason to maintain the restriction to real values when the untimate goal is the identification of deterministic relations between physical properties. As discussed above, the complex weak value is a valid quantification of the intersection of the orbits $\mid \psi \rangle$ and $\mid m \rangle$ in terms of the quantity defined by the operator $\hat{A}$. By extending the estimate to complex values, it is always possible to obtain the error free value $\tilde{A}_m = A(\psi, m)$ from a maximally precise measurement. Since the value $A(\psi, m)$ is error free, it can serve as a deterministic expression of the relation between the value of A and the precisely defined conditions ψ and m which holds for all quantum states $\hat{\rho}$. We can verify that this is indeed correct by using the joint statistics of ψ and m defined by the Dirac distribution of $\hat{\rho}$,

$$\rho(m, \psi) = \langle \psi \mid \hat{\rho} \mid m \rangle \langle m \mid \psi \rangle. \qquad (21)$$

Note that here, the quantum state is given by $\hat{\rho}$, whereas ψ is merely a basis state used to characterize the statistics of $\hat{\rho}$. The expectation value of $\hat{A}$ in $\hat{\rho}$ can now be explained as an average of the deterministic values $A(\psi, m)$ of $\hat{A}$ determined by the combinations of ψ and m,

$$\langle \hat{A} \rangle = \sum_{m,\psi} A(\psi, m) \rho(m, \psi). \qquad (22)$$

We can therefore conclude that error free relations between $\hat{A}$, m and ψ provide a state independent description of deterministic relations between physical properties [20, 21].

6 Empirical Objectivity and Non-classical Correlations

The central merit of the Hilbert space formalism is that it provides an objective description of the quantum system. In quantum mechanics, this presents a problem because we cannot simply neglect the role of the environment in the physical processes used to prepare and measure the system. In popular discussions of quantum physics, it has often been suggested that quantum physics involves some mysterious influence of the observer on the result, implying the complete absence of objective laws of causality. It is therefore important to stress that the Hilbert space formalism does not allow any such "external" effects. Even the description of preparation and measurement is entirely objective. The problem arises only from the possible choice between different state preparations or measurement procedures. However, these procedures are all defined by physical interactions with the object, and the effects of these physical interactions can then be described objectively by using the Hilbert space algebra.

We need to understand the algebra of Hilbert space as a description of causality that relates a physical object to the evidence of its existence found outside of the system. Objectivity is only possible if we can apply rules of causality to eliminate the unavoidable contextuality introduced by external devices. Quantum physics shows that the most fundamental elements of reality are processes, not properties. Processes can be objectified as orbits described by Hilbert space projectors. The result is a proper causal description of the system, where the self-adjoint operators describing physical properties can be used to evaluate the quantitative effects observed at finite sensitivities.

It is somewhat unfortunate that quantum physics is rarely applied properly to systems that behave in a nearly classical fashion. It is important to remember that the classical description of such systems is merely approximate, no matter how large they are. In most cases, the observation of objects involves fluctuations that are much larger than the quantum limit. Just as an extreme example, we can consider the motion of the moon around the earth. At a distance of about 400 000 Km from the center of the earth, a single photon of visible light scattered by the surface of the moon will change the angular momentum by about $5 \times 10^{15}\hbar$. Since no classical description of the orbit of the moon can take into account every single photon scattered by the moon, it is obvious that classical physics is no more than a very crude approximation - except by relative standards, of course, where we should consider that the total angular momentum of the moon going around the earth is about $3 \times 10^{68}\hbar$. The motion of the moon is therefore quite robust against the disturbances caused by the light we need to see it by. We should just avoid the misconception that the moon has a reality independent of its interaction with light and matter. The fact that the moon is continuously immersed in interactions with its environment makes the moon real, just as all other objects are only real as a source of their physical interactions.

The detailed investigations of non-classical correlations we have recently performed indicate that we should take imaginary correlations seriously [26, 28]. This is a direct consequence of the relation between unitary transformations and statistics in the operator algebra. Specifically, the imaginary correlations of two non-commuting operators $\hat{A}$ and $\hat{B}$ is given by the expectation value of the commutation relation,

$$\text{Im}\left(\langle \hat{A}\hat{B}\rangle\right) = -\frac{i}{2}\langle[\hat{A}, \hat{B}]\rangle. \tag{23}$$

Therefore, the time evolution of any physical property $\hat{A}$ is evidence of an imaginary correlation between $\hat{A}$ and the energy $\hat{H}$,

$$\text{Im}\left(\langle \hat{A}\hat{H}\rangle\right) = \frac{\hbar}{2}\frac{d}{dt}\langle \hat{A}\rangle. \tag{24}$$

Importantly, it is possible to experimentally observe the imaginary correlation between $\hat{A}$ and $\hat{H}$ in weak measurements or in any other experimental reconstruction of the Dirac distribution $\rho(A, H)$. Oppositely, it is not possible to observe the change in $\hat{A}$ without changing the energy $\hat{H}$ as a result of the necessary interactions. Therefore, the identification of the rate of change with an imaginary correlation does not contradict our experience. Rather, the assumption that we observe physical properties as exact real numbers is at odds with the empirical evidence. We can quickly confirm that the limit placed by Eq. (24) on our ability to estimate both the energy and the value of $\hat{A}$ from the evidence is not unrealistically high. After all, $\hbar$ is a very small action. For example, the imaginary correlation between position and energy achieves its maximal possible value at the speed of light, where it is a mere 1.58×10^{-26} Jm. The lesson we should learn from such considerations is that the assumption that we could hypothetically control physical systems with absolute precision is mostly a fantasy based on sloppy thinking. Quantum mechanics reveals that we need to make corrections to the artificially precise laws of motion once we approach the limit where small actions do matter. Nevertheless the laws of motion remain objective and consistent. The origin of the randomness observed in quantum experiments is explained by the limitations of control that these deterministic laws of motion impose on the possible interactions with the system. We can understand these limitations once we realize that the mathematical formalism describes dynamics and causality, and not the static realities represented by the classical phase space algebra.

7 Conclusions

In quantum theory, Hilbert space is used to describe the deterministic relations between physical properties that allow us to trace external effects of a system back to causes within the system. Objective reality emerges as a result of the causality relations between observations of the same object made at different times, or, in the spirit of ergodic theory, between observations made on identically prepared objects of the same type. Importantly, the physical properties of an object are known only through the effects of interactions - by "touch and sight." It is a serious mistake to assume that reality is accessible by abstract thought. The elements of the theory do not represent platonic realities. Each one of them needs to be justified by actual effects observed in the laboratory. This demand may seem overly restrictive, and it should not be taken as an attempt to reject speculations about possible observations that have not been realized due to technical limitations - but the present discussion of quantum mechanics suffers from unnecessary confusion because scientists cling to concepts of reality that are clearly at odds with the observed phenomena. A more careful distinction between the observable world and unobservable figments of the imagination may therefore be helpful. In particular, we should be more humble in admitting that our knowledge of reality is limited to our actual experience, and that the extrapolations of our personal experience to possible experiences beyond our technical capabilities may result in delusions about the real world. Science should provide a tool by which we can reach an agreement on questions about the external reality, and this can only be achieved if there is a shared experience of the world that we can all relate to.

The intention of the analysis of quantum theory developed here and in a number of related works [20, 21, 25] is to provide an empirical foundation of quantum physics that explains how the formalism describes the observable laws of physics that shape our experience of the world around us. At the center is the realization that objects obtain their reality by their appearance, and the properties of the object are the quantities that determine the possible effects of the object that determine its appearance, both in the laboratory and in nature. The abstraction of the "state" should really be understood in terms of this experience, where the projection on a Hilbert space vector actually represents an interaction that randomizes the dynamics of the system in the course of the interaction by which the object causes an observable effect. The strangeness of quantum statistics originates from the peculiar role played by the laws of motion that determine this dynamical randomization. Specifically, our approximate separation between reality and dynamics - the assumption of a static reality - breaks down when the interaction is sensitive to actions of $\hbar$ or less. This is no different from the breakdown of the independence of time and motion when velocities approach the speed of light. It may therefore be possible to gain a better fundamental understanding of physics by noticing that nothing in our experience indicates that the reality of objects is static and can be frozen in time. Quantum mechanics simply shows that this unnecessary assumption is wrong, and that dynamics forms an essential part of objective reality in the limit of small actions.

References

1. Y. Aharonov and L. Vaidman, J. Phys. A: Math. Gen. **24**, 2315 (1991).

2. Y. Aharonov, D. Z. Albert, and L. Vaidman, Phys. Rev. Lett. **60** 1351 (1988).
3. Y. Aharonov, S. Popescu, D. Rohrlich, and P. Skrzypczyk, New J. Phys. **15**, 113015 (2013).
4. S.-Y. Baek, F. Kaneda, M. Ozawa, and K. Edamatsu, Sci. Rep. **3**, 2221 (2013).
5. C. Bamber and J. S. Lundeen, Phys. Rev. Lett. **112**, 070405 (2014)
6. J. S. Bell, Physics **1**, 195 (1964).
7. P. Busch, P. Lahti, and R. F. Werner, Phys. Rev. Lett. **111**, 160405 (2013).
8. P. Busch, P. Lahti, and R. F. Werner, Rev. Mod. Phys. **86**, 1261 (2014).
9. T. Denkmayr, H. Geppert, S. Sponar, H. Lemmel, A. Matzkin, J. Tollaksen, and Y. Hasegawa, Nat. Commun. **5**, 4492 (2014).
10. P. A. M. Dirac, *Rev. Mod. Phys.* **17** 195 (1945).
11. J. Dressel and F. Nori, Phys. Rev. A **89**, 022106 (2014).
12. J. Erhart, S. Sponar, G. Sulyok, G. Badurek, M. Ozawa, and Y. Hasegawa, Nat. Phys. **8**, 185 (2012).
13. M. E. Goggin, M. P. Almeida, M. Barbieri, B. P. Lanyon, J. L. O'Brien, A. G. White, and G.J. Pryde, Proc. Natl. Acad. Sci. U. S. A. **108** 1256 (2011).
14. M. J. W. Hall, Phys. Rev. A **69**, 052113 (2004).
15. L. Hardy, Phys. Rev. Lett. **68**, 2981 (1992).
16. W. Heisenberg, *Quantentheorie und Philosophie* (Reclam, Stuttgart, 1979)
17. M. Hiroishi and H.F. Hofmann, J. Phys. A: Math. Theor. **46**, 245302 (2013).
18. H. F. Hofmann, New J. Phys. **13**, 103009 (2011).
19. H. F. Hofmann, Phys. Rev. Lett. **109**, 020408 (2012).
20. H. F. Hofmann, New J. Phys. **14**, 043031 (2012).
21. H. F. Hofmann, Phys. Rev. A **89**, 042115 (2014).
22. H. F. Hofmann, New J. Phys. 16, 063056 (2014).
23. H. F. Hofmann, Quantum Stud. : Math. Found. **1**, 39 (2014).
24. H. F. Hofmann, Phys. Rev. A **91**, 062123 (2015).
25. H. F. Hofmann, Eur. Phys. J. D **70**, 118 (2016).
26. M. Iinuma, Y. Suzuki, T. Nii, R. Kinoshita, and H. F. Hofmann, Phys. Rev. A **93**, 032104 (2016).
27. A. N. Jordan, A. N. Korotkov, and M. Büttiker, Phys. Rev. Lett. **97**, 026805 (2006).
28. S. Kino, T. Nii, and H. F. Hofmann, Phys. Rev. A **92**, 042113 (2015).
29. J. G. Kirkwood, *Phys. Rev.* **44**, 31 (1933).
30. S. Kochen and E.P. Specker E P, J. Math. Mech. **17**, 59 (1967).
31. A. J. Leggett and A. Garg, Phys. Rev. Lett. **54**, 857 (1985).
32. J. S. Lundeen and C. Bamber, Phys. Rev. Lett. **108**, 070402 (2012).
33. J. S. Lundeen and A. M. Steinberg, Phys. Rev. Lett. **102**, 020404 (2009).
34. J. S. Lundeen, B. Sutherland, A. Patel, C. Stewart, and C. Bamber, *Nature* **474**, 188 (2011).
35. N. H. McCoy, *Proc. Natl. Acad. Sci. U. S. A.* **18**, 674 (1932).
36. T. Nii, M. Iinuma, and H. F. Hofmann, Quantum Stud.: Math. Found. **5**, 229 (2018)
37. M. Ozawa, Phys. Rev. A **67**, 042105 (2003).
38. K. J. Resch, J. S. Lundeen, and A. M. Steinberg, Phys. Lett. A **324**, 125 (2004).
39. M. Ringbauer, D. N. Biggerstaff, M. A. Broome, A. Fedrizzi, C. Branciard, and A. G. White, Phys. Rev. Lett. **112**, 020401 (2014).
40. L. A. Rozema, A. Darabi, D. H. Mahler, A. Hayat, Y. Soudagar, and A. M. Steinberg, Phys. Rev. Lett. **109**, 100404 (2012).
41. J. Z. Salvail, M. Agnew, A. S. Johnson, E. Bolduc, J. Leach, and R. W. Boyd, Nature Photon **7**, 316 (2013).
42. Y. Suzuki, M. Iinuma, and H. F. Hofmann, New J. Phys. **14**, 103022 (2012).
43. J. Tollaksen, J. Phys. A: Math. Gen. **40**, 9033 (2007).
44. G. Vallone and D. Dequal, Phys. Rev. Lett. **116**, 040502 (2016).
45. Y. Watanabe, T. Sagawa, and M. Ueda, Phys. Rev. A **84**, 042121 (2011).
46. M. M. Weston, M. J. W. Hall, M. S. Palsson, H. M. Wiseman, and G. J. Pryde, Phys. Rev. Lett. **110**, 220402 (2013).
47. N. S. Williams and A. N. Jordan, Phys. Rev. Lett. **100**, 026804 (2008).
48. S. Wu, Sci. Rep. **3**, 1193 (2013).
49. K. Yokota, T. Yamamoto, M. Koashi, and N. Imoto, New J. Phys. **11**, 033011 (2009).
50. P. Zou, Z. Zhang, and W. Song, Phys. Rev. A **91**, 052109 (2015).

A General Framework of Quasi-probabilities and the Statistical Behaviour of Non-commuting Quantum Observables

Jaeha Lee and Izumi Tsutsui

Abstract We propose a general framework of the quantum/quasi-classical transformations by introducing the concept of quasi-joint-spectral distribution (QJSD). Specifically, we show that the QJSDs uniquely yield various pairs of quantum/quasi-classical transformations, including the Wigner–Weyl transform. We also discuss the statistical behaviour of combinations of generally non-commuting quantum observables by introducing the concept of quantum correlations and conditional expectations defined analogously to the classical counterpart. Based on these, Aharonov's weak value is given a statistical interpretation as one realisation of the quantum conditional expectations furnished in our formalism.

Keywords Quasi-probability · Wigner-Weyl transform · Weak value · Quantum correlation · Conditional expectation · Quantisation

1 Introduction

Since the advent of quantum theory founded nearly a century ago, non-commutativity of quantum observables has undoubtedly been in the centrepiece of the theory marking its departure from classical theory. The hallmark of this is Heisenberg's uncertainty relation [1], which has later been elaborated from operational viewpoints by taking account of the measurement device by Ozawa [2, 3]. At the same time, the non-commutativity has been one of the major sources of troubles we face when we try to interpret their measurement outcomes in a sensible manner. This has naturally led to various attempts of 'quantisation' of classical systems, most notably in

J. Lee
National Institute of Informatics, 2-1-2 Hitotsubashi, Chiyoda-ku, Tokyo 101-8430, Japan
e-mail: jlee@post.kek.jp

I. Tsutsui (✉)
Theory Center, Institute of Particle and Nuclear Studies, High Energy Accelerator Research Organization (KEK), 1-1 Oho, Tsukuba, Ibaraki 305-0801, Japan
e-mail: izumi.tsutsui@kek.jp

M. Ozawa et al. (eds.), *Reality and Measurement in Algebraic Quantum Theory*, Springer Proceedings in Mathematics & Statistics 261,
https://doi.org/10.1007/978-981-13-2487-1_9

terms of non-commuting Hilbert space operators, or conversely of 'quasi-classical' interpretation of quantum systems in terms of commuting quantities familiar to us in classical theory.

The study on quantum and quasi-classical transformations has a long history dating back to the early days of quantum mechanics. Wigner and Weyl were among the prominent figures who have made much contribution in this effort bearing the theory of Wigner–Weyl transform [4, 5]. Historically, however, all these contributions in this area have been made more or less in a heuristic manner, and apparently their systematic treatment is still underdeveloped, not to mention a transparent overview of the relations among the various proposals of the transformations made so far.

On the other hand, in recent years we have witnessed the rise of interest in an issue related, at the roots, to the interpretation of measurement outcomes under non-commutativity. It is the novel quantity called the *weak value*,

$$A_w := \frac{\langle \psi', A\psi \rangle}{\langle \psi', \psi \rangle} \tag{1}$$

which has been proposed by Aharonov and co-workers [6] based on their time-symmetric formulation of quantum mechanics [7]. The weak value is a physical quantity that characterises the value of the observable A in the process specified by an initial (pre-selected) state $|\psi\rangle$ and a final (post-selected) state $|\psi'\rangle$. Unlike the standard measurement outcomes given by one of the eigenvalues of an observable A obtained in an ideal measurement, the weak value admits a definite value and is considered to be meaningful even for a set of non-commutable observables. The relation between the weak value and the quasi-classical transformations has been argued earlier, specifically with the Kirkwood–Dirac distribution [8, 9].

One of the aims of our present paper, expounded in Sect. 2, is to propose a general framework of the quantum/quasi-classical transformations by introducing the concept of *quasi-joint-spectral distribution* (QJSD). Specifically, we show that the QJSDs, of which definition shortly follows, uniquely yield various pairs of quantum/quasi-classical transformations, and that notable previous proposals of the transformations belong to this framework as special cases. Another aim, to which Sect. 3 is devoted, is to discuss the statistical behaviour of combinations of *generally non-commuting* quantum observables. Specifically, we introduce the concept of quantum correlations and conditional expectations, which are defined in analogue to the classical counterpart, and see how these concepts play together. Based on these, we finally endow Aharonov's weak value with a statistical interpretation as one realisation of the quantum conditional expectations furnished in our formalism.

Mathematical Notations Employed

Throughout this paper, we denote by $\mathbb{K}$ either the real field $\mathbb{R}$ or the complex field $\mathbb{C}$. Since our primary interest is on quantum mechanics, Hilbert spaces are always assumed to be complex. Conforming to the convention in physical literature, we denote the complex conjugate of a complex number $c \in \mathbb{C}$ by c^*, and an inner product

$\langle \cdot, \cdot \rangle$ defined on a complex linear space is anti-linear in its first argument and linear in the second. For simplicity, we adopt the natural units where we specifically have $\hbar = 1$, unless stated otherwise.

2 Quantisation and Quasi-Classicalisation via QJSDs

For commuting quantum observables, a 'trivial' method of quantum and quasi-classical transformation is available, where the former is known as the functional calculus whereas the latter is known as the Born rule. These maps are both known to be characterised by the joint-spectral measure (JSM) of the observables concerned, and they are understood to be adjoint operations to each other. On the other hand, the problem becomes non-trivial when non-commuting observables are put in to consideration, primarily due to the lack of the JSM.

In this section, we first propose a novel approach to the problem of quantum/quasi-classical transformations by introducing the concept of quasi-joint-spectral distributions (QJSDs), which are intended as non-commuting generalisations to the JSMs of commuting observables. Just as the JSM induces a unique adjoint pair of quantum/quasi-classical transformation for commuting observables, QJSDs induce various adjoint pairs for non-commuting observables. Specifically, we see that there exists inherent indefiniteness in the possible definition of QJSDs, each leading to different possible transformations, in which the Wigner–Weyl transform belongs as a special case.

2.1 Preliminary Observations

As a prelude to our study, we first review some basic facts in quantum theory regarding the spectral theorem of self-adjoint operators, the functional calculus and the Born rule. In what follows, we consider a finite combination of simultaneously measurable observables, and observe that both the functional calculus and the Born rule can respectively be understood as the trivial realisation of *quantisation* and *quasi-classicalisation*, in the sense that the former allows us to map real functions (i.e., classical observables) to self-adjoint Hilbert space operators (i.e., quantum observables), and the latter defines a map from density operators (i.e., quantum states) to probability distributions (i.e., classical states). We then point out that the functional calculus and the Born rule are *adjoint* notions to each other. These rather trivial observations shall be the guiding line for our further study in considering the non-trivial problem of quantisation and quasi-classicalisation for the general case involving combination of non-commuting quantum observables.

2.1.1 Spectral Theorem for Self-adjoint Operators

A basic but important result of functional analysis (i.e., linear algebra for arbitrary-dimensions) regarding self-adjoint operators is the *spectral theorem*, which states that for a self-adjoint operator A on a Hilbert space $\mathscr{H}$, there corresponds a unique *spectral measure* E_A such that

$$A = \int_{\mathbb{R}} a \, dE_A(a) \tag{2}$$

holds. In simple terms, the spectral measure

$$E_A(a) = \begin{cases} 0 & (a \text{ is not an eigenvalue of } A) \\ P_a & (a \text{ is an eigenvalue of } A) \end{cases}, \quad a \in \mathbb{R} \tag{3}$$

of A is a map from the eigenvalues a of A to the orthogonal projections P_a on the corresponding eigenspaces. For the case where the eigenvalues are all non-degenerate, the spectral measure is simply nothing but $E_A(a) = P_a = |a\rangle\langle a|$, where $|a\rangle\langle a|$ is the orthogonal projection on the 1-dimensional subspace of $\mathscr{H}$ spanned by the eigenspace corresponding to the eigenvalue a. In this case, the integral (2) formally reduces to the familiar form

$$A = \int_{\mathbb{R}} a|a\rangle\langle a| \, da. \tag{4}$$

For the case in which the Hilbert space $\mathscr{H}$ under consideration is moreover finite-dimensional, the spectral theorem is nothing but the eigendecomposition theorem

$$A = \sum_{i=1}^{n} a_i |a_i\rangle\langle a_i|$$

valid for Hermitian matrices A, and the spectral measure $E_A(a_i) = |a_i\rangle\langle a_i|$ reduces to the collection of orthogonal projections corresponding to the eigenvectors a_i of A. Simply put, spectral theorem is thus a generalisation of the eigendecomposition theorem for the infinite dimensional case, and the spectral measures are in turn the generalisation of orthogonal projections onto the corresponding eigenspaces. The primary advantage of this generalisation becomes apparent when infinite dimensional Hilbert spaces must be taken into consideration.[1]

Joint-Spectral Measures

Now, suppose one is given an ordered combination

[1] A self-adjoint operator defined on infinite dimensional Hilbert spaces may sometimes fail to have any eigenvalues in the sense that $A|\psi\rangle = a|\psi\rangle$ holds for some non-zero vector $|\psi\rangle \in \mathscr{H}$, as most famously exemplified by the position operator $\hat{x}$ and the momentum operator $\hat{p}$ of a free particle.

$$\boldsymbol{A} := (A_1, \dots, A_n), \quad 1 \leq n < \infty \tag{5}$$

of a finite number of pairwise strongly commuting[2] distinct self-adjoint operators (i.e., simultaneously measurable quantum observables) on $\mathscr{H}$. An important fact regarding strongly commuting self-adjoint operators is that, one may uniquely construct the *joint-spectral measure* (JSM) of the combination

$$E_{\boldsymbol{A}}(\boldsymbol{a}) = \prod_{i=1}^{n} E_{A_i}(a_i), \quad \boldsymbol{a} := (a_1, \dots, a_n) \in \mathbb{R}^n \tag{6}$$

that fully describes their joint behaviour, where each E_{A_i} is the unique spectral measure corresponding to A_i, $(1 \leq i \leq n)$. One then trivially has

$$A_i = \int_{\mathbb{R}^n} a_i \, dE_{\boldsymbol{A}}(\boldsymbol{a}), \quad 1 \leq i \leq n, \tag{7}$$

if one is to reclaim the original self-adjoint operators.

2.1.2 Functional Calculus

An important fact regarding spectral measures is that it induces a map from the space of functions to Hilbert space operators. Indeed, under the same situation as above, the JSM of the ordered combination (5) induces a map that maps a function f defined on $\mathbb{R}^n$ to the operator

$$f_{E_{\boldsymbol{A}}} := \int_{\mathbb{R}^n} f(\boldsymbol{a}) \, dE_{\boldsymbol{A}}(\boldsymbol{a}) \tag{8}$$

on $\mathscr{H}$. The map $f \mapsto f_{E_{\boldsymbol{A}}}$ is one realisation of the *functional calculus*, which is a general term that points to a map from functions to operators satisfying certain algebraic properties. In fact, under some appropriate conditions, one finds that there is a one-to-one correspondence between functional calculi and spectral measures. In our context, we view the functional calculus as the trivial way to quantise classical observables (i.e., real functions) into quantum observables (i.e., self-adjoint operators). In what follows, we may occasionally write the image of the functional calculus by either of the following notations: $f_{E_{\boldsymbol{A}}} = f_{\boldsymbol{A}} = f(\boldsymbol{A}) = f(A_1, \dots, A_n)$.

---

[2] We say that a pair of self-adjoint operators A and B strongly commutes, if and only if they commute in the level of spectral measures. In a laxer notation, this is to say that

$$E_A(a)E_B(b) = E_B(b)E_A(a), \quad a, b \in \mathbb{R}$$

holds. Strictly speaking, strong commutativity is generally stronger than mere commutativity when unbounded operators are concerned, but we do not intend to delve into the intricacies, which are not essential for our discussion.

2.1.3 Born Rule

The Born rule is the corner stone of the probabilistic interpretation of quantum measurements, which states that, given a density operator (i.e., mixed quantum state) ρ on $\mathscr{H}$ and a combination (5) of simultaneously measurable observables, the joint behaviour of the measurement outcomes is described by the *joint-probability distribution*

$$\rho_{E_A}(\boldsymbol{a}) := \mathrm{Tr}[E_A(\boldsymbol{a})\rho] \tag{9}$$

defined for the combination of simultaneously measurable observables concerned on the state. In our context, we view the Born rule as the trivial realisation of quasi-classicalisation of quantum states (i.e., density operators) into classical states (i.e., probability distributions). In what follows, we occasionally denote the resulting probability distribution by either of the following notations: $\rho_{E_A} = \rho_A = \rho_{(A_1,\ldots,A_n)}$.

2.1.4 Adjointness of the Functional Calculus and the Born Rule

An important observation we point out here that is crucial for our further discussion is that, both quantisation (functional calculus) and quasi-classicalisation (Born rule) are *adjoint* notions.

Dual Pair

To see this point, we first prepare some necessary terminologies and notations. Let $L(\mathscr{H})$ denote the space of all bounded linear operators on the Hilbert space $\mathscr{H}$, and let $N(\mathscr{H})$ denote the space of all nuclear operators (or, better known as trace-class operators) on $\mathscr{H}$. Bounded quantum observables $A \in L(\mathscr{H})$ and quantum states $\rho \in N(\mathscr{H})$ belong to the respective spaces. On the product space $L(\mathscr{H}) \times N(\mathscr{H})$ is defined a bilinear form

$$\langle X, N\rangle_Q := \mathrm{Tr}[XN], \quad X \in L(\mathscr{H}),\ N \in N(\mathscr{H}) \tag{10}$$

that maps a pair of bounded linear operator and a nuclear operator to the trace of their product. In mathematics, a triple consisting of a pair of linear spaces X, Y and a bilinear form $\langle\cdot\,,\,\cdot\rangle : X \times Y \to \mathbb{K}$ satisfying the conditions

$$\begin{aligned} &\forall x \in X \setminus \{0\},\ \exists y \in Y \quad \langle x, y\rangle \neq 0, \\ &\forall y \in Y \setminus \{0\},\ \exists x \in X \quad \langle x, y\rangle \neq 0, \end{aligned} \tag{11}$$

is called a *dual pair*. The triple $\big(L(\mathscr{H}), N(\mathscr{H}), \langle\cdot\,,\,\cdot\rangle_Q\big)$ is one typical realisation of a dual pair, which are the familiar tools we use to describes quantum measurements.

On the other hand, let $\mathscr{S}(\mathbb{R}^n)$, $(1 \leq n < \infty)$ denote the n-dimensional *Schwartz space*, which is the space of all smooth functions that, even being multiplied by any polynomials after being differentiated arbitrarily many times, they 'vanish at infinity'. Also, let $\mathscr{S}'(\mathbb{R}^n)$ denote the continuous dual of the Schwartz space, which is called the space of *tempered distributions*. The space of tempered distributions is an extension of the familiar space of density functions or complex measures: every probability density function is a complex measure, and every complex measure is a tempered distribution, while the converses are not always true.[3] It reveals that the spaces of density functions or that of complex measures is not sufficient in properly handling quasi-probability distributions for non-commuting observables.

Now, if we allow ourselves some abuse of notation, we may formally treat a tempered distribution $\varphi \in \mathscr{S}'(\mathbb{R}^n)$ as a function $\varphi(\boldsymbol{x})$ on $\mathbb{R}^n$, and define the bilinear form

$$\langle f, \varphi \rangle_C := \int_{\mathbb{R}^n} f(\boldsymbol{x})\varphi(\boldsymbol{x})\, dm_n(\boldsymbol{x}), \quad f \in \mathscr{S}(\mathbb{R}^n),\ \varphi \in \mathscr{S}'(\mathbb{R}^n) \tag{12}$$

where we have introduced the renormalised Lebesgue–Borel measure

$$dm_n(\boldsymbol{x}) := (2\pi)^{-n/2} dx^n. \tag{13}$$

For brevity, we occasionally write $dm_1 = dm$ whenever there is no risk for confusion. This renormalisation is mostly of aesthetic purpose, whose advantage becomes apparent when we introduce the Fourier transformation later in our discussion. Equipped with the bilinear form, the triple $\left(\mathscr{S}(\mathbb{R}^n), \mathscr{S}'(\mathbb{R}^n), \langle\,\cdot\,,\,\cdot\,\rangle_C\right)$ also qualifies as a dual pair that becomes a tool in describing classical measurements.

Adjointness of the Transformations

Given an ordered combination of simultaneously measurable quantum observables (5), let

$$\Phi_{E_A} : \mathscr{S}(\mathbb{R}^n) \to L(\mathscr{H}),\ f \mapsto f_{E_A} \tag{14}$$

denote the functional calculus[4] of the ordered combination (5) of self-adjoint operators defined in (8), and in turn let

$$\Phi'_{E_A} : N(\mathscr{H}) \to \mathscr{S}'(\mathbb{R}^n),\ \rho \mapsto \rho_{E_A} \tag{15}$$

[3] The space of density functions on $\mathbb{R}^n$ is a proper subspace of the space of complex measures. An example of this is the delta measure, which is well-defined as a measure but not as a density function. The space of complex measures on $\mathbb{R}^n$ is also a proper subspace of the space of tempered distributions. An example for this is the derivative of the delta measure, which is well-defined as a tempered distribution but not as a complex measure.

[4] Note that the image of a Schwartz function under the functional calculus (14) is a bounded operator with the operator norm $\|f(A)\| \leq \sup_{x\in\mathbb{R}^n} |f(x)|$, hence the map is well-defined.

denote the Born rule[5] described in (9). It can be demonstrated that both (14) and (15) are continuous linear maps between the respective spaces equipped with the usual topologies. One then readily observes by the following straightforward computation

$$\begin{aligned}
\langle \Phi_{E_A}(f), \rho \rangle_Q &:= \mathrm{Tr}[f_{E_A}\rho] \\
&= \int_{\mathbb{R}^n} f(\boldsymbol{a})\, d\mathrm{Tr}[E_A(\boldsymbol{a})\rho] \\
&= \int_{\mathbb{R}^n} f(\boldsymbol{a})\rho_{E_A}(\boldsymbol{a})\, dm_n(\boldsymbol{a}) \\
&= \langle f, \Phi'_{E_A}(\rho) \rangle_C, \quad f \in \mathscr{S}(\mathbb{R}^n),\ \rho \in N(\mathscr{H}) \qquad (16)
\end{aligned}$$

that the functional calculus (14) and the Born rule (15) are *adjoint maps* to each other. This relation can be illustrated by the following diagram:

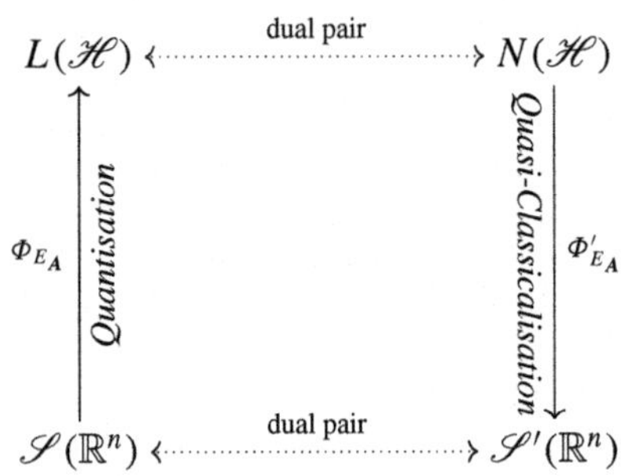

Here, the top row denotes the dual pair of quantum observables and quantum states, whereas the bottom row depicts the classical counterpart. The left column consists of (quantum and classical) observables, whereas the right column consists of (quantum and classical) states.

2.2 *Quantisation and Quasi-classicalisation via QJSDs*

In the previous subsection, we have reviewed the very basics of the spectral theorem, the functional calculus and the Born rule defined for commuting observables, and have seen that the functional calculus (i.e., quantisation) and the born rule (i.e., quasi-classicalisation) are adjoint operations to each other. The next step is to generalise our whole arguments into the case for *non-commuting* observables.

[5]The image of a nuclear operator under the Born rule (15) belongs to the space of complex measures, which can be uniquely embedded into the space of tempered distributions. In this sense, we extend the codomain of the map (15) and understand their images to be tempered distributions, rather than complex measures, for later convenience.

2.2.1 Introducing Quasi-joint-Spectral Distributions (QJSDs)

The key observation to make here is that, it was the JSM that uniquely gave rise to the adjoint pair of the desired maps. A straightforward idea for our current problem would be thus to introduce non-commuting analogues to the JSM.

Strong Commutativity in the Fourier Space

As a preparation to our further discussion, we review the characterisation for strong commutativity of spectral measures in their Fourier spaces. Let A be self-adjoint, and let E_A be its spectral measure. We call the operator valued function

$$(\mathscr{F} E_A)(s) := \int_{\mathbb{R}} e^{-isa}\, dE_A(a) = e^{-isA}, \quad s \in \mathbb{R}, \tag{17}$$

the *Fourier transform* of E_A. It is a basic fact of functional calculus that one may characterise the strong commutativity of self-adjoint operators by the Fourier transforms of their spectral measures: a pair of self-adjoint operators A and B strongly commutes if and only if the Fourier transforms of the respective spectral measures

$$e^{itA}e^{isB} = e^{isB}e^{itA}, \quad s, t \in \mathbb{R} \tag{18}$$

commute.

Fourier Transform of JSM

Let us first compute the Fourier transform of the JSMs. Given the JSM $E_{\boldsymbol{A}}$ of an ordered combination of strongly commuting self-adjoint operators, let

$$\begin{aligned}(\mathscr{F} E_{\boldsymbol{A}})(\boldsymbol{s}) &:= \int_{\mathbb{R}^n} e^{-i\langle \boldsymbol{s}, \boldsymbol{a}\rangle}\, dE_{\boldsymbol{A}}(\boldsymbol{a}) \\ &= e^{-i\langle \boldsymbol{s}, \boldsymbol{A}\rangle} \\ &= \prod_{i=1}^{n} e^{-is_i A_i}, \quad \boldsymbol{s} \in \mathbb{R}^n \end{aligned} \tag{19}$$

denote the *Fourier transform* of $E_{\boldsymbol{A}}$. Here, $\langle \boldsymbol{s}, \boldsymbol{a}\rangle := \sum_{i=1}^{n} s_i a_i$ denotes the standard inner product on $\mathbb{R}^n$, and $\langle \boldsymbol{s}, \boldsymbol{A}\rangle := \sum_{i=1}^{n} s_i A_i$. We also note that the last line of the above equality is due to the iterated application of the Lie–Trotter–Kato product formula.

Hashed Operators

From now on, we generalise the use of the notation $\boldsymbol{A}$ so that it may also admit ordered combinations of generally *non-commuting* distinct self-adjoint operators.

To this end, we introduce, in a less formal way, the *hashed operator*

$$\hat{\#}_{\boldsymbol{A}}(\boldsymbol{s}) := \big\{\text{a suitable `mixture' of the `disintegrated'} \\ \text{components of the unitary groups } e^{-is_1 A_1}, \ldots, e^{-is_n A_n}\big\}, \quad \boldsymbol{s} \in \mathbb{K}^n \quad (20)$$

of the operators concerned. Here, by 'disintegration' we mean breaking each of the unitary operators $e^{-is_k A_k}$ into chunks of operator valued functions $T_{k,\iota_k}(s_k)$, $(\iota_k \in I_k)$ such that

$$\prod_{\iota_k \in I_k} T_{k,\iota_k}(s_k) = e^{-is_k A_k} \tag{21}$$

holds. A straightforward example of this is provided by

$$T_{k,\iota_k}(s_k) := e^{-is_k \lambda_{\iota_k} A}, \tag{22}$$

where $\lambda_{\iota_k} \in \mathbb{R}$ are real numbers satisfying the normalisation $\sum_{\iota_k} \lambda_{\iota_k} = 1$. Then, by 'mixture' we imply that a hashed operator $\hat{\#}_{\boldsymbol{A}}(\boldsymbol{s})$ is constructed as a product of all components $T_{k,\iota_k}(s_k)$, $(\iota_k \in I_k, k = 1, \ldots, n)$ in arbitrary orders. We may also allow it to be given by convex combinations of two hashed operators $\hat{\#}_{\boldsymbol{A}}$ and $\hat{\#}'_{\boldsymbol{A}}$ as

$$\hat{\#}''_{\boldsymbol{A}}(\boldsymbol{s}) = \lambda\, \hat{\#}_{\boldsymbol{A}}(\boldsymbol{s}) + (1 - \lambda)\, \hat{\#}'_{\boldsymbol{A}}(\boldsymbol{s}). \tag{23}$$

Examples of the hashed operators pertaining to the simplest case $\boldsymbol{A} = (A, B)$ are thus given by

$$\hat{\#}_{\boldsymbol{A}}(s,t) = \begin{cases} e^{-itB} e^{-isA}, \\ e^{-isA} e^{-itB}, \\ \frac{1+\alpha}{2} \cdot e^{-itB} e^{-isA} + \frac{1-\alpha}{2} \cdot e^{-isA} e^{-itB}, \quad \alpha \in \mathbb{C} \\ \prod_{k=1}^{N} e^{-is\lambda_k A} e^{-it\mu_k B}, \quad \left(\sum_{k=1}^{N} \lambda_k = 1,\ \sum_{k=1}^{N} \mu_k = 1\right), \\ \left(e^{-isA/N} e^{-itB/N}\right)^N, \\ e^{-i\overline{(sA+tB)}} = \lim_{N\to\infty} \left(e^{-isA/N} e^{-itB/N}\right)^N. \end{cases} \tag{24}$$

In general, either or both of the parameters s, t can be made to even admit complex numbers. One such example is given by

$$\hat{\#}^{\kappa}_{\boldsymbol{A}}(s,t) = e^{-i\langle \frac{1-\kappa}{2}, s\rangle A} e^{-itB} e^{-i\langle \frac{1+\kappa}{2}, s\rangle A}, \quad \kappa \in \mathbb{C},\ s \in \mathbb{C},\ t \in \mathbb{R}, \tag{25}$$

where $\langle \kappa, s\rangle = \kappa_1 s_1 + \kappa_2 s_2$ is understood as the standard inner product defined on $\mathbb{R}^2$, where $\kappa = \kappa_1 + i\kappa_2$ $(\kappa_1, \kappa_2 \in \mathbb{R})$ and $s = s_1 + is_2$ $(s_1, s_2 \in \mathbb{R})$ are complex numbers identified as vectors on $\mathbb{R}^2$.

The precise definition as to what types of construction one may allow for hashed operators will depend on the context and the properties that one would like them to retain. For example, the evaluation of the operator norm of the hashed operator

$$\left\| \hat{\#}_{A}(s) \right\| = 1 \tag{26}$$

can be guaranteed if we restrict ourselves to the specific type of construction that only allows hashed operators to be defined in the form of products of (22), since in that case $\hat{\#}_{A}(s)$ becomes a unitary operator. If we moreover consider their convex combinations (23), the equality in (26) generally becomes an inequality $\left\| \hat{\#}_{A}(s) \right\| \leq 1$. If we moreover allow complex numbers to be chosen as λ in (23), the operator norm of $\hat{\#}_{A}(s)$, while still being bounded, may exceed 1.

Hashed Operator of Commuting Observables

One readily realises that the hashed operators (20) are, while differing in their representations, unique *if and only if* the self-adjoint operators concerned are all simultaneously measurable. In that case, it is easy to see that the hashed operators all reduce to the same Fourier transform (19) of the JSM.

Quasi-joint-Spectral Distributions

Under the same conditions as above, let us choose any hashing $\hat{\#}_{A}$ introduced in (20), and introduce the *quasi-joint-spectral distribution*[6] (QJSD) of the ordered pair $\boldsymbol{A}$ defined by its inverse Fourier transform

$$\begin{aligned} \#_{A}(\boldsymbol{a}) &:= (\mathscr{F}^{-1}\, \hat{\#}_{A})(\boldsymbol{a}) \\ &:= \int_{\mathbb{K}^n} e^{i\langle \boldsymbol{a}, \boldsymbol{s}\rangle}\, \hat{\#}_{A}(\boldsymbol{s})\, dm_n(\boldsymbol{s}), \quad \boldsymbol{a} \in \mathbb{K}^n. \end{aligned} \tag{27}$$

Due to the bijectivity of the Fourier transformation, to each QJSD corresponds a unique hashed operator, and hence QJSDs are highly non-unique in the case a given ordered combination $\boldsymbol{A}$ is non-commutative. The QJSD is unique if and only if $\boldsymbol{A}$ admits simultaneous measurability, and in such case, the unique QJSD actually reduces to the JSM itself

$$\#_{A} = E_{A}. \tag{28}$$

By construction and the observations made above, one may surmise that the QJSDs serve as generalisations of the JSM to generally non-commuting observables. Indeed, QJSDs share some of the basic properties one finds in common with the standard JSM. The primary fact we mention is the normalisation property: the total integration

[6]The reason for our choice of the nomination quasi-joint-spectral *distributions*, rather than *measures*, lies in the fact that, in contrast to the JSMs, QJSDs does not necessarily lie in the space of operator valued measures (OVMs). In fact, we understand them as members of the operator valued distributions (OVDs), which is an operator analogue of generalised functions (distributions). The space of OVDs is larger than the space of OVMs, and the latter can be embedded into the former.

of any QJSD reduces to the identity Id, as one readily finds through the following formal computation

$$\begin{aligned}\int_{\mathbb{K}^n} \#_{\boldsymbol{A}}(\boldsymbol{a})\, dm_n(\boldsymbol{a}) &= \int_{\mathbb{K}^n} e^{-i\langle \boldsymbol{0},\boldsymbol{a}\rangle} \#_{\boldsymbol{A}}(\boldsymbol{a})\, dm_n(\boldsymbol{a}) \\ &= (\mathscr{F}\#_{\boldsymbol{A}})(\boldsymbol{0}) \\ &= \hat{\#}_{\boldsymbol{A}}(\boldsymbol{0}) = \mathrm{Id}. \end{aligned} \tag{29}$$

In fact, this is actually a corollary to a more stronger property regarding the marginals

$$\int_{\mathbb{K}} \#_{\boldsymbol{A}}(\boldsymbol{a})\, dm(a_k) = \#_{\boldsymbol{A}_k}(a_1, \ldots, a_{k-1}, a_{k+1}, \ldots, a_n), \tag{30}$$

where $\boldsymbol{A}_k := (A_1, \ldots, A_{k-1}, A_{k+1}, \ldots, A_n)$, $1 \le k \le n$, denotes the ordered combination of the self-adjoint operators that lacks the kth component of the original ordered combination $\boldsymbol{A}$, and $\#_{\boldsymbol{A}_k}$ denotes the QJSD of $\boldsymbol{A}_k$ defined by

$$\hat{\#}_{\boldsymbol{A}_k}(s_1, \ldots, s_{k-1}, s_{k+1}, \ldots, s_n) := \hat{\#}_{\boldsymbol{A}}(s_1, \ldots, s_{k-1}, 0, s_{k+1}, \ldots, s_n), \tag{31}$$

which corresponds to the hashing constructed by 'taking away' all the disintegrated components of the kth member $e^{-is_kA_k}$ from the original hashing $\hat{\#}_{\boldsymbol{A}}$. To see this, let H_k denote the l.h.s. of (30). The Fourier transform of H_k then reads

$$\begin{aligned}&(\mathscr{F}H_k)(s_1, \ldots, s_{k-1}, s_{k+1} \ldots, s_n) \\ &= \int_{\mathbb{K}^{n-1}} \sum_{i=1}^{k-1} \sum_{i=k+1}^{n} e^{-is_ia_i} \left(\int_{\mathbb{K}} \#_{\boldsymbol{A}}(\boldsymbol{a})\, dm(a_k) \right) dm_{n-1}(a_1, \ldots, a_{k-1}, a_{k+1} \ldots, a_n) \\ &= \int_{\mathbb{K}^n} \sum_{i=1}^{k-1} \sum_{i=k+1}^{n} e^{-is_ia_i} e^{-i0\cdot a_n} \#_{\boldsymbol{A}}(\boldsymbol{a})\, dm_n(\boldsymbol{a}) \\ &= \hat{\#}_{\boldsymbol{A}}(s_1, \ldots, s_{k-1}, 0, s_{k+1} \ldots, s_n) \\ &= (\mathscr{F}\#_{\boldsymbol{A}_k})(s_1, \ldots, s_{k-1}, s_{k+1} \ldots, s_n), \end{aligned} \tag{32}$$

and by the injectivity of the Fourier transformation, one concludes $H_k = \#_{\boldsymbol{A}_k}$.

Reclaiming the JSMs

A straightforward but important corollary to the above property is the following observation. Let

$$\boldsymbol{B} = (A_{i_1}, \ldots, A_{i_k}) \tag{33}$$

be an order-preserving (i.e., $1 \leq i_1 < \cdots < i_k \leq n$) subset of $\boldsymbol{A}$ consisting of k ($1 \leq k \leq n$) numbers of *pairwise strongly commuting* distinct members, and let

$$\boldsymbol{B}^c := (A_{j_1}, \ldots, A_{j_{n-k}}) \tag{34}$$

denote its order-preserving (i.e., $1 \leq j_1 < \cdots < j_{n-k} \leq n$) complement consisting of $n - k$ numbers of the members of $\boldsymbol{A}$ that do not belong to $\boldsymbol{B}$. Then, an iterated application of (30) leads to

$$E_{\boldsymbol{B}}(\boldsymbol{b}) = \int_{\mathbb{K}^{n-k}} \#_{\boldsymbol{A}}(\boldsymbol{a})\, dm_{n-k}(\boldsymbol{b}^c), \tag{35}$$

where $\boldsymbol{b} := (a_{i_1}, \ldots, a_{i_k})$ and $\boldsymbol{b}^c := (a_{j_1}, \ldots, a_{j_{n-k}})$ denotes the variables corresponding to the respective order-preserving subsets.[7] This implies that, if one 'integrate-outs' all the variables corresponding to the complement $\boldsymbol{B}^c$, one may reclaim the authentic JSM of B.

2.2.2 Quantisation and Quasi-classicalisation

Now that we have constructed the QJSDs, which could be understood as non-commutative analogues to the standard JSM, we shall now embark on the construction of quantisation and quasi-classicalisation regarding combination of observables that may fail to be measured simultaneously.

Quantisation of Classical Observables

For an ordered combination of (generally non-commuting) distinct quantum observables $\boldsymbol{A}$, let $\#_{\boldsymbol{A}}$ be any QJSD of one's choice. Guided by a straightforward analogy of the functional calculus originally defined for the commutative case, we thus define the map

$$\Phi_{\#_{\boldsymbol{A}}} : f \mapsto f_{\#_{\boldsymbol{A}}} := \int_{\mathbb{K}^n} f(\boldsymbol{a})\, \#_{\boldsymbol{A}}(\boldsymbol{a})\, dm_n(\boldsymbol{a}) \tag{36}$$

[7] Here, we adopt the convention

$$\int_{\mathbb{K}^{n-k}} \#_{\boldsymbol{A}}(\boldsymbol{a})\, dm_{n=k}(\boldsymbol{b}^c) = \#_{\boldsymbol{A}}(\boldsymbol{a})$$

for the case $k = n$.

that maps a Schwartz function $f \in \mathscr{S}(\mathbb{K}^n)$ to a bounded[8] linear operator $f_{\#_A} \in L(\mathscr{H})$. We call the map (36) the *quantisation* pertaining to the QJSD $\#_A$, and in turn call the image $\Phi_{\#_A}(f)$ the quantisation of f. Occasionally, we denote the quantisation of f by either of the following notations: $f_{\#_A} = f(\#_A)$. We may sometimes even omit $\boldsymbol{A}$ and write $f_{\#}$, when the observables concerned are obvious from the context.

Quasi-classicalisation of Quantum States

Conversely, we also allow ourselves to be guided by a straightforward analogy of the Born rule and intend to extend it to the non-commutative case. We thus define the map

$$\Phi'_{\#_A} : \rho \mapsto \rho_{\#_A}(\boldsymbol{a}) := \mathrm{Tr}\,[\,\#_A(\boldsymbol{a})\rho\,]\,, \quad \boldsymbol{a} \in \mathbb{K}^n \tag{37}$$

that maps a density operator $\rho \in N(\mathscr{H})$ to a tempered distribution $\rho_{\#_A} \in \mathscr{S}'(\mathbb{K}^n)$. We call the (image $\rho_{\#_A}$ of the) map (37) the *quasi-classicalisation* (of ρ) pertaining to the QJSD $\#_A$. Specifically, for a density operator $\rho \in N(\mathscr{H})$, i.e., a positive nuclear operator with the normalisation condition $\mathrm{Tr}[\rho] = 1$, we occasionally call the distribution $\rho_{\#_A}$ the *quasi-joint-probability* (QJP) distribution of $\boldsymbol{A}$ on ρ pertaining to the QJSD $\#_A$. Note that, in general, the corresponding QJP distributions may be negative or even complex valued.

Adjointness of Quantisation and Quasi-classicalisation

As one may surmise, quantisation (36) and quasi-classicalisation (37) are *adjoint* operations to each other, as one may readily check by the formal computation

$$\begin{aligned}
\langle \Phi_{\#_A}(f), \rho\rangle_Q &:= \mathrm{Tr}[f_{\#_A}\rho] \\
&= \int_{\mathbb{K}^n} f(\boldsymbol{a})\mathrm{Tr}\,[\,\#_A(\boldsymbol{a})\rho\,]\; dm_n(\boldsymbol{a}) \\
&= \int_{\mathbb{K}^n} f(\boldsymbol{a})\rho_{\#_A}(\boldsymbol{a})\; dm_n(\boldsymbol{a}) \\
&= \langle f, \Phi'_{\#_A}(\rho)\rangle_C, \quad f \in \mathscr{S}(\mathbb{R}^n),\ \rho \in N(\mathscr{H}).
\end{aligned} \tag{38}$$

[8]The quantisation of a Schwartz function f is formally defined as a unique bounded map $f_{\#_A}$ such that the equality

$$\mathrm{Tr}\left[f_{\#_A}\rho\right] := \int_{\mathbb{R}^n} \check{f}(\boldsymbol{s})\mathrm{Tr}\left[\hat{\#}_A(\boldsymbol{s})\rho\right] dm_n(\boldsymbol{s})$$

holds for all $\rho \in N(\mathscr{H})$, where $\check{f}$ denotes the inverse Fourier transform of f. The r. h. s of the above equation is well-defined, since

$$\left|\int_{\mathbb{R}^n} \check{f}(\boldsymbol{s})\mathrm{Tr}\left[\hat{\#}_A(\boldsymbol{s})\rho\right] dm_n(\boldsymbol{s})\right| \le \|\check{f}\|_1 \cdot M\|\rho\|_{\mathrm{nuk}}$$

is finite. Here, $\|\cdot\|_1$ and $\|\cdot\|_{\mathrm{nuk}}$ respectively denote the L^1 norm of integrable functions and the nuclear norm (*alias* trace norm) of nuclear operators, and $\|\hat{\#}_A(\boldsymbol{s})\| \le M$, $\boldsymbol{s} \in \mathbb{K}^n$ is the upper bound of the operator norm of the hashed operator. The existence and uniqueness of such an operator $f_{\#_A} \in L(\mathscr{H})$ is due to the fact that the space of continuous linear functionals on $N(\mathscr{H})$ is isomorphic to the space of bounded operators $N(\mathscr{H})' \cong L(\mathscr{H})$. The continuity of the linear functional $\rho \mapsto \mathrm{Tr}[f_{\#_A}\rho]$ follows directly from the above evaluation.

This relation can be illustrated by the following diagram:

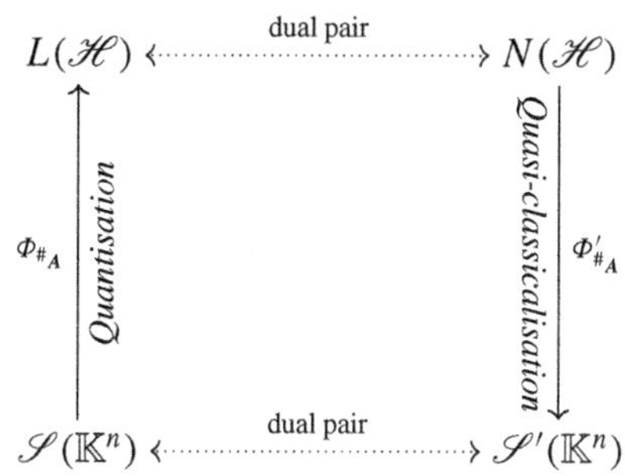

Note again that, as maps, quantisation (36) and quasi-classicalisation (37) are uniquely dictated by the choice of the QJSD $\#_A$. Even for the same classical observable $f \in \mathscr{S}(\mathbb{K}^n)$, its quantisation generally differs $f_{\#_A} \neq f_{\tilde{\#}_A}$ given a distinct choice of the QJSD $\#_A \neq \tilde{\#}_A$, and the same is also true for the quasi-classicalisation of quantum states $\rho \in N(\mathscr{H})$.

2.3 *Quantum/Quasi-classical Representations*

Since quantisation and quasi-classicalisation are adjoint notions to each other, they are different facets of a single entity. In this sense, we occasionally use the term *quantum/quasi-classical representations* or *transformations* referring to the adjoint pair. In general, these representations are non-unique, and each of the representations can be specified by the choice of the QJSD, whose indefiniteness originates directly from the non-commutative nature of the observables concerned.

2.3.1 Transformation of Representations

We may define transformations of QJSDs in various manners. In some cases, it may occur that a group of quantisation/quasi-classical representations could be understood as being *equivalent* to each other in the sense that they can be mutually transformed into one another. Given an ordered combination $\boldsymbol{A}$ of quantum observables, it is an interesting question to ask ourselves how many QJSDs there are (or in other words, the way of ordering of non-commuting observables) that are essentially distinct to each other up to isomorphisms.

The Simplest Case

The simplest case for this is when two QJSDs, while being distinct in their form as hashed operators, are identical. Needless to say, this is trivially always the case when all the members of the ordered combination $\boldsymbol{A}$ pairwise strongly commute. As an example of less trivial cases, let $\boldsymbol{X} = (Q, P)$ be an ordered pair of self-adjoint operators such that the members satisfy the Weyl representation

$$e^{-isQ}e^{-itP} = e^{-ist}e^{-itP}e^{-isQ}, \quad s, t \in \mathbb{R} \tag{39}$$

of the canonical commutation relation (CCR). Then, by a simple observation, one verifies that the hashed operators of the form

$$\begin{aligned} e^{-i\frac{t}{2}P}e^{-isQ}e^{-i\frac{t}{2}P} &= e^{-i\frac{s}{3}Q}e^{-i\frac{t}{2}P}e^{-i\frac{s}{3}Q}e^{-i\frac{t}{2}P}e^{-i\frac{s}{3}Q} \\ &= e^{-i\frac{t}{4}P}e^{-i\frac{s}{3}Q}e^{-i\frac{t}{4}P}e^{-i\frac{s}{3}Q}e^{-i\frac{t}{4}P}e^{-i\frac{s}{3}Q}e^{-i\frac{t}{4}P} \\ &= \cdots \\ &= e^{-i\overline{(sQ+tP)}} \\ &= \cdots \\ &= e^{-i\frac{s}{4}Q}e^{-i\frac{t}{3}P}e^{-i\frac{s}{4}Q}e^{-i\frac{t}{3}P}e^{-i\frac{s}{4}Q}e^{-i\frac{t}{3}P}e^{-i\frac{s}{4}Q} \\ &= e^{-i\frac{t}{3}P}e^{-i\frac{s}{2}Q}e^{-i\frac{t}{3}P}e^{-i\frac{s}{2}Q}e^{-i\frac{t}{3}P} \\ &= e^{-i\frac{s}{2}Q}e^{-itP}e^{-i\frac{s}{2}Q}, \quad s, t \in \mathbb{R}, \end{aligned} \tag{40}$$

are all identical. Here, the equality in the center is due to the product formula proven by Lie–Trotter–Kato, and the overline on the operator $sQ + tP$ denotes its self-adjoint extension. Although they differ in their representation as mixtures, their corresponding QJSDs are identical, and thus all the quantisation/quasi-classical representations induced could be naturally understood as being equivalent.

Affine Transformations

As another example, let $T : \mathbb{K}^n \to \mathbb{K}^n$ be a linear map, and consider the affine map

$$T_b : \boldsymbol{a} \mapsto T\boldsymbol{a} + \boldsymbol{b} \tag{41}$$

defined for $\boldsymbol{b} \in \mathbb{K}^n$. We then introduce the *affine transform* of a QJSD $\#_{\boldsymbol{A}}$ with respect to the affine map (41) by

$$(T_b\#_{\boldsymbol{A}})(\boldsymbol{a}) := \#_{\boldsymbol{A}}\left(T_b^{-1}\boldsymbol{a}\right), \quad \boldsymbol{a} \in \mathbb{K}^n. \tag{42}$$

If T is a bijection (i.e., $\det T \neq 0$), both $T_b\#_{\boldsymbol{A}}$ and $\#_{\boldsymbol{A}}$ are invertible to one another and thus essentially contain the same information of the combination $\boldsymbol{A}$ of the observables concerned. It is to be noted that the affine transform of a QJSD is generally not a member of the QJSDs. Indeed, while the total integration of (42) reduces to the identity Id by definition, it most importantly fails to satisfy the marginal properties (30) in general. Even so, affine transforms of a QJSD give rise to adjoint pairs of quantisation and quasi-classicalisation in an extended sense, which are respectively defined by

$$\Phi_{(T_b\#_A)} : f \mapsto f_{(T_b\#_A)} := \int_{\mathbb{K}^n} f(\boldsymbol{a}) \, (T_b\#_A)\,(\boldsymbol{a}) \, dm_n(\boldsymbol{a}) \tag{43}$$

and

$$\Phi'_{(T_b\#_A)} : \rho \mapsto \rho_{(T_b\#_A)}(\boldsymbol{a}) := \mathrm{Tr}[(T_b\#_A)\,(\boldsymbol{a})\rho]. \tag{44}$$

The adjointness of these operations

$$\langle f_{(T_b\#_A)}, \rho \rangle_Q = \langle f, \rho_{(T_b\#_A)} \rangle_C \tag{45}$$

can be readily confirmed by a simple computation. To see how the quantum/quasi-classical representations corresponding to affine transforms relate to the original representation, we first observe

$$\begin{aligned} f_{(T_b\#_A)} &:= \int_{\mathbb{K}^n} f(\boldsymbol{a}) \, (T_b\#_A)\,(\boldsymbol{a}) \, dm_n(\boldsymbol{a}) \\ &= \int_{\mathbb{K}^n} f(T_b\boldsymbol{a})\#_A(\boldsymbol{a}) \, dm_n(\boldsymbol{a}) \\ &= (f \circ T_b)_{\#_A} \\ &=: (T_b^* f)_{\#_A} \end{aligned} \tag{46}$$

and

$$\begin{aligned} \rho_{(T_b\#_A)}(\boldsymbol{a}) &= \rho_{\#_A}\left(T_b^{-1}\boldsymbol{a}\right), \\ &=: T_{b*}\rho_{\#_A} \end{aligned} \tag{47}$$

where

$$T_b^* f := f \circ T_b, \tag{48}$$

$$T_{b*}\rho := \rho \circ T_b^{-1}, \tag{49}$$

respectively denote the pullback of a function f and the pushforward of a distribution ρ by the affine map T_b. The relation can thus be illustrated by the following diagram

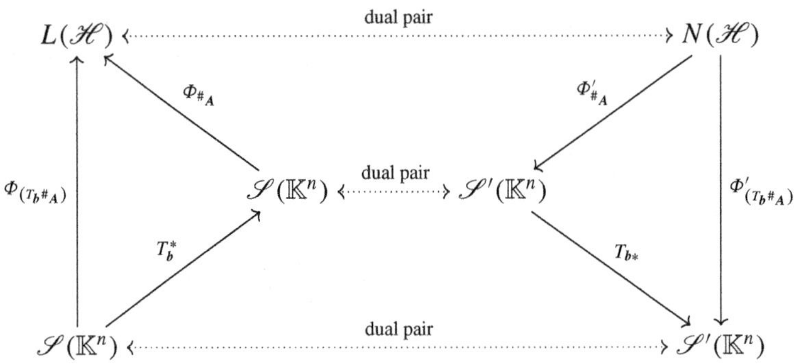

As a simple concrete example, we consider the simplest case $\boldsymbol{A} = (A, B)$. Below, we see that all the members $\#^{\kappa}_{\boldsymbol{A}}$ of the subfamily of the QJSDs of the form (25) are linear transforms of $\#^{i}_{\boldsymbol{A}}$ for the specific choice $\kappa = i$. To see this, consider the matrix

$$\tilde{T}_\kappa := \begin{pmatrix} 1 & \kappa_1 \\ 0 & \kappa_2 \end{pmatrix} \tag{50}$$

defined for each $\kappa = \kappa_1 + i\kappa_2$, $(\kappa_1, \kappa_2 \in \mathbb{R})$, and the linear transformation $T_\kappa := \tilde{T}_\kappa \times \mathrm{Id}$ on $\mathbb{C} \times \mathbb{R}$ defined by $T_\kappa(a, b) := (\tilde{T}_\kappa a, b)$. Then, a simple computation yields

$$\begin{aligned} \mathscr{F}\left(T_\kappa \#^{i}_{\boldsymbol{A}}\right)(s, t) &= \int_{\mathbb{C}\times\mathbb{R}} e^{-i\langle s, T_\kappa a\rangle} e^{-itb}\, \#^{i}_{\boldsymbol{A}}(a, b)\, dm_2(a, b) \\ &= \hat{\#}^{i}_{\boldsymbol{A}}(T'_\kappa s, t) \\ &= \mathscr{F}\#^{\kappa}_{\boldsymbol{A}}(s, t), \end{aligned} \tag{51}$$

where T'_κ denotes the adjoint matrix of T_κ. We thus conclude

$$T_\kappa \#^{i}_{\boldsymbol{A}} = \#^{\kappa}_{\boldsymbol{A}}. \tag{52}$$

Since $\det T_\kappa = \kappa_2$, it is straightforward to see that all the members $\#^{\kappa}_{\boldsymbol{A}}$ of the QJSDs for the choice $\operatorname{Im} \kappa \neq 0$ are equivalent to one another by linear transformations.

Convolutions

As another important class of transformations, we consider the *convolution*

$$(h * \#_{\boldsymbol{A}})(\boldsymbol{a}) := \int_{\mathbb{K}^n} h(\boldsymbol{a} - \boldsymbol{a}')\, \#_{\boldsymbol{A}}(\boldsymbol{a}')\, dm_k(\boldsymbol{a}') \tag{53}$$

of a QJSD $\#_{\boldsymbol{A}}$ and a function h with the total integration of unity. The Fourier transform of the convolution reads

$$
\begin{aligned}
\mathscr{F}(h * \#_{A})(s) &= \hat{h}(s)\,\hat{\#}_{A}(s) \\
&= \int_{\mathbb{K}^n} h(\boldsymbol{k}) e^{-i\langle s, \boldsymbol{k}\rangle}\, \hat{\#}_{A}(s)\, dm_n(\boldsymbol{k}) \\
&= \int_{\mathbb{K}^n} h(\boldsymbol{k})\, \hat{\#}_{A+k}(s)\, dm_n(\boldsymbol{k}),
\end{aligned} \tag{54}
$$

where $\boldsymbol{A} + \boldsymbol{k} := (A_1 + k_1 \cdot \mathrm{Id}, \ldots, A_n + k_n \cdot \mathrm{Id})$ denotes the ordered combination of the normal operators defined as the parallel translation of $\boldsymbol{A}$ towards the direction $\boldsymbol{k} \in \mathbb{K}^n$. From this, the convolution

$$
h * \#_{A} = \int_{\mathbb{K}^n} h(\boldsymbol{k})\, \#_{A+k}\, dm_n(\boldsymbol{k}) \tag{55}
$$

could be understood as the 'weighted average' of the family of QJSDs $\#_{A+k}$ of the ordered combination $\boldsymbol{A} + \boldsymbol{k}$ of normal operators, or equivalently, the parallel translation

$$
\#_{A+k} = \tau_k \#_{A}, \quad (\tau_k \boldsymbol{a} := \boldsymbol{a} + \boldsymbol{k}) \tag{56}
$$

of the original QJSD $\#_{A}$, with respect to the 'weight function' h.

It is important to note that, in general, the convolution (53) itself is not necessarily a QJSD of $\boldsymbol{A}$. Indeed, consider the most extreme case where all the members of $\boldsymbol{A}$ pairwise strongly commute. In such case, the unique QJSD of $\boldsymbol{A}$ is the JSM E_{A}, so in order for the convolution $h * E_{A} = E_{A}$ to be the unique QJSD of $\boldsymbol{A}$, we must have $h = \delta_0 \Leftrightarrow \hat{h} = 1$. In a more general setting, a necessary condition for $h * \#_{A}$ to satisfy the marginal condition (35) is given by[9]

$$
\int_{\mathbb{K}^{n-k}} h(\boldsymbol{a})\, dm_{n-k}(\boldsymbol{b}^c) = \delta_0(\boldsymbol{b}), \tag{57}
$$

where $\boldsymbol{B}$ is an order-preserving subset (33) of $\boldsymbol{A}$ consisting of k $(1 \le k \le n)$ numbers of its pairwise strongly commuting distinct members, $\boldsymbol{B}^c$ the order-preserving complement (34) of $\boldsymbol{B}$, and $\boldsymbol{b} \in \mathbb{K}^k$, $\boldsymbol{b}^c \in \mathbb{K}^{n-k}$ are their corresponding variables. Here, $\delta_{\boldsymbol{a}}$, denotes the *delta distribution* centred at $\boldsymbol{a} \in \mathbb{K}^n$, which is a generalised function symbolically defined as

$$
\delta_{\boldsymbol{a}}(\boldsymbol{x}) = \begin{cases} \infty, & (\boldsymbol{x} = \boldsymbol{a}) \\ 0, & (\boldsymbol{x} \neq \boldsymbol{a}) \end{cases} \tag{58}
$$

in the usual manner.

---

[9]Here, we adopt the convention

$$
\int_{\mathbb{K}^{n-k}} h(\boldsymbol{a})\, dm_{n-k}(\boldsymbol{b}^c) = h(\boldsymbol{a})
$$

for the case $k = n$.

On the other hand, the convolution still gives rise to the adjoint pair of representations in an extended sense in a similar manner to the affine transforms of a QJSD. In order to see its relation to the original representation, we first observe that

$$\begin{aligned} f_{(h*\#_A)} &:= \int_{\mathbb{K}^n} f(\boldsymbol{a})(h * \#_A)(\boldsymbol{a})\, dm_n(\boldsymbol{a}) \\ &= \int_{\mathbb{K}^n} (\tilde{h} * f)(\boldsymbol{a})\#_A(\boldsymbol{a})\, dm_n(\boldsymbol{a}) \\ &=: (\tilde{h} * f)_{\#_A}, \end{aligned} \tag{59}$$

where $\tilde{h}(\boldsymbol{a}) := h(-\boldsymbol{a})$ denotes the transpose of h. One also finds

$$\rho_{(h*\#_A)} = h * \rho_{\#_A}. \tag{60}$$

The adjointness of these operations

$$\langle f_{(h*\#_A)}, \rho \rangle_Q = \langle f, \rho_{(h*\#_A)} \rangle_C \tag{61}$$

can be readily confirmed by a simple computation. The diagram

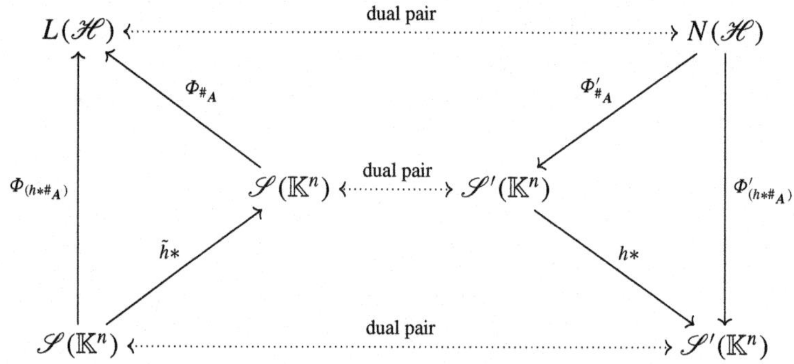

gives a visual summary as to how the quantum/quasi-classical representation corresponding to the convolution of the QJSD $\#_A$ with the function h relates to the original representation. One readily sees that the two representations corresponding to $\#_A$ and $h * \#_A$ are equivalent if the map $h* : f \mapsto h * f$ is a bijection.

2.3.2 Faithfulness of the Representations

For a fixed $\boldsymbol{A}$, we say that the quantum/quasi-classical representation pertaining to a given choice of QJDS $\#_{\boldsymbol{A}}$ is *faithful*, if either of the following equivalent[10] conditions are met:

1. The quantisation $\Phi_{\#_{\boldsymbol{A}}}$ has a dense range, i.e.,

$$\overline{\mathrm{ran}\, \Phi_{\#_{\boldsymbol{A}}}} = L(\mathscr{H}). \tag{62}$$

2. The quasi-classicalisation $\Phi'_{\#_{\boldsymbol{A}}}$ is injective, i.e.,

$$\rho_{\#_{\boldsymbol{A}}} = \sigma_{\#_{\boldsymbol{A}}} \quad \Leftrightarrow \quad \rho = \sigma, \quad (\rho, \sigma \in N(\mathscr{H})) \tag{63}$$

In physical terms, this is to say that every quantum operator $X = \lim f_i(\#_{\boldsymbol{A}})$ can be represented by limits of quantised operators if and only if no two quantum states give rise to the same QJP distribution. In general, the larger the number of observables belonging to the ordered combination $\boldsymbol{A}$ becomes, the closer the representation approaches to faithfulness.

2.3.3 Realness of the Representations

Among the various candidates of representations, 'real' representations are oftentimes prized. To state the precise definition, we first need some preparations. Given a QJSD $\#_{\boldsymbol{A}}$, the *conjugate* of $\#_{\boldsymbol{A}}$ is formally defined by

$$\#^*_{\boldsymbol{A}}(\boldsymbol{a}) := \#_{\boldsymbol{A}}(\boldsymbol{a})^*, \quad \boldsymbol{a} \in \mathbb{K}^n, \tag{64}$$

where the asterisk on the r. h. s. denotes the adjoint of $\#_{\boldsymbol{A}}(\boldsymbol{a})$. The Fourier transform of the conjugate of a QJSD reads

$$\begin{aligned}(\mathscr{F}\#^*_{\boldsymbol{A}})(\boldsymbol{s}) &:= \int_{\mathbb{K}^n} e^{-i\langle \boldsymbol{s}, \boldsymbol{a}\rangle} \#_{\boldsymbol{A}}(\boldsymbol{a})^* \, dm_n(\boldsymbol{a}) \\ &= \left(\int_{\mathbb{K}^n} e^{-i\langle -\boldsymbol{s}, \boldsymbol{a}\rangle} \#_{\boldsymbol{A}}(\boldsymbol{a}) \, dm_n(\boldsymbol{a}) \right)^* \\ &= \hat{\#}_{\boldsymbol{A}}(-\boldsymbol{s})^* \\ &=: \hat{\#}^{\dagger}_{\boldsymbol{A}}(\boldsymbol{s}),\end{aligned} \tag{65}$$

[10]The proof for the equivalence of the conditions can be carried out by applying the Hahn-Banach theorem on locally convex spaces.

where we have introduced the *involution* $\hat{\#}_{A}^{\dagger}$ of the hashed operator $\hat{\#}_{A}$ in the last equality. It is not difficult to see that involutions of hashed operators are again hashed operators, hence a conjugate of a QJSD is a QJSD.

We say that a QJSD $\#_{A}$ is *real* if $\#_{A}^{*} = \#_{A}$ holds. By the bijectivity of the Fourier transformation, a QJSD is real if and only if its Fourier transform (i.e., the corresponding hashed operator) is a self-involution. Examples of real QJSDs of $A = (A, B)$ are those whose corresponding hashed operators read

$$\hat{\#}(s,t) = \begin{cases} \frac{1}{2} \cdot (e^{-isB} e^{-isA} + e^{-isA} e^{-itB}) \\ e^{-i\frac{s}{2}A} e^{-itB} e^{-i\frac{s}{2}A} \\ e^{-i\frac{t}{2}B} e^{-isA} e^{-i\frac{t}{2}B} \\ e^{-i\frac{s}{3}A} e^{-i\frac{t}{2}B} e^{-i\frac{s}{3}A} e^{-i\frac{t}{2}B} e^{-i\frac{s}{3}A} \\ e^{-i\frac{t}{4}B} e^{-i\frac{s}{3}A} e^{-i\frac{t}{4}B} e^{-i\frac{s}{3}A} e^{-i\frac{t}{4}B} e^{-i\frac{s}{3}A} e^{-i\frac{t}{4}B} \\ e^{-i\overline{(sA+tB)}} \end{cases} \qquad s,\ t \in \mathbb{R},$$

were the overline on the operator $sA + tB$ denotes its unique self-adjoint extension. Colloquially speaking, a QJSD is real if its corresponding hashed operator is 'symmetric' in its form. We call the quantum/quasi-classical representation real, if the corresponding QJSD is real. Real representations have the following convenient properties:

1. The quantisation $f_{\#_A}$ of a real function $f \in \mathscr{S}(\mathbb{K}^n)$ is always self-adjoint.
2. The quasi-classicalisation $\rho_{\#_A}$ of a density operator $\rho \in N(\mathscr{H})$ is always real.

Real representations thus have a formal advantage of taking a classical observables into self-adjoint operators, and a quantum state into real QJP distribution, which some may find favourable above non-real representations.

2.3.4 Relation to Some Prior Works

The study on quantum-classical transformation has a long history. In this passage, we investigate the relation of the formalism of QJSDs presented in this paper to some of the prior works on this topic. In this passage, we are specifically interested in the choice $X = (Q, P)$, where Q and P are operators satisfying the Weyl representation (39) of the CCR.

The Theory of Wigner and Weyl

We first point out that the Weyl map and the Wigner map are respectively the quantisation $\Phi_{\#_X}$ and the quasi-classicalisation $\Phi'_{\#_X}$ pertaining to the QJSD W_X of X corresponding to the hashed operator of the form

$$\hat{W}_X(s,t) := e^{-i\overline{(sQ+tP)}}. \tag{66}$$

In fact, since the hashed operator (66) is equivalent to any choice of the form (40), the Wigner–Weyl transformation is the quantum/quasi-classical representation pertaining to any of the choices. It is easy to see from the self-involutive form of the hashed operator (66) that the Wigner–Weyl transformation is real. It is widely known that the Wigner map is injective, which is equivalent for its adjoint map (i.e., the Weyl map) to have a dense range: in the terminology of this paper, the Wigner–Weyl transformation is faithful.

In order to confirm the claim, we first compute the quantisation of functions with respect to the QJSD corresponding to (66). The quantisation of $f \in \mathscr{S}(\mathbb{R}^2)$ reads

$$
\begin{aligned}
f_{W_X} &:= \int_{\mathbb{R}^2} f(q,p)\, W_X(q,p) dm_2(q,p) \\
&= \int_{\mathbb{R}^4} f(q,p) e^{i(qs+pt)}\, \hat{W}_X(s,t) dm_2(q,p,s,t) \\
&= \int_{\mathbb{R}^4} f(q,p) e^{-is(Q-q)} e^{-it(P-p)} e^{ist/2} dm_2(q,p,s,t) \\
&=: W_X(f),
\end{aligned} \tag{67}
$$

where the last equality is precisely the definition of the Weyl quantisation of f. Here, note that we have used the relation

$$
e^{-itP/2} e^{-isQ} e^{-itP/2} = e^{ist/2} e^{-isQ} e^{-itP} \tag{68}
$$

in order to obtain the third equality. As for the quasi-classical representation of a quantum state, we assume without loss of generality that $\mathscr{H} = L^2(\mathbb{R})$, and that $Q = \hat{x}$, $P = \hat{p}$ are respectively the familiar position and momentum operators. For better readability, we moreover restrict ourselves to the case that $\rho = |\psi\rangle\langle\psi|$ for some wave-function $\psi \in L^2(\mathbb{R})$. Then, the Fourier transform of ρ_{W_X} reads

$$
\begin{aligned}
\left(\mathscr{F}\rho_{W_X}\right)(s,t) &= \mathrm{Tr}\left[\hat{W}_X(s,t)\rho\right] \\
&= \langle e^{itP/2}\psi, e^{-isQ} e^{-itP/2}\psi\rangle \\
&= \int_{\mathbb{R}} e^{-isq}\, \psi^*(q+t/2)\psi(q-t/2)\, dm(x) \\
&= \int_{\mathbb{R}} e^{-i(sq+tp)} \left(\int_{\mathbb{R}} \psi^*(q+t/2)\psi(q-t/2) e^{ipt}\, dm(t)\right) dm_2(q,p) \\
&= \left(\mathscr{F} W_X^{\rho}\right)(s,t),
\end{aligned} \tag{69}
$$

where

$$
W_X^{\rho}(q,p) := \int_{\mathbb{R}} \psi^*(q+t/2)\psi(q-t/2) e^{ipt}\, dm(t) \tag{70}
$$

is the Wigner function of the wave function $\psi \in L^2(\mathbb{R})$. Due to the injectivity of the Fourier transformation, we conclude $\rho_{W_X} = W_X^{\rho}$. The proof for the general case goes essentially the same.

QSJDs Generated by Convolution

We next consider the problem of generating a family of QJSDs of $\boldsymbol{X}$ by means of the convolution of $W_{\boldsymbol{X}}$ with functions. This setting has been previously examined by Cohen et al. [10] with the intention of constructing a family of generalised phase space distribution functions. For the convolution $h * W_{\boldsymbol{X}}$ to be a QJSD of $\boldsymbol{X}$, one sees from the result (57) that the condition $\hat{h}(0,t) = \hat{h}(s,0) = 1$, $s, t \in \mathbb{R}$ is necessary. Specifically, we demonstrate below that, under the assumption that the Fourier transform

$$\hat{h}(s,t) = \hat{g}\,(st/2)\,, \quad s,\, t \in \mathbb{R} \tag{71}$$

of the function h can be represented by a function $g : \mathbb{R} \to \mathbb{R}$ with the total integration of unity, the convolution (55) becomes a QJSD of the pair $\boldsymbol{X}$. Indeed, observe that since

$$\mathscr{F}(h * W_{\boldsymbol{X}})(s,t) = \int_{\mathbb{R}} g(\kappa) e^{-i\kappa st/2}\, \hat{W}_{\boldsymbol{X}}(s,t)\, dm(\kappa), \tag{72}$$

we have

$$h * W_{\boldsymbol{X}} = \int_{\mathbb{R}} g(\kappa)\, \#_{\boldsymbol{X}}^{\kappa}(s,t)\, dm(\kappa), \tag{73}$$

where we have used the parametrised family $\#_{\boldsymbol{X}}^{\kappa}$, $\kappa \in \mathbb{R}$, of the QJSD of $\boldsymbol{X}$ corresponding to the hashed operators of the form

$$\begin{aligned} \hat{\#}_{\boldsymbol{X}}^{\kappa}(s,t) &:= e^{-i\kappa st/2}\, \hat{W}_{\boldsymbol{X}}(s,t) \\ &= e^{-i\frac{1-\kappa}{2}sQ} e^{-itP} e^{-i\frac{1+\kappa}{2}sQ}, \end{aligned} \tag{74}$$

which was originally introduced in (25) for general $\boldsymbol{A} = (A, B)$. Hence, as the 'weighted average' of the parametrised families of the QJSDs of $\boldsymbol{X}$, the convolution (73) itself is a QJSD of $\boldsymbol{X}$. Each choice of the function g yields distinct representation. Among the well-known representations are those proposed by:

1. Weyl [4] and Wigner [5]

$$g(k) = \delta_0(k) \quad \Leftrightarrow \quad \hat{g}(\omega) = 1 \tag{75}$$

2. Kirkwood [11] and Dirac [12]

$$g(k) = \delta_{\pm 1}(k) \quad \Leftrightarrow \quad \hat{g}(\omega) = e^{\mp i\omega} \tag{76}$$

3. Margenau and Hill [13]

$$g(k) = \frac{(\delta_{-1}(k) + \delta_{+1}(k))}{2} \quad \Leftrightarrow \quad \hat{g}(\omega) = \cos\omega \tag{77}$$

4. Born and Jordan [14]

$$g(k) = \begin{cases} \frac{1}{2}, & (|k| \leq 1) \\ 0, & (|k| > 1) \end{cases} \quad \Leftrightarrow \quad \hat{g}(k) = \frac{\sin\omega}{\omega} \tag{78}$$

which all belongs to the same class generated by convolutions.

The Theory of Hushimi and Glauber–Sudarshan

In the previous paragraph, we have considered the problem of constructing various QJSDs of $\boldsymbol{X}$ by means of convolution. In a broader perspective, however, the convolution $h * W_X$ itself need not be a QJSD of $\boldsymbol{X}$ in the sense that it gives rise to the adjoint pair of representations in an extended sense. If the map $h* : f \mapsto h * f$ is bijective, both the original QJSD and the convolution contain the same amount of information of $\boldsymbol{X}$, and a sufficient condition for this is $\hat{h} > 0$.

As for the choice of the function h, which may from the result (55) be interpreted as the 'weight function' of the family of QJSDs, we typically consider the normal distribution (i.e., Gaußian function)

$$G(\boldsymbol{x}) := e^{-\|\boldsymbol{x}\|^2/2^n}, \quad \boldsymbol{x} \in \mathbb{R}^n, \tag{79}$$

where $\|\boldsymbol{x}\| := \sqrt{\sum_{i=1}^{n} |x_i|^2}$ denotes the Euclidean norm of an n-dimensional vector $\boldsymbol{x} \in \mathbb{R}^n$. Since the normal distribution never satisfies the condition (57), the convolution $G * \#_X$ of a normal distribution with a QJSD of $\boldsymbol{X}$ is not a QJSD of $\boldsymbol{X}$ in general. However, since the Fourier transform of a normal distribution is another normal distribution, hence $\hat{G} > 0$, the original QJSD $\#_X$ and the convolution $G * \#_X$ are equivalent to each other. We thus introduce two operator valued distributions (OVDs) H_X and GS_X that are uniquely specified through the relations

$$H_X = G * W_X, \tag{80}$$

$$W_X = G * GS_X. \tag{81}$$

By the above argument, we see that all W_X, H_X and GS_X contain the same information in the sense that they can be transformed to one another by convolution with normal distributions, and thus the adjoint pairs of quantum/quasi-classical representations which they yield are all equivalent. In this paper, we casually call the QJSD W_X the *Wigner–Weyl* type, and the OVDs H_X and GS_X the *Hushimi* and the *Glauber–Sudarshan* type, respectively. From the result (60), one sees that the quasi-classical representation of a density operator $\rho \in N(\mathscr{H})$ pertaining to the respective representations are related to each other by

$$H_X^\rho = G * W_X^\rho, \tag{82}$$
$$W_X^\rho = G * GS_X^\rho. \tag{83}$$

As we have seen in the previous passage that $W_X^\rho(q,p)$ is the Wigner function of ρ, we here learn that the distributions $H_X^\rho(q,p)$ and $GS_X^\rho(q,p)$ are precisely the *Hushimi Q-function* [15] and the *Glauber–Sudarshan P-function* [16, 17] of the quantum state ρ, respectively.

3 Some Applications: Quantum Correlations, Conditional Expectations and the Weak Value

We now seek application of the formalisms we have constructed in the previous chapter. By its nature, the framework of quantisation/quasi-classicalisation by QJSDs should become useful in analysing problems where non-commutativity of quantum observables is concerned. In this section, we specifically focus on 'correlations' and 'conditioning' between (generally non-commuting) quantum observables induced by QJSDs. Due to the non-commuting nature of quantum observables, there are various candidates of quantum correlations and conditional expectations. Specifically, we see that Aharonov's weak value [7] can be identified as one realisation among various candidates of quantum conditional expectations.

Complex Parametrised Subfamily

In handling relatively abstract objects as QJSDs of quantum observables, concrete examples are always of use. To this, we occasionally consider the simplest case where only two self-adjoint operators $\boldsymbol{A} = (A, B)$ are concerned, and make use of the complex parametrised subfamily of hashed operators

$$\hat{\#}_{\boldsymbol{A}}^{\alpha}(s,t) := \frac{1+\alpha}{2} \cdot e^{-itB} e^{-isA} + \frac{1-\alpha}{2} \cdot e^{-isA} e^{-itB}, \quad \alpha \in \mathbb{C}, \tag{84}$$

and the resulting subfamily $\#_{\boldsymbol{A}}^{\alpha} := \mathscr{F}^{-1} \hat{\#}_{\boldsymbol{A}}^{\alpha}$ of QJSDs for demonstration.[11]

3.1 *Correlations of Quantum Observables*

In classical probability theory, *correlation* has been an important quantity in various aspects. The definition of the quantum counterpart, however, is not so obvious, when

[11] Do not confuse the subfamily introduced above with that of (25). We have used different superscript characters κ, α as parameters to make the distinction more easier.

non-commutative observables are taken into account. In what follows, we define a family of quantum correlation based on our framework of QJSDs of quantum observables, and observe their very basic properties.

3.1.1 Sesquilinear Forms Induced by QJP Distributions

As usual, let $\boldsymbol{A} = (A_1, \ldots, A_n)$, $n \geq 1$, be an ordered combination of self-adjoint operators on a Hilbert space $\mathscr{H}$, and choose a QJSD $\#_{\boldsymbol{A}}$ and a density operator $\rho \in N(\mathscr{H})$. In what follows, in order to refrain ourselves from dealing with unessential mathematical intricacies, we assume that the resulting QJP distribution $\rho_{\#_{\boldsymbol{A}}}$ can be represented by a density function.[12] This allows us to introduce a sesquilinear form

$$\langle g, f \rangle_{\rho_{\#_{\boldsymbol{A}}}} := \int_{\mathbb{K}^n} g^*(\boldsymbol{a}) f(\boldsymbol{a})\, \rho_{\#_{\boldsymbol{A}}}(\boldsymbol{a}) dm_n(\boldsymbol{a}), \quad f, g \in L^2(\rho_{\#_{\boldsymbol{A}}}) \tag{85}$$

defined on the space of square-integrable functions with respect to the complex density function $\rho_{\#_{\boldsymbol{A}}}$. By definition, we have

$$\langle f, g \rangle_{\rho_{\#_{\boldsymbol{A}}}} = \langle g^*, f^* \rangle_{\rho_{\#_{\boldsymbol{A}}}} \tag{86}$$

and

$$\langle f, g \rangle_{\rho^*_{\#_{\boldsymbol{A}}}} = \langle g, f \rangle^*_{\rho_{\#_{\boldsymbol{A}}}}, \tag{87}$$

where the superscript asterisk denotes the complex conjugate. The following observations are direct consequences of the above properties.

1. The quantum correlation (85) is *symmetric* (Hermitian)

$$\langle f, g \rangle_{\rho_{\#_{\boldsymbol{A}}}} = \langle g, f \rangle^*_{\rho_{\#_{\boldsymbol{A}}}} \tag{88}$$

   if and only if the QJP distribution $\rho_{\#_{\boldsymbol{A}}}$ is *real*. This is guaranteed for every $\rho \in N(\mathscr{H})$ if and only if the choice of the QJPD $\#_{\boldsymbol{A}}$ is self-adjoint.
2. The quantum correlation (85) is *positive definite*[13]

$$\begin{aligned} \forall f \in L^2 \quad \langle f, f \rangle_{\rho_{\#_{\boldsymbol{A}}}} &\geq 0, \\ \langle f, f \rangle_{\rho_{\#_{\boldsymbol{A}}}} = 0 \quad &\Leftrightarrow \quad f = 0 \end{aligned} \tag{89}$$

   if and only if the QJP distribution $\rho_{\#_{\boldsymbol{A}}}$ is *positive*. This is guaranteed for every $\rho \in N(\mathscr{H})$ if and only if the QJPD $\#_{\boldsymbol{A}}$ is positive.

---

[12]We say that a tempered distribution $u \in \mathscr{S}'(\mathbb{K}^n)$ admits representation by a density function if $u(x)$ is actually an integrable function.

[13]Here, the equality $f = 0$ in the second line of (89) is meant to hold $\rho_{\#_{\boldsymbol{A}}}$-almost everywhere.

Note that the second condition (i.e., positive definiteness) is stronger that the first condition (i.e., symmetricity), for indeed the positiveness of the QJP distributions trivially implies its realness, and in parallel, the positiveness of the QJPDs implies its self-adjointness.

3.1.2 Quantum Correlations

We next introduce the concept of quantum correlations based on the sesquilinear forms defined above. In what follows, in order to ease our arguments, we confine ourselves to the simplest case $\boldsymbol{A} = (A, B)$ without loss of generality. We also write $\# := \#_{\boldsymbol{A}}$ for better readability. Now, let $f(A)$, $g(B)$ be operators respectively defined from the functions $f(a)$ and $g(b)$ by means of the functional calculus. We then define the *quantum correlations* or *quasi-correlations* between the operators $f(A)$, $g(B)$ by

$$\langle g(B), f(A) \rangle_{\rho_{\#}} := \int_{\mathbb{K}^2} g^*(b) f(a)\ \rho_{\#}(a, b) dm_2(a, b) \tag{90}$$

By construction, quantum correlations are dependent on the choice of the QJP distributions $\rho_{\#}$, which is generally non-unique due to the indefiniteness of the QJSDs. When A and B happen to be simultaneously measurable, indefiniteness of the QJSDs vanishes, and the quantum correlation reduces to the unique classical correlation in the standard sense.

3.1.3 Quantum Covariances

Now that we have introduced the concept of quantum correlations, we next introduce the concept of quantum covariances. Under the same assumptions, we introduce the *quantum covariance* of the pair with respect to the QJP distribution $\rho_{\#}$ defined as

$$\begin{aligned} \mathbb{CV}[f(A), g(B); \rho_{\#}] &:= \langle g(B) - \mathbb{E}[g(B); \rho],\ f(A) - \mathbb{E}[f(A); \rho] \rangle_{\rho_{\#}} \\ &= \langle g(B),\ f(A) \rangle_{\rho_{\#}} - \mathbb{E}[f(A); \rho] \cdot \mathbb{E}[g(B); \rho], \end{aligned} \tag{91}$$

where $\mathbb{E}[X; \rho] := \mathrm{Tr}[X\rho]$ denotes the expectation value of $X \in L(\mathscr{H})$ on a density operator $\rho \in N(\mathscr{H})$ as usual. The quantum covariance serves as a natural extension to the standard covariance in classical probability theory, and they indeed coincide when the pair of self-adjoint operators $f(A)$ and $g(B)$ strongly commute.

Example

As an example, let $\#^{\alpha}$ be a complex parametrised QJSD for $\alpha \in \mathbb{C}$ introduced in (84). By a simple computation, the quantum correlation of the operators A and B reads

$$
\begin{aligned}
\langle B, A \rangle_{\rho_\#} &:= \int_{\mathbb{R}^2} ba \, \rho_\#(a, b) \, dm_2(a, b) \\
&= (i\partial_{s_1})(i\partial_{s_2})(\mathscr{F}\rho_\#)(s)\big|_{s=0} \\
&= \mathrm{Tr}\left[\frac{\{A, B\}}{2}\rho\right] + i\,\alpha\, \mathrm{Tr}\left[\frac{[A, B]}{2i}\rho\right],
\end{aligned}
\tag{92}
$$

where $\{X, Y\} := XY + YX$ and $[X, Y] := XY - YX$ respectively denotes the anti-commutator and the commutator of X and Y as usual. This computation leads to

$$
\mathbb{CV}[A, B; \rho_{\#^\alpha}] = \mathbb{CV}_{\mathrm{S}}[A, B; \rho] + i\,\alpha\,\mathbb{CV}_{\mathrm{A}}[A, B; \rho], \tag{93}
$$

where we have introduced the *standard symmetric* and *standard anti-symmetric* quantum covariances

$$
\mathbb{CV}_{\mathrm{S}}[A, B; \rho] := \mathrm{Tr}\left[\frac{\{A, B\}}{2}\rho\right] - \mathbb{E}[A; \rho] \cdot \mathbb{E}[B; \rho], \tag{94}
$$

$$
\mathbb{CV}_{\mathrm{A}}[A, B; \rho] := \mathrm{Tr}\left[\frac{[A, B]}{2i}\rho\right], \tag{95}
$$

for better readability.

3.2 *Conditioning by Quantum Observables*

In the previous passage, we have introduced quantum analogues of correlations by means of QJP distributions. Closely related to these are quantum analogues of conditional expectations.

3.2.1 Introducing Quantum Conditional Expectations

In what follows, in order to avoid distraction by unessential mathematical intricacies, we impose an additional assumption that, for the given choice of the density operator $\rho \in N(\mathscr{H})$, the probability of finding the outcome of B is always positive $\rho_B(b) > 0$ on its spectrum.[14] The quantum correlation of both the operators $f(A)$ and $g(B)$ then reads

[14]The spectrum of a self-adjoint operator A is defined as the largest closed subset $J \subset \mathbb{R}$ such that $E_B(J) = \mathrm{Id}$ holds.

$$\begin{aligned}\langle g(B),\ f(A)\rangle_{\rho_\#} &:= \int_{\mathbb{K}^2} g^*(b) f(a)\ \rho_\#(a,b) dm_2(a,b) \\ &= \int_{\mathbb{K}} g^*(b)\mathbb{E}[f(A)|B=b;\rho_\#]\ \rho_B(b) dm(b) \\ &= \langle g(B),\ \mathbb{E}[f(A)|B;\rho_\#]\rangle_{\rho_B}, \end{aligned} \tag{96}$$

where we have introduced the *quantum conditional expectation* or *conditional quasi-expectation*

$$\mathbb{E}[f(A)|B=b;\rho_\#] := \frac{\int_{\mathbb{K}} f(a)\ \rho_\#(a,b) dm(a)}{\rho_B(b)}. \tag{97}$$

of the operator $f(A)$ given the outcome of B under the QJP distribution $\rho_\#$. Note that the quantum conditional expectation $\mathbb{E}[f(A)|B;\rho_\#]$ is defined as an (equivalence class of) complex function(s) rather than a scalar. The normal operator

$$\mathbb{E}[f(A)|B;\rho_\#] := \int_{\mathbb{K}} \mathbb{E}[f(A)|B=b;\rho_\#](b)\ dE_B(b) \tag{98}$$

in the last equation is the image of the functional calculus of the (equivalence class of) function(s) (97).

Some 'Statistical' Properties and its Interpretation

The key observation to make here is that the quantum correlation of an operator $f(A)$ with any operator $g(B)$ generated by B can be reproduced by the *authentic* correlation of the quantum conditional expectation $\mathbb{E}[f(B)|B;\rho_\#]$ with $g(B)$, which we reiterate for emphasis as:

$$\langle g(B),\ \mathbb{E}[f(A)|B;\rho_\#]\rangle_{\rho_B} = \langle g(B),\ f(A)\rangle_{\rho_\#},\quad \forall g \in L^2(\rho_B). \tag{99}$$

Also, by taking the constant function $g = 1$, we have

$$\mathbb{E}\left[\mathbb{E}[f(A)|B;\rho_\#];\rho\right] = \mathbb{E}\left[f(A);\rho\right]. \tag{100}$$

The above two equalities show that the quantum conditional expectation serve as the 'approximation' of the original operator $f(A)$ by operators generated by B, and that it is unique in the sense that it precisely reproduces the quantum correlation with any other operators generated by B in place of the original $f(A)$.

In physical terms, the quantum conditional expectation can be interpreted as the quantum analogue of conditional expectations of the operator $f(A)$ given the outcome b of B under the hypothetical 'joint' distribution (i.e., QJP distribution) $\rho_\#$. If the combination of the observables $\boldsymbol{A} = (A, B)$ happens to be simultaneously measurable, the quantum conditional expectation simply reduces to the conditional expectation in the classical sense that is familiar to us. On the other hand, if some of the pair of observables fail to admit simultaneous measurability, the quantum conditional expectation becomes a hypothetical quantity, whose definition becomes non-unique due to the indefiniteness of the choice of the QJPDs.

Examples

We next provide some concrete examples to actually compute the quantum conditional expectations. To this end, let $\#^{\alpha}$ be the parametrised subfamily of QJSDs of the ordered pair $\boldsymbol{A} = (A, B)$ of self-adjoint operators introduced in (84). For a given function $f(a)$, let

$$f(a) = \sum_{m=0}^{\infty} f_m a^m \tag{101}$$

denote its Taylor expansion. If we let $\varphi(b)$ denote the numerator of (97), we then have

$$\begin{aligned}
(\mathscr{F}\varphi)(t) &= \int_{\mathbb{R}} e^{-itb} \left(\int_{\mathbb{R}} \sum_{m=0}^{\infty} f_m a^m \ \rho_{\#^{\alpha}}(a, b) dm(a) \right) dm(b) \\
&= \sum_{m=0}^{\infty} f_m \left(\int_{\mathbb{R}^2} e^{-i(0a+tb)} a^m b^0 \ \rho_{\#^{\alpha}}(a, b) dm_2(a, b) \right) \\
&= \sum_{m=0}^{\infty} f_m \left((i\partial_s)^m (i\partial_t)^0 \ (\mathscr{F}\rho_{\#^{\alpha}})(0, t) \right) \\
&= \sum_{m=0}^{\infty} f_m \left(\frac{1+\alpha}{2} \cdot \mathrm{Tr}\left[e^{-itB} A^m e^{-i0A} \rho \right] + \frac{1-\alpha}{2} \cdot \mathrm{Tr}\left[A^m e^{-i0A} e^{-itB} \rho \right] \right) \\
&= \frac{1+\alpha}{2} \cdot \mathrm{Tr}\left[e^{-itB} f(A) \rho \right] + \frac{1-\alpha}{2} \cdot \mathrm{Tr}\left[f(A) e^{-itB} \rho \right].
\end{aligned} \tag{102}$$

Observing that the Fourier transform of the function $b \mapsto \mathrm{Tr}\,[E_B(b) f(A)\rho]$ and $b \mapsto \mathrm{Tr}\,[f(A) E_B(b)\rho]$ are respectively $\mathrm{Tr}\left[e^{-itB} f(A)\rho\right]$ and $\mathrm{Tr}\left[f(A) e^{-itB}\rho\right]$, one finds that the injectivity and the linearity of the Fourier transformation leads to

$$\int_{\mathbb{R}} f(a) \ \rho_{\#^{\alpha}}(a, b) dm(a) = \frac{1+\alpha}{2} \cdot \mathrm{Tr}\,[E_B(b) f(A)\rho] + \frac{1-\alpha}{2} \cdot \mathrm{Tr}\,[f(A) E_B(b)\rho] \tag{103}$$

by combining the results. We thus finally have

$$\begin{aligned}
\mathbb{E}^{\alpha}[A|B=b; \rho] &:= \mathbb{E}[f(A)|B=b; \rho_{\#^{\alpha}}] \\
&= \frac{1+\alpha}{2} \cdot \frac{\mathrm{Tr}\,[E_B(b) f(A)\rho]}{\mathrm{Tr}[E_B(b)\rho]} + \frac{1-\alpha}{2} \cdot \frac{\mathrm{Tr}\,[f(A) E_B(b)\rho]}{\mathrm{Tr}[E_B(b)\rho]} \\
&= \mathrm{Re}\left[\frac{\mathrm{Tr}\,[E_B(b) f(A)\rho]}{\mathrm{Tr}[E_B(b)\rho]} \right] + i\,\alpha\,\mathrm{Im}\left[\frac{\mathrm{Tr}\,[E_B(b) f(A)\rho]}{\mathrm{Tr}[E_B(b)\rho]} \right],
\end{aligned} \tag{104}$$

where we have introduced an abbreviated symbol (the first line) for better readability.

3.3 The Weak Value as a Quantum Conditional Expectation

As the simplest case of the examples provided in the previous Sect. 3.2, let us consider the quantum conditional expectation of the function $f(a, b) = a$. Based on the formula (104), one obtains

$$\begin{aligned}\mathbb{E}^{\alpha}[A|B = b; \rho] &= \frac{1+\alpha}{2} \cdot A_w^{\rho}(b) + \frac{1-\alpha}{2} \cdot A_w^{\rho}(b)^* \\ &= \mathrm{Re}\left[A_w^{\rho}(b)\right] + i\,\alpha\,\mathrm{Im}\left[A_w^{\rho}(b)\right], \quad \alpha \in \mathbb{C},\end{aligned} \tag{105}$$

where we have introduced the Aharonov's weak value

$$A_w^{\rho}(b) := \mathbb{E}^1[A|B = b; \rho] = \frac{\mathrm{Tr}\,[E_B(b)A\rho]}{\mathrm{Tr}[E_B(b)\rho]}.$$

The quantity (105), defined as the complex convex combination of the weak value, was initially proposed in [18], in which it was called the *two state value*. Provided that the density operator $\rho = |\psi\rangle\langle\psi|$ is an orthogonal projection onto the 1-dimensional linear subspace spanned by a unit vector $|\psi\rangle \in \mathscr{H}$, weak value reduces to its familiar form

$$A_w^{\psi}(b) = \frac{\langle b|A|\psi\rangle}{\langle b|\psi\rangle}. \tag{106}$$

In this respect, the weak value admits an interpretation as one manifestation of the possible family of quantum conditional expectations corresponding to the specific choice $\#^{\alpha}$, $\alpha = 1$, of the subfamily of the QJSDs.

The interpretation of the weak value, as one of the possible candidates of quantum analogues of classical conditional expectations, has been proposed earlier in several literatures. Relevant to our framework presented here is [19], in which the inherent non-uniqueness of the QJP distributions for non-commuting observables is particularly emphasised, a fact which is oftentimes overlooked in discussing the weak value. There, quantum conditional expectations are computed not only for the Kirkwood–Dirac distribution (76), but also for several other types of QJP distributions, including the Wigner–Weyl distribution (75) and the Margenau–Hill distribution (77).

4 Summary and Discussion

In the former part of this paper (Sect. 2), we focused on the problem of *quantisation* of classical observables and *quasi-classicalisation* of quantum states. For the simplest case in which the observables concerned are all simultaneously measurable, we reviewed that the joint-spectral measure (JSM) uniquely attributed to the commuting observables gives rise to the unique pair of functional analysis and the Born rule, which could respectively be considered as the trivial realisation of quantisation and quasi-classicalisation. Specifically, by taking the duality relation between observables and states into account, we saw that quantisation and quasi-classicalisation are actually adjoint to one another as maps, and thus the JSM, quantisation and quasi-classicalisation are all equivalent as an entity, although they may differ in concept. In this sense, we occasionally referred to the pair of maps as quantum/quasi-classical transformations or representations.

We next considered the general case in which observables concerned are arbitrary. To this, we let ourselves be guided by the observation above, and introduced the concept of *quasi-joint-spectral distributions* (QJSDs), which could be interpreted as non-commutative analogues to the standard JSM. In contrast to the commutative case, QJSDs attributed to a given set of non-commuting observables are non-unique, and thus they give rise to various distinct pairs of quantisations and quasi-classicalisations. We also discussed the basic properties of QJSDs and the transformations between them. An important implication of this framework is that, although there may be countless possible ways to construct quantisation and quasi-classicalisation, there is a precise one-to-one correspondence between them. These realisations help us to understand the relation between various proposals made historically, including those proposed by Wigner–Weyl, Kirkwood–Dirac, Margenau–Hill, Born–Jordan, Hushimi and Glauber–Sudarshan.

As an application to this framework, the latter portion of this paper (Sect. 3) focused on the problem of constructing quantum analogues to the classical concept of correlation and conditioning. We proposed a framework to this problem by means of QJSDs introduced earlier, and demonstrated that some of the statistical properties familiar in classical probability theory are still preserved even under the quantum counterpart, especially the relation between correlation and conditional expectation. We finally mentioned that Aharonov's weak value could be interpreted as one manifestation of quantum conditional expectations. One of the virtues of this interpretation is that it reveals a novel aspect of the uncertainty relations [20], in the sense that the weak value appears as the optimal choice of approximation: quantum conditional expectations are best approximations of an observable by another observable, just as classical conditional expectation is the best approximation of a random variable by means of another.

The framework of the quantum/quasi-classical transformation proposed in this paper may find a variety of applications. In fact, it should be obvious from our arguments that one can always draw an analogy to various concepts and results in classical probability theory when one considers the quantum counterparts obtained by this method. Naturally, this will allow for an intuitive treatment of the latter based on the statistical and geometric structures present in classical probability theory.

References

1. W. K. Heisenberg. Über den anschaulichen Inhalt der Quantenmechanik. *Z. Phys.* 43:172, 1927.
2. M. Ozawa. Universally valid reformulation of the Heisenberg uncertainty principle on noise and disturbance in measurement. *Phys. Rev. A* 67:042105, 2003.

3. M. Ozawa. Uncertainty relations for joint measurements of noncommuting observables. *Phys. Lett. A* 320:367, 2004.
4. H. Weyl. Quantenmechanik und Gruppentheorie. *Zeitschrift für Physik* 46(1):1–46, 1927.
5. E. Wigner. On the quantum correction for thermodynamic equilibrium. *Phys. Rev.* 40:749, 1932.
6. Y. Aharonov, D. Z. Albert, and L. Vaidman. How the result of a measurement of a component of the spin of a spin-1/2 particle can turn out to be 100. *Phys. Rev. Lett.* 60:1351, 1988.
7. Y. Aharonov, P. G. Bergmann, and L. Lebowitz. Time symmetry in the quantum process of measurement. *Phys. Rev.* 134:B1410, 1964.
8. M. Ozawa. Universal uncertainty principle, simultaneous measurability, and weak values. *AIP Conf. Proc.* 1363:53–62, 2011.
9. H. F. Hofmann. Reasonable conditions for joint probabilities of non-commuting observables. *Quantum Stud.: Math. Found.* 1:39, 2014.
10. M. O. Scully and L. Cohen. *The Physics of Phase Space Nonlinear Dynamics and Chaos Geometric Quantization, and Wigner Function*, volume 278 of *Lecture Notes in Physics*, pages 253–260. Springer, 1987.
11. J. G. Kirkwood. Quantum statistics of almost classical assemblies. *Phys. Rev.* 44:31, 1933.
12. P. A. M. Dirac. On the analogy between classical and quantum mechanics. *Rev. Mod. Phys* 17:195–199, 1945.
13. H. Margenau and N. R. Hill. Correlations between measurements in quantum theory. *Prog. Theoret. Phys.* 26:722, 1961.
14. M. Born and P. Jordan. Zur Quantenmechanik. *Z. Phys.* 34:858, 1925.
15. K. Husimi. Some formal properties of the density matrix. *Proc. Phys. Math. Soc. Jap.* 22:264, 1940.
16. R. J. Glauber. Coherent and incoherent states of the radiation field. *Phys. Rev.* 131:2766, 1963.
17. E. C. G. Sudarshan. Equivalence of semiclassical and quantum mechanical descriptions of statistical light beams. *Phys. Rev. Lett.* 10:277–279, 1963.
18. T. Morita, T. Sasaki, and I. Tsutsui. Complex probability measure and Aharonov's weak value. *Prog. Theor. Exp. Phys.* 2013(053A02), 2013.
19. Justin Dressel. Weak values as interference phenomena. *Phys. Rev. A* 91(032116), 2015.
20. J. Lee and I. Tsutsui. Uncertainty relations for approximation and estimation. *Phys. Lett. A* 380:2045, 2016.

A New Quantum Version of f-Divergence

Keiji Matsumoto

Abstract This paper proposes and studies new quantum version of f-divergences, a class of functionals of a pair of probability distributions including Kullback–Leibler divergence, Renyi-type relative entropy and so on. distance. There are several quantum versions so far, including the one by Petz (Rev Math Phys 23:691–747, 2011, [1]). We introduce another quantum version ($D_f^{\max}$, below), defined as the solution to an optimization problem, or the minimum classical f-divergence necessary to generate a given pair of quantum states. It turns out to be the largest quantum f-divergence. The closed formula of $D_f^{\max}$ is given either if f is operator convex, or if one of the state is a pure state. Also, concise representation of $D_f^{\max}$ as a pointwise supremum of linear functionals is given and used for the clarification of various properties of the quality. Using the closed formula of $D_f^{\max}$, we show: Suppose f is operator convex. Then the maximum f-divergence of the probability distributions of a measurement under the state ρ and σ is strictly less than $D_f^{\max}(\rho\|\sigma)$. This statement may seem intuitively trivial, but when f is not operator convex, this is not always true. A counter example is $f(\lambda) = |1-\lambda|$, which corresponds to total variation distance. We mostly work on finite dimensional Hilbert space, but some results are extended to infinite dimensional case.

Keywords f-divergence · Kullback–Leibler divergence · Renyi-type relative entropy · Operator convex · Total variation distance

1 Introduction

This paper proposes and studies a new quantum version of f-divergence:

$$D_f(p\|q) := \sum_x q(x) f(p(x)/q(x)),$$

K. Matsumoto (✉)
National Institute of Informatics, 2-1-2 Hitotsubashi, Chiyoda-ku, Tokyo 101-8430, Japan
e-mail: keiji@nii.ac.jp

M. Ozawa et al. (eds.), *Reality and Measurement in Algebraic Quantum Theory*,
Springer Proceedings in Mathematics & Statistics 261,
https://doi.org/10.1007/978-981-13-2487-1_10

where p and q are probability distributions. Several important quantities in information theory and statistics are in this class. For example, $\mathrm{D}_{\lambda \ln \lambda}$ and $\mathrm{D}_{\lambda^{\alpha}}$ correspond to Kullback–Leibler divergence and Renyi-type relative entropy, respectively, which are extensively used in asymptotic analysis of error probability of decoding, hypothesis test, and so on.

Other f-divergences than these have at least one operational meaning. If f is a convex function and satisfies some moderate conditions, $\mathrm{D}_f(p\|q)$ is the optimal gain of a certain Bayes decision problem: for each f, there is a pair of functions w_1 and w_2 on decision space representing a gain of decision d with

$$\mathrm{D}_f(p\|q) = \sup_{d(\cdot)} \sum_x (w_1(d(x))\, p(x) + w_2(d(x))\, q(x)). \tag{1}$$

Conversely, for each $(w_1(\cdot), w_2(\cdot))$, there is a convex function f with this identity. Also, by (1) and the celebrated randomization criterion [2], there is a Markov map which sends (p, q) to (p', q') iff $\mathrm{D}_f(p\|q) \geq \mathrm{D}_f(p'\|q')$ holds for any convex function f with above mentioned properties.

In quantum information theory, a series of works by Petz (see [1] and references therein) is most impressive, and his version of quantum divergence have been widely studied and applied. Also, recent development of theory of quantum Renyi entropy is significant.

In this paper, we introduce another quantum version, the maximal quantum f-divergence $\mathrm{D}_f^{\max}(\rho\|\sigma)$. This quantity is defined as the solution to the following optimization problem: given a pair of quantum states $\{\rho, \sigma\}$, consider a (completely) positive trace preserving map Γ that sends probability distributions $\{p, q\}$ to $\{\rho, \sigma\}$. The triple $(\Gamma, \{p, q\})$ (*reverse test*, here after), is optimized to minimize $\mathrm{D}_f(p\|q)$, and this infimum is $\mathrm{D}_f^{\max}(\rho\|\sigma)$. The name comes from the fact that $\mathrm{D}_f^{\max}$ is the largest of the all possible quantum f-divergences.

Some historical remarks are in order. When f is $\lambda \ln \lambda$ and σ is invertible,

$$\mathrm{D}_{\lambda \ln \lambda}^{\max}(\rho\|\sigma) = \mathrm{tr}\, \rho \log \rho^{1/2}\sigma^{-1}\rho^{1/2}.$$

This RHS quantity had been studied by several authors from operator theoretic point of view [3–5] . Also, some authors had pointed out this quantity is path dependent divergence [6] of RLD quantum Fisher metric [7], which plays an important role in quantum statistical estimation theory [8], along e- and m- geodesic connecting ρ and σ [9–11]. However, its characterization as the solution to the optimization problem and the largest quantum version is first pointed out by the present author [10]. In [12], the present author studied $\mathrm{D}_{\lambda^{1/2}}^{\max}$ rather intensively, and briefly treated the case when f is operator monotone decreasing.

Below, we summarize our main results. When f is operator convex, a series of rich results are available. First, we can write down the value of $\mathrm{D}_f^{\max}$ and the operation achieving the minimum explicitly: Suppose σ and ρ are

$$\rho = \begin{bmatrix} \rho_{11} & \rho_{12} \\ \rho_{21} & \rho_{22} \end{bmatrix}, \ \sigma = \begin{bmatrix} \sigma_{11} & 0 \\ 0 & 0 \end{bmatrix},$$

then

$$\mathrm{D}_f^{\max}(\rho||\sigma) = \mathrm{tr}\, \sigma\, f\left(\sigma^{-1}\tilde{\rho}\right) + \mathrm{tr}\,(\rho - \tilde{\rho}) \lim_{\varepsilon \downarrow 0} \varepsilon f(1/\varepsilon). \tag{2}$$

where $\tilde{\rho} := \rho_{11} - \rho_{12}(\rho_{22})^{-1}\rho_{21}$. The operation achieves the minimum is obtained using spectral decomposition of $\sigma^{-1/2}\tilde{\rho}\sigma^{-1/2}$, and the same reverse test is optimal for all operator convex function f 's . Uniqueness of optimal operation modulo trivial redundancy is also shown.

Based on these analysis, we had shown, for example: Suppose f is operator convex. Then the maximum f-divergence of the probability distributions of a measurement under the state ρ and σ is strictly less than $\mathrm{D}_f^{\max}(\rho\|\sigma)$. Thus, once encoded into non- commutative quantum states, some amount of classical f-divergence is irrecoverably lost. (This statement may seem intuitively trivial, but when f is not operator convex, this is not always true.)

After the detailed analysis of the case of f is operator convex, we study the case where such an assumption is not true. One of the motivation is much of the results in the former case generalizes.

First, when one of the states are a pure state, (2) generalize to all the convex functions, and the optimal reverse test is also the same, and also unique. Also, $\mathrm{D}_f^{\max}$ is strictly larger than measured f-divergence, unless two states commute.

Next, we analyzed $f(\lambda) = |1 - \lambda|$, since this corresponds to total variation distance, which is quite often used in statistics, information theory, and so on. Though we failed to obtain the closed formula, the optimization problem is reduced to quite simple linear semidefinite program. Using this, we had shown that (2) is not true in this case, and the optimal reverse test is no the same either.

In addition, when $\{\rho, \sigma\}$ satisfies some conditions, it turns out

$$\mathrm{D}_{|1-\lambda|}^{\max}(\rho||\sigma) = \|\rho - \sigma\|_1 . \tag{3}$$

Since the RHS equals the measured total variation distance, this means the total variation sometimes does not decrease by embedding into non - commutative quantum states. The condition for (3) is not too restrictive. If either two states are close enough, or either if $\rho\sigma + \sigma\rho \geq 0$, this identity holds. The numerical check for 2-dim case shows fairly large area of Bloch sphere satisfies (3).

Besides from these case studies, we had shown the dual expression of $\mathrm{D}_f^{\max}$,

$$\mathrm{D}_f^{\max}(\rho\|\sigma) = \sup\left\{\mathrm{tr}\,(\rho W_1 + \sigma W_2)\ ;\ pW_1 + qW_2 \leq g_f(p, q)\mathbf{1}, (p, q) \in [0, 1]^{\times 2}\right\}, \tag{4}$$

where $g_f(p, q)$ is essentially $pf(p/q)$. This shows that $\mathrm{D}_f^{\max}$ is the pointwise supremum of linear functionals, thus it is lower semi- continuous. Thus, $\mathrm{D}_f^{\max}$ behaves extremely nicely at the edge of the domain. In fact, if $(\rho_\varepsilon, \sigma_\varepsilon)$ is an arbitrary line segment connecting (ρ, σ) and an interior point of the domain,

$\lim_{\varepsilon\downarrow 0} \mathrm{D}_f^{\max}(\rho_\varepsilon\|\sigma_\varepsilon) = \mathrm{D}_f^{\max}(\rho\|\sigma)$ holds. (4) is also valid, with certain restrictions, even when the underlying Hilbert space is separable infinite dimensional space.

Except for the last section, we will work on a finite dimensional Hilbert space $\mathscr{H}$. In most cases, the underlying Hilbert space is not mentioned unless it is confusing. The space of trace class operators, and bounded operators on $\mathscr{H}$ is denoted by $\mathscr{B}_1(\mathscr{H})$, and $\mathscr{B}(\mathscr{H})$, respectively, and the space of their self-adjoint elements are denoted by $\mathscr{B}_{1,sa}(\mathscr{H})$, and $\mathscr{B}_{sa}(\mathscr{H})$. In most cases, specification of underlying Hilbert space is dropped, thus $\mathscr{B}_{sa}$ in stead of $\mathscr{B}_{sa}(\mathscr{H})$, for example. When $\dim \mathscr{H} < \infty$ (thus in most of the paper,) to denote the space of all linear operators, we use $\mathscr{B}(\mathscr{H})$.

Also, for each positive operator X, denote by $\operatorname{supp} X$ the its support, and by π_X the projection onto $\operatorname{supp} X$. The projection onto the space $\mathscr{K}$ is denoted by $\pi_{\mathscr{K}}$. Orthogonal complement of the projector π is denoted by $\pi^\perp$. In this paper, in most part, probability distributions or positive measures are defined on the finite set $\mathscr{X}$. These are easily identified with commutative elements of $\mathscr{B}_{sa}\left(\mathbb{C}^{|\mathscr{X}|}\right)$. Note the support of the measures μ is also denoted by $\operatorname{supp}\mu$.

2 Classical f-Divergence

This section explains the definition and known useful facts about classical f-divergence, and convex analysis.

The definition of D_f in the introduction obviously cannot be used when $q(x) = 0$ for some x. Convex analysis supplies useful tools to cope with such continuity issue. As in [13], we suppose that h is a map from $\mathbb{R}^n$ to $\mathbb{R}\cup\{\pm\infty\}$. Instead of saying that h is not defined on a certain set, we say that $h(\lambda) = \infty$ on that set. The *effective domain* of h, denoted by $\operatorname{dom} h$, is the set of all λ's with $h(\lambda) < \infty$. h is said to be convex iff its *epigraph*, or the set $\operatorname{epi} h := \{(\lambda_1, \lambda_2)\,;\, \lambda_2 \geq h(\lambda_1)\}$ is convex. A convex function h is *proper* iff h is nowhere $-\infty$ and not ∞ everywhere, and is *closed* iff the set $\{\lambda_1; \lambda_2 \geq h(\lambda_1)\}$ is closed for any λ_2, or equivalently, iff its epigraph is closed (Theorem 7.1 of [13]) , or equivalently, iff it is lower semi-continuous.

Given a convex function h, its *closure* $\operatorname{cl} h$ is the greatest closed, or equivalently, lower semi-continuous (not necessarily finite) function majorized by h. The name comes from the fact that $\operatorname{epi}(\operatorname{cl} h) = \operatorname{cl}(\operatorname{epi} h)$. $\operatorname{cl} h$ coincide with h except perhaps at the relative boundary points of its effective domain. If h is proper and convex, so is $\operatorname{cl} h$ (Theorem 7.4, [13]).

The following Proposition will be intensively used later.

Proposition 1 (Theorem 10.2, [13]) *If h is closed, proper and convex, it is continuous on any simplex in* $\operatorname{dom} h$.

From here, unless otherwise mentioned, f , which is used to define f-divergence, is supposed to satisfy the following condition.

(FC) f is a proper closed convex function with $\operatorname{dom} f \supset (0, \infty)$. Also, $f(0) = 0$.

Now we are in the position to define the classical f-divergence D_f between the positive measures p and q over the finite set $\mathscr{X}$. It is defined in the following manner, so that the function $(p, q) \to \mathrm{D}_f(p||q)$ is closed: Namely,

$$\mathrm{D}_f(p\|q) := \sum_{x\in\mathscr{X}} g_f(p(x), q(x)),$$

where $g_f(\lambda_1, \lambda_2)$ is the closure of $\lambda_2 f\left(\frac{\lambda_1}{\lambda_2}\right)$ (see pp. 35 and 67 of [13]),

$$g_f(\lambda_1, \lambda_2) := \begin{cases} \lambda_2 f(\lambda_1/\lambda_2), & \text{if } \lambda_1 \in \mathrm{dom}\, f,\ \lambda_2 > 0 \\ \lim_{\lambda_2\downarrow 0} \lambda_2 f(\lambda_1/\lambda_2), & \text{if } \lambda_1 \in \mathrm{dom}\, f,\ \lambda_2 = 0, \\ 0, & \text{if } \lambda_1 = \lambda_2 = 0, \\ \infty, & \text{otherwise.} \end{cases} \tag{5}$$

It is easy to check that

$$\mathrm{D}_f(p\|q) = \sum_{x\in \mathrm{supp}\, q} q(x) f\left(\frac{p(x)}{q(x)}\right) + \sum_{x\in\mathscr{X}/\mathrm{supp}\, q} p(x) \lim_{\varepsilon\downarrow 0} \varepsilon f\left(\frac{1}{\varepsilon}\right).$$

Remark 1 Though p and q have to be probability for D_f to have operational meanings, we extend the domain of D_f to pairs of positive finite measures on finite set for the sake of mathematical convenience.

Observe also g_f is in addition positively homogeneous, or

$$\forall a \geq 0,\ \ g_f(a\lambda_1, a\lambda_2) = a g_f(\lambda_1, \lambda_2).$$

Since it is positively homogeneous, proper, closed and convex, by Corollary 13.5.1 of [13], it is the pointwise supremum of linear functions,

$$g_f(\lambda_1, \lambda_2) = \sup_{(w_1,w_2)\in\mathscr{W}_f} w_1\lambda_1 + w_2\lambda_2, \tag{6}$$

where the set $\mathscr{W}$ is convex and unbounded from below.

Therefore,

$$\mathrm{D}_f(p\|q) = \sup\left\{\sum_{x\in\mathscr{X}} w_1(x) p(x) + w_2(x) q(x);\ (w_1(x), w_2(x)) \in \mathscr{W}_f\right\}. \tag{7}$$

This in turn shows, by Corollary 13.5.1 of [13], D_f is positively homogeneous, proper, closed and convex.

Remark 2 Equation (7) indicates (1). To see this, use $\mathscr{W}_f$ as a decision space.

If f satisfies (FC), the function

$$\hat{f}(\lambda) := g_f(\lambda, 1) \tag{8}$$

also satisfies (FC) and $g_{\hat{f}}(\lambda_2, \lambda_1) = g_f(\lambda_1, \lambda_2)$. This identity implies

$$\mathrm{D}_f(p||q) = \mathrm{D}_{\hat{f}}(q||p). \tag{9}$$

Also,

$$\lim_{\varepsilon\downarrow 0} \varepsilon f(1/\varepsilon) = \hat{f}(0), \ f(0) = \lim_{\varepsilon\downarrow 0} \varepsilon \hat{f}(1/\varepsilon). \tag{10}$$

Introduction of $\hat{f}$ often simplifies the argument, allowing to switch the first and the second variables.

3 Reverse Test and Maximal f-Divergence

In this section we define maximal f-divergence $\mathrm{D}_f^{\max}$ as the solution to an operationally defined minimization problem.

A *reverse test* of a pair $\{\rho, \sigma\}$ of positive definite operators is a triple $(\Gamma, \{p, q\})$. Here, Γ is a trace preserving positive linear map from positive measures over some a finite set $\mathscr{X}$ (or commutative algebra with dimension $|\mathscr{X}|$) to Hermitian operators, and p and q are positive measures over $\mathscr{X}$, with

$$\Gamma(p) = \rho, \ \Gamma(q) = \sigma.$$

(Note Γ is necessarily completely positive.)

For a function f satisfying above (FC), we define *maximal* f*-divergence*

$$\mathrm{D}_f^{\max}(\rho\|\sigma) = \inf_{(\Gamma,\{p,q\})} \mathrm{D}_f(p\|q), \tag{11}$$

where the infimum is taken over all the reverse tests. The name comes from the fact that $\mathrm{D}_f^{\max}(\rho\|\sigma)$ is the largest quantum version of $\mathrm{D}_f(p\|q)$; here, quantum version of $\mathrm{D}_f(p\|q)$ is any $\mathrm{D}_f^Q(\rho\|\sigma)$ such that

(D1) $\mathrm{D}_f^Q(\Lambda(\rho)\|\Lambda(\sigma)) \leq \mathrm{D}_f^Q(\rho\|\sigma)$ holds for any completely positive trace preserving (CPTP) map, any density operators ρ, σ on finite dimensional Hilbert spaces.

(D2) $\mathrm{D}_f^Q(p\|q) = \mathrm{D}_f(p\|q)$ for any probability distributions p, q over any finite sets.

Here p is identified $\sum_{x\in\mathscr{X}} p(x)|e_x\rangle\langle e_x|$, for example, where $\{|e_x\rangle; x\in\mathscr{X}\}$ is a CONS. Choice of a particular CONS is not important, since

$$\mathrm{D}_f^Q\left(U\rho U^\dagger\|U\sigma U^\dagger\right)=\mathrm{D}_f^Q(\rho\|\sigma)$$

for any unitary operator U due to (D1).

We also consider the following stronger condition.

(D1') $\mathrm{D}_f^Q(\Lambda(\rho)\,||\Lambda(\sigma))\le\mathrm{D}_f^Q(\rho||\sigma)$ holds for any trace preserving positive map Λ, any any density operators ρ,σ, on finite dimensional Hilbert spaces.

Lemma 1 *If (FC) is satisfied,* $\mathrm{D}_f^{\max}$ *satisfies above (D1), (D1') and (D2). Also, if a two point functional* D_f^Q *satisfies satisfies both of (D1) and (D2), or both of (D1') and (D2),*

$$\mathrm{D}_f^Q(\rho\|\sigma)\le\mathrm{D}_f^{\max}(\rho\|\sigma).$$

Proof Let Λ be a trace preserving positive map. Then,

$$\begin{aligned}
&\mathrm{D}_f^{\max}(\Lambda(\rho)\,\|\Lambda(\sigma))\\
&=\inf_{(\Gamma,\{p,q\})}\left\{\mathrm{D}_f(p\|q);(\Gamma,\{p,q\})\ \text{: a reverse test of }\{\Lambda(\rho),\Lambda(\sigma)\}\right\}\\
&\le\inf_{(\Gamma,\{p,q\})}\left\{\mathrm{D}_f(p\|q);\Gamma=\Gamma'\circ\Lambda,\left(\Gamma',\{p,q\}\right)\ \text{: a reverse test of }\{\rho,\sigma\}\right\}\\
&=\mathrm{D}_f^{\max}(\rho\|\sigma).
\end{aligned}$$

Hence, $\mathrm{D}_f^{\max}$ satisfies (D1'), and thus (D1) also. Also,

$$\begin{aligned}
\mathrm{D}_f^{\max}(p\|q)&=\inf\left\{\mathrm{D}_f(p'\|q');p=\Gamma(p'),q=\Gamma(q'),\Gamma\text{: stochastic map}\right\}\\
&\ge\inf\left\{\mathrm{D}_f(\Gamma(p')\,\|\Gamma(q'));p=\Gamma(p'),q=\Gamma(q'),\Gamma\text{ : stochastic map}\right\}\\
&=\mathrm{D}_f(p\|q).
\end{aligned}$$

The opposite inequality is trivial. so we have $\mathrm{D}_f^{\max}(p\|q)=\mathrm{D}_f(p\|q)$. Thus, $\mathrm{D}_f^{\max}$ satisfies (D2).

Suppose D_f^Q satisfies (D1) (or (D1')) and (D2), and let $(\Gamma,\{p,q\})$ be a reverse test of $\{\rho,\sigma\}$. Then,

$$\mathrm{D}_f^Q(\rho\|\sigma)=\mathrm{D}_f^Q(\Gamma(p)\,\|\Gamma(q))\le\mathrm{D}_f^Q(p\|q)=\mathrm{D}_f(p\|q).$$

Therefore, taking infimum over all the reverse tests of $\{\rho,\sigma\}$, we have $\mathrm{D}_f^Q(\rho\|\sigma)\le\mathrm{D}_f^{\max}(\rho\|\sigma)$.

Remark 3 In defining reverse test, we had assumed the cardinality of $\mathscr{X}$, where $\{p,q\}$ are defined, is finite for mathematical simplicity. But, this restriction is not essential as long as $\dim\mathscr{H}<\infty$: By usual argument using Caratheodory's theorem, we can show $|\mathscr{X}|\le 2(\dim\mathscr{H})^2+1$ is enough for minimization.

Also, the infimum can be replaced by the minimum if $g_f(\lambda_1, 0) = \lim_{\varepsilon\downarrow 0} \varepsilon f(1/\varepsilon) < \infty$ and $g_f(0, \lambda_2) = f(0) < \infty$. In this case, the map

$$F\left(\{p(x), q(x), \Gamma(\delta_x)\}_{x\in\mathscr{X}}\right) := \mathrm{D}_f(p\|q) = \sum_{x\in\mathscr{X}} g_f(p(x), q(x))$$

is bounded and continuous. Since the constrains on $\{p(x), q(x), \Gamma(\delta_x)\}_{x\in\mathscr{X}}$ is continuous, the domain of F is a closed subset of the compact set $[0, \mathrm{tr}\,\rho]^{|\mathscr{X}|} \times [0, \mathrm{tr}\,\sigma]^{|\mathscr{X}|} \times \{X; X \geq 0, \mathrm{tr}\,X = 1\}^{|\mathscr{X}|}$. Thus, the domain of F is compact. Therefore, it has the minimum.

4 Representations of Reverse Tests

4.1 A Representation

Denote by δ_x the delta distribution at x, let $\mathscr{X}' := \mathrm{supp}\, q$, and define

$$S_x := \begin{cases} q(x)\,\Gamma(\delta_x), & \text{if } x \in \mathscr{X}', \\ p(x)\,\Gamma(\delta_x), & \text{otherwise,} \end{cases} \quad r_x := \begin{cases} p(x)/q(x), & \text{if } x \in \mathscr{X}', \\ \infty, & \text{otherwise,} \end{cases}$$

Then it should satisfy

$$\sum_{x\in\mathscr{X}'} r_x S_x + \sum_{x\notin\mathscr{X}'} S_x = \rho, \quad \sum_{x\in\mathscr{X}'} S_x = \sigma. \tag{12}$$

Conversely, to each such $\{S_x, r_x; x \in \mathscr{X}\}$, there corresponds a reverse test with $\Gamma(\delta_x) = \frac{1}{\mathrm{tr}\, S_x} S_x$ and

$$\{p(x), q(x)\} = \begin{cases} \{r_x \mathrm{tr}\, S_x, \mathrm{tr}\, S_x\} & \text{if } x \in \mathscr{X}', \\ \{\mathrm{tr}\, S_x, 0\}, & \text{otherwise.} \end{cases}$$

So $\{S_x, r_x; x \in \mathscr{X}\}$ is a bijective representation of a reverse test. In this representation, the value of classical divergence is

$$\mathrm{D}_f(p\|q) = \sum_{x\in\mathscr{X}'} f(r_x)\,\mathrm{tr}\, S_x + \lim_{\varepsilon\downarrow 0} \varepsilon f(1/\varepsilon) \sum_{x\notin\mathscr{X}'} \mathrm{tr}\, S_x.$$

Note that x affect the value of $\mathrm{D}_f(p\|q)$ through r_x and S_x. Thus, when the map $x \to r_x$ is not injective, one can always find "simpler" reverse test $\left\{S'_x, r'_y; y \in \mathscr{Y}\right\}$ without changing the value of D_f:

$$S'_y = \sum_{x;r_x=r_y} S_x,\ S_{y_0} := \sum_{x \notin \mathscr{X}'} S_x.$$

Then, elementary computation verifies that the induced revere test $\left(\Gamma', \left\{p', q'\right\}\right)$ does not change f-divergence. Thus, without changing the infimum, we can safely assume that $x \to r_x$ is injective.

Since $x \to r_x$ is injective, we may omit x in representation; Let $S(\cdot)$ be the map from $[0, \infty]$ into positive operators such that

$$\sum_{r \in [0,\infty)} r S(r) + S(\infty) = \rho,\ \sum_{r \in [0,\infty)} S(r) = \sigma. \tag{13}$$

Here, $S(\cdot)$ is non-zero only on the finite set $\{r_x; x \in \mathscr{X}\}$. By this representation, $\mathrm{D}_f^{\max}$ is the solution to the linear optimization problem

$$\mathrm{D}_f^{\max}(\rho\|\sigma) = \inf\left\{\sum_{r \in [0,\infty)} f(r)\,\mathrm{tr}\,S(r) + \lim_{\varepsilon \downarrow 0} \varepsilon f\left(\frac{1}{\varepsilon}\right) \mathrm{tr}\,S(\infty)\,;\, S(\cdot) \text{ with } (13)\right\}. \tag{14}$$

Observe this optimization can be done in the two stages; Fixing $S(\infty)$, or equivalently

$$\rho_* := \rho - S(\infty) = \sum_{r \in [0,\infty)} r\mathrm{tr}\,S(r) \le \rho, \tag{15}$$

optimize $\{S(r)\,;\, r \in [0, \infty)\}$ to minimize

$$\sum_{r \in [0,\infty)} f(r)\,\mathrm{tr}\,S(r) = \sum_{x:\mathrm{supp}\,q(x)} q(x) f\left(\frac{p(x)}{q(x)}\right)$$
$$= \mathrm{D}_f(\tilde{p}\|q),$$

where $\tilde{p}$ is restriction of p to $\mathrm{supp}\,q$. Since $(\Gamma, \{\tilde{p}, q\})$ is a reverse test of $\{\rho_*, \sigma\}$, the minimum of $\mathrm{D}_f(\tilde{p}\|q)$ equals $\mathrm{D}_f^{\max}(\rho_*\|\sigma)$.

After this is done, we optimize $\rho_*$to minimize

$$\mathrm{D}_f^{\max}(\rho_*\|\sigma) + \lim_{\varepsilon \downarrow 0} \varepsilon f\left(\frac{1}{\varepsilon}\right) \mathrm{tr}\,S(\infty) = \mathrm{D}_f^{\max}(\rho_*\|\sigma) + \mathrm{tr}\,(\rho - \rho_*) \lim_{\varepsilon \downarrow 0} \varepsilon f\left(\frac{1}{\varepsilon}\right).$$

Note that $\mathrm{supp}\,\rho_* \subset \mathrm{supp}\,\sigma$ holds by (15) and (13), and $0 \le \rho_* \le \rho$ by its definition (15). Therefore,

$$\mathrm{D}_f^{\max}(\rho\|\sigma) = \inf_{\rho_*}\left\{\mathrm{D}_f^{\max}(\rho_*\|\sigma) + \mathrm{tr}\,(\rho - \rho_*) \lim_{\varepsilon \downarrow 0} \varepsilon f\left(\frac{1}{\varepsilon}\right); 0 < \rho_* \le \rho,\, \mathrm{supp}\,\rho_* \subset \mathrm{supp}\,\sigma\right\}. \tag{16}$$

4.2 Representation and Radon–Nikodym Derivative

To make the list of all $\{S_x, r_x\}$'s with (12), or almost equivalently, $S(\cdot)$'s with (13), commutative Radon–Nikodym derivative is useful. Given a positive operator ρ_* with

$$\mathrm{sppp}\,\rho_* \subset \mathrm{sppp}\,\sigma, \tag{17}$$

the commutative Radon–Nikodym derivative of ρ_* with respect to σ is defined by

$$d(\rho_*, \sigma) := \sigma^{-1/2}\rho_*\sigma^{-1/2}. \tag{18}$$

(σ^{-1} is generalized inverse.) Suppose

$$\rho_* \leq \rho \tag{19}$$

and let $\{M_x\}$ be a resolution of identity into positive operators with

$$d(\rho_*, \sigma) = \sum_{x\in\mathscr{X}'} r_x M_x, \ \sum_{x\in\mathscr{X}'} M_x = \mathbf{1}.$$

Then

$$S_x = \begin{cases} \sigma^{1/2}M_x\sigma^{1/2}, & \text{if } r_x < \infty, \\ \frac{1}{\mathrm{tr}(\rho-\rho_*)}(\rho-\rho_*), & \text{if } r_x = \infty. \end{cases}$$

Therefore,

$$\{q(x), p(x)\} = \begin{cases} \{\mathrm{tr}\,\sigma M_x, r_x q_x\}, & \text{if } r_x < \infty, \\ \{0, \mathrm{tr}\,(\rho-\rho_*)\}, & \text{if } r_x = \infty, \end{cases} \tag{20}$$

and

$$\Gamma(\delta_x) := \begin{cases} \frac{1}{q(x)}\sigma^{1/2}M_x\sigma^{1/2}, & \text{if } r_x < \infty, \\ \frac{1}{\mathrm{tr}\,(\rho-\rho_*)}(\rho-\rho_*), & \text{if } r_x = \infty. \end{cases} \tag{21}$$

Thus, $\{M_x, r_x\}$ and ρ_* specifies a reverse test.

Define

$$\tilde{\rho} := \rho_{11} - \rho_{12}\rho_{22}{}^{-1}\rho_{21}, \tag{22}$$

where

$$\rho = \begin{bmatrix} \rho_{11} & \rho_{12} \\ \rho_{21} & \rho_{22} \end{bmatrix}, \ \sigma = \begin{bmatrix} \sigma_{11} & 0 \\ 0 & 0 \end{bmatrix}.$$

When M_x's are projectors and $\rho_* = \tilde{\rho}$, we say the corresponding reverse test is itminimal. The minimal reverse test turns out to be optimal under certain natural conditions (namely, the condition (F) in Sect. 3) on f. $\rho_* = \tilde{\rho}$ is chosen because it is the largest positive operator with (17) and (19);

Lemma 2 *Suppose $\rho_* \geq 0$ is supported on* $\operatorname{supp}\sigma$ *and $\rho_* \leq \rho$. Then, $\tilde{\rho} \geq \rho_*$. Also, $0 \leq \tilde{\rho} \leq \rho$ and* $\operatorname{supp}\tilde{\rho} \subset \operatorname{supp}\sigma$.

Proof By Proposition 9, $\tilde{\rho} \geq 0$. (in fact, $\tilde{\rho}^{-1} = \pi_\sigma \rho^{-1} \pi_\sigma$.) $\tilde{\rho} \leq \rho$ and $\operatorname{supp}\tilde{\rho} \subset \operatorname{supp}\sigma$ are obvious by definition.

Since $\rho_* \leq \rho$,

$$\begin{bmatrix} \rho_{11} - \rho_* & \rho_{12} \\ \rho_{21} & \rho_{22} \end{bmatrix} \geq 0,$$

Therefore, by Proposition 9, we should have $\rho_{11} - \rho_1 \geq \rho_{12}\rho_{22}^{-1}\rho_{21}$, or equivalently,

$$\tilde{\rho} = \rho_{11} - \rho_{12}\rho_{22}^{-1}\rho_{21} \geq \rho_*.$$

5 Properties of $\mathrm{D}_f^{\max}$

Theorem 1 *When f satisfies (FC), $\mathrm{D}_f^{\max}$ has the following properties.*

(i) $\mathrm{D}_f^{\max}$ is jointly convex: if $\rho = \sum_i c_i\rho_i$, $\sigma = \sum_i c_i\sigma_i$, $\sum_i c_i = 1$ ($c_i \geq 0$),

$$\mathrm{D}_f^{\max}(\rho\|\sigma) \leq \sum_i c_i \mathrm{D}_f^{\max}(\rho_i\|\sigma_i), \tag{23}$$

(ii) If $f(0) = 0$ in addition, it is monotone decreasing in the second argument:

$$\mathrm{D}_f^{\max}(\rho\|X) \leq \mathrm{D}_f^{\max}(\rho\|\sigma), \quad X \geq \sigma \tag{24}$$

(iii) $\mathrm{D}_f^{\max}$ is positively homogeneous.

$$\mathrm{D}_f^{\max}(c\rho\|c\sigma) = c\mathrm{D}_f^{\max}(\rho\|\sigma), \ c \geq 0. \tag{25}$$

In particular,

$$\mathrm{D}_f^{\max}(0\|0) = 0. \tag{26}$$

(iv) Direct sum property:

$$\mathrm{D}_f^{\max}(\rho_0 \oplus \rho_1\|\sigma_0 \oplus \rho_1) = \mathrm{D}_f^{\max}(\rho_0\|\sigma_0) + \mathrm{D}_f^{\max}(\rho_1\|\sigma_1), \tag{27}$$

where ρ_i, σ_i are supported on $\mathscr{H}_i$ ($i = 0, 1$), and $\mathscr{H}_0 \perp \mathscr{H}_1$.

Proof (i): let $(\Gamma_i, \{p_i, q_i\})$ be a reverse tests of $\{\rho_i, \sigma_i\}$, where p_i, q_i are positive measures over the finite set $\mathscr{X}_i$. Then 'mixture' of these reverse tests with probability c_i, compose a reverse test $(\Gamma, \{p, q\})$ of $\{\rho, \sigma\}$: let $\mathscr{X} = \bigcup_i \mathscr{X}_i$, and define

$$p_i(x) := c_i p_0,\ q_i(x) := c_i q_i(x),\ \Gamma(\delta_x) = \Gamma_i(\delta_x),\ (x \in \mathscr{X}_i).$$

Then,

$$\mathrm{D}_f\left(p\|q\right)=\sum_i c_i \mathrm{D}_f\left(p_i\|q_i\right).$$

Therefore, minimizing over all the reverse tests of $\{\rho,\sigma\}$, we obtain (23).

(ii): let $X':=X-\sigma\geq 0$. Then

$$\begin{aligned}
&\mathrm{D}_f^{\max}\left(\rho\|X\right)=\mathrm{D}_f^{\max}\left(\rho\|\sigma+X'\right)\\
&\leq\inf\left\{\mathrm{D}_f\left(p\|q+q'\right);\varGamma\left(p\right)=\rho,\varGamma\left(q\right)=\sigma,\varGamma\left(q'\right)=X,\ \mathrm{supp}\,q'\text{ is disjoint with }\mathrm{supp}q,\mathrm{supp}p\right\}\\
&=\inf\left\{\mathrm{D}_f\left(p\|q\right);\varGamma\left(p\right)=\rho,\varGamma\left(q\right)=\sigma,\varGamma\left(q'\right)=X,\ \mathrm{supp}\,q'\text{ is disjoint with }\mathrm{supp}q,\mathrm{supp}p\right\}\\
&=\mathrm{D}_f^{\max}\left(\rho\|\sigma\right),
\end{aligned}$$

where the identity in the fourth line is due to the following calculation: let $\mathscr{X}_1=\mathrm{supp}q\cup\mathrm{supp}p$, and $\mathscr{X}_2=\mathrm{supp}\,q'$. Since these two sets are disjoint,

$$\sum_{x\in\mathscr{X}_1\cup\mathscr{X}_2}g_f\left(p\left(x\right),q\left(x\right)+q'\left(x\right)\right)=\sum_{x\in\mathscr{X}_1}g_f\left(p\left(x\right),q\left(x\right)\right)+\sum_{x\in\mathscr{X}_2}g_f\left(0,q'\left(x\right)\right).$$

Since $f\left(0\right)=0$ by the condition (FC), we have the identity.

(iii): Let $c>0$. Then to each reverse test $(\varGamma,\{p,q\})$ of $\{\rho,\sigma\}$, corresponds the reverse test $(\varGamma,\{cp,cq\})$ test of $\{c\rho,c\sigma\}$, and vice versa. Hence, due to the fact that D_f positively homogeneous, we have the identity. When $c=0$, we only have to show the LHS is 0. In fact, if $(\varGamma,\{p,q\})$ is an arbitrary reverse test of $\{0,0\}$, supp p and supp q are empty, and $\mathrm{D}_f\left(p\|q\right)=0$. Thus $\mathrm{D}_f^{\max}\left(0\|0\right)=0$.

(iv): "$\leq$" is trivial. Thus, we show "$\geq$". Let $(\varGamma,\{p,q\})$ be a reverse test of $\{\rho,\sigma\}=\{\rho_0\oplus\rho_1,\sigma_0\oplus\sigma_1\}$, and define

$$\begin{aligned}
\varGamma_i\left(\delta_x\right)&:=\frac{1}{\mathrm{tr}\,\pi_{\mathscr{H}_i}\varGamma\left(\delta_x\right)}\pi_{\mathscr{H}_i}\varGamma\left(\delta_x\right)\pi_{\mathscr{H}_i},\\
p_i\left(x\right)&:=p\left(x\right)\mathrm{tr}\,\pi_{\mathscr{H}_i}\varGamma\left(\delta_x\right),\ q_i\left(x\right):=p\left(x\right)\mathrm{tr}\,\pi_{\mathscr{H}_i}\varGamma\left(\delta_x\right).
\end{aligned}$$

Then $(\varGamma_i,\{p_i,q_i\})$ is a reverse test of $\{\rho_i,\sigma_i\}$ $(i=0,1)$. Also, since g_f is positively homogeneous,

$$\begin{aligned}
&\mathrm{D}_f\left(p_0\|q_0\right)+\mathrm{D}_f\left(p_1\|q_1\right)\\
&=\sum_{x\in\mathscr{X}}g_f\left(p\left(x\right)\mathrm{tr}\,\pi_{\mathscr{H}_0}\varGamma\left(\delta_x\right),q\left(x\right)\mathrm{tr}\,\pi_{\mathscr{H}_0}\varGamma\left(\delta_x\right)\right)\\
&\quad+\sum_{x\in\mathscr{X}}g_f\left(p\left(x\right)\mathrm{tr}\,\pi_{\mathscr{H}_1}\varGamma\left(\delta_x\right),q\left(x\right)\mathrm{tr}\,\pi_{\mathscr{H}_1}\varGamma\left(\delta_x\right)\right)\\
&=\sum_{x\in\mathscr{X}}\left\{\mathrm{tr}\,\pi_{\mathscr{H}_0}\varGamma\left(\delta_x\right)+\mathrm{tr}\,\pi_{\mathscr{H}_1}\varGamma\left(\delta_x\right)\right\}g_f\left(p\left(x\right),q\left(x\right)\right)\\
&=\sum_{x\in\mathscr{X}}g_f\left(p\left(x\right),q\left(x\right)\right)=\mathrm{D}_f\left(p\|q\right).
\end{aligned}$$

Thus,

$$\inf \mathrm{D}_f\left(p\|q\right)=\inf\left\{\mathrm{D}_f\left(p_0\|q_0\right)+\mathrm{D}_f\left(p_1\|q_1\right)\right\}\geq \inf \mathrm{D}_f\left(p_0\|q_0\right)+\inf \mathrm{D}_f\left(p_1\|q_1\right),$$

which leads to the asserted inequality. After all, we have (27).

Lemma 3 *Suppose f satisfies (FC). Then the convex function $(\rho,\sigma)\rightarrow \mathrm{D}_f^{\max}\left(\rho||\sigma\right)$ is proper. Thus, it is nowhere $-\infty$.*

Proof An improper convex function is necessarily infinite except perhaps at relative boundary points of its effective domain (Theorem 7.2 of [13]). But

$$\mathrm{D}_f^{\max}\left(p\|p\right)=\mathrm{D}_f\left(p\|p\right)=\sum_{x\in\mathscr{X}}p\left(x\right)f\left(1\right)$$

is finite. Thus $\mathrm{D}_f^{\max}$ cannot be improper.

Theorem 2 *Suppose f satisfies (FC). Then $\mathrm{D}_f^{\max}\left(\rho||\sigma\right)<\infty$ only in the following four cases.*

(i) $\lim_{\varepsilon\downarrow 0}\varepsilon f\left(1/\varepsilon\right)<\infty$ *and* $f\left(0\right)<\infty$*;*
(ii) $\lim_{\varepsilon\downarrow 0}\varepsilon f\left(1/\varepsilon\right)<\infty$, $f\left(0\right)=\infty$, *and* $\operatorname{supp}\rho\supset\operatorname{supp}\sigma$*;*
(iii) $\lim_{\varepsilon\downarrow 0}\varepsilon f\left(1/\varepsilon\right)=\infty$, $f\left(0\right)<\infty$, *and* $\operatorname{supp}\rho\subset\operatorname{supp}\sigma$*;*
(iv) $\lim_{\varepsilon\downarrow 0}\varepsilon f\left(1/\varepsilon\right)=\infty$, $f\left(0\right)=\infty$, *and* $\operatorname{supp}\rho=\operatorname{supp}\sigma$.

Proof (i): Consider a reverse test with $\mathscr{X}=\{0,1\}$, $p=\delta_0$, $q=\delta_1$, $\Gamma\left(\delta_0\right)=\rho$, and $\Gamma\left(\delta_1\right)=\sigma$. In other words, we generate ρ or σ deterministically. Then

$$\mathrm{D}_f^{\max}\left(\rho||\sigma\right)\leq \mathrm{D}_f\left(p||q\right)=f\left(0\right)+\lim_{\varepsilon\downarrow 0}\varepsilon f\left(1/\varepsilon\right)<\infty.$$

(ii): Suppose $\operatorname{supp}\rho\supset\operatorname{supp}\sigma$, and, let $\mathscr{X}:=\{0,1\}$,

$$\Gamma\left(\delta_0\right):=\frac{1}{\operatorname{tr}\left(\rho-\alpha\sigma\right)}\left(\rho-\alpha\sigma\right),\ \Gamma\left(\delta_1\right):=\sigma,$$
$$p\left(0\right):=\operatorname{tr}\left(\rho-\alpha\sigma\right),\ p\left(1\right):=\alpha,\ q=\delta_1,$$

where $\alpha>0$ is a constant satisfying $\rho-\alpha\sigma\geq 0$. Then the corresponding classical divergence is finite,

$$\mathrm{D}_f^{\max}\left(\rho\|\sigma\right)\leq \mathrm{D}_f\left(p\|q\right)=\lim_{\varepsilon\downarrow 0}\varepsilon f\left(1/\varepsilon\right)\operatorname{tr}\left(\rho-\alpha\sigma\right)+1\cdot f\left(\frac{\alpha}{1}\right)<\infty.$$

On the other hand, if $\operatorname{supp}\rho\not\supset\operatorname{supp}\sigma$, for any reverse test $(\Gamma,\{p,q\})$, there is $x_0\in\mathscr{X}$ with $p\left(x_0\right)=0$ and $q\left(x_0\right)\neq 0$. To show this, suppose the otherwise, or equivalently, for any x with $q\left(x\right)\neq 0$, $p\left(x\right)\neq 0$. Then the support of $\rho=\sum_x p\left(x\right)\Gamma\left(\delta_x\right)$

contains the support of $\sigma = \sum_x q(x)\,\Gamma(\delta_x)$, contradicting with the assumption. Therefore, there is $x_0 \in \mathscr{X}$ with $p(x_0) = 0$ and $q(x_0) \neq 0$, and therefore

$$\mathrm{D}_f^{\max}(\rho\|\sigma) = \inf \mathrm{D}_f(p\|q) = \inf \left\{ q(x_0)\, f\left(\frac{0}{q(x_0)}\right) + \sum_{x \neq x_0} g_f(p(x), q(x)) \right\} = \infty.$$

(iii): This case is reduced to (ii) using $\hat{f}$ defined by (8), which also satisfies (FC). The reduction uses

$$\mathrm{D}_{\hat{f}}^{\max}(\sigma||\rho) = \mathrm{D}_f^{\max}(\rho||\sigma)$$

which is true by (9), and also (10).

(iv): Suppose $\mathrm{supp}\,\rho = \mathrm{supp}\,\sigma$. Then considering the minimal reverse test, we can verify $\mathrm{D}_f^{\max}(\rho||\sigma) \leq \infty$. On the other hand, if $\rho \neq \mathrm{supp}\,\sigma$, the same argument as (ii) or (iii) proves $\mathrm{D}_f^{\max}(\rho||\sigma) = \infty$.

6 When f is Operator Convex

6.1 Closed Formula

In this section, we suppose that f is operator convex and $f(0) = 0$ in addition to satisfying (FC):

(F) f is proper, closed, and operator convex. In addition, $\mathrm{dom}\, f(x) = [0, \infty)$ and $f(0) = 0$.

If this is true and $\lim_{\varepsilon \downarrow 0} \varepsilon\, f\left(\frac{1}{\varepsilon}\right) < \infty$, by Proposition 7,

$$f(\lambda) = a\lambda + f_0(\lambda), \tag{28}$$

where $f_0(\lambda)$ satisfies (F) and is operator monotone decreasing.

When $\mathrm{supp}\,\sigma \supset \mathrm{supp}\,\rho$, by the correspondence (20) and (21),

$$\mathrm{D}_f^{\max}(\rho||\sigma) = \inf_{\{M_x\},\{r_x\}_*} \left\{ \sum_{x \in \mathscr{X}} f(r_x)\,\mathrm{tr}\,\sigma M_x \;;\; \sum_{x \in \mathscr{X}} r_x M_x = d(\rho, \sigma),\ \sum_{x \in \mathscr{X}} M_x = \mathbf{1} \right\} \tag{29}$$

Here we use Naimark extension. Denoting the extended space by $\mathscr{H}'$, and letting V be an isometry from $\mathscr{H}$ (,where ρ, σ, etc. are living in) into $\mathscr{H}'$, there is a tuple of mutually orthogonal projectors $\{E_x\}$ in $\mathscr{H}'$ with $VE_xV^{\dagger} = M_x$. Therefore,

$$\begin{aligned}\sum_{x\in\mathscr{X}} f(r_x)\operatorname{tr}\sigma M_x &= \operatorname{tr}\sigma V f\left(\sum_{x\in\mathscr{X}} r_x E_x\right)V^{\dagger}\\ &\geq \operatorname{tr}\sigma f\left(\sum_{x\in\mathscr{X}} r_x V E_x V^{\dagger}\right) = \operatorname{tr}\sigma f\left(\sum_{x\in\mathscr{X}} r_x M_x\right) = \operatorname{tr}\sigma f\left(d\left(\rho,\sigma\right)\right),\end{aligned}$$

where the inequality in the second line is by Jensen's inequality, Proposition 6 (Note $X \to VXV^{\dagger}$ is a positive unital map into $\mathscr{B}(\mathscr{H})$). The identity is true if M_x's are mutually orthogonal projectors and $\mathscr{H}' = \mathscr{H}$, i.e., if the reverse test is minimal. Thus,

$$\mathrm{D}_f^{\max}(\rho\|\sigma) = \operatorname{tr}\sigma f\left(d\left(\rho,\sigma\right)\right)$$

and the identity is achieved by the minimal test.

Let us proceed to the case where $\operatorname{supp}\sigma \not\supset \operatorname{supp}\rho$. By (16),

$$\begin{aligned}&\mathrm{D}_f^{\max}(\rho\|\sigma)\\ &= \inf_{\rho_*}\left\{\mathrm{D}_f^{\max}(\rho_*\|\sigma) + \operatorname{tr}(\rho-\rho_*)\lim_{\varepsilon\downarrow 0}\varepsilon f\left(\frac{1}{\varepsilon}\right);\ \rho\geq\rho_*\geq 0,\ \operatorname{supp}\sigma\supset\operatorname{supp}\rho_*\right\}\\ &= \inf_{\rho_*}\left\{\operatorname{tr}\sigma f\left(d\left(\rho_*,\sigma\right)\right) + \operatorname{tr}(\rho-\rho_*)\lim_{\varepsilon\downarrow 0}\varepsilon f\left(\frac{1}{\varepsilon}\right);\ \rho\geq\rho_*\geq 0,\ \operatorname{supp}\sigma\supset\operatorname{supp}\rho_*\right\}.\end{aligned}$$

By Theorem 2, $\mathrm{D}_f^{\max}(\rho\|\sigma) = \infty$ if $a := \lim_{\varepsilon\downarrow 0}\varepsilon f(1/\varepsilon) = \infty$. Hence, suppose $a < \infty$. In this case, recall that we have the decomposition (28) due to Proposition 7, that is, $f(\lambda) = a\lambda + f_0(\lambda)$ and f_0 is operator monotone decreasing. Therefore,

$$\begin{aligned}&\operatorname{tr}\sigma\ f\ \left(d\left(\rho_*,\sigma\right)\right) + \operatorname{tr}\left(\rho-\rho_*\right)\lim_{\varepsilon\downarrow 0}\varepsilon f\left(1/\varepsilon\right)\\ &= \operatorname{tr}\sigma\ \left\{ad\left(\rho_*,\sigma\right) + f_0\left(d\left(\rho_*,\sigma\right)\right)\right\} + \operatorname{tr}\left(\rho-\rho_*\right)\lim_{\varepsilon\downarrow 0}\varepsilon\left\{a/\varepsilon + f_0\left(1/\varepsilon\right)\right\}\\ &= a + \operatorname{tr}\sigma\ f_0\ \left(d\left(\rho_*,\sigma\right)\right) + \operatorname{tr}\left(\rho-\rho_*\right) f_0\left(0\right)\\ &\underset{(a)}{\geq} a + \operatorname{tr}\sigma\ f_0\ \left(d\left(\tilde{\rho},\sigma\right)\right) + \operatorname{tr}\left(\rho-\tilde{\rho}\right) f_0\left(0\right)\\ &= \operatorname{tr}\sigma\ f\ \left(d\left(\tilde{\rho},\sigma\right)\right) + \operatorname{tr}\left(\rho-\tilde{\rho}\right)\lim_{\varepsilon\downarrow 0}\varepsilon f\left(1/\varepsilon\right),\end{aligned}$$

where $\tilde{\rho}$ is defined by (22), and the inequality (a) is by Lemma 2.

After all:

Theorem 3

$$\mathrm{D}_f^{\max}(\rho||\sigma) = \mathrm{tr}\,\sigma\, f\,(d\,(\tilde{\rho},\sigma)) + \mathrm{tr}\,(\rho - \tilde{\rho}) \lim_{\varepsilon\downarrow 0} \varepsilon f\,(1/\varepsilon) \tag{30}$$
$$= \mathrm{tr}\,\sigma\, f\,\left(\sigma^{-1}\tilde{\rho}\right) + \mathrm{tr}\,(\rho - \tilde{\rho}) \lim_{\varepsilon\downarrow 0} \varepsilon f\,(1/\varepsilon)$$

holds if (F) is true. (If $\lim_{\varepsilon\downarrow 0} \varepsilon f\left(\frac{1}{\varepsilon}\right) = \infty$*, both ends are* ∞ *and the identity holds.) The minimum is achieved by the minimal reverse test of* $\{\rho,\sigma\}$*.*

6.2 Examples

Throughout this subsection, we suppose $\mathrm{supp}\,\sigma \supset \mathrm{supp}\,\rho$. With $f_{\mathrm{KL}}(\lambda) := \lambda \log \lambda$,

$$\begin{aligned}\mathrm{D}_{f_{\mathrm{KL}}}^{\max}(\rho||\sigma) &= \mathrm{tr}\,\sigma\sigma^{-1}\rho\left(\log \sigma^{-1}\rho\right)\\ &= \mathrm{tr}\,\rho \log \sigma^{-1}\rho\\ &= \mathrm{tr}\,\rho \log \rho^{1/2}\sigma^{-1}\rho^{1/2}.\end{aligned}$$

This quantity, corresponding to Kullback–Leibler divergence, had been studied by various authors [3–5, 9, 11]. The relation to the reverse test problem is first pointed out by [10].

Define

$$f_\alpha(\lambda) := (\pm)\,\lambda^\alpha,$$

where the sign is chosen so that the function is convex on the positive half- line. This is operator convex if $\alpha \in [-1, 1]\,/\,\{0\}$, and

$$\mathrm{D}_{f_\alpha}^{\max}(\rho||\sigma) = \mathrm{tr}\,\sigma f_\alpha\left(\sigma^{-1}\rho\right).$$

One can check the following identity, which is a special case of $\mathrm{D}_f^{\max} = \mathrm{D}_{\hat{f}}^{\max}$:

$$\mathrm{D}_{f_\alpha}^{\max}(\rho||\sigma) = \mathrm{D}_{f_{1-\alpha}}^{\max}(\sigma||\rho).$$

6.3 Operator Inequality Versions of the Properties of $\mathrm{D}_f^{\max}$

When (F) is true, the following operator valued quantity $g_f(\rho,\sigma)$, which is self-adjoint and

$$\mathrm{tr}\, g_f(\rho,\sigma) = \mathrm{D}_f^{\max}(\rho\|\sigma),$$

satisfies some operator version of properties of $\mathrm{D}_f^{\max}$ has:

$$g_f(\rho,\sigma) := \begin{cases} \sqrt{\sigma} f\left(\sigma^{-1/2}\rho\sigma^{-1/2}\right)\sqrt{\sigma}, & \text{if } \operatorname{supp}\sigma \supset \operatorname{supp}\rho, \\ g_f(\tilde{\rho},\sigma) + a(\rho-\tilde{\rho}), & \text{if } \operatorname{supp}\sigma \not\supset \operatorname{supp}\rho,\ a := \lim_{\varepsilon\downarrow 0}\varepsilon f(1/\varepsilon) < \infty, \\ \text{undefined.} & \text{otherwise.} \end{cases}$$

In the second case, since $a := \lim_{\varepsilon\downarrow 0}\varepsilon f\left(\frac{1}{\varepsilon}\right) < \infty$, $.f(\lambda) = f_0(\lambda) + a\lambda$, with operator monotone decreasing f_0. Therefore,

$$\begin{aligned} g_f(\rho,\sigma) &= g_f(\tilde{\rho},\sigma) + a(\rho-\tilde{\rho}) \\ &= g_{f_0}(\tilde{\rho},\sigma) + a\rho \\ &= \inf\left\{g_{f_0}(\rho_*, X) + a\rho;\ 0 \le \rho_* \le \rho, \operatorname{supp} X \supset \operatorname{supp}\rho_*\right\}. \end{aligned} \tag{31}$$

The most important one, which is used later, is operator version of (D1):

Lemma 4 *(i) For any positive trace preserving map Λ, we have*

$$\Lambda\left(g_f(\rho,\sigma)\right) \ge g_f(\Lambda(\rho), \Lambda(\sigma)). \tag{32}$$

(ii) If $g_f(\Lambda(\rho), \Lambda(\sigma)) = \Lambda\left(g_f(\rho,\sigma)\right)$, then $\Lambda(\tilde{\rho})$ is the largest positive operator supported on $\Lambda(\sigma)$ and majorized by $\Lambda(\rho)$. Thus,

$$\Lambda\left(g_f(\tilde{\rho},\sigma)\right) = g_f(\Lambda(\tilde{\rho}), \Lambda(\sigma)). \tag{33}$$

Proof For a given positive trace preserving map Λ, define

$$\Lambda_\sigma(X) := \{\Lambda(\sigma)\}^{-1/2}\Lambda\left(\sigma^{1/2}X\sigma^{1/2}\right)\{\Lambda(\sigma)\}^{-1/2}, \tag{34}$$

which is a positive unital map into $\mathscr{B}(\operatorname{supp}\Lambda(\sigma))$:

$$\Lambda_\sigma(\mathbf{1}) = \pi_{\Lambda(\sigma)}, \tag{35}$$

and

$$\Lambda_\sigma(d(\rho,\sigma)) = d(\Lambda(\rho), \Lambda(\sigma)). \tag{36}$$

If $\operatorname{supp}\sigma \supset \operatorname{supp}\rho$, since Λ is positive, $\Lambda(\rho)$ is supported on $\operatorname{supp}\Lambda(\sigma)$ and $d(\Lambda(\rho), \Lambda(\sigma))$ exists. Also,

$$\begin{aligned} g_f(\Lambda(\rho), \Lambda(\sigma)) &\underset{(a)}{=} \Lambda(\sigma)^{1/2}\, f\left(\Lambda_\sigma(d(\rho,\sigma))\right)\Lambda(\sigma)^{1/2} \\ &\underset{(b)}{\le} \Lambda(\sigma)^{1/2}\,\Lambda_\sigma\left(f(d(\rho,\sigma))\right)\Lambda(\sigma)^{1/2} = \Lambda\left(\sigma^{1/2}f(d(\rho,\sigma))\sigma^{1/2}\right), \end{aligned}$$

where (a) and (b) is by (36) and Proposition 6, respectively. If $\operatorname{supp}\sigma \not\supset \operatorname{supp}\rho$ and $a := \lim_{\varepsilon\downarrow 0}\varepsilon f(1/\varepsilon) < \infty$,

$$
\begin{aligned}
& g_f(\Lambda(\rho), \Lambda(\sigma)) \\
& = \inf_{\rho_*} \left\{ g_{f_0}(\rho_*, \Lambda(\sigma)) + a\Lambda(\rho);\ \Lambda(\rho) \geq \rho_* \geq 0,\ \operatorname{supp} \Lambda(\sigma) \supset \operatorname{supp} \rho_* \right\} \\
& \leq g_{f_0}(\Lambda(\tilde{\rho}), \Lambda(\sigma)) + a\Lambda(\rho) \\
& \leq \Lambda\left(g_{f_0}(\tilde{\rho}, \sigma)\right) + a\Lambda(\rho) = \Lambda\left(g_{f_0}(\tilde{\rho}, \sigma) + a\rho\right) \\
& = \Lambda\left(g_f(\rho, \sigma)\right),
\end{aligned}
$$

where the inequality in the fourth line is due to the inequality for the case of $\operatorname{supp} \sigma \supset \operatorname{supp} \rho$ (Recall $\tilde{\rho}$ as of (22) is supported on $\operatorname{supp} \sigma$).

Therefore, if $g_f(\Lambda(\rho), \Lambda(\sigma)) = \Lambda\left(g_f(\rho, \sigma)\right)$, $\Lambda(\tilde{\rho})$ should achieve the infimum in the second line. Thus the first statement of (ii) is true. Then,

$$
g_f(\Lambda(\rho), \Lambda(\sigma)) = g_f(\Lambda(\tilde{\rho}), \Lambda(\sigma)) + a\Lambda(\rho - \tilde{\rho}).
$$

Equating this to $\Lambda\left(g_f(\rho, \sigma)\right) = \Lambda\left(g_f(\tilde{\rho}, \sigma)\right) + a\Lambda(\tilde{\rho})$, we have (33).

Operator versions of (25) and (27) are trivial. Thus next we show the operator versions of (23):

$$
g_f(\rho, \sigma) \leq \sum_i c_i\, g_f(\rho_i, \sigma_i), \tag{37}
$$

where $\rho := \sum_i c_i \rho_i, \sigma := \sum_i c_i \sigma_i, \sum_i c_i = 1$, and $c_i \geq 0$. Equation (37) for the case $\operatorname{supp} \sigma \supset \operatorname{supp} \rho$ is known [14]. if $a := \lim_{\varepsilon \downarrow 0} \varepsilon f\left(\frac{1}{\varepsilon}\right) < \infty$,

$$
\begin{aligned}
& \sum_i c_i\, g_f(\rho_i, \sigma_i) = \sum_i c_i\, g_f(\tilde{\rho}_i, \sigma_i) + a \sum_i c_i (\rho_i - \tilde{\rho}_i) \\
& \geq g_{f_0}\left(\sum_i c_i\, \tilde{\rho}_i, \sum_i c_i \sigma_i\right) + a \sum_i c_i \rho_i
\end{aligned}
$$

Above, since $\operatorname{supp} \sigma = \operatorname{span}\{\bigcup \operatorname{supp} \sigma_i\}$, $\sum_i c_i\, \tilde{\rho}_i$ is supported on $\operatorname{supp} \sigma$. Thus the last end is well-defined. Also, $\sum_i c_i \tilde{\rho}_i \leq \sum_i c_i \rho_i = \rho$. Thus, The last end is bounded from below by

$$
\begin{aligned}
& \inf \left\{ g_{f_0}(\rho_*, \sigma) + a\rho;\ \rho \geq \rho_* \geq 0, \operatorname{supp} \sigma \supset \operatorname{supp} \rho_* \right\} \\
& = g_{f_0}(\tilde{\rho}, \sigma) + a\rho = g_f(\rho, \sigma),
\end{aligned}
$$

concluding (37).

Lastly, the analogue of (24) is

$$
g_f(\rho, \sigma) \leq g_f(\rho, X), \quad X \geq \sigma. \tag{38}
$$

This is proved as follows. Since $X \geq \sigma$, there is an operator C such that

$$CX^{1/2} = \sigma^{1/2}, \ \ \|C\| \leq 1.$$

Since $\operatorname{supp} X \supset \operatorname{supp} \sigma$, it follows that

$$C = \sigma^{1/2} X^{-1/2},$$

where $X^{-1/2}$ stands for the generalized inverse of $X^{1/2}$. If $\operatorname{supp} X \supset \operatorname{supp} \sigma \supset \operatorname{supp} \rho$,

$$\begin{aligned} g_f(\rho, \sigma) &= X^{1/2} C^{\dagger} f(d(\rho, \sigma)) X^{1/2} C \\ &\geq X^{1/2} f\left(C^{\dagger} d(\rho, \sigma) C\right) X^{1/2} \\ &= X^{1/2} f\left(X^{-1/2} \rho X^{-1/2}\right) X^{1/2} \\ &= g_f(\rho, X), \end{aligned}$$

where the inequality in the second line is due to Proposition 5. If $a := \lim_{\varepsilon \downarrow 0} \varepsilon f\left(\frac{1}{\varepsilon}\right) < \infty$,

$$\begin{aligned} g_f(\rho, \sigma) &= g_{f_0}(\tilde{\rho}, \sigma) + a\rho \\ &\geq g_{f_0}(\tilde{\rho}, X) + a\rho \\ &\geq \inf\left\{g_{f_0}(\rho_*, X) + a\rho; 0 \leq \rho_* \leq \rho, \operatorname{supp} X \supset \operatorname{supp} \rho_*\right\} \\ &= g_f(\rho, X). \end{aligned}$$

Remark 4 Here, $\tilde{\rho}$ is the largest element of the set $\{\rho_*; 0 \leq \rho_* \leq \rho, \operatorname{supp} \sigma \supset \operatorname{supp} \rho_*\}$ but not necessarily the largest element of the set

$$\{\rho_*; 0 \leq \rho_* \leq \rho, \operatorname{supp} X \supset \operatorname{supp} \rho_*\}.$$

Thus, in general, the equality between the second and third line does not hold.

6.4 Relation to RLD Fisher Metric

$\mathrm{D}_f^{\max}$ is closely related to RLD Fisher metric

$$J_\rho^R(X, Y) := \operatorname{tr} X \rho^{-1} Y,$$

where X and Y are self-adjoint operators living in the support of $\rho > 0$, with $\operatorname{tr} X = \operatorname{tr} Y = 0$. This quantity plays important role in quantum statistical estimation theory [8], and is the largest monotone metric on the space of density operators [7]. Also, this quantity is the solution to infinitesimal version of reverse test [10]; The triple (Γ, p, v) of the positive trace preserving map Γ from positive measures to self-adjoint operators, the probability distribution over the finite set $\mathscr{X}$, and the real valued function over $\mathscr{X}$ with $\sum_{x\in\mathscr{X}} v(x)$, is said to be *reverse estimation of* $\{\rho, X\}$ iff

$$\Gamma(p) = \rho, \Gamma(v) = X.$$

Then

$$J_\rho^R(X, X) = \inf\left\{\sum_{x\in\mathscr{X}} \frac{\{v(x)\}^2}{p(x)}; (\Gamma, p, v) \text{ is a reverse estimation of } \{\rho, X\}\right\}.$$

Here, the function minimized is called *Fisher information* and plays significant roll in point estimation.

This problem reduces to our reverse test problem of $\{\rho + \varepsilon X, \rho\}$, where ε is chosen so that $\rho + \varepsilon X \geq 0$, and $f(\lambda) = (1 - \lambda)^2$. Since this is operator convex and $f(\lambda) - 1$ satisfies (F), (30) shows the above identity. Also, by this argument, the optimal reverse estimation is the one such that $(\Gamma, \{p + \varepsilon v, p\})$ is the minimal reverse test of $\{\rho + \varepsilon X, \rho\}$.

Based on this observation, we prove

$$f''(1) J_\rho^R(X, X), = \left.\frac{d^2}{d\varepsilon^2} D_f^{\max}(\rho + \varepsilon X \| \rho)\right|_{\varepsilon=0} = \left.\frac{d^2}{d\varepsilon^2} D_f^{\max}(\rho \| \rho + \varepsilon X)\right|_{\varepsilon=0} \tag{39}$$

when f''' exists and uniformly bounded in the sense that

$$\left|f'''(x)\right| < c, \ \exists\varepsilon > 0 \forall x \in (1 - \varepsilon, 1 + \varepsilon).$$

(Differentiating (30) twice, one can also obtain (39).

The key observation is the following. For any $\varepsilon > 0$ with $\rho + \varepsilon X \geq 0$, $\{\rho + \varepsilon X, \rho\}$ are the same in its minimal reverse test. Therefore,

$$D_f^{\max}(\rho + \varepsilon X || \rho) = D_f(p + \varepsilon v || p).$$

Then differentiation of both sides, well-known relation between Fisher information and f-divergence leads to the first identity.

To obtain the second identity, we use Corollary 1 that will be shown later; the minimal reverse test of $\{\rho + \varepsilon X, \rho\}$ and $\{\rho, \rho + \varepsilon X\}$ are identical. Then, the rest of the argument is almost parallel.

6.5 Essential Uniqueness of Optimal Reverse Test

In this section, we show that any optimal reverse test is essentially identical to the minimal reverse test, provided that (F) is satisfied. First, we show some technical lemmas. (They themselves are of interest. We show another application of them in the next section)

A function f with (F), by Theorem 8.1 of [1] , is written as

$$f(\lambda) = c\lambda + b\lambda^2 + \int_{(0,\infty)} \left(\frac{\lambda}{1+t} + \psi_t(\lambda) \right) \mathrm{d}\mu(t), \tag{40}$$

where c is a real number, $b > 0$, μ is a positive Borel measure with $\int_{(0,\infty)} \frac{\mathrm{d}\mu(t)}{(1+t)^2} < \infty$, and $\psi_t(\lambda) := -\frac{\lambda}{t+\lambda}$.

In what follows, we suppose

$$(0,\infty) \subset \operatorname{supp}\mu \cup \{0\}, \tag{41}$$

where $\operatorname{supp}\mu$ is the set of all points λ having property that $\mu(U) > 0$ for any open set U containing λ (see Theorem 2.2.1 and Definition 2.2.1 of [15].) $\lambda \ln \lambda$, $(\pm 1)\lambda^{\alpha}$ $(-1 \leq \alpha \leq 1, \alpha \neq 1)$ satisfies (41) (see Example 8.3, [1]).

Lemma 5 *Suppose that* $t \in (0,\infty) \to h(t) \in [0,\infty)$ *is continuous. Suppose also*

$$\int_{(0,\infty)} h(t)\,\mathrm{d}\mu(t) = 0,$$

where μ *is a positive measure over Borel* σ*-field. Then, if* $t_0 \in \operatorname{supp}\mu$*,*

$$h(t_0) = 0.$$

Proof Suppose $t_0 \in \operatorname{supp}\mu$ and $h(t_0) > 0$. Then, by continuity of h, there is an $\varepsilon > 0$ such that $h(t) > 0$ for any $t \in [t_0 - \varepsilon, t_0 + \varepsilon]$. Therefore,

$$\begin{aligned} \int_{(0,\infty)} h(t)\,\mathrm{d}\mu(t) &\geq \int_{[t_0-\varepsilon, t_0+\varepsilon]} h(t)\,\mathrm{d}\mu(t) \\ &\geq \mu((t_0-\varepsilon, t_0+\varepsilon)) \min_{t\in[t_0-\varepsilon,t_0+\varepsilon]} h(t) > 0. \end{aligned}$$

This contradicts with $\int_{(0,\infty)} h(t)\,\mathrm{d}\mu(t) = 0$. Therefore, we have $h(t_0) = 0$.

Lemma 6 *Suppose* f *satisfies (F) and (41). Let* Λ *be a positive trace preserving map. Then,*

$$\mathrm{D}_f^{\max}(\rho||\sigma) = \mathrm{D}_f^{\max}(\Lambda(\rho)\,||\Lambda(\sigma)) \tag{42}$$

and $\mathrm{D}_f^{\max}(\rho||\sigma) < \infty$ *implies*

$$\Lambda_\sigma\left(h\left(d\left(\tilde{\rho},\sigma\right)\right)\right)=h\left(\Lambda_\sigma\left(d\left(\tilde{\rho},\sigma\right)\right)\right). \tag{43}$$

Here, Λ_σ is a subunital positive map defined by (34), and h is an arbitrary function on $[0,\infty)$.

Conversely, if (43) holds, (42) holds for any f with (F). In fact, for any function h on $[0,\infty)$,

$$\operatorname{tr}\sigma h(d\left(\tilde{\rho},\sigma\right))=\operatorname{tr}\Lambda\left(\sigma\right)h(d\left(\Lambda\left(\tilde{\rho}\right),\Lambda\left(\sigma\right)\right)). \tag{44}$$

Proof By (32), (42) and $\mathrm{D}_f^{\max}\left(\rho||\sigma\right)<\infty$ implies

$$\Lambda\left(g_f\left(\rho,\sigma\right)\right)=g_f\left(\Lambda\left(\rho\right),\Lambda\left(\sigma\right)\right). \tag{45}$$

First, suppose $\operatorname{supp}\rho\subset\operatorname{supp}\sigma$ holds. Observe, by (40),

$$g_f\left(\rho,\sigma\right)=c\rho+b\,g_{f_2}\left(\rho,\sigma\right)+\int_{(0,\infty)}\left(\frac{\rho}{1+t}+g_{\psi_t}\left(\rho,\sigma\right)\right)\mathrm{d}\mu\left(t\right) \tag{46}$$

where $f_2\left(\lambda\right):=\lambda^2$ and $\psi_t\left(\lambda\right)=-\frac{\lambda}{\lambda+t}$. Since $\psi_t\left(\lambda\right)$ satisfies (F), by (32) and Lemma 5, (45) is possible only if

$$\Lambda\left(g_{\psi_t}\left(\rho,\sigma\right)\right)=g_{\psi_t}\left(\Lambda\left(\rho\right),\Lambda\left(\sigma\right)\right),\ \forall t\geq 0.$$

This, by the assumption (41) and Proposition 8, implies

$$\Lambda\left(g_h\left(\rho,\sigma\right)\right)=g_h\left(\Lambda\left(\rho\right),\Lambda\left(\sigma\right)\right), \tag{47}$$

which, using (36), implies (43).

Next, suppose $\operatorname{supp}\rho\not\subset\operatorname{supp}\sigma$. Then for $\mathrm{D}_f^{\max}\left(\rho||\sigma\right)<\infty$ to be true, $a:=\lim_{\varepsilon\downarrow 0}\varepsilon f\left(1/\varepsilon\right)<0$ should hold. Therefore, by (33),

$$\Lambda\left(g_f\left(\tilde{\rho},\sigma\right)\right)=g_f\left(\Lambda\left(\tilde{\rho}\right),\Lambda\left(\sigma\right)\right).$$

Therefore, using the parallel argument as above, we have (43).

The second assertion of the theorem is proved by straightforward computation.

Lemma 7 *Suppose f satisfies (F) and (41). Let Λ be a positive trace preserving map. Also, let $\left(\Gamma,\{p,q\}\right)$ and $\left(\Gamma',\{p',q'\}\right)$ be the minimal reverse test of $\{\rho,\sigma\}$ and $\{\Lambda\left(\rho\right),\Lambda\left(\sigma\right)\}$, respectively. Then, (42) holds iff $\mathrm{D}_f\left(p||q\right)<\infty$ and*

$$\Lambda\left(\Gamma\left(\delta_x\right)\right)=\Gamma'\left(\delta_x\right),\ \{p,q\}=\{p',q'\}. \tag{48}$$

Proof Since 'if' is trivial, we prove 'only if'. Recall the minimal reverse test is given by

$$\Gamma\left(\delta_{x_0}\right) = \frac{1}{\mathrm{tr}\ (\rho - \tilde{\rho})}\,(\rho - \tilde{\rho}),$$
$$\Gamma\left(\delta_x\right) = \frac{1}{\mathrm{tr}\,\sigma\,P_x}\sqrt{\sigma}\,P_x\sqrt{\sigma},\ (x \neq x_0),$$

where $d\left(\tilde{\rho},\sigma\right) = \sum_x d_x P_x$, where d_x and P_x is an eigenvalue and projection onto eigenspace.

Let $P_x' := \Lambda_\sigma\left(P_x\right)$, and applying (43) with

$$h_0\left(\lambda\right) := \begin{cases} 1, & (\lambda = d_x), \\ 0, & \text{otherwise.} \end{cases}$$

we have

$$P_x' = \Lambda_\sigma\left(P_x\right) = \Lambda_\sigma\left(h_0\left(d\left(\tilde{\rho},\sigma\right)\right)\right) = h_0\left(\Lambda_\sigma\left(d\left(\tilde{\rho},\sigma\right)\right)\right).$$

Since eigenvalues of $h_0\left(\Lambda_\sigma\left(d\right)\right)$ are either 0 or 1, P_x' is a projector. Since

$$d\left(\Lambda\left(\tilde{\rho}\right),\Lambda\left(\sigma\right)\right) = \Lambda_\sigma\left(d\left(\tilde{\rho},\sigma\right)\right) = \sum_x d_x \Lambda_\sigma\left(P_x\right) = \sum_x d_x P_x',$$

P_x's are the projectors onto the eigenspaces of d', and $\mathrm{spec}\, d = \mathrm{spec}\, d'$. Therefore, if $x \neq x_0$,

$$\Lambda\left(\Gamma\left(\delta_x\right)\right) = \Lambda\left(\sqrt{\sigma}\,P_x\sqrt{\sigma}\right) = \sqrt{\Lambda\left(\sigma\right)}\Lambda_\sigma\left(P_x\right)\sqrt{\Lambda\left(\sigma\right)}$$
$$= \sqrt{\Lambda\left(\sigma\right)}P_x'\sqrt{\Lambda\left(\sigma\right)} = \Gamma'\left(\delta_x\right).$$

This relation is easily checked for $x = x_0$,

$$\Gamma'\left(\delta_{x_0}\right) = \frac{1}{\mathrm{tr}\,\Lambda\left(\rho - \tilde{\rho}\right)}\Lambda\left(\rho - \tilde{\rho}\right)$$
$$= \Lambda\left(\frac{1}{\mathrm{tr}\ (\rho - \tilde{\rho})}\,(\rho - \tilde{\rho})\right) = \Lambda\left(\Gamma\left(\delta_{x_0}\right)\right),$$

where we had used Λ is trace preserving. Therefore, if $\left(\Gamma',\left\{p',q'\right\}\right)$ is the minimal reverse test of $\left\{\Lambda\left(\rho\right),\Lambda\left(\sigma\right)\right\}$, $\Gamma' = \Lambda \circ \Gamma$.

Having specified the map Γ', next we specify $\left\{p',q'\right\}$. For $\left(\Lambda \circ \Gamma,\left\{p',q'\right\}\right)$ to be a reverse test of $\left\{\Lambda\left(\rho\right),\Lambda\left(\sigma\right)\right\}$,

$$\sum_{x \neq x_0} q'\left(x\right)\Lambda \circ \Gamma\left(\delta_x\right) = \Lambda\left(\sigma\right) = \sum_{x \neq x_0} q\left(x\right)\Lambda \circ \Gamma\left(\delta_x\right).$$

Thus we have to have, for all $x \neq x_0$,

$$\sum_x q'(x)\sqrt{\Lambda(\sigma)}P'_x\sqrt{\Lambda(\sigma)} = \sum_x q(x)\sqrt{\Lambda(\sigma)}P'_x\sqrt{\Lambda(\sigma)}.$$

Since P'_x is supported on $\operatorname{supp} d' = \operatorname{supp} \Lambda(\sigma)$, this is equivalent to

$$\sum_x q'(x)\, P'_x = \sum_x q(x)\, P'_x.$$

Since P'_xs are orthogonal projectors, we have $q' = q$.

In the same way, we can prove that $p'(x) = p(x)$, for all $x \neq x_0$. Then by trace preserving nature of Λ, obviously $p'(x_0) = p(x_0)$, concluding $p' = p$. Thus we have the assertion.

In the following, we extend the notion of the minimal reverse test to the pair $\{p, q\}$of positive measures, by identifying p with the diagonal matrix, $\sum_x p(x)\,|e_x\rangle\langle e_x|$, where $\{|e_x\rangle\}$ is a CONS. Let $(\Gamma_0, \{p_0, q_0\})$ be the minimal reverse test of $\{p, q\}$, where $\{p_0, q_0\}$ are positive measures over the finite set $\mathscr{Y}$. Then Γ_0 is in fact stochastic map, but at the same time viewed as the positive trace preserving map sending diagonal density matrices to diagonal density matrices.

With this correspondence, the equivalence $\tilde{p}$ of $\tilde{\rho}$ (see (22)) is in fact restriction of p to supp$q$, and

$$d(\tilde{p}, q) = \sum_{x \in \operatorname{supp} q} \frac{p(x)}{q(x)}\,|e_x\rangle\langle e_x| = \sum_{y \in \mathscr{Y}\setminus\{y_0\}} r_y P_y,$$
$$\operatorname{supp} P_y = \operatorname{span}\{|e_x\rangle\,;\, p(x)/q(x) = r_y\}.$$

In the case that $y \neq y_0$, transition $y \to x$ by Γ_1 occurs iff $p(x)/q(x) = r_y$. y_0 is mapped to $x \in (\operatorname{supp} q)^c$, to produce $p - \tilde{p}$. (Detailed form of the transition probability is not important now.)

Lemma 8 *Let $(\Gamma_0, \{p_0, q_0\})$ be the minimal reverse test of $\{p, q\}$, where $\{p_0, q_0\}$ are positive measures over the finite set $\mathscr{Y}$. Then, there is a positive trace preserving map Γ_0^- that invert Γ_0,*

$$\Gamma_0^-(p) = p_0,\ \Gamma_0^-(q) = q_0.$$

In addition, Γ_0^- is deterministic. Therefore,

$$\Gamma_0^- \circ \Gamma_0(\delta_y) = \delta_y.$$

Proof Γ_1^- corresponds to the following deterministic map from $\mathscr{X}$ to $\mathscr{Y}$: in the case of $x \in \operatorname{supp} q$, $x \to y$ occurs iff and $p(x)/q(x) = r_y$. If $x \in (\operatorname{supp} q)^c$, x is mapped to y_0. By this construction, Γ_1^- is deterministic.

Remark 5 In the statistician's term, r_y is likelihood ratio, and thus y is minimal sufficient statistic of the family $\{p, q\}$ [2]. That roughly means y contains all the information about the family $\{p, q\}$, and the smallest one among those having the same property. Thus, $\{p_0, q_0\}$ is a kind of "compression" of $\{p, q\}$. In fact, the map from $\{p, q\}$ to $\{p_0, q_0\}$ is deterministic, while the inverse is noisy.

Lemmas 7 and 8 indicate that the optimal reverse test is essentially unique.

Theorem 4 *Suppose* $\mathrm{D}_f^{\max}(\rho||\sigma) < \infty$, *where* f *is a function with (F), and* μ *defined by (40) satisfies (41). Let* $(\Gamma, \{p, q\})$ *be an optimal reverse test,*

$$\mathrm{D}_f^{\max}(\rho||\sigma) = \mathrm{D}_f(p||q). \tag{49}$$

If $(\Gamma_1, \{p_1, q_1\})$ *is the minimal reverse test of* $\{\rho, \sigma\}$, *there is a CPTP map* Γ_0 *and* Γ_0^- *with*

$$\{p, q\} \underset{\Gamma_0^-}{\overset{\Gamma_0}{\leftrightarrows}} \{p_1, q_1\}. \tag{50}$$

Therefore,

$$\Gamma = \Gamma_1 \circ \Gamma_0^-, \quad \Gamma_1 = \Gamma \circ \Gamma_0. \tag{51}$$

In addition, Γ_1^- *is deterministic:* $\Gamma_0^- \circ \Gamma_0(\delta_y) = \delta_y$.

Before proving this, let us see its implication. Equation (51) intuitively means, given a physical device implementing Γ, one can implement Γ_1 by preprocessing of the classical input data to Γ, and vice versa. This is what essential "uniqueness" meant to say.

Proof Let $(\Gamma_0, \{p_0, q_0\})$ be the minimal reverse test of $\{p, q\}$. Then, taking recourse to Lemma 7, we have $\{p_1, q_1\} = \{p_0, q_0\}$.

Therefore,

$$p = \Gamma_0(p_1), \ q = \Gamma_0(q_1).$$

Also, by Lemma 8, there is a positive trace preserving map Γ_0^- with

$$p_0 = \Gamma_0^-(p), q_0 = \Gamma_0^-(q), \tag{52}$$

and $\Gamma_0^- \circ \Gamma_0(\delta_y) = \delta_y$.

The following simple statement is not easy to prove directly, but quite easy if the theorem is given.

Corollary 1 *The minimal reverse test of* $\{\rho, \sigma\}$ *and* $\{\sigma, \rho\}$ *are identical*

Proof Let us consider an operator convex function $f(\lambda) = -\sqrt{\lambda}$, that satisfies (F) and (41) (Example 8.3, [1]). Since $\hat{f}$ as of (8) satisfies $\hat{f}(\lambda) = -\sqrt{\lambda} = f(\lambda)$, by (10), we have

$$\mathrm{D}_f^{\max}(\sigma\|\rho) = \mathrm{D}_{\hat{f}}^{\max}(\rho\|\sigma) = \mathrm{D}_f^{\max}(\rho\|\sigma).$$

The left most end and the right most end is achieved by the minimal reverse test of $\{\sigma, \rho\}$ and $\{\rho, \sigma\}$, respectively. Therefore, Theorem 4 above shows the assertion.

6.6 Invertible Reverse Tests

Corollary 2 *Suppose* $\mathrm{D}_f^{\max}(\rho||\sigma) < \infty$*, where* f *is a function with (F), and* μ *defined by (40) satisfies (41). Then if there is a measurement* M *taking values on the finite set* $\mathscr{X}$ *such that*

$$\mathrm{D}_f^{\max}(\rho||\sigma) = \mathrm{D}_f\left(P_\rho^M \,||\, P_\sigma^M\right),$$

ρ *and* σ *commute.*

Intuitively, this result seems trivial: It is not possible to retrieve classical information imbedded in quantum states perfectly, unless the quantum states are in fact commutative (classical). However, as later turns out, this result is generally not true if f is not operator convex A counter example is $f = |1 - \lambda|$, and this corresponds to the total variation distance, which is one of the most commonly used distance measure.

Proof Suppose $\left\{P_\rho^M, P_\sigma^M\right\}$ is a positive measures over the finite set $\mathscr{X}$. Let $(\Gamma, \{p, q\})$ and $(\Gamma_0, \{p_0, q_0\})$ be the minimal reverse test of $\{\rho, \sigma\}$ and $\left\{P_\rho^M, P_\sigma^M\right\}$, respectively. Then by Lemma 7, we have $\{p_0, q_0\} = \{p, q\}$ and

$$P_{\Gamma(\delta_x)}^M = \Gamma_0(\delta_x) \tag{53}$$

Since $\left\{P_\rho^M, P_\sigma^M\right\}$ are probability distributions, by Lemma 8, there is a positive trace preserving map Γ_0^- with

$$\Gamma_0^-(\Gamma_0(\delta_x)) = \delta_x.$$

Composing them, we obtain

$$\Gamma_0^-\left(P_{\Gamma(\delta_x)}^M\right) = \delta_x.$$

The composition of the measurement M followed by the data processing Γ_0^- can be viewed as a measurement, which is denoted by $\tilde{M}$. Then, this can be rewritten as

$$P_{\Gamma(\delta_x)}^{\tilde{M}} = \mathrm{tr}\,\Gamma(\delta_x)\,\tilde{M}_x = \delta_x,\; x \in \mathscr{X}.$$

This means that the support of $\Gamma(\delta_x)$ and $\Gamma(\delta_{x'})$ $(x' \neq x)$ do not overlap, and M' is a projective measurement. Therefore, $\rho = \sum_{x\in\mathscr{X}} p(x)\,\Gamma(\delta_x)$ and $\sigma = \sum_{x\in\mathscr{X}} q(x)\,\Gamma(\delta_x)$ commute.

$$
\begin{array}{ccccc}
 & & \mathrm{span}\,\{\Gamma(\delta_x)\}_{x\in\mathscr{X}} & & \\
 & \nearrow_{\Gamma} & \circlearrowleft & \searrow^{M} & \\
\mathrm{span}\,\{\delta_x\}_{x\in\mathscr{X}} & & \begin{array}{c} \Gamma_0 \longrightarrow \\ \longleftarrow_{\Gamma_0^-} \end{array} & & \mathrm{span}\,\{\delta_z\}_{z\in\mathscr{Z}}
\end{array}
$$

Proposition 2 *If f satisfies (F) and $\left|f'''(x)\right| < c$, $\exists\varepsilon > 0 \forall x \in (1-\varepsilon, 1+\varepsilon)$. Let $\rho_\varepsilon := \rho + \varepsilon X > 0$. Then if that there is a measurement M_ε for each ε taking values on the finite set $\mathscr{Z}$ such that*

$$\mathrm{D}_f^{\max}(\rho\|\rho_\varepsilon) = \mathrm{D}_f(P_{\rho_\varepsilon}^{M_\varepsilon} \,\|\, P_{\rho_\varepsilon}^{M_\varepsilon}),$$

then ρ and X commute.

Proof By (39) and by Section 9 of [16], this is equivalent to the existence of the measurement M with
$$J_\rho^R(X, X) = J_{p^M}\left(v^M, v^M\right),$$

where $J_p(v, v)$ is the Fisher information of $p^M := P_\rho^M$, $v^M = \frac{1}{\varepsilon}\left(P_{\rho+\varepsilon X}^M - P_\rho^M\right)$. But this is impossible unless ρ and X commute (see, for example, [17]).

6.7 Relation to Comparison of State Families

Let $\left\{\left|\hat{\varphi}_x\right\rangle; x \in \mathscr{X}\right\}$ be family of linearly independent state vectors. Also let $\{\tau_x; x \in \mathscr{X}\}$ be a family of density operators. The necessary and sufficient condition for the existence of CPTP map Λ with

$$\Lambda(\tau_x) = \left|\hat{\varphi}_x\right\rangle\left\langle\hat{\varphi}_x\right|, \ \forall x \in \mathscr{X} \tag{54}$$

have been studied by several authors. Especially, if $\tau_x = |\varphi_x\rangle\langle\varphi_x|$, it is expressed in the following very simple form.

$$\exists A \geq 0 \quad \langle\varphi_x|\varphi_{x'}\rangle = A_{x,x'}\left\langle\hat{\varphi}_x|\hat{\varphi}_{x'}\right\rangle$$

(see [18, 19]).

Here we show this in fact is equivalent to

$$\Lambda(\rho) = \hat{\rho}, \Lambda(\sigma) = \hat{\sigma}, \tag{55}$$

where

$$\rho := \sum_x p(x)\,\tau_x,\ \sigma := \sum_x q(x)\,\tau_x,$$
$$\hat{\rho} := \sum_x p(x)\,|\hat{\varphi}_x\rangle\langle\hat{\varphi}_x|,\ \hat{\sigma} := \sum_x q(x)\,|\hat{\varphi}_x\rangle\langle\hat{\varphi}_x|,$$

and $\{p, q\}$ is probability distributions over $\mathscr{X}$. In addition, we suppose, with $r_x := p(x)/q(x)$,

$$r_x < \infty,\ r_x \neq r_{x'},\ (x \neq x'). \tag{56}$$

That (54) implies (55) is trivial. To show the opposite implication, we take recourse to Lemma 7.

First, observe $(\Gamma, \{p, q\})$ and $\left(\tilde{\Gamma}, \{p, q\}\right)$, where $\Gamma(\delta_x) := \tau_x$ and $\hat{\Gamma}(\delta_x) := |\hat{\varphi}_x\rangle\langle\hat{\varphi}_x|$, is a reverse test of $\{\rho, \sigma\}$ and $\{\hat{\rho}, \hat{\sigma}\}$, respectively. In addition, the latter one is minimal, since we had supposed that $|\hat{\varphi}_x\rangle$'s are linearly independent and that $\{p, q\}$ satisfies (56).

(Compute the minimal reverse test as follows. Define $N := \sum_{x\in\mathscr{X}} |\hat{\varphi}_x\rangle\langle e_x|$, $D_p := \sum_{x\in\mathscr{X}} p(x)\,|e_x\rangle\langle e_x|$, $D_q := \sum_{x\in\mathscr{X}} q(x)\,|e_x\rangle\langle e_x|$. Then, $\hat{\sigma} = N D_q N^{\dagger}$. Therefore, there is a unitary U with

$$\hat{\sigma}^{1/2} = N D_q^{1/2} U.$$

Therefore,

$$\hat{\sigma}^{-1/2}\hat{\rho}\hat{\sigma}^{-1/2} = \sum_{x\in\mathscr{X}} r_x U^{\dagger}\,|e_x\rangle\langle e_x|\,U.$$

Therefore, the minimal reverse test maps δ_x to the constant multiple of $\hat{\sigma}^{1/2}U^{\dagger}\,|e_x\rangle\langle e_x|\,U\hat{\sigma}^{1/2} = q(x)\,|\hat{\varphi}_x\rangle\langle\hat{\varphi}_x|$.)

Therefore,

$$\begin{aligned}\mathrm{D}_f(p\|q) &= \mathrm{D}_f^{\max}\left(\hat{\rho}\|\hat{\sigma}\right) = \mathrm{D}_f^{\max}\left(\Lambda(\rho)\,\|\,\Lambda(\sigma)\right)\\ &\leq \mathrm{D}_f^{\max}(\rho\|\sigma) \leq \mathrm{D}_f(p\|q),\end{aligned}$$

indicating

$$\mathrm{D}_f^{\max}(\rho\|\sigma) = \mathrm{D}_f^{\max}\left(\hat{\rho}\|\hat{\sigma}\right) = \mathrm{D}_f(p\|q).$$

By Theorem 4, the minimal reverse test of $\{\rho, \sigma\}$ should be essentially identical to $(\Gamma, \{p, q\})$. But by the assumption (56), $(\Gamma, \{p, q\})$ has to be the minimal reverse test. Therefore, by Lemma 7, (55) implies (54).

7 When One of the Argument is a Pure State

From this section, again we remove the assumption of operator convexity and $f(0) = 0$, and come back to our initial assumption (FC). To start, we treat the case where one of the argument is rank-1.

Without loss of generality, we suppose σ is rank-1. (The other case, ρ is rank-1, is reduce to this case by replacing f by $\hat{f}$ defined by (8).)

and obtain the following:

$$\mathrm{D}_f^{\max}(\rho\|\sigma) = \sigma_{11} f\left(\sigma_{11}^{-1}\tilde{\rho}\right) + (\operatorname{tr}\rho - \tilde{\rho}) \lim_{\varepsilon\downarrow 0} \varepsilon f\left(\frac{1}{\varepsilon}\right), \tag{57}$$

where $\tilde{\rho}$ is defined by (22) (Note now $\tilde{\rho}$ is scalar, since σ 's rank is 1.) Though this coincide with (30), it holds irrespective of the assumption of operator convexity.

Especially, if ρ is also rank-1, $\tilde{\rho} = 0$, thus

$$\mathrm{D}_f^{\max}(\rho\|\sigma) = \sigma_{11} f(0) + \rho_{11} \lim_{\varepsilon\downarrow 0} \varepsilon f(1/\varepsilon).$$

(When $0 \notin \operatorname{dom} f$, $f(0) = \infty$, following the convention used by [13].)

If $(\Gamma, \{p, q\})$ is a reverse test of $\{\rho, \sigma\}$, we should have

$$\Gamma(\delta_x) = \sigma, \ \forall x \in \operatorname{supp} q,$$

since σ is rank-1 and positive. Therefore,

$$\rho = \Gamma(p) = \sum_{x\in\operatorname{supp} q} p(x)\,\sigma + \sum_{x\notin\operatorname{supp} q} p(x)\,\Gamma(\delta_x),$$

which implies $\rho - \sum_{x\in\operatorname{supp} q} p(x)\,\sigma \geq 0$, or equivalently,

$$\sum_{x\in\operatorname{supp} q} p(x) \leq \tilde{\rho}. \tag{58}$$

If $a := \lim_{\varepsilon\downarrow 0} \varepsilon f(1/\varepsilon) = \infty$, then both ends of (57) is ∞. Thus, suppose $a < \infty$. Then, there is a monotone decreasing convex function h with (FC), such that

$$f(\lambda) = a\lambda + f_0(\lambda), \tag{59}$$

To prove this, we take recourse to Corollary 8.5.1 of [13], which claims for any $\lambda' \in \operatorname{dom} f$

$$f(\lambda) \leq f\left(\lambda'\right) + a\left(\lambda - \lambda'\right),$$

(to understand this intuitively, observe $a = \lim_{\lambda\to\infty} f(\lambda)/\lambda$ is the slope at ∞) or equivalently,

$$f_0(\lambda) = f(\lambda) - a\lambda \leq f(\lambda') - a\lambda' = f_0(\lambda').$$

Thus $h(\cdot)$ is monotone decreasing. That $h(\cdot)$ satisfies (FC) is obvious.

By (59),

$$\begin{aligned}
\mathrm{D}_f(p\|q) &= \sum_{x\in\operatorname{supp} q} q(x) f\left(\frac{p(x)}{q(x)}\right) + a \sum_{x\notin\operatorname{supp} q} p(x) \\
&= \sum_{x\in\operatorname{supp} q} q(x) h\left(\frac{p(x)}{q(x)}\right) + a \sum_{x\in\operatorname{supp} q} p(x) + a \sum_{x\notin\operatorname{supp} q} p(x) \\
&\underset{(i)}{\geq} \sum_{x\in\operatorname{supp} q} q(x) f_0\left(\sum_{x\in\operatorname{supp} q} \frac{q(x)}{\sum_{x'\in\operatorname{supp} q} q(x')} \frac{p(x)}{q(x)}\right) + a \sum_{x\in\mathscr{X}} p(x) \\
&= \sigma_{11} f_0\left(\sigma_{11}^{-1} \sum_{x\in\operatorname{supp} q} p(x)\right) + a \sum_{x\in\mathscr{X}} p(x) \\
&\underset{(ii)}{\geq} \sigma_{11} f_0(\sigma_{11}^{-1}\tilde{\rho}) + a \sum_{x\in\mathscr{X}} p(x) \\
&= \sigma_{11} f(\sigma_{11}^{-1}\tilde{\rho}) + a(\operatorname{tr}\rho - \tilde{\rho}).
\end{aligned}$$

So the LHS of (57) is surely the lower bound to the RHS. In the chain of inequalities, the identity in (i) is achieved if $p(x) = cq(x)$ for all $x \in \operatorname{supp} q$, and the one in (ii) is achieved if $\sum_{x\in\operatorname{supp} q} p(x)$ is set to the maximum indicated by (58).

8 Total Variation Distance

8.1 Set Up and a General Formula

The divergence corresponding to $f(\lambda) = |1-\lambda|$ is called total variation distance,

$$\mathrm{D}_{|1-\lambda|}(p\|q) = \|p - q\|_1.$$

This is one of most frequently used distance measures between two measures. Frequently used quantum version of this is,

$$\|\rho - \sigma\|_1 = \sup_M \left\| P_\rho^M - P_\sigma^M \right\|_1,$$

where P_ρ^M is the distribution of the outcome of the measurement M under ρ. This quantum version in fact is smallest of all the quantum versions satisfying (D1') and (D2):

$$\mathrm{D}^Q_{|1-\lambda|}(\rho\|\sigma) \geq \|\rho - \sigma\|_1. \tag{60}$$

(D1') and (D2) imply

$$\mathrm{D}^{Q}_{|1-\lambda|}\left(\rho\|\sigma\right) \geq \left\|P^{M}_{\rho} - P^{M}_{\sigma}\right\|_{1},$$

whose maximization about M leads to the assertion.

In this section we study $\mathrm{D}^{\max}_{|1-\lambda|}\left(\rho\|\sigma\right)$.

Given $(\Gamma, \{p, q\})$ be a reverse test of $\{\rho, \sigma\}$, we define $\left(\Gamma', \left\{p', q'\right\}\right)$, where $\left\{p', q'\right\}$ are probability distributions on $\{0, 1, 2\}$:

$$\Gamma'\left(\delta_0\right) := c_0 \sum_{x\in\mathscr{X}} \min\left\{p\left(x\right), q\left(x\right)\right\} \Gamma\left(\delta_x\right),$$
$$\Gamma'\left(\delta_1\right) := c_1 \sum_{x:p(x)\geq q(x)} \left(p\left(x\right) - q\left(x\right)\right) \Gamma\left(\delta_x\right),$$
$$\Gamma'\left(\delta_2\right) := c_2 \sum_{x:p(x)\leq q(x)} \left(q\left(x\right) - p\left(x\right)\right) \Gamma\left(\delta_x\right),$$

where c_i are chosen so that $\Gamma'\left(\delta_i\right)$ will be normalized, and

$$p\left(0\right) = c_0,\ p\left(1\right) = c_1,\ p\left(2\right) = 0.$$
$$p\left(0\right) = 0,\ p\left(1\right) = c_1,\ p\left(2\right) = c_2.$$

Then $\left(\Gamma', \left\{p', q'\right\}\right)$ is another $\left(\Gamma', \left\{p', q'\right\}\right)$ of $\{\rho, \sigma\}$ with $\left\|p' - q'\right\|_1 = \|p - q\|_1$.

(Intuitively, $\Gamma'\left(\delta_0\right)$ takes care of the common part of two states, and $\Gamma'\left(\delta_1\right)$ and $\Gamma'\left(\delta_2\right)$ compensates the difference.)

Therefore, without loss of generality, we may restrict ourselves to the one in the following form: Let A be an operator with

$$A \geq 0, \rho - A \geq 0, \sigma \geq A,$$

(where A corresponds to $\sum_{x\in\mathscr{X}} \min\left\{p\left(x\right), q\left(x\right)\right\} \Gamma\left(\delta_x\right)$ above,)

$$\begin{aligned}
&\Gamma\left(\delta_0\right) := \frac{1}{\mathrm{tr}\, A} A,\ \Gamma\left(\delta_1\right) := \frac{\rho - A}{\mathrm{tr}\, \left(\rho - A\right)},\ \Gamma\left(\delta_2\right) := \frac{\sigma - A}{\mathrm{tr}\, \left(\sigma - A\right)}, \\
&p\left(0\right) := \mathrm{tr}\, A,\ \ p\left(1\right) := \mathrm{tr}\, \left(\rho - A\right),\ \ p\left(2\right) := 0, \\
&q\left(0\right) := 0,\ \ q\left(1\right) := \mathrm{tr}\, \left(\sigma - A\right),\ \ q\left(2\right) := \mathrm{tr}\, A.
\end{aligned} \tag{61}$$

Therefore, we have:

$$\mathrm{D}^{\max}_{|1-\lambda|}\left(\rho\|\sigma\right) = \inf\left\{\mathrm{tr}\, \left(\rho + \sigma - 2A\right);\, A \geq 0, \rho \geq A, \sigma \geq A\right\}. \tag{62}$$

8.2 Invertible Reverse Test

From here, for a while $\mathrm{tr}\,\rho = \mathrm{tr}\,\sigma = 1$. We study the conditions for

$$\mathrm{D}^{\max}_{|1-\lambda|}(\rho\|\sigma) = \|\rho - \sigma\|_1 \tag{63}$$

holds. This implies that any quantum version of statistical distance $\mathrm{D}^{Q}_{|1-\lambda|}(\rho\|\sigma)$ equals to $\|\rho-\sigma\|_1$. Intuitively, this means some aspects of classical information encoded into quantum states can be completely retrieved. As stated in Sect. 6.6, such complete retrieval of f-divergence scarcely occurs if f is operator convex and ρ and σ do not commute. Especially, if ρ and σ are close enough, Proposition 2 states its impossibility. But $|1-\lambda|$ is very different from operator convex functions in this respect.

If we drop the constraint $A \geq 0$ and suppose $\mathrm{tr}\,\rho = \mathrm{tr}\,\sigma$,

$$\begin{aligned}\mathrm{D}^{\max}_{|1-\lambda|}(\rho\|\sigma) &\geq \inf\{\mathrm{tr}\,(\rho+\sigma-2A)\,;\,\rho\geq A, \sigma\geq A\}\\ &= \inf\{2\mathrm{tr}\,(\rho-A)\,;\,\rho\geq A,\sigma\geq A\}\\ &= \inf\{2\mathrm{tr}\,(\rho-A)\,;\,\rho-A\geq 0, \rho-A\geq\rho-\sigma\}\\ &= 2\mathrm{tr}\,[\rho-\sigma]_+ = \|\rho-\sigma\|_1\,.\end{aligned}$$

Here, the minimum in the third line is achieved if $\rho - A = [\rho-\sigma]_+$. ($[X]_+$ is the positive part of the self-adjoint operator X.)

Therefore, (63) holds iff

$$A = \rho - [\rho-\sigma]_+ = \frac{1}{2}(\rho+\sigma-|\rho-\sigma|) \geq 0. \tag{64}$$

(Here, $|X| := \sqrt{X^\dagger X}$.) Another necessary and sufficient condition is the existence of A, Δ_1, $\Delta_2 \geq 0$ with

$$\rho = A + \Delta_1, \sigma = A + \Delta_2, \tag{65}$$

$$\Delta_1\Delta_2 = 0. \tag{66}$$

To see this, observe

$$\|\Delta_1 - \Delta_2\|_1 = \|\rho-\sigma\|_1 \leq \mathrm{D}^{\max}_{|1-\lambda|}(\rho\|\sigma) = \min\,(\mathrm{tr}\,\Delta_1 + \mathrm{tr}\,\Delta_2)\,,$$

where the minimum is taken for all the Δ_1, Δ_2 with (65), not necessarily $\Delta_1\Delta_2 = 0$. For (63) to hold, existence of Δ_1, Δ_2 with $\mathrm{tr}\,\Delta_1 + \mathrm{tr}\,\Delta_2 = \|\Delta_1 - \Delta_2\|_1$ or equivalently, is necessary and sufficient. Thus $\Delta_1\Delta_2 = 0$.

Of course, in general, (64) is not true. For example, if ρ is a pure state and $\rho \neq c\sigma$, it is not positive. (Let $f = |1-\lambda|$ in the formula (57). Then what we obtain is very much different from $\|\rho-\sigma\|_1$.) However, if ρ and σ are very close so that

$$\| \, |\rho - \sigma| \, \| \leq \text{ minimum eigenvalue of } \rho + \sigma,$$

it is positive. (Compare this with Proposition 2.)

Another sufficient condition is

$$(\rho - \sigma)^2 = |\rho - \sigma|^2 \leq (\rho + \sigma)^2 .$$

Taking the square root of both sides, we have (64). (Recall $\sqrt{\cdot}$ is operator monotone. This condition is not necessary, since λ^2 is not operator monotone.) This condition can be rewritten as

$$\rho\sigma + \sigma\rho \geq 0. \tag{67}$$

These sufficient conditions are enough to see the difference from the case where f is operator convex, indicated by Corollary 2 and Proposition 2. Especially, the latter's claims was: $\mathrm{D}_f^{\max}(\rho \| \rho + \varepsilon X) = \mathrm{D}_f^{\min}(\rho \| \rho + \varepsilon X)$ for all the small $\varepsilon > 0$ leads to commutativity of ρ and $\rho + \varepsilon X$, while $\mathrm{D}_{|1-\lambda|}^{\max}(\rho \| \rho + \varepsilon X) = \| \rho - (\rho + \varepsilon X) \|_1$ for all the small $\varepsilon > 0$, for any X.

8.3 2-Dimensional Case

For mathematical simplicity, we assume $\dim \mathscr{H} = 2$. First, numerical computation shows, for each fixed ρ, the set $\{\sigma; (63)\}$ is fairly large. For example, the largest eigenvalue of ρ is ≤ 0.85, this occupies more than the half of the volume of the Bloch sphere. From various circumstantial evidences, the present author conjectures the set is spheroid with one of its focus being ρ.

If

$$\rho = \begin{bmatrix} a & \bar{c} \\ c & b \end{bmatrix}, \sigma = \begin{bmatrix} a & -\bar{c} \\ -c & b \end{bmatrix}, \quad (a \geq b)$$

the minimization problem (62) is solved explicitly. With $Z := \operatorname{diag}(1, -1)$, $\sigma = Z\rho Z^{\dagger}$, $\rho = Z\sigma Z^{\dagger}$. Thus, if A satisfies constrains of (62), so does $\frac{1}{2}\left(ZAZ^{\dagger} + A\right)$, and $\operatorname{tr} A = \operatorname{tr} \frac{1}{2}\left(ZAZ^{\dagger} + A\right)$. Therefore, without loss of generality, we suppose A is diagonal. After some elementary analysis, the optimal A turns out to be

$$A = \begin{cases} \operatorname{diag}(a - |c|, b - |c|), & (a \geq b \geq |c|) \\ \operatorname{diag}\left(a - \frac{|c|^2}{b}, 0\right), & (a \geq |c| \geq b) \end{cases}$$

and we have

$$\mathrm{D}_{|1-\lambda|}^{\max}(\rho \| \sigma) = \begin{cases} 4|c| = \| \rho - \sigma \|_1, & (a \geq b \geq |c|) \\ 2\left(b + \frac{|c|^2}{b}\right). & (a \geq |c| \geq b). \end{cases}$$

9 Divergence as the Supremum of Linear Functionals

In this section we give the dual of (11), or representation of $\mathrm{D}_f^{\max}$ by maximization of a linear functional. Since some properties are preserved by taking supremum, this representations useful for the investigation of the properties of $\mathrm{D}_f^{\max}$. Our main tool is the celebrated following proposition.

Proposition 3 (See Problem 8.8.7, Theorem 8.6.1 of [20]) *Let F be a real-valued convex functional defined on a convex subset Ω of a vector space $\mathfrak{X}$, and let G be a convex mapping of $\mathfrak{X}$ into a partially ordered normed space $\mathfrak{Z}$. Also, let $H(\xi) = A\xi + \eta_0$ is an affine map of $\mathfrak{X}$ into the finite - dimensional normed space $\mathfrak{Y}$. Suppose there exists an ξ_1 such that $G(\xi_1) < 0$ and $H(\xi_1) = 0$. Suppose also that 0 is an interior point of $\{\eta; \eta = H(\xi), \exists\xi \in \Omega\}$, and*

$$\mu_0 := \inf\{F(\xi); \xi \in \Omega, G(\xi) \le 0, H(\xi) = 0\} < \infty.$$

Then there is $\zeta_0^* \ge 0$, and η_0^* such that

$$\mu_0 = \inf\{F(\xi) + \langle\zeta_0^*, G(\xi)\rangle + \langle\eta_0^*, H(\xi)\rangle; \xi \in \Omega\}.$$

Since for any $\zeta_0^* \ge 0$, and η_0^*,

$$\mu_0 \ge \inf\{F(\xi) + \langle\zeta_0^*, G(\xi)\rangle + \langle\eta_0^*, H(\xi)\rangle; \xi \in \Omega\},$$

this means

$$\mu_0 = \max_{\zeta_0^* \ge 0, \eta_0^*} \inf\{F(\xi) + \langle\zeta_0^*, G(\xi)\rangle + \langle\eta_0^*, H(\xi)\rangle; \xi \in \Omega\}. \tag{68}$$

In this section, we suppose f satisfies (FC), and apply this proposition to (14). Let

$$\begin{aligned}\mathfrak{X} &:= \{\{S(r)\}_{r\in[0,\infty]}; S(r) \ge 0, \forall r \in [0,\infty]\}\\ \mathfrak{Y} &:= \mathscr{B}_{sa}(\operatorname{supp}\rho) \oplus \mathscr{B}_{sa}(\operatorname{supp}\sigma),\\ H(S(\cdot)) &:= \left(\sum_{r\in[0,\infty)} rS(r) + S(\infty) - \rho, \sum_{r\in[0,\infty)} S(r) - \sigma\right),\end{aligned}$$

where $\mathscr{B}_{sa}(\mathscr{K})$ denotes the space of (bounded) self-adjoint linear operators on the Hilbert space $\mathscr{K}$.

Below, we show $H(\mathfrak{X})$ contains $(0,0)$ in its interior. If $\varepsilon > 0$ is small enough,

$$\rho + A \ge 0, \sigma + B \ge 0$$

holds for any $A \in \mathscr{B}_{sa}(\operatorname{supp}\rho)$ and $B \in \mathscr{B}_{sa}(\operatorname{supp}\sigma)$ with $\|A\| \leq \varepsilon$ and $\|B\| \leq \varepsilon$. Thus, $S(\cdot)$ induced by the minimal reverse test of $\{\rho + A, \sigma + B\}$ satisfies $H(S(\cdot)) = (A, B)$. This means that ε- ball is in the set $H(\mathfrak{X})$, implying the assertion.

Here, the choice of $\mathfrak{Y}$ is important for the proof to work. If the larger space, for example $\mathscr{B}(\mathscr{H}) \oplus \mathscr{B}(\mathscr{H})$, was chosen, $(0, 0)$ was not an interior point of $H(\mathfrak{X})$.

After all, we can use Proposition 3, if $\mathrm{D}_f^{\max}(\rho\|\sigma) < \infty$.

Now we compute our dual problem. If $a := \lim_{\varepsilon\downarrow 0} \varepsilon f\left(\frac{1}{\varepsilon}\right) < \infty$, with Lagrange multipliers $W_1, W_2 \in \mathscr{B}_{sa}(\mathscr{H})$, the following gives an relaxation to the initial problem:

$$\begin{aligned}
&\inf_{S(\cdot)\geq 0}\left\{\sum_{r\in[0,\infty)} f(r)\operatorname{tr} S(r) + aS(\infty) - \operatorname{tr} W_1\left(\sum_{r\in[0,\infty)} rS(r) + S(\infty) - \rho\right) - \operatorname{tr} W_2\left(\sum_{r\in[0,\infty)} S(r) - \sigma\right)\right\}\\
&= \inf_{S(\cdot)\geq 0}\left\{\sum_{r\in[0,\infty)} \operatorname{tr} S(r)\left(f(r)\mathbf{1} - rW_1 - W_2\right) + \operatorname{tr} S(\infty)\left(-W_1 + a\mathbf{1}\right) + \operatorname{tr} W_1\rho + \operatorname{tr} W_2\sigma\right\}\\
&= \begin{cases} \operatorname{tr} W_1\rho + \operatorname{tr} W_2\sigma, & \text{if } f(r)\mathbf{1} - rW_1 - W_2 \geq 0,\ \forall r \in [0,\infty),\ W_1 \leq a\mathbf{1}\\ -\infty, & \text{otherwise.}\end{cases}\\
&\leq \mathrm{D}_f^{\max}(\rho\|\sigma).
\end{aligned}$$

By Proposition 3, taking supremum by moving (W_1, W_2) over $\mathscr{B}_{sa}(\operatorname{supp}\rho) \oplus \mathscr{B}_{sa}(\operatorname{supp}\sigma)$, the equality achieved. Thus, supremum over $\mathscr{B}_{sa}(\mathscr{H}) \oplus \mathscr{B}_{sa}(\mathscr{H})$ can only be larger than $\mathrm{D}_f^{\max}(\rho\|\sigma)$. But as pointed out above, the inequality '$\leq$' is still valid on $\mathscr{B}_{sa}(\mathscr{H}) \oplus \mathscr{B}_{sa}(\mathscr{H})$. Thus, extending the domain of (W_1, W_2) does not change the supremum, which is $\mathrm{D}_f^{\max}(\rho\|\sigma)$.

After all, taking supremum of W_1, W_2 over $\mathscr{B}_{sa}(\mathscr{H})$, if $\lim_{\varepsilon\downarrow 0} \varepsilon f\left(\frac{1}{\varepsilon}\right) < \infty$ and $\mathrm{D}_f^{\max}(\rho\|\sigma) < 0$,

$$\begin{aligned}
&\mathrm{D}_f^{\max}(\rho\|\sigma)\\
&= \sup\left\{\operatorname{tr} W_1\rho + \operatorname{tr} W_2\sigma;\ W_1 \leq a\mathbf{1},\ f(r)\mathbf{1} - rW_1 - W_2 \geq 0, \forall r \in [0,\infty)\right\}\\
&= \sup\left\{\operatorname{tr} W_1\rho + \operatorname{tr} W_2\sigma;\ rW_1 + W_2 - g_f(p,q) \leq 0, \forall p, q \in [0,\infty)\right\}. \qquad (69)
\end{aligned}$$

(To see the last identity, let $r = p/q$). The proof for the case of $\lim_{\varepsilon\downarrow 0} \varepsilon f\left(\frac{1}{\varepsilon}\right) = \infty$ and $\mathrm{D}_f^{\max}(\rho\|\sigma) < 0$ is almost parallel.

We close this section with a comment of the case where f is piecewise linear function, that is, with $0 = b_0 < b_1 < b_2 < .. < b_k < b_{k+1} = \infty$, f is linear on each $\left[b_i, b_{i+1}\right]$. Then the constraint on W_1 and W_2 reduces to

$$\begin{aligned}
b_i W_1 + W_2 - f(b_i)\mathbf{1} \leq 0,\ (i = 1, \cdots, k),&\\
W_1 \leq \lim_{\varepsilon\downarrow 0} \varepsilon f(1/\varepsilon)\mathbf{1},\ W_2 \leq f(0)\mathbf{1},& \qquad (70)
\end{aligned}$$

where $a := \lim_{\varepsilon\downarrow 0} \varepsilon f(1/\varepsilon)$. Since this is a semi definite program with finite number of variables, numerical solution can be obtained quite efficiently. Also, $f(\lambda) := |1 - \lambda|$ is a special case of this $(k = 1)$; In fact, (62) is the dual of (70).

9.1 Another Proof and Another Characterization

In this subsection we give another proof of (69) when $\operatorname{supp}\rho = \operatorname{supp}\sigma$. A motivation is for the double- checking, but also the present author conjectures that this proof can be generalized to the infinite dimensional case with minor modifications.

Stated differently, we give another proof that the closure $\mathrm{cl}\,\mathrm{D}_f^{\max}$ of $\mathrm{D}_f^{\max}$ equals the RHS of (69). Since $\mathrm{cl}\,\mathrm{D}_f^{\max}$ is proper, positively homogeneous, convex and closed, it is pointwise supremum of linear functionals.

Consider D_f^Q satisfying (D1'), (D2) and having the form

$$\mathrm{D}_f^Q(\rho\|\sigma) = \sup_{(W_1,W_2)\in\mathscr{W}_f^Q} \mathrm{tr}\,(\rho W_1 + \sigma W_2),$$

where $\mathscr{W}_f^Q$ is, without loss of generality, convex and unbounded from below.

Since D_f^Q satisfies (D2),

$$\begin{aligned}
&\mathrm{D}_f^Q\left(\sum_x p(x)\,|e_x\rangle\langle e_x| \,\middle\|\, \sum_x q(x)\,|e_x\rangle\langle e_x|\right)\\
&= \sup_{(W_1,W_2)\in\mathscr{W}_f^Q} \sum_{x\in\mathscr{X}} \left(p(x)\,\langle e_x|\,W_1\,|e_x\rangle + q(x)\,\langle e_x|\,W_2\,|e_x\rangle\right)\\
&= \mathrm{D}_f(p\|q)
\end{aligned}$$

holds for any orthonormal system of vectors $\{|e_x\rangle\}$. By (7), this implies

$$p\,\langle e|\,W_1\,|e\rangle + q\,\langle e|\,W_2\,|e\rangle \le g_f(p,q),\ \forall p,q\ge 0,\ \forall\,|e\rangle,$$

or equivalently,

$$pW_1 + qW_2 \le g_f(p,q)\,\mathbf{1},\ \forall p,q\ge 0.$$

Denote by $\mathscr{W}_f^{\max}$ the set of all the pairs (W_1, W_2) satisfying this. Then, the above argument has shown that $\mathscr{W}_f^Q \subset \mathscr{W}_f^{\max}$, or equivalently,

$$\mathrm{D}_f^Q(\rho\|\sigma) \le \sup_{(W_1,W_2)\in\mathscr{W}_f^{\max}} \mathrm{tr}\,(\rho W_1 + \sigma W_2).$$

On the other hand, it is easy to verify the RHS satisfies (D1') and (D2). Therefore, it is the largest of all D_f^Q's having above mentioned properties. Thus,

$$\mathrm{cl}\,\mathrm{D}_f^{\max}(\rho\|\sigma) = \sup_{(W_1,W_2)\in\mathscr{W}_f^{\max}} \mathrm{tr}\,(\rho W_1 + \sigma W_2). \tag{71}$$

Recall $\mathscr{W}_f^{\max}$ was specified only by (D2), and (D1') just followed. Thus we have:

Proposition 4 *If* D_f^Q *satisfies (D2), convexity, positive homogeneity and lower semi-continuity, then* $\mathrm{D}_f^Q(\rho\|\sigma) \leq \mathrm{D}_f^{\max}(\rho\|\sigma)$.

10 On Continuity

In this section, we study continuity of $\mathrm{D}_f^{\max}$. Since it is proper and convex, $\mathrm{D}_f^{\max}$ is continuous in the interior of the effective domain (i.e., if $\rho > 0$ and $\sigma > 0$). Thus our focus is on its behavior at the boundary of the effective domain.

By (69), on $\mathrm{dom}\,\mathrm{D}_f^{\max} = \left\{(\rho,\sigma)\,;\,\mathrm{D}_f^{\max}(\rho\|\sigma) < \infty\right\}$, $\mathrm{D}_f^{\max}$ is pointwise supremum of linear functionals. Therefore, it is proper and lower semi- continuous. Thus, our concern is the points (ρ,σ) on the boundary of $\mathrm{dom}\,\mathrm{D}_f^{\max}$ with $\mathrm{D}_f^{\max}(\rho\|\sigma) = \infty$. The following lemma is quite useful for our present purpose.

Lemma 9 *If there is a straight line* $\{(\rho_\varepsilon,\sigma_\varepsilon)\}_{\varepsilon>0}$ *in* $\mathrm{dom}\,\mathrm{D}_f^{\max}$ *such that* $\lim_{\varepsilon\downarrow 0}(\rho_\varepsilon,\sigma_\varepsilon) = (\rho,\sigma)$ *and*

$$\lim_{\varepsilon\downarrow 0}\mathrm{D}_f^{\max}(\rho_\varepsilon\|\sigma_\varepsilon) = \mathrm{D}_f^{\max}(\rho\|\sigma), \tag{72}$$

we have

$$\mathrm{D}_f^{\max}(\rho\|\sigma) = \mathrm{cl}\,\mathrm{D}_f^{\max}(\rho\|\sigma).$$

Proof Recall that $\mathrm{D}_f^{\max}$ and $\mathrm{cl}\,\mathrm{D}_f^{\max}$ coincide except at the relative boundary of its effective domain. Therefore,

$$\lim_{\varepsilon\downarrow 0}\mathrm{D}_f^{\max}(\rho_\varepsilon\|\sigma_\varepsilon) = \lim_{\varepsilon\downarrow 0}\mathrm{cl}\,\mathrm{D}_f^{\max}(\rho_\varepsilon\|\sigma_\varepsilon) = \mathrm{cl}\,\mathrm{D}_f^{\max}(\rho\|\sigma),$$

where the second identity is by Proposition 1. Therefore, (72), combined with this, leads to the assertion.

Thus, our focus is to show (72) for the points (ρ,σ) on the boundary of $\mathrm{dom}\,\mathrm{D}_f^{\max}$ with $\mathrm{D}_f^{\max}(\rho\|\sigma) = \infty$. Let π_1, and π_2 be the projector onto $(\mathrm{supp}\,\rho)^\perp \cap \mathrm{supp}\,\sigma$ and $\mathrm{supp}\,\rho \cap (\mathrm{supp}\,\sigma)^\perp$, respectively. Define also $\pi_0 := \mathbf{1} - \pi_1 - \pi_2$. Then they are all disjoint, $\pi_i\pi_j = 0$, if $i \neq j$. Define

$$\rho_\varepsilon := \rho + \varepsilon\pi_1,\ \sigma_\varepsilon := \sigma + \varepsilon\pi_2,$$

and CPTP map Λ by

$$\Lambda(A) := \pi_0 A\pi_0 + \pi_1 A\pi_1 + \pi_2 A\pi_2.$$

Then,

$$\begin{aligned}\mathrm{D}_f^{\max}\left(\rho_\varepsilon\|\sigma_\varepsilon\right) &\geq \mathrm{D}_f^{\max}\left(\Lambda\left(\rho_\varepsilon\right)\|\Lambda\left(\sigma_\varepsilon\right)\right)\\ &= \mathrm{D}_f^{\max}\left(\varepsilon\pi_1\|\pi_1\sigma\pi_1\right) + \mathrm{D}_f^{\max}\left(\pi_2\rho\pi_2\|\varepsilon\pi_2\right) + \mathrm{D}_f^{\max}\left(\pi_0\rho\pi_0\|\pi_0\sigma\pi_0\right)\end{aligned}$$

By definition, $\pi_1\sigma\pi_1$ ($\pi_2\rho\pi_2 = 0$, resp.) has non - trivial kernel iff $\pi_1 = 0$ (if $\pi_2 = 0$, resp.). Also, $\mathrm{supp}\,\pi_0\rho\pi_0 = \mathrm{supp}\,\pi_0\sigma\pi_0$. Therefore, by Theorem 2, all the three terms in the last line are finite, if $\varepsilon > 0$.

To see the limit $\lim_{\varepsilon\downarrow 0}$, observe $\pi_1\sigma\pi_1$ commute with $\varepsilon\pi_1$ and $\pi_2\rho\pi_2$ commute with $\varepsilon\pi_2$, and thus the first and the second term can be written explicitly.

First, suppose $f(0) = \infty$ and $\mathrm{supp}\,\rho \not\supset \mathrm{supp}\,\sigma$. Then $\pi_1 \neq 0$ and, letting b_x's be the eigenvalues of $\pi_1\sigma\pi_1$ (note they are positive),

$$\lim_{\varepsilon\downarrow 0} \mathrm{D}_f^{\max}\left(\varepsilon\pi_1\|\pi_1\sigma\pi_1\right) = \lim_{\varepsilon\downarrow 0}\sum_x b_x f\left(\varepsilon/b_x\right) = \infty.$$

Second, suppose $\lim_{\varepsilon\downarrow 0}\varepsilon f\left(1/\varepsilon\right) = \infty$ and $\mathrm{supp}\,\rho \not\subset \mathrm{supp}\,\sigma$. Then $\pi_2 \neq 0$, and

$$\lim_{\varepsilon\downarrow 0} \mathrm{D}_f^{\max}\left(\pi_2\rho\pi_2\|\varepsilon\pi_2\right) = \lim_{\varepsilon\downarrow 0}\sum_x \varepsilon f\left(c_x/\varepsilon\right) = \infty.$$

Thus, whenever $\mathrm{D}_f^{\max}\left(\rho\|\sigma\right) = \infty$,

$$\lim_{\varepsilon\downarrow 0} \mathrm{D}_f^{\max}\left(\rho_\varepsilon\|\sigma_\varepsilon\right) = \infty = \mathrm{D}_f^{\max}\left(\rho\|\sigma\right).$$

After all, by the lemma above, we have $\mathrm{cl}\,\mathrm{D}_f^{\max} = \mathrm{D}_f^{\max}$.

By Proposition 1, this implies that along any straight line, $\mathrm{D}_f^{\max}$ converges to the same value. Summarizing arguments so far, we have:

Theorem 5 *Suppose f satisfies (FC). Then, the positively homogeneous convex function $(\rho,\sigma) \to \mathrm{D}_f^{\max}\left(\rho\|\sigma\right)$ is proper and closed. Thus*

$$\lim_{\varepsilon\downarrow 0} \mathrm{D}_f^{\max}\left(\rho_\varepsilon\|\sigma_\varepsilon\right) = \mathrm{D}_f^{\max}\left(\rho\|\sigma\right), \tag{73}$$

where $\rho_\varepsilon = \rho + \varepsilon X > 0$, $\sigma_\varepsilon = \varepsilon Y > 0$ for $\varepsilon \in (0, \varepsilon_0)$. (In fact, it suffices that $\{(\rho_\varepsilon, \sigma_\varepsilon)\}_{\varepsilon>0}$ is a straight line in the effective domain of $\mathrm{D}_f^{\max}$).

$\{(\rho_\varepsilon, \sigma_\varepsilon)\}_{\varepsilon>0}$ cannot be arbitrary curve for (73) to hold. To see this, suppose σ is a pure state, and use (57). Suppose $\lim_{\varepsilon\downarrow 0}\varepsilon f\left(1/\varepsilon\right) < \infty$, and let

$$\sigma = \begin{bmatrix} a & 0\\ 0 & 0\end{bmatrix},\ \rho_\varepsilon = \begin{bmatrix} b & \sqrt{\varepsilon}C^\dagger\\ \sqrt{\varepsilon}C & \varepsilon D\end{bmatrix},$$

and $\rho_1 \geq 0$, tr $\rho_\varepsilon = 1$. Then

$$\tilde{\rho}_\varepsilon = b - \sqrt{\varepsilon} C\,(\varepsilon D)^{-1} \sqrt{\varepsilon} C^\dagger = b - C D^{-1} C^\dagger = \tilde{\rho}_1,$$

is constant of ε, and

$$\begin{aligned}\lim_{\varepsilon\downarrow 0} \mathrm{D}_f^{\max}(\rho_\varepsilon\|\sigma) &= \mathrm{D}_f^{\max}(\tilde{\rho}_1\|\sigma) + (1 - \tilde{\rho}_1) \lim_{\varepsilon\downarrow 0} \varepsilon f\,(1/\varepsilon) \\ &= \mathrm{D}_f^{\max}(\rho_1\|\sigma) \neq \mathrm{D}_f^{\max}(\rho_0\|\sigma).\end{aligned}$$

11 Infinite Dimensional Separable Hilbert Space

So far, we had supposed that the dimension of the underlying Hilbert space is finite. Some of them, namely Sects. 5 and 7, are obviously generalized to separable infinite dimensional case. In this section, we consider the existence of the minimum, and also lower semi - continuity. under the assumption

$$\lim_{\varepsilon\downarrow 0} \varepsilon f\,(1/\varepsilon) < \infty\,,\ \ f\,(0) < \infty.$$

This is done by justifying the dual expression (69). Here, ρ and σ are positive element of $\mathscr{B}_{1,sa}$, the space of all the self-adjoint trace class operators (operators with finite trace), and W_1 and W_2 are self-adjoint bounded operators (whose collection is denoted by $\mathscr{B}_{sa}$.)

For that, we have to admit positive measures over infinite set as input to the reverse test. To state that object, operator valued functions and their integrals are used, see Section 2.3, [21] for these concepts. In this new definition, the reverse test is specified by a regular finite measure ν over the Borel sets of $[0, 1]^{\times 2}$, a ν- measurable function $Z_{p,q}$ from $(p, q) \in [0, 1]^{\times 2}$ into $\mathscr{B}_{1,sa}$ with

$$\mathrm{tr}\, Z_{p,q} = 1,\ \nu - \text{ a.e.}, \tag{74}$$

$$\int_{[0,1]^{\times 2}} p\, Z_{p,q} \mathrm{d}\nu = \rho,\ \int_{[0,1]^{\times 2}} q Z_{p,q} \mathrm{d}\nu = \sigma, \tag{75}$$

where the integrals of operator valued functions are understood as short for

$$\int_{[0,1]^{\times 2}} p\mathrm{tr}\ Z_{p,q} W \mathrm{d}\nu = \mathrm{tr}\, \rho W,\ \forall W \in \mathscr{B}_{sa}.$$

This defines a positive map from measures which are absolutely continuous relative to ν into $\mathscr{B}_{1,sa}$ such that

$$\Gamma(R) = \int_{[0,1]^{\times 2}} \frac{\mathrm{d}R}{\mathrm{d}\nu} Z_{p,q} \mathrm{d}\nu.$$

This is 'trace preserving' in the sense $R\left([0,1]^{\times 2}\right) = \operatorname{tr}\Gamma(R)$ .Thus the pair ν and $\left\{Z_{p,q}\right\}$ represents a "reverse test" $(\Gamma, \{P, Q\})$, where P and Q are positive measures with density p and q, respectively.

Γ is defined only on the regular measures, and though it is not in accordance with the definition of reverse tests in the beginning of the present paper, it is analogous enough, to be called its generalization.

Therefore, our new definition is

$$\mathrm{D}_f^{\max}(\rho\|\sigma) := \inf\left\{\int_{[0,1]^{\times 2}} g_f(p,q) \operatorname{tr} Z_{p,q} \mathrm{d}\nu; \text{(74) and (75)}\right\} \tag{76}$$

Remark 6 A function $Z_{p,q}$ is ν- measurable iff its norm - approximable by simple functions, but the definition is equivalent to that the scalar valued function $(p,q) \to \operatorname{tr} Z_{p.q} W$ is ν-measurable (Proposition 2.15, [21]). Equation (75) may be understood in the "weak" sense as stated, but since $\left\|pZ_{p,q}\right\|_1 \leq \operatorname{tr} Z_{p,q}$ is ν-integrable, also can be understood as Bochner integral, or the limit of integral of approximate simple functions in norm.

Theorem 6 *Suppose $\mathscr{H}$ a separable Hilbert space and* $\lim_{\varepsilon\downarrow 0} \varepsilon f(1/\varepsilon) < \infty$ *and* $f(0) < \infty$*,Then,*

(i) Equation (69) holds if W_1 *and* W_2 *ranges over* $\mathscr{B}_{sa}$.

(ii) inf *in (76) can be replaced by* min.

(iii) $\mathrm{D}_f^{\max}$ *is lower semi-continuous.*

In the proof, we take recourse to Proposition 3, which asserts the strong duality and the existence of the optimal solution to the dual problem, seeing (69) as the primal problem. (In application, exchange supremum and infimum by adding minus sign to the function to be optimized.) To proceed, we need to define a proper mathematical framework.

Consider the space $\mathscr{C}$ of continuous real valued functions on $[0,1]^{\times 2}$ and the space $\mathscr{B}_{sa}$ of the space of the bounded self-adjoint linear operators on the Hilbert space $\mathscr{H}$. Endorse $\mathscr{C}$ and $\mathscr{B}_{sa}$ with the norm $\|h\| := \sup_{(p,q)\in[0,1]^{\times 2}} |h(p,q)|$ and the oprator norm $\|W\|$, respectively. From these two spaces, we compose the linear space,

$$\left\{\sum_{i=1}^{n} h_i W_i\,;\, h_i \in \mathscr{C},\, W_i \in \mathscr{B}_{sa}\right\},$$

and its completion with respect to the projective norm

$$\|z\|_\pi := \inf\left\{\sum_{i=1}^{n} \|h_i\|\,\|W_i\|\,;\, z = \sum_{i=1}^{n} h_i W_i\right\}$$

is denoted by $\mathfrak{Z}$. In fact, $\mathfrak{Z}$ is projective tensor product $\mathscr{C}\hat{\otimes}_{\pi}\mathscr{B}_{sa}$ (that $\|\cdot\|_{\pi}$ is a norm and $\|hW\|_{\pi} = \|h\|\,\|W\|$ is known [21]).

Then for each $z \in \mathfrak{Z}$, there exist bounded sequences $\{h_i\}$ and $\{W_i\}$ with $z = \sum_{i=1}^{\infty} h_i W_i$ and

$$\|z\|_{\pi} = \inf\left\{\sum_{i=1}^{\infty} \|h_i\|\,\|W_i\|\,;\ z = \sum_{i=1}^{\infty} h_i W_i\right\}.$$

One can endorse the partial order $\geq$ in $\mathfrak{Z}$ by

$$\begin{aligned} & z_1 \geq z_2 \\ & \Leftrightarrow \sum_{i=1}^{\infty} h_{i,1}(p,q)\,W_{i,1} \geq \sum_{i=1}^{\infty} h_{i,2}(p,q)\,W_{i,2}, \forall (p,q) \in [0,1]^{\times 2}. \end{aligned}$$

The strict inequality '>' is defined similarly.

Any bounded linear functional ζ on $\mathfrak{Z}$ is the linearization of bilinear form on $\mathscr{C}$ and $\mathscr{B}_{sa}$ (see Section 2.2, [21]):

$$\zeta(hW) = \zeta(h)(W),$$

where $\zeta(h)(\cdot)$ and $\zeta(\cdot)(W)$ is an element of $\mathscr{B}_{sa}^{*}$ and $\mathscr{C}^{*}$, respectively. After these preparations, now we are in the position to state the proof.

Proof We prove that the RHS of (76) is the dual problem of the RHS of (69), and that satisfies premises of Proposition 3. Then, (i) and (ii) are simultaneously proved. Since (i) means $\mathrm{D}_f^{\max}$ is the pointwise supremum of linear functionals, (iii) follows.

With the Lagrange multiplier $\zeta \in \mathfrak{Z}^{*}$, $\zeta \geq 0$, the dual function is

$$\begin{aligned} & \sup_{W_1, W_2} \left\{\operatorname{tr}\rho W_1 + \operatorname{tr}\sigma W_2 - \zeta\left(pW_1 + qW_2 - g_f(p,q)\,\mathbf{1}\right)\right\} \\ & = \sup_{W_1, W_2} \left\{(\operatorname{tr}\rho W_1 - \zeta(p)(W_1)) + (\operatorname{tr}\sigma W_2 - \zeta(q)(W_2)) + \zeta\left(g_f\right)(\mathbf{1})\right\} \\ & = \begin{cases} \zeta\left(g_f\right)(\mathbf{1}), & \text{if } \zeta(p)(W) = \operatorname{tr}\rho W \text{ and } \zeta(q)(W) = \operatorname{tr}\sigma W, \\ \infty, & \text{otherwise,} \end{cases} \end{aligned}$$

and the dual problem is

$$\inf\left\{\zeta\left(g_f\right)(\mathbf{1})\,;\ \zeta(p)(W) = \operatorname{tr}\rho W\,,\ \zeta(q)(W) = \operatorname{tr}\sigma W,\ \forall W \in \mathscr{B}_{sa}\right\}. \quad (77)$$

First we show this equals to the RHS of (69), using Proposition 3 (Now our primal problem is (69)). Define $G(W_1, W_2)$ from $\mathscr{B}_{sa} \times \mathscr{B}_{sa}$ to $\mathfrak{Z}$ by

$$G(W_1, W_2) = pW_1 + qW_2 - g_f(p,q)\,\mathbf{1}.$$

Then this map is linear and convex. Thus, by we are done if we can show there is $\left(W_{1,*}, W_{2,*}\right) \in \mathscr{B}_{sa} \times \mathscr{B}_{sa}$ with

$$G\left(W_{1,*}, W_{2,*}\right) < 0.$$

Observe, by (6), arbitrary interior point $\left(w_{1,*}, w_{2,*}\right)$ of $\mathscr{W}_f$ satisfies

$$w_{1,*}p + w_{2,*}q < g_f(p,q), \forall (p,q) \in [0,1]^{\times 2}.$$

Thus $\left(W_{1,*}, W_{2,*}\right) = \left(w_{1,*}\mathbf{1}, w_{2,*}\mathbf{1}\right)$ satisfies $G\left(W_{1,*}, W_{2,*}\right) < 0$. Therefore, by Proposition 3, (77) has minimum.

Suppose ζ is in the following form:

$$\zeta(h)(W) = \int_{[0,1]^{\times 2}} h(p,q) \operatorname{tr} Z_{p,q} W \mathrm{d}\nu, \tag{78}$$

where ν is a finite regular measure over the Borel set of $[0,1]^{\times 2}$, and $Z_{p,q}$ is a ν-measurable function into $\mathscr{B}_{1,sa}$, satisfying (74) and (75). Then it is a bounded bilinear form satisfying constraints $\zeta(p)(W) = \operatorname{tr}\rho W$ and $\zeta(q)(W) = \operatorname{tr}\sigma W$, and

$$\zeta(h)(\mathbf{1}) = \int_{[0,1]^{\times 2}} g_f(p,q) \operatorname{tr} Z_{p,q} \mathrm{d}\nu.$$

Therefore, if we show that ζ can be limited to those with this form, (77) is identical to (75) and the proof is done.

More precisely, we show the following: for each given ζ, there is $Z_{p,q}$ and ν that satisfying the constraints (74), (75), and improves the value of the function to be minimized,

$$\zeta(h)(\mathbf{1}) \geq \int_{[0,1]^{\times 2}} g_f(p,q) \operatorname{tr} Z_{p,q} \mathrm{d}\nu. \tag{79}$$

Observe $h \to \zeta(h)(W)$ is a bounded functional on $\mathscr{C}$. Therefore, by Riesz–Markov representation theorem,

$$\zeta(h)(W) = \int h \mathrm{d}\nu_W,$$

where ν_W is a regular measure over the Borel sets of $[0,1]^{\times 2}$. Since ζ is positive, $|\nu_W(B)| \leq \|W\| \, \nu_1(B)$ and ν_W is absolutely continuous relative to ν_1.Thus

$$\eta_{p,q}(W) := \frac{\mathrm{d}\nu_W}{\mathrm{d}\nu_1}$$

exists and

$$\left|\eta_{p,q}(W)\right| \leq \|W\|, \nu_1 - \text{ a.e.}. \tag{80}$$

Observe there is $Z_{p,q} \in \mathscr{B}_{1,sa}$ with $\operatorname{tr} Z_{p,q} W = \eta_{p,q}(W)$ for any W with finite rank. Then, by positivity of ζ, $Z_{p,q} \geq 0$, ν_1-a.e.. By (80),

$$\operatorname{tr} Z_{p,q} \leq 1, \nu_1 - \text{ a.e..} \tag{81}$$

Also,

$$\operatorname{tr} Z_{p,q} W \leq \eta_{p,q}(W), \; W \in \mathscr{B}_{sa}, W \geq 0, \nu_1 - \text{ a.e..} \tag{82}$$

By (79), (82) follows immediately. Thus, we only have to show this satisfies the constraints (75) and (81).

Suppose $W \geq 0$, and let $\{W_k\}$ be the sequence of positive finite rank operators such that $W_k = \pi_k W \pi_k$, where π_k is the projector onto k-dimensional subspace. Then $0 \leq W_k \leq W$ and as $k \to \infty$, for ν_1-a.e.,

$$p\operatorname{tr} Z_{p,q} W_k \nearrow p\operatorname{tr} Z_{p,q} W, \; \forall A \geq 0, \operatorname{tr} A \in \mathscr{B}_{1,sa}.$$

Since the function $(p, q) \to p\operatorname{tr} Z_{p,q} W$ is ν_1-integrable, by Lebesgue convergence theorem or by monotone convergence theorem,

$$\begin{aligned} \operatorname{tr} \rho W &= \lim_{k\to\infty} \operatorname{tr} \rho W_k = \lim_{k\to\infty} \int p\operatorname{tr} Z_{p,q} W_k \mathrm{d}\nu_1 \\ &= \int \lim_{k\to\infty} p\operatorname{tr} Z_{p,q} W_k \mathrm{d}\nu_1 \\ &= \int p\operatorname{tr} Z_{p,q} W \mathrm{d}\nu_1. \end{aligned}$$

When $W \in \mathscr{B}_{sa}$ is not positive, decompose it into its positive and negative part, and apply the above argument to each part. The other identity of (75) is shown exactly in the same manner. Thus we only have to show (74).

Since (81) holds, we define new Z and ν by

$$\begin{aligned} Z'_{p,q} &:= \begin{cases} \frac{1}{\operatorname{tr} Z_{p,q}} Z_{p,q}, & \text{if } Z_{p,q} \neq 0, \\ 0, & \text{if } Z_{p,q} = 0, \end{cases} \\ \nu(B) &:= \int_B \operatorname{tr} Z_{p,q} \mathrm{d}\nu_1. \end{aligned}$$

Then this satisfies (75) and (74), and does not change the value of the function to be minimized.

After all, ζ can be limited to those in the form (78), and the dual problem of (77) is identical to (75), and the proof is complete.

12 Discussions

We had introduced the maximal f-divergence as the solution to an optimization problem, reverse test, and shown its closed formula in some important cases. Next step is to consider asymptotic version of the problem, in the hope that this close the gap between the maximum and minimum quantum divergence. The present author's long standing project is to characterize all the possible quantum f-divergence, as [7] had characterized all the quantum Fisher information.

Appendix 1: Some Backgrounds from Matrix Analysis

Proposition 5 (Theorem V.2.3 of [22]) *Let f be a continuous function on $[0,\infty)$. Then, if f is operator convex and $f(0)\le 0$, for any positive operator X and an operator C such that $\|C\|\le 1$, $f\left(C^{\dagger}XC\right)\le C^{\dagger}f(X)C$.*

Proposition 6 (Equation (2.43) of [23]) *Let f be a operator convex function defined on $[0,\infty)$. Let $\Lambda^{\dagger}$ be a unital positive map. Then*

$$f\left(\Lambda^{\dagger}(A)\right)\le\Lambda^{\dagger}(f(A))$$

holds for any $A\ge 0$.

Proposition 7 (Proposition 8.4 of [1]) *Let f be a continuous operator convex function on $[0,\infty)$. Then, if*

$$a:=\lim_{\varepsilon\downarrow 0}\varepsilon f(1/\varepsilon)=\lim_{x\to\infty}f(\lambda)/\lambda<\infty,$$

there is a real number a and a positive Borel measure μ such that

$$f(\lambda)=f(0)+a\lambda+\int_{(0,\infty)}\psi_t(\lambda)\,\mathrm{d}\mu(t),\quad \psi_t(\lambda):=-\frac{\lambda}{\lambda+t},$$

and $\int_{(0,\infty)}\frac{\mathrm{d}\mu(t)}{1+t}<\infty$. Since ψ_t is operator monotone decreasing, this means that $f(\lambda)$ is sum of linear function and operator monotone decreasing function.

Proposition 8 (Lemma 5.2 of [1]) *If f is a complex-valued function on finitely many points $\{x_i;i\in I\}\subset[0,\infty)$, then for any pairwise different positive numbers $\{t_i;i\in I\}$ there exist complex numbers $\{c_i;i\in I\}$ such that $f(x_i)=\sum_{j\in I}\frac{c_j}{x_i+t_j}$, $i\in I$.*

Proposition 9 (Exercise 1.3.5 of [23]) *Let X, Y be a positive definite matrices. Then,*

$$\begin{bmatrix} X & C \\ C^{\dagger} & Y \end{bmatrix} \geq 0 \tag{83}$$

implies

$$X \geq CY^{-1}C^{\dagger}, \quad Y \geq C^{\dagger}X^{-1}C. \tag{84}$$

References

1. Hiai, F., Mosonyi, M., Petz D., and Beny, C.: *Quantum f-divergences and error corrections*. Rev. Math. Phys. 23, 691–747 (2011)
2. Strasser, H.: *Mathematical Theory of Statistics—Statistical Experiments and Asymptotic Decision Theory*. Walter de Gruyter, Berlin (1985).
3. Belavkin, V. P.: *On Entangled Quantum Capacity*. In: *Quantum Communication, Computing, and Measurement*, vol. 3, pp.325–333. Kluwer, Boston (2001)
4. Hammersley, S. J., Belavkin, V. P.: *Information Divergence for Quantum Channels, Infinite Dimensional Analysis*. In: *Quantum Information and Computing, Quantum Probability and White Noise Analysis*, pp.149–166, World Scientific, Singapore (2006)
5. Hiai, F., Petz, D.: *The proper formula for relative entropy and its asymptotics in quantum probability*. Comm. Math. Phys. 143, 99–114 (1991)
6. Amari, S., Nagaoka, H.: *Methods of Information Geometry*. AMS (2001)
7. Petz, D.: *Monotone Metrics on Matrix Spaces*. Linear Algebra and its Applications, 224, 81–96 (1996)
8. Holevo, A. S.: *Probabilistic and Statistical Aspects of Quantum Theory*, North-Holland, Amsterdam, (1982) (in Russian, 1980)
9. Hayashi, M.: *Characterization of Several Kinds of Quantum Analogues of Relative Entropy*. Quantum Information and Computation, Vol. 6, 583–596 (2006)
10. Matsumoto, K.: *Reverse estimation theory, Complementarity between RLD and SLD, and monotone distances*. arXiv:quant-ph/0511170 (2005)
11. Jencova, A.: *Affine connections, duality and divergences for a von Neumann algebra*. arXiv:math-ph/0311004 (2003)
12. Matsumoto, K.: *Reverse test and quantum analogue of classical fidelity and generalized fidelity*, arXiv:quant-ph/1006.0302 (2010)
13. Rockafellar, R. T.: *Convex Analysis*. Princeton (1970)
14. Ebadian, A., Nikoufar, I., and Gordjic, M.: *Perspectives of matrix convex functions*. Proc. Natl Acad. Sci. USA, 108(18), 7313–7314 (2011)
15. Parthasarathy, K.: *Probability and Measures on Metric Spaces*. Academic Press (1967)
16. Matsumoto, K. : *On maximization of measured f-divergence between a given pair of quantum states*, arXiv:1412.3676 (2014)
17. Matsumoto, K.: *A Geometrical Approach to Quantum Estimation Theory*, doctoral dissertation, University of Tokyo (1998)
18. Chefles, A.: *Deterministic quantum state transformations*. Phys. Lett A 270, 14 (2000)
19. Uhlmann, A.: *Eine Bemerkung uber vollstandig positive Abbildungen von Dichteopera-toren*. Wiss. Z. KMU Leipzig, Math.-Naturwiss. R. 34(6), 580–582 (1985).
20. Luenberger, D. G.: *Optimization by vector space methods*. Wiley, New York (1969)
21. Ryan, R. A.: *Introduction to tensor products of Banach spaces*. Springer, Berlin (2002)
22. Bhatia, R.: *Matrix Analysis*. Springer, Berlin (1996)
23. Bhatia, R.: *Positive Definite Matrices*. Princeton (2007)

Part III
Quantum Measurement

Peaceful Coexistence: Examining Kent's Relativistic Solution to the Quantum Measurement Problem

Jeremy Butterfield

Abstract Can there be 'peaceful coexistence' between quantum theory and special relativity? Thirty years ago, Shimony hoped that isolating the culprit (i.e. the false assumption) in proofs of Bell inequalities as Outcome Independence would secure such peaceful coexistence: or, if not secure it, at least show a way—maybe the best or only way—to secure it. In this paper, I begin by being sceptical of Shimony's approach, urging that we need a relativistic solution to the quantum measurement problem (Sect. 2). Then I analyse Outcome Independence in Kent's realist one-world Lorentz-invariant interpretation of quantum theory (Sects. 3 and 4). Then I consider Shimony's other condition, Parameter Independence, both in Kent's proposal and more generally, in the light of recent remarkable theorems by Colbeck, Renner and Leegwater (Sect. 5). For both Outcome Independence and Parameter Independence, there is a striking analogy with the situation in pilot-wave theory. Finally, I will suggest that these recent theorems make some kind of peaceful coexistence mandatory for someone who, like Shimony, endorses Parameter Independence.

Keywords Kent · Shimony · Outcome independence · Parameter independence Measurement problem

1 Introduction

My topic is the assumption of proofs of Bell inequalities that is usually considered 'the culprit', i.e. considered to be shown false by the experimental violation of the Bell inequality in question. Thirty years ago, [59, 60] argued that denying this assumption ('condemning this culprit'), which he called 'Outcome Independence' (i.e. accepting

Dedicated to the memory of Abner Shimony (1928–2015).

J. Butterfield (✉)
Trinity College, Cambridge, Cambridge CB2 1TQ, UK
e-mail: jb56@cam.ac.uk

M. Ozawa et al. (eds.), *Reality and Measurement in Algebraic Quantum Theory*,
Springer Proceedings in Mathematics & Statistics 261,
https://doi.org/10.1007/978-981-13-2487-1_11

its negation, Outcome Dependence) would secure a 'peaceful coexistence' between quantum non-locality and relativity theory.

My discussion will proceed in two main stages: first, Sect. 2, and then Sects. 3–5. The stages are linked by a common theme, viz. that Outcome Independence, and so also its negation, Outcome Dependence, use a merely *schematic* notion of 'outcome'; (and similarly, their contrasted notion of 'parameter' is schematic). This will mean that Outcome Dependence does not itself give the detailed physical account that peaceful coexistence needs (Sect. 2). But this negative verdict prompts a positive project: to examine a detailed physical account—I take that of Adrian Kent—and assess whether Outcome Dependence holds in it. I do this in Sects. 3 and 4. This project then prompts, in Sect. 5, a final discussion of whether the obvious 'alternative culprit'—the condition Shimony called 'Parameter Independence'—holds in Kent's proposal. The main message of this final discussion will be to underline the importance of theorems by Colbeck and Renner (made rigorous by Landsman), and by Leegwater.

The rest of this Introduction will spell out this plan in more detail.

Thus I begin in Sect. 2 by urging that our predicament is not as fortunate as Shimony hoped. Outcome Dependence does not—at least by itself—secure peaceful coexistence, for someone (such as Shimony and myself) seeking a 'realist' and 'one-world' interpretation of quantum theory. For the schematic notion of 'outcome' in Outcome Dependence leads inevitably into the quantum measurement problem. So if one rejects solving this problem by adopting some kind of 'instrumentalism', or by saving 'realism' with some kind of 'many worlds' solution, one needs a realist one-world—and relativistic—solution. Once given such a solution, one can then define 'outcome' (unschematically!) and assess whether Outcome Independence fails.

So the gist of Sect. 2 is scepticism about Shimony's hope: forgive me, Abner! But I note that in his final years, he himself came to doubt the proposal (2009: 489; 2009a: Section 7, (1)). It is not just that he was a long-standing advocate of some process of dynamical reduction, i.e. of non-unitary evolution of isolated quantum systems.[1] He also cites Bell, who seems to have doubted the proposal. Although (so far as I know) Bell did not explicitly discuss Shimony's distinction between Outcome Independence and Parameter Independence (cf. Sect. 2 for definitions), he of course resisted making a fundamental interpretative distinction between outcomes (i.e. measurement results) and parameters (i.e. apparatus-settings). His viewpoint was that a pointer-reading (outcome) and a knob-setting (parameter) are both macrophysical facts, and so surely on equal terms, as regards whether a curious (i.e. unscreenable-off) correlation between examples of them at spacelike separation violates relativity theory. As he might put it: 'surely Nature, in her causal structure, does not care whether a macrophysical fact is 'controllable' by humans, in the way a knob-setting, but not a pointer-reading, is—or at least seems to be?' This viewpoint is espe-

[1]For the dynamical reduction programme, cf. e.g. [3] and Pearle's essay (2009) in honour of Shimony. For Shimony's advocacy, cf. e.g. his (1990) and (2009a: Sect. 7, (2)). Excellent philosophical discussions include [49, 50, 52].

cially clear in his last essays, 'Against measurement' (2004: pp. 213–7, 227–30) and 'La nouvelle cuisine' (2004: pp. 237–8, 244–6): Shimony cites this last reference.

From Sect. 3 onwards, I turn to the positive project. Section 3 introduces my chosen realist and relativistic one-world interpretation, namely Kent's (2014, 2015, 2017). Indeed, my aim is in part simply to *advertise* Kent's proposal. For I think philosophers' discussions of the interpretation of quantum theory, especially the measurement problem, focus too much on the 'usual suspects', especially dynamical reduction, 'many worlds' and the pilot-wave.[2] Then in Sect. 4, I use the fact that Kent's proposed beable—Bell's jargon for preferred quantity, whose extra values solve the measurement problem—makes precise, and unschematic, the idea of an outcome, to investigate whether his interpretation satisfies Outcome Dependence.

The situation will be interestingly analogous to that for the pilot-wave theory. It is well-known to satisfy Outcome Independence at the 'micro-level', i.e. the level of its hidden variables (particles' positions), while recovering Outcome Dependence at the observable level of experimental statistics by averaging over the hidden variables. I will urge that despite Kent's proposal being otherwise very different from the pilot-wave theory, the situation is analogous: Outcome Independence at the 'micro-level', and Outcome Dependence at the observable level.

This verdict prompts the question: what about the other much-discussed locality condition or 'possible culprit'—the condition Shimony called 'Parameter Independence'? Does it hold in Kent's proposal? This question is hard to answer, and must wait for another occasion. But in Sect. 5, I briefly address it. I will make two main points, both arising from some recent theorems. The first is about Kent, the second about peaceful coexistence. (I owe the first to discussion with Guido Bacciagaluppi and Gijs Leegwater: to whom my thanks.)

First: Theorems by Colbeck and Renner, and by Leegwater, give us powerful tools for addressing our question about Parameter Independence. For they say (roughly speaking!) that, under some apparently natural assumptions: any theory that supplements orthodox quantum theory must violate either a 'no conspiracy' assumption, or Parameter Independence. It is clear, even allowing for rough speaking and for the unmentioned assumptions, that this result is important for understanding quantum theory, quite apart from evaluating Kent's proposal. After all: as is well-known, Bell was prompted to prove his first non-locality theorem by his awareness that the pilot-wave theory was non-local in the way we now call 'Parameter Dependent', so that he naturally asked himself whether any supplementation of quantum theory had to be non-local.[3] Since then, we have learnt—again, following Bell's lead—to prove Bell inequalities for stochastic rather than deterministic hidden variables, and to distin-

---

[2]My (2015) is another such effort. But there are several other proposals deserving more attention from philosophers, such as those of [1], and Landsman and co-authors (2013, 2013a, 2017 Chapters 10.1–3, 11.4).

[3]As is also well-known, he asked himself this question in print, in the closing paragraph of his ground-breaking first paper on hidden variables: a paper which, due to an editorial oversight at *Reviews of Modern Physics*, was only published in 1966, i.e. after he had given a 'Yes' answer to the question for deterministic hidden variables, in his 1964 proof of a Bell inequality.

guish subtly different 'locality' assumptions: in particular, we have learnt, following Shimony, to distinguish between Outcome Dependence and Parameter Dependence.

So the Colbeck, Renner and Leegwater theorems are remarkable—one might say, ironic—for leading us back, after all these years, to see Parameter Independence, which Bell long ago saw as violated by the pilot-wave theory, as *the* central locality notion—the notion that, on pain of some 'conspiracy', any supplementation of quantum theory must deny. So here again, my aim will be in part to *advertise* the theorems to philosophers—just as I wish to advertise Kent's proposal.

So the obvious question arising here is whether Kent's proposal is again (as for Outcome Independence) analogous to pilot-wave theory: which obeys 'no conspiracy' and of course violates Parameter Independence. At first sight, it seems that the answer is Yes, despite Kent's significant differences from pilot-wave theory. (These differences go well beyond his proposal being relativistic. For example, it does not use position as its preferred quantity (beable); and it invokes a final condition of its quantity, rather than, as in the pilot-wave theory, an initial condition.) But we will see that on reflection, the analogy falters. For Kent's proposal may violate 'no conspiracy': though as I shall stress, 'conspiracy' is an unfair label, since there is nothing conspiratorial (or suspicious or 'spooky') about the violation. So the upshot will be that, despite these theorems, Kent's proposal may yet obey Parameter Independence, even while supplementing orthodox quantum theory—and solving the measurement problem!

My second, and final, point will turn on the fact that these theorems say that Parameter Independence and 'no conspiracy' lead to 'unsupplemented' quantum theory. (Leegwater's theorem is especially impressive, in being free of assumptions beyond these two.) This means in effect, that they make some kind of peaceful coexistence mandatory for someone who, like Shimony, endorses Parameter Independence.

To sum up this Introduction:—The upshot of the paper will be twofold:

(1): On the one hand, Shimony's hope for peaceful coexistence is alive and well. Indeed, recent theorems in a sense make it 'mandatory'.

(2): But on the other hand: peaceful coexistence needs more than a judicious or subtle choice of which assumption of a Bell theorem (which 'locality condition') to deny. It will probably need no less than an agreed relativistic solution to the quantum measurement problem. And in seeking such a solution—in particular, in developing Kent's solution—we need to bear in mind whatever constraints on the solution are implied by general theorems like those of Colbeck, Renner and Leegwater.

2 Does Outcome Dependence Secure 'Peaceful Coexistence'?

In this Section, I will (i) introduce Shimony's proposal that Outcome Dependence promises peaceful coexistence (Sect. 2.1); then (ii) report the details of the proposal (Sect. 2.2), and (iii) give some reasons for scepticism about this promise (Sect. 2.3).

2.1 Shimony's Proposal

Shimony proposed (1984, pp. 131–136; 1986, 146–154; following [35]) that denying Outcome Independence led, or at least promised to lead, to some sort of 'peaceful coexistence' between quantum nonlocality and relativity theory. This proposal was based on Jarrett's insight: that the assumption that had hitherto been the main one in proofs of Bell inequalities (then usually called 'factorizability' or 'conditional stochastic independence') is a conjunction of two conditions—so that one could deny one, but not the other. (Details in Sect. 2.2.) Shimony labelled them 'Outcome Independence' (OI) and 'Parameter Independence' (PI).

Both assumptions concern the traditional two-wing Bell experiment (but can be generalized to set-ups with three or more wings/'parties'). Roughly speaking: Outcome Independence says that, conditional on sufficient information (including the choice of the two quantities to be measured), the outcome in one wing is stochastically independent of the outcome in the other wing; while Parameter Independence says that the outcome in one wing is (conditional on sufficient information: including the quantity chosen in that wing, and the outcome in the other wing) stochastically independent of the choice of quantity to be measured in the other wing. (So here, the quantity chosen is dubbed 'parameter', though 'setting' would be clearer.)

Thus Jarrett and Shimony proposed that in the light of Bell inequalities being violated, it is OI, not PI, that we should deny. It will be convenient to have labels for the negations of these assumptions: so we speak of 'Parameter Dependence' (PD), and 'Outcome Dependence' (OD).

Jarrett's and Shimony's reason for denying OI instead of PI, and saying that OD combined with PI led to, or at least promised, peaceful coexistence with relativity theory centred around the ideas that:

(i): since experimenters could choose parameters i.e. settings, PD would enable one experimenter to signal to the other her choice of parameter, which for spacelike-related regions would amount to superluminal signalling; while on the other hand

(ii): experimenters could not choose (nor, apparently: influence) which outcome occurred, so that OD did not threaten superluminal signalling: it instead reflected, or at least suggested, a 'holism' of the quantum correlations—which one might well hope relativity could accommodate.[4]

This viewpoint was also supported by analysing which of the conditions, OI and PI, are obeyed by the two 'obvious' theories one might consider: on the one hand, orthodox quantum theory; and on the other, its well-known rival—the pilot-wave theory. In the metaphor of the court-room: it was supported by these two theories' verdicts about these conditions.

The overall point here is that one can consider different theories' verdicts, because the conditions are *schematic*. For they are equations about probabilities, with nota-

[4]From a large subsequent literature, let me pick out just: an early collection, [20], Jarrett's essay [36] in honour of Shimony, Myrvold's recent (2016) statement of a position similar to Shimony's—and a more sceptical position of my own (2007: especially Sects. 3.1 and 3.3, pp. 825–832, 846–851).

tions for outcomes, for apparatus-settings (i.e. quantity-choices, 'parameters'), and a 'hidden variable' or 'complete state' of the pair: they say nothing physical about outcomes etc. So given a theory, it is a matter of judgment exactly which of its notions to take the schema's notations to refer to, so as to get a verdict on whether the condition is obeyed. But for these two theories, applied to the two-wing Bell experiment, there are very natural judgments about the interpretation of the schema's notations. And applying these, we find:

(i): orthodox quantum theory obeys PI, but not OI;

(ii): the pilot-wave theory obeys OI, but not PI.

Section 2.2 gives details. Then Sect. 2.3 urges scepticism.

2.2 The Conditions and their Diverse Verdicts

2.2.1 Parameter Independence and Outcome Independence

Consider stochastic models of the usual two-wing Bell experiment. We represent the two possible choices of measurement on the left (L) wing by a_1, a_2; and on the right (R) wing, by b_1, b_2. The idea is that a complete state ("hidden variable") $\lambda \in \Lambda$ encodes all the factors that influence the measurement outcomes that are settled before the particles enter the apparatuses, and that are therefore not causally or stochastically dependent on the measurement choices. So λ specifies probabilities for outcomes ± 1 of the various single and joint measurements:

$$pr_{\lambda,a_i}(\pm 1)\ ,\ pr_{\lambda,b_j}(\pm 1)\ ,\ \text{and}\ \ pr_{\lambda,a_i,b_j}(\pm 1 \& \pm 1)\ ;\ i,j = 1,2. \tag{1}$$

We also represent outcomes by $A_i, B_i, i = 1, 2$, where $A_i = \pm 1$ is the event that measuring a_i yields ± 1. We will also use x as a variable over a_1, a_2; X as a variable over A_1, A_2 and their negations (i.e., outcomes ∓ 1); and for the right wing, we similarly use y and Y.

Observable probabilities are predicted by averaging over λ. For example, the observable left wing single probability for $A_1 = +1$ is:

$$pr(A_1 = +1) := \int_\Lambda pr_{\lambda,a_1}(+1)\, d\rho\ . \tag{2}$$

Here, the use of the same measure ρ irrespective of which quantity, e.g. a_1 versus a_2, is chosen to be measured, encodes a locality assumption. Namely, that there is no correlation between (i) the causal factors influencing which quantity, e.g. a_1 versus a_2, is measured, and (ii) the causal factors influencing which value of λ is realized.

This assumption seems very reasonable, especially if one takes the causal factors (i) to be localized in the wings of the experiment, and the causal factors (ii) to be localized in the central source of the emitted particle-pairs. One thinks: would it not be a *conspiracy* if there was a correlation between such surely disparate causal factors? So we shall grant this assumption without demur—for *most* of this paper. But at the end of Sect. 5.1, we shall see that it can be denied for wholly unproblematic reasons, if λ encodes facts about a final boundary condition, and thereby can encode information in that boundary condition about traces (records) of earlier choices of measurement. (So it would be misleading to think of such a λ as localized in the central source.) Besides: since the notion of a 'hidden variable' λ is schematic (as stressed at the start of Sect. 1), considering such a λ is wholly legitimate. On the other hand, a 'reassurance': invoking such a final boundary condition does not necessarily imply that any probability measure over the final boundary conditions must depend on which quantity has been previously measured.

In any case, the assumption of locality used in Bell's theorem, that is traditionally most focussed on, is: 'factorizability' or 'conditional stochastic independence'. This says: the joint probabilities prescribed by each value of λ factorize into the corresponding single probabilities.[5] In our notation:

$$\forall\lambda; \forall x, y; \forall X, Y = \pm 1: \quad pr_{\lambda,x,y}(X\&Y) = pr_{\lambda,x}(X) \cdot pr_{\lambda,y}(Y) . \tag{3}$$

Jarrett's and Shimony's main formal point is that Eq. 3 is the conjunction of two apparently disparate independence conditions for a probability of a "local" i.e. "this-wing" result.[6] The first is, roughly speaking, independence from the measurement choice in the other wing; called 'Parameter Independence' (PI) (where 'parameter' means 'apparatus-setting'):

$$\forall\lambda; x, y; X, Y = \pm 1: \quad pr_{\lambda,x}(X) = pr_{\lambda,x,y}(X) := pr_{\lambda,x,y}(X\&Y) + pr_{\lambda,x,y}(X\&\neg Y) ; \tag{4}$$

and similarly for R-probabilities. The second condition is, roughly, independence from the outcome obtained in the other wing: 'Outcome Independence' (OI):

$$\forall\lambda, x, y; X, Y = \pm 1: \quad pr_{\lambda,x,y}(X\&Y) = pr_{\lambda,x,y}(X) \cdot pr_{\lambda,x,y}(Y) . \tag{5}$$

Then Bell's theorem states that any stochastic model obeying Eq. 3, or equivalently, Eqs. 4 and 5 is committed to a Bell inequality governing certain combinations

---

[5]This condition is found in, for example: Bell's 1971 paper (2004: pp. 36–38), Clauser and Horne ([16]: Eq. (2'), p. 528), and Shimony ([64]: Sect. 2 Eq. (10), and Sect. 4 Eq. (37)). The facts that this condition (i) occurs in both the Bell, and the Clauser and Horne, papers, and (ii) suffices, together with our earlier 'no conspiracy' assumption (that ρ in Eq. 2 is independent of which quantity is measured), for a Bell inquality, were first clarified by Shimony, Horne, Clauser and Bell in a famous 1976 exchange. It is reprinted as [5], and as Chap. 12 of [62].

[6]Reference [35] is the most thorough early source for this 'conjunction' point. But we should recall van Fraassen's brief but masterly exposition, which dubs Parameter Independence 'Hidden Locality', and Outcome Independence 'Causality' (1982: principles IV and III, respectively, p. 31). Incidentally, Shimony himself came to agree that 'parameter' was too general a word for 'distant setting'. His (2009a: Eqs. 8 and 9) replaces the label 'Parameter Independence' by 'Remote Context Independence', and correspondingly 'Outcome Independence' by 'Remote Outcome Independence'. But I shall stick to using the established labels.

of probabilities: which is experimentally violated. On the other hand, quantum theory is not committed to such an inequality—and its predictions for these combinations of probabilities are confirmed.[7]

2.2.2 The Orthodox Verdict: The 'No Signalling Theorem'

Given that Bell inequalities are violated by quantum theory and by experiment: the usual verdict is that OI is false. In particular, quantum theory obeys PI but not OI. More precisely: let us exploit the schematic nature of Sect. 2.2.1's conditions, and so put a quantum mechanical state for each λ, and take the probabilities at a given λ to be given by the orthodox Born rule. Then we infer that:—

(a): PI holds. It is now a statement of the orthodox quantum no-signalling theorem (e.g. [25], Shimony ([59]: 134–6), Redhead ([57]: Sect. 4.6, 113–116)). This theorem says that single-wing probabilities are not affected by any distant-wing *non-selective* measurement (i.e. measurement with no outcome selected). To prove it, the measurement process is often modelled as a projective or POVM measurement, i.e. with the "projection postulate". Thus in non-selective projective measurement of a quantity Q with pure discrete spectrum and spectral decomposition $Q = \Sigma\, q_n \Pi_n$, the initial density matrix ρ is changed by measurement to $\Sigma\, \Pi_n \rho \Pi_n$. Then the theorem is immediately proven in the density matrix formalism, using the cyclicity of trace and the commutation of the L- and R-quantities (cf. [37]).

(b): OI fails, as a mathematical triviality except in the special case of the quantum state being a product state, which of course makes the relevant probabilities factorize. (In terms of density matrices: for any projectors Π_L, Π_R representing outcomes on the left and right respectively, a product state $\rho_L \otimes \rho_R$ gives $\mathrm{tr}(\Pi_L \otimes \Pi_R \cdot \rho_L \otimes \rho_R)$ $= \mathrm{tr}(\Pi_L \rho_L)\mathrm{tr}(\Pi_R \rho_R)$.) Of course, as Shimony and Jarrett admit, and much of the literature (cf. footnotes 5–8) stresses, this mathematical triviality should not blind us to the intuitive plausibility of factorizability, Eq. 3. One feels that, given the measurement choices (parameters): if λ is the *complete* state of the pair, the probability conditional on it of a L-outcome must be unaffected by conditioning on further information about the R-outcome. It is from this intuition that the mysteriousness of—and historically, the surprise at—violations of Bell inequalities, arises.

2.2.3 The Heterodox Verdict: The Pilot-Wave

If we turn to the pilot-wave theory (e.g. [8, 9, 31]), then the orthodox quantum verdicts, (a) and (b), in Sect. 2.2.2 are *reversed*, once we identify the "hidden variable" λ occurring in PI and OI with what in the pilot-wave theory, one naturally considers the complete, or total, state. In the most-studied versions of pilot-wave theory, the preferred quantity (beable) whose extra values solve the measurement problem, is the

[7] Shimony ([64]: Sects. 3–5) surveys experimental aspects. Fine recent discussions of the contents of the various versions of the theorem include [10, 71].

position of point-particles. So the natural notion of complete state is: the conjunction of (or: the ordered pair comprising) the quantum state (especially: the wave-function on configuration space) and the particles' possessed positions.

This reversal of verdicts is normally stated for the non-relativistic pilot-wave theory, and its description of experiments that use Stern–Gerlach magnets to measure spin; and I will follow this. (But it also holds for relativistic versions, which retain an absolute simultaneity structure.). These descriptions take the wave-packet of one of the two particles, say the left particle, to be incident on a bifurcation plane across the magnet: the interaction with the magnetic field then splits the wave packet in two, the two halves being swept away from the plane—and the point-particle gets swept along within whichever 'half-packet' it happens to be in. That is; it gets swept away from the plane without crossing it. So the pilot-wave theory makes precise the two possible outcomes—spin being 'up' or 'down' in the direction concerned—in as straightforward a way as you could wish for: in the point-particle being, indeed, 'up' or 'down' relative to the plane. (Cf. e.g. Dewdney et al. ([21]: Sects. 3–5, pp. 4721–4730), Holland ([31], Sects. 11.2–11.3, pp. 465–476), Barrett ([2], 127–132), Bricmont ([9], Sect. 5.14, 141–150).)

Thus the situation is:—

(a): PI fails. In each individual run of the experiment, there is action-at-a-distance; or, phrased more cautiously: instantaneous functional dependence of the value of a quantity 'here', on the choice of a measurement setting at a distant location. The reason, in short, is that point-particles' possessed positions evolve according to the guidance equation—which for an entangled state of two particles, makes the velocity of each particle instantaneously dependent on the position of the other. In more detail: the momentum of each particle, $i = 1, 2$ is given by $\mathbf{p}_i = \nabla_i S$, where $S \equiv S(\mathbf{x}_1, \mathbf{x}_2)$ is the phase of the wave function in configuration space. This means that the possessed position of particle 1 (respectively, 2) contributes to *where* in configuration space the gradient is taken, for determining the momentum of particle 2 (respectively 1). So in a Bell experiment, the position of, for example, the R-particle, swept upwards from its bifurcation plane, contributes to determining the momentum, and so the later position, of the L-particle. The orthodox quantum probabilities, and in particular the no-signalling theorem, are then recovered at the observable level, by averaging over the "hidden variables", i.e. the possessed positions, using the Born-rule distribution. (And more generally: the much-celebrated empirical equivalence with orthodox quantum theory is obtained by such averaging. But if another distribution is used, signalling *would* be possible. Besides, this point generalizes to other deterministic hidden variable theories that reproduce quantum theoretic statistics by a 'quantum equilibrium' distribution of hidden variables; ([66, 67].)

(b): OI holds. The reason, in short, is that Outcome Independence is trivial for a deterministic theory. Given the measurement choices (parameters), and a state rich enough to determine an L-outcome, conditioning on a distant R-outcome gives no further information. And of course, the pilot-wave theory, with λ taken to contain not just the quantum state but also the possessed positions (or corresponding "beables" in other versions), *is* deterministic. (Agreed, there are subtleties about this last statement: (thanks to Bryan Roberts for stressing this). The existence and uniqueness of solutions for the combined Schrödinger and guidance equations is guaranteed only under certain, albeit reasonable, conditions. But I shall set these subtleties aside: [6] is a review, and details are in [7].)

2.3 *Too Easy?*

Taken together, the verdicts in Sects. 2.2.2 and 2.2.3 seem to support the idea that OD, with PI, makes for peaceful coexistence between quantum theory and relativity. For orthodox quantum theory seems 'at peace' with relativity, thanks to its no-signalling theorem; while the pilot-wave theory being 'at war' is explicitly shown by its violating PI—a 'state of war' that its obeying OI does not calm.[8]

But on examination, this support is questionable. It is not just that, as we have stressed, the conditions OD etc. are schematic. More specifically:

(1): Note that the no-signalling theorem uses only commutation of the two quantities that get measured. Nothing is assumed about the spacetime location of the measurements. Indeed, the theorem is often presented (e.g. [57], pp. 113–116; [8], pp. 139–140) in a wholly non-relativistic quantum formalism that, "notoriously", allows superluminal propagation: in particular, wave-packets with an initially compact spatial support spreading instantaneously. So, as some authors (e.g. [48], p. 242) point out: for this formalism, the no-signalling theorem is really a 'coincidence', since nothing in the conceptual framework of the formalism suggests a non-selective measurement must be forbidden from affecting the statistics of a distant measurement.[9]

(2): Mentioning outcomes, and saying that they cannot be influenced ('controlled' as Shimony puts it) puts one face-to-face with the measurement problem: how does quantum theory represent a definite experimental outcome? Or perhaps better: how *can* it? Or: how *should* it? This problem was of course repeatedly pressed by Bell in his condemnation of orthodox quantum theory's 'shifty split' (i.e. its vagueness about the quantum-classical transition), and its shifty replacement of the 'and' of superposition by the 'or' of an ignorance-interpretable mixture (2004: pp. 93–4, 117–8, 155–6, 213–7, 245–6). Even setting aside all issues about quantum nonlocality, this problem still has no agreed answer—and the best of authorities continue to press the problem (e.g. Isham ([33], Sects. 8.5, 9.4), [24, 46], Landsman ([44], Chap. 11)). Besides: considering quantum nonlocality only aggravates the problem. For we have no agreed relativistic description of quantum measurement processes: in particular, no agreed relativistic formulation of the 'collapse of the wave-packet' (whether the

[8]Similarly, some authors ([32, 65]) suggested that the moral of OD was that quantum theory exhibited a species of holism or non-separability. Cf. Morganti ([47]: Sect. 2) as an example of the ongoing philosophical discussion. But as the sequel indicates, I concur with Henson ([30]: pp. 1021–1028, Sect. 3.1–3.4) that this moral does not, as Henson puts it, 'relieve the problem of Bell's theorem'.

[9]This point echoes Bell's viewpoint, mentioned in Sect. 1. But perhaps one should say, so as to reflect one's hope of reconciling the quantum with relativity: not 'coincidence', but 'manna from Heaven'.

collapse is treated as a fundamental physical process, or as in some way effective or even subjective). And we have no consensus about how to obtain one. Again, various authors have pressed the problem. Among discussions of how OD in particular bears on these issues, earlier work includes Butterfield ([12]: Sect. 7, p. 72f.), [17, 48]; and recent work includes [30, 54, 55]. All these authors are sceptical that OD secures 'peaceful coexistence'.

Agreed, I have stated the measurement problem—as do Bell and the other authors cited—in terms that set aside solutions that are 'instrumentalist' rather than 'realist', or that save 'realism' with some kind of 'many worlds' solution.[10] And I frankly admit to hoping, as Shimony did, for a realist one-world—and relativistic—solution. Hence my positive project, from Sect. 3 onwards, to focus on one such proposal.

To sum up this Section: we cannot expect peaceful coexistence to be readily established: in particular, not just by appealing to OD. The word 'outcome' leads to the measurement problem; and correlatively, so does the contrast made by OI and PI between outcomes and parameters. The apparent neatness of the contrast in the equations of OI and PI belies controversial issues about what the quantum state represents, and how measurement processes unfold—especially in a relativistic spacetime.

3 Kent's Proposal for a Realist One-World Lorentz-Invariant Interpretation

This Section presents Kent's proposal (2014, 2015, 2017): first its strategy (Sect. 3.1), then its details (Sect. 3.2). The details involve three stages, of which the third is the main one (Sect. 3.2.1). Then we can see how Kent recovers both the empirical success of orthodox quantum theory, and a single actual quasiclassical history (Sects. 3.2.2 and 3.2.3).

3.1 The Strategy

We imagine, to begin with, that we are given a Lorentz-invariant quantum theory defined on Minkowski spacetime, which is able to rigorously describe interactions, in particular measurements, and which describes the total system as evolving unitarily.

Agreed: we in fact—notoriously—do not have a rigorous formulation of an empirically adequate Lorentz-invariant quantum theory describing interactions. (Not even such a theory of the basic interactions between microphysical entities like electrons and photons, let alone measurement interactions.) But Kent's proposal for how we should augment such a theory can be understood, and assessed, without knowing all

[10] Fine recent versions of 'instrumentalism' include Friederich ([23], Chaps. 7–10) and Healey ([29]: Chap. 4, especially Sect. 4.6 pp. 72–74, and Chap. 10, especially Sect. 10.6, pp. 179–183: building on his previous (2012, 2013, 2014)). For 'many worlds', [70] is nowadays the *locus classicus*.

the details of such a theory. For although Kent's proposal must refer to interactions, he endorses (and indeed uses) the conventional though non-rigorous Lorentz-invariant physics of interactions. And there is every reason to think that what he postulates additionally, about the values of a preferred physical quantity (traditionally 'observable'; but better: 'beable') and their probabilities, will *not* conflict with this conventional physics: nor with a rigorous formulation of it, were we ever to have such.[11]

The aim is to augment this theory in a precise way so as to give a realist one-world Lorentz-invariant interpretation of quantum theory (modulo the issues just mentioned about describing interactions both rigorously and empirically adequately).[12]

Kent augments this theory by specifying—all in a suitably Lorentz-invariant way:

(i) an appropriate physical quantity (traditionally 'observable'; but better: 'beable');

(ii) possible temporal sequences of values for it; and

(iii) probabilities for those sequences:

such that the one real world corresponds to one such sequence (taken together, of course, with the total history of the orthodox unitary evolution of the quantum state, according to the given theory).

In other words: Kent specifies within the given theory a beable, and thereby a sample space of its possible values, and histories of values; on which he then defines a probability measure (in terms of a sequence of final conditions, each given a conventional Born-rule probability), so as to recover both:

(a) the successful standard quantum description of microphysics (in particular: the principle of superposition and Lorentz-invariance), and

(b) the successful standard classical description of macrophysics, i.e. the emergence of a quasiclassical history.

Thus each total history of the universe is given fundamentally by the conjunction of

(a') the history throughout time of the quantum state, which evolves unitarily and Lorentz-invariantly (so (a') might be called the 'orthodox part of the history'—which Everettians claim is the whole history); and

(b') the history throughout time of the beable's actual possessed values: i.e. one specific trajectory through the sample space.

Remark: So far, the broad 'natural philosophy' of the proposal seems to be like the pilot-wave theory: a unitary quantum evolution is conjoined with the history of the

---

[11] As we will see, the spacetime need not be Minkowski: any appropriately causally well-behaved spacetime will do. There is a more substantial constraint, arising from his appeal to a final boundary condition: viz. that there should be a well-defined limit to a sequence of probability distributions, associated with a sequence of successively later and later spacelike hypersurfaces. But as Kent discusses, there is good reason to think this will be satisfied in favoured cosmological models.

[12] Remark: Kent's focus is on the measurement problem, which he prefers to call *the reality problem*, 'since few physicists now believe that the fundamental laws of nature involve measuring devices per se or that progress can be made by analysing them' (2014, 012107–1). Hence the title of his paper. I agree with his preference for 'reality problem' over 'measurement problem': but I will keep to the traditional term, reflecting his wider concern by talking about seeking an 'interpretation of quantum theory'.

actual possessed values of the beable. Besides, as in the pilot-wave theory: there is no dynamical back-reaction from the actual possessed value to the quantum state and-or its evolution. But as we will see in Sect. 3.2, there are substantial differences. Overall, the traditional language of 'hidden variables' sits ill with Kent's ideas (hence his use of 'beable'), and it will be clearest to forget it until we return to Outcome Dependence in Sect. 4.3. More specifically, the principal differences from pilot-wave theory are:

(i) fundamental Lorentz-invariance (of course);

(ii) a different choice of the beable (not: position), and a different prescription for how its actual possessed value evolves (not: a deterministic 'guidance equation');

(iii) a different prescription for the probabilities of the various possible possessed values (not: the Born-rule applied to the initial values of the beable, and proven equivariant for the unitary evolution).

3.2 *The Details*

The details of Kent's proposals vary between his three papers. In particular, the first includes (in its Section II) an extended presentation of an analogous proposal for a non-relativistic spacetime; and its relativistic proposal differs significantly from the second and third papers. The second paper gives toy models of photons scattering off a massive quantum system whose initial wave-function is an archetypal 'two hump' superposition (in one spatial dimension) of 'being on the left' and 'being on the right'. In these models, the photons are treated as point-like objects propagating along light-like curves, and interacting with the massive quantum system by reflection. The third paper's models are more realistic: Kent treats the photons as well as the massive systems quantum mechanically, using the formalism of photon wave mechanics. But in all three papers, his proposal secures his desired result: that there is, under appropriate circumstances, an 'effective collapse' onto one or other location—cf. (b) and (b') at the end of Sect. 3.1.

I will concentrate on the *second* paper: discussing first, the detailed proposal in three stages (Sect. 3.2.1), and then the recovery of a quasiclassical history (Sects. 3.2.2 and 3.2.3).

3.2.1 Three Stages

The main idea of the choice of beable is, in a word, that it should be mass-energy. And the main idea about the probabilities of its various values unfolds in three stages:

(i) to consider the orthodox Born-rule probabilities, prescribed by the quantum state, for mass-energy's possible values at a suitable late time (i.e. at all points on a suitable late spacelike hypersurface), though there is no assumption that an actual physical measurement is made anywhere on this hypersurface; and then

(ii) to take one possible distribution of mass-energy on this hypersurface as a final boundary condition; and

(iii) to evaluate probabilities for values of mass-energy at an earlier spacetime point by (a) using the quantum state (for the appropriate time), *but also* (b) conditioning on the final boundary condition given by (ii).

More precisely:—Kent first recalls that the Tomonaga-Schwinger formalism enables us, given a quantum state $|\psi_0\rangle$ prescribed on some initial spacelike hypersurface S_0, to define formally the evolved state $|\psi_S\rangle$ on any hypersurface S in the future of S_0 via a unitary operator $U_{S_0 S}$. This formalism enables him to fulfil stage (i) above. Thus we are to envisage a world in which physics plays out between two hypersurfaces S_0 and S, and a quantum state is given on S_0. We consider the local mass-energy density operators $T_S(x)$ for $x \in S$. (So as usual, $T_S(x) = T_{\mu\nu}(x)\eta^\mu(x)\eta^\nu(x)$ where $T_{\mu\nu}(x)$ is the stress-energy tensor at x, and η is the future-directed unit 4-vector orthogonal to S at x.) Then the quantum state $|\psi_S\rangle \equiv U_{S_0 S}|\psi_0\rangle$ prescribes orthodox Born-rule probabilities for the various possible distributions of values of all these operators. But of course, there is no need to suppose that a joint measurement of these operators in fact occurs on S.[13]

Kent then proposes that one possible mass-energy distribution $t_S(x)$ on S is randomly selected, using the Born-rule probability distribution prescribed by $|\psi_S\rangle$. That is: physical reality includes one such distribution. This is stage (ii).

As for stage (iii), i.e. proposing probabilities for the beable mass-energy at a spacetime point y between S_0 and S, Kent proposes that these should be calculated conditionally on—not the whole final boundary condition—but only on that part of it that lies outside the future light-cone of the spacetime point y. The effect of this, as we will see in Sect. 3.2.2, is that there can be an 'effective collapse' of appropriate superpositions of values of mass-energy at intermediate points y (thus securing the desired definiteness of macroscopic quantities), thanks to photons scattering differently off the components of the superposition and then later registering differently on part of the surface S, and so contributing to the final boundary condition. Note that this collapse is by construction Lorentz-invariant: roughly speaking, a 'collapse along the light-cone'.

But before discussing that, here is a summary of stages (i)–(iii), in Kent's own words (2015, Sect. 2 (a)):

> We wish to define a generalized expectation value for the stress-energy tensor at a point y between S_0 and S, using post-selected final data $t_S(x)$ on S. More precisely ... we will use the post-selected data $t_S(x)$ for all points $x \in S$ outside the future light cone of y, *and only for those points*. [Kent labels the set of all these points $S^1(y)$.]
>
> In words, our recipe is to take the expectation value for $T_{\mu\nu}(y)$ given that the initial state was $|\psi_0\rangle$ on S_0, conditioned on the measurement outcomes for $T_S(x)$ being $t_S(x)$ for x outside the future light cone of y. So, for any given point y, our calculation ignores the outcomes $t_S(x)$ for x inside the future light cone of y.

---

[13] After expounding stages (i)–(iii), Kent discusses taking S ever later in spacetime, i.e. letting S go to future infinity; and therefore his proposal needs the probability distributions associated to ever later S to have an appropriate limit: cf. footnote 11. But I shall not discuss this aspect.

[Kent then defines a sequence $S_i(y)$ of spacelike hypersurfaces that: (i) include almost all of $S^1(y)$, i.e. almost all of the part of S outside the future light cone of y; and (ii) include y; and (iii) as $i \to \infty$, get ever closer to that part of the future light cone of y that lies to the past of S.]

Now for any of the $S_i(y)$, we can consider the Born rule probability distribution of outcomes of joint measurements of $T_S(x)$ (for all $x \in S \cap S_i(y)$) and of $T_{\mu\nu}(y)$. These are calculated in the standard way, taking the initial state $|\psi_0\rangle$ on S_0, unitarily evolving to $S_i(y)$, and applying the measurement postulate there. . . . By taking the limit as $i \to \infty$ we obtain a joint probability density function $P(t_S(x), t_{\mu\nu}(y))$ [for $x \in S^1(y)$]. From this, we can calculate conditional probabilities and conditional expectations for $t_{\mu\nu}(y)$, conditioned on any set of outcomes for $t_S(x)$ (for $x \in S^1(y)$), in the standard way.

Our mathematical description of reality, in a hypothetical world in which physics takes place only between S_0 and S and in which the outcomes $t_S(x)$ were randomly selected, is then given by the set of conditional expectations $\langle t_{\mu\nu}(y)\rangle$ for each y between S_0 and S, calculated as above. We stress that the calculations for the beables $\langle t_{\mu\nu}(y)\rangle$ at each point y all use the same final outcome data $t_S(x)$. However, different subsets of these data are used in these calculations: for each y, the relevant subset is $\{t_S(x) : x \in S^1(y)\}$.

3.2.2 Recovering Quantum Theory's Empirical Success, and A Quasiclassical History

Kent now proceeds to recover both:

(a) the empirical success of standard quantum theory in microphysics; and

(b) a single quasiclassical history in macrophysics: (cf. (a) and (b) at the end of Sect. 3.1).

To do so, he relies on the existence of "environmental" 'particles, or wave packets, or field perturbations, that travel at light speed' (2015, Sect. 1). But as we shall see, Kent's appeal to the "environment" is judiciously different from a *mistaken* (though all too common!) appeal to decoherence: in short, he does not make the error of thinking that an improper mixture is ignorance-interpretable.

Kent's main idea here can be usefully divided into two parts. The first part can be briefly stated; the second will occupy the rest of this Subsection. It will require discussion of decoherence; and this discussion will lead to Kent's needing two constraints to hold good—discussed in Sect. 3.2.3. (Kent argues, by considering toy models (2015, Sect. 3; 2017, Examples 1–5), that there is good reason to think the constraints do hold good.)

The first part:—The first part just assumes our universe has such lightlike propagations (I will say 'photons', for short) and then applies Sect. 3.2.1 to them. So the first part says that these photons, propagating from some point y in the spacetime, arrive on the later spacelike hypersurface S, and:

(i) by registering there, these photons contribute to the actual values (outcomes of Kent's notional measurements) $t_S(x)$ for $x \in S^1(y)$, i.e. to the actual mass-energy density distribution on the hypersurface S;

(ii) by being correlated, according to the unitarily evolving quantum state, with other degrees of freedom at y, these photons function as *records* of those degrees of freedom; in particular, the relevant subset $\{t_S(x) : x \in S^1(y)\}$ of the actual values determines the beable $\langle t_{\mu\nu}(y)\rangle$ at the point y, i.e. the expectation of $T_{\mu\nu}(y)$ conditional on $\{t_S(x) : x \in S^1(y)\}$—cf. the end of Sect. 3.2.1.

The second part:—The second part is less general, and explicitly directed at recovering (a) and (b) above. So it is, inevitably, less systematic than the first part: Kent supports it with some analyses of toy models (2015, Sect. 3; 2017, Examples 1–5). To understand this second part, I propose that we think of it as a reconstrual, within Kent's framework, of the insight (nowadays universally accepted) of decoherence theory, that:

(i): when a quantum system is *very* well isolated, so that the very fast, efficient and ubiquitous process of decoherence, arising when a quantum system interacts with its environment (e.g. photons, or air molecules), can be avoided or at least postponed: the interference terms, that are characteristic of the system being in a superposition rather than a mixture, will persist and characteristic quantum phenomena (like the iconic interference patterns in the two-slit experiment) will occur; whereas

(ii): when the quantum system is *not* well isolated from its environment, decoherence will rapidly "diffuse" the interference terms out into the environment: so that the system's reduced state is a mixture—and accordingly, one is tempted to say that a component of the mixture represents a quasiclassical history (more precisely: an instantaneous slice, or member, of such a history).

But note: Kent's postulate of a specific beable and its actual values $t_S(x)$, and so of actual values $\langle t_{\mu\nu}(y)\rangle$ make this insight, about (i) versus (ii), play out differently, as regards conceptual aspects (though not numerical aspects), from the way it usually plays out when decoherence is invoked in discussions of the measurement problem. To better understand Kent's proposal, it will be helpful to spell out these contrasts: (and as presaged at the end of Sect. 3.1, Kent's proposal will be similar in some respects to the pilot-wave theory). Spelling out these contrasts will also prepare us for Sect. 4 discussion of Outcome Dependence.

Contrasts with 'decoherence as usual':—So recall the idea of decoherence: plausible Hamiltonians for the interaction between a quantum system that is comparatively massive—paradigmatically, the pointer of an apparatus—and another system or systems that is/are comparatively light—paradigmatically, the air molecules or photons scattering off the pointer—imply that after the interaction, the reduced state (i.e. density matrix) of the pointer is nearly diagonal in a variable that is a collective variable encoding information about mass and position. Thus in some models, it is nearly diagonal in the centre of mass of the pointer. (The reason for the implication is, broadly speaking, that interaction Hamiltonians are local in position.) This result prompts two striking and much-discussed suggestions in relation to the measurement problem. Both will give a contrast with Kent's proposal (and with the pilot-wave theory).

First: Notice that these models give a dynamical explanation of the salience, or 'selection', of a quantity such as the centre of mass of the pointer. The quantity is not given a special role—e.g., that of always having a value—by some general *ab initio* postulate: it is just made salient, or selected, by the nature of the interaction in question. This marks a contrast with Kent, who postulates a special role for mass-energy density and other components of stress-energy. (And as mentioned at the end of Sect. 3.1: it marks a contrast with the pilot-wave theory which postulates (in its usual version) a special role for the position of point-particles.)

Second: The pointer being in a mixture for a quantity such as the centre of mass suggests the measurement problem is solved! It is tempting to exclaim: surely, it represents the desired mixture, as against superposition, of macroscopically distinguishable configurations! As is nowadays well known, this is a chimera. In d'Espagnat's ([15], Chap. 6.2) hallowed terminology: the mixture is 'improper' i.e. it is not ignorance-interpretable—as it would have to be, in order to solve the measurement problem at one fell swoop, along the lines envisaged.[14]

But Kent makes no such error. The postulate of a randomly selected final condition, i.e. the postulated fact of one specific mass-energy distribution $t_S(x)$ on the later spacelike hypersurface S, gives a single, definite value to the conditional expectation $\langle t_{\mu\nu}(y) \rangle$. The point here is again similar to the situation for the pilot-wave theory. It also makes no such error. For according to it (in the usual version), just one component of the density matrix, in the fundamental position representation, has the relevant point-particles in its support: thus giving a single, definite position. (And again, for both Kent and the pilot-wave theory: there is no back-reaction from the actual, randomly selected, value of the beable to the evolving universal quantum state.)

To sum up this Subsection: Kent proposes we can achieve the two goals (a) and (b)—we can recover both (a) the empirical success of standard quantum theory in microphysics, and (b) a single quasiclassical history in macrophysics—and we can do so in a Lorentz-invariant way. To do so, we invoke—not: position as a beable—but mass-energy as a beable at a late time (i.e. spacelike hypersurface S) together with its orthodox Born-rule correlations to the expectation values of components of stress-energy at earlier spacetime points. Lorentz-invariance is respected by appropriately restricting which part of the entire distribution $t_S(x)$ on S is conditioned on, for determining the expectation value at an earlier spacetime point y.

[14]Though much could be said about this topic, this is not the place: except for two short points, the first historical and the second conceptual. (1): Though I duly cited d'Espagnat's clear and influential presentation of the point, and nowadays the decoherence literature often says it clearly (e.g. Zeh, Joos et al. (2003, p. 36, 43); Janssen (2008, Sects. 1.2.2, 3.3.2)), it is humbling to recall that Schrödinger already was clear on it in his amazing 1935 papers: cf. especially the analogy with a school examination (1935, Sect. 13, p. 335 f.). (2): Although I thus condemn this 'one fell swoop' solution (joining, of course, much wiser authorities including Bell): I of course agree that the result here—obtaining an improper mixture nearly diagonal in a quantity you 'want' to have definite values—can form an important part of a principled, and clear-headed, solution to the measurement problem. The obvious example is the modal interpretation, in its early versions from the mid-1990s (cf. [22]): which, roughly speaking, proposed as a postulate, going beyond quantum theory, that the eigenprojections of any system's density matrix have definite values.

3.2.3 The Recovery Needs Two Constraints to be Satisfied

Combining Sect. 3.2.1 discussion of Kent's three stages, and Sect. 3.2.2's discussion, we can now see that Kent's proposal depends on two constraints holding good. I will call them, (α) and (β). They correspond, respectively, to the goals (a), 'accurate microphysics', and (b), 'a single macrohistory', listed at the start of Sect. 3.2.2 (and repeated in its final paragraph). But I should of course note that those goals (and so also the constraints below) are connected, e.g. because we use the statistics of macrophysical pointers, i.e. facts about the quasiclassical world, to confirm quantum theory's description of microphysics.[15]

(α): Quantum theory's empirically successful descriptions of microphysical phenomena, e.g. interference patterns in the two-slit experiment:

(i) are recorded in (the expectation values of) mass-energy and other components of stress-energy at appropriate points y in spacetime, e.g. in the positions of massive pointers in front of a calibrated dial; and

(ii) are thus recorded with statistics that are close to the orthodox Born-rule probabilities prescribed by the quantum textbook, i.e. by the quantum state ascribed in the standard manner to the measured system (so that indeed, orthodox applications of quantum theory are vindicated, by Kent's lights); and

(iii) these textbook probabilities are equal, or close enough, to the probabilities prescribed in Kent's stage (iii) of Sect. 3.2.1. Namely: equal or close to a probability derived by combining:

[i] the correlation (in simple cases: strict or near-strict correlation) of the relevant component(s) of stress-energy at y with appropriate features of the final condition $t_S(x)$ on S; with

[ii] the orthodox Born-rule probability of those appropriate features of $t_S(x)$: i.e. the probability prescribed by $|\psi_S\rangle$, the unitary time-evolute on S of the universe's initial state $|\psi_0\rangle$ on S_0.

Three comments on this constraint, (α), before I turn to the second constraint.

First: Note that all three of (α)'s clauses (i)-(iii) are needed in order to link Kent's proposed beables, and his proposed probabilities of their values, with quantum theory's empirical success, and with how we know it (i.e. how we confirm quantum theory by collecting experimental statistics).

Second: In (α)'s clause (iii), we are to consider some single actual value (final condition: outcome of a notional measurement) $t_S(x)$: conditioning on which defines the beable $\langle t_{\mu\nu}(y)\rangle$ at the earlier spacetime point y; (cf. Sect. 3.2.1's exposition of Kent's stage (iii)). The idea is: the state of the environmental particles (for short: photons) encodes a value of the beable at y, and later on this gives a contribution to (i.e. a non-zero component of) the quantum state $|\psi_S\rangle$; and, we can suppose, this contribution survives in the randomly selected actual final condition $t_S(x)$ (i.e. $t_S(x)$ actually includes this contribution). Thus the random selection on S actually

[15]For clarity and simplicity, I will again suppress the need to let the late spacelike hypersurface S go to future infinity. Cf. footnotes 11 and 13.

including this contribution, combined with this contribution's strict correlation to the value of the beable at the earlier point y, *makes it true* that the beable has that value at y. Cf. also the discussion of constraint (β) below.[16]

Third: Let us ask: How plausible is it that (α) holds good? Of course, a conclusive assessment is not possible, in the present state of knowledge. After all, recall from the beginning of Sect. 3.1 that, quite apart from Kent's proposals, we lack a rigorous Lorentz-invariant quantum field theoretic account of interactions. So we have no rigorous relativistic theory of quantum measurement, and we cannot now rigorously test clauses (i) and (ii), even though they concern only Born-rule probabilities prescribed by the quantum textbook, i.e. by the standard quantum state of the measured system. A rigorous test of clause (iii) is even more difficult: it corresponds to Kent's stage (iii) of Sect. 3.2.1, i.e. the calculational algorithm that needs appropriate correlations between beables to be encoded in the universal state $|\psi_S\rangle$. So testing (α)'s clause (iii) amounts to combining, in a realistic, relativistic setting:

[i] the sorts of ideas and formalism used in the physics of measurement-processes and decoherence (and we must again note sadly that hitherto, this physics is almost exclusively studied in a non-relativistic setting ...); with

[ii] the sorts of ideas and formalism used in pilot-wave theory's discussions of effective quantum states (and conditional wave-functions) whereby one justifies attributing a quantum state (i.e. a wave-function, not merely an improper mixture) to a subsystem of the universe, in terms of both the universal state and the actual values of the subsystem's beables: ideas and formalism that would then be adapted to Kent's postulated beables.

Clearly, in the present state of knowledge, the best we can do by way of testing (α) is to look at various toy models of measurement: for example, with photons scattering off some massive object (thought of as a pointer) and later registering on a hypersurface. This, Kent proceeds to do (2015, Sect. 3; 2017, Examples 1–5). These models give detail to the ideas I have sketched here, and in my second comment above. Cf. also the second constraint that Kent needs, (β)—to which I now turn.

(β): The actual single quasiclassical history in macrophysical terms—the sequence through time of the values of countless macrophysical quantities (such as whether the centre of mass of Erwin's cat's tail is above the floor and moving ('alive'!) or on the floor and still ('dead')), with the sequence obeying approximately classical laws of motion, and so attesting to the emergent validity of classical physics—is *picked out* by the actual mass-energy distribution $t_S(x)$ on the late hypersurface S. Here, 'picked out' means: the expectation values of components of stress-energy at the various points y throughout spacetime, that encode the actual single quasiclassical history, have strict, or nearly strict, Born-rule correlations, according to the calculational algorithm in Kent's stage (iii), with the actual mass-energy distribution $t_S(x)$

[16]So in terms of the formalism: Kent's idea is *like* an appeal to one 'branch' of the state of an environmental particle, i.e. one component of its improper mixture, to pick out as factual the corresponding (strictly correlated) component of the improper mixture of the system of interest. But only 'like', not 'identical with'! As I stressed in Sect. 3.2.2's *Contrasts with 'decoherence as usual'*: the difference is that Kent *avoids* the error of assuming an improper mixture is ignorance-interpretable. He knowingly postulates the beables, and their probabilities, that secure an actual quasiclassical history, while avoiding this error.

on the late hypersurface S. More precisely, so as to respect the light cone structure: 'picked out' means that each expectation value at a point y has such a correlation with the actual mass-energy distribution $t_S(x)$ in that part of S that is outside y's future light cone.

Finally, we need to address a question about this constraint, (β): a question which relates back to my second and third comments on constraint (α). One is bound to ask:

What, if anything, does Kent's proposal need to say about the fact that we cannot now know the actual mass-energy distribution $t_S(x)$ on the hypersurface S; and relatedly, what does it need to say about the fact that we cannot now (or ever) know more than a tiny fraction of the actual single quasiclassical history?

This question—these two pools of ignorance—prompts two remarks: the first conceptual, the second empirical.

(1): *How to represent a definite perception?*:— Agreed, these pools of ignorance do not cause any immediate problem for Kent. Nothing in the exposition above (or in Kent's papers) requires us (or anyone) to know the actual mass-energy distribution $t_S(x)$, or even anything substantive about it. Nor, correspondingly, are we or anyone required to know facts about the actual single quasiclassical history. But . . . there is an issue here, about how we should conceive the representation *in that quasiclassical history* of our knowledge of it—partial, indeed tiny, though that knowledge no doubt is. Thus suppose that in the actual history, the centre of mass of Erwin's cat's tail is above the floor and moving (i.e. the infernal device did not kill the cat—the actual world is better than it might have been ...), and Erwin knows this, since he sees the tail vertical and moving. Then this definite perception, and the knowledge it engenders, are also part of the actual definite quasiclassical history—and so presumably, Kent proposes that they are represented by appropriate values of appropriate beables. I do not wish here to foist on Kent an account of the relation between mental and physical states, or even require that he should have some such account. But clearly, there is an issue to consider. Namely: do definite perceptions (and their consequent states of belief and knowledge) correspond to values of components of stress-energy, i.e. of the *same* beable that Kent has already proposed as sufficient to secure a definite *inanimate* macroscopic realm? Or are the subtleties of the mental/physical nexus such that they correspond to values of some other beable?[17]

(2): *How to assess the constraint?*:—But these pools of ignorance *do* cause a problem for efforts to assess whether this constraint (β) for 'recovering a quasiclassical

[17] Again, there is an interesting comparison with the pilot-wave theory. For my question to Kent is the analogue of a question sometimes raised about the pilot-wave theory. Namely: does the 'psycho-physical parallelism' ([69], Chap. VI.1 p. 418f.) between some mental states, such as states of perceptual knowledge, and some physical states of our sensory organs (e.g. depression of a touch receptor in my fingertip, or photons impinging on my retina) mean that a perception being definite—one way rather than another—involves a point-particle being in one wave-packet rather than another? (For example, cf. Brown and Wallace (2005: Sect. 7, pp. 533–537).) But I think that on this topic, Kent's situation is *much* more comfortable than the pilot-wave theorist's. For according to modern psychophysics, it is much more plausible that mental states being one way rather than another correspond to (i) values of components of stress-energy at locations in the brain being one way rather another, than to (ii) point-particles being in one location rather than another.

history' in fact holds good or fails. If we are to assess (β), despite our ignorance, we will have to somehow simplify and-or idealize, as regards both (1) the final actual mass-energy distribution and (2) the details of the actual single quasiclassical history. For both (1) and (2), the obvious strategy is to look at a toy model, including very simple idealizing assumptions about both the environmental particles and the quasiclassical history. This is of course what Kent does (2015, Sect. 3; 2017, Examples 1–5). For example, his simpler models assume that 'photons' scatter off macroscopic systems without any recoil by the latter; and that the quasiclassical history consists just of the locations of one or more macroscopic massive quantum systems that have zero self-Hamiltonian.

Thus his simplest example, which uses just one spatial dimension, goes roughly as follows. I will develop this example in Sect. 4.1, so as to assess Outcome Dependence.

(i) A macroscopic massive quantum system with zero self-Hamiltonian is initially stationary in a two-peak superposition of two locations, call them 'left' and 'right';

(ii) it then scatters a photon off one peak—*and* the other, i.e. the photon's quantum state after the interaction is itself a two-peak wave-packet (more precisely: an improper mixture with two equi-weighted components) thus encoding both possible locations of reflection;

(iii) later on, the photon registers on the hypersurface S where—we can suppose—the randomly selected actual final condition encodes that the photon's location is such as to record that it had earlier scattered off the massive system's *left* peak, not its right one—so that

(iv) applying Kent's calculational algorithm (especially stage (iii) of Sect. 3.2.1), the macroscopic massive quantum system is localized on the left: that is, the beable $\langle t_{\mu\nu}(y)\rangle$ is substantially non-zero for $y =$ 'left', and is zero, or close to zero, for $y =$ 'right'.

So much by way of briefly expounding Kent's proposal. Obviously, there is a great deal one could explore here: for example, the detail of Kent's toy models, and his suggestions for developing them, or for varying the postulates so as to get 'cousin' theories that could differ empirically from quantum theory. But I will now confine myself to assessing the fate, in Kent's proposal, of the conditions, Outcome Independence and Parameter Independence. As announced in Sect. 1, there will be an interesting analogy with the pilot-wave theory: similar verdicts on Outcome Independence, and—I believe!—on Parameter Independence. And we will see a connection with some recent no-go, i.e. 'no hidden-variable' theorems.

4 Outcome Independence?

I now investigate whether Kent's proposal satisfies Outcome Dependence. I will argue that the situation is analogous to that for the pilot-wave theory. As we saw in Sect. 2.2.3, the pilot-wave theory satisfies Outcome Independence. That is: it is satisfied at the 'micro-level', i.e. the level of hidden variables (particles' positions).

But by averaging over the hidden variables using, indeed, the orthodox Born-rule probability distribution, the theory recovers orthodox quantum theory's Outcome Dependence at the observable level of experimental statistics (cf. Sect. 2.2.2, (b)).

Thus I will argue that despite Kent's proposal being otherwise very different from the pilot-wave theory, the situation is analogous:

(i) Outcome Independence at the 'micro-level', i.e. using probabilities conditioned on a specific value of Kent's beable, i.e. specific values of the mass-energy distribution on (appropriate parts of) the late hypersurface S, $\{t_S(x) : x \in S\}$; while on the other hand, there is:

(ii) Outcome Dependence at the observable level, after averaging over these values, using the orthodox Born-rule probabilities prescribed by the quantum state $|\psi_S\rangle \equiv U_{S_0S}|\psi_0\rangle$ on S.

To make this argument, I need to adapt the ideas of Kent's toy models, as summarized at the end of Sect. 3.2.3, to a Bell experiment. I will first quote Kent's own presentation of his first toy model (Sect. 4.1). Here I will emphasize that Kent's invocation of a final boundary condition is conceptually unproblematic. Then I adapt the ideas to (a very simple model of) a Bell experiment (Sect. 4.2). Then in the last Subsection (Sect. 4.3), I conclude, as announced in (i) and (ii) above, that Kent and the pilot-wave theory give similar verdicts about whether Outcome Independence is obeyed: Yes at the micro-level, No at the observable level. So this last Subsection will return us to the language of 'hidden variables', which we set aside in order to expound Kent's ideas; (cf. the *Remark* at the end of Sect. 3.1).

4.1 Kent's First Toy Model

Kent presents his first toy model as follows (2015, Sect. 3). The algebra in this quotation—in particular, the arguments in the δ-functions describing the positions of the point-like photons—can be checked by looking at the spacetime diagram (thanks to Bryan Roberts): where X is the position coordinate of the photon.

> We consider a toy version of "semi-relativistic" quantum theory, in which a non-relativistic system interacts with a small number of "photons". We treat the photons as following light-like path segments. We model their interactions with the system as bounces, which alter the trajectory of the photon. For simplicity, we neglect the effect of these interactions on the non-relativistic system, and also neglect its wave function spread and self-interaction, so that in isolation its Hamiltonian $H_{\text{sys}} = 0$. We simplify further by working in one spatial dimension, and we take $c = 1$.
>
> We suppose that the initial state of the system is a superposition of two separate localized states, $\psi_0^{\text{sys}} = a\psi_1^{\text{sys}} + b\psi_2^{\text{sys}}$. Here $|a|^2 + |b|^2 = 1$ and the ψ_i^{sys} are states localized around the points $x = x_i$, with $x_2 > x_1$. For example, the ψ_i^{sys} could be taken to be Gaussians (but recall that we are neglecting changes in their width over time). We take $|x_1 - x_2|$ to be large compared to the regions over which the wave functions are non-negligible. We thus have a crude model of a superposition of two well separated beams, or of a macroscopic object in a superposition of two macroscopically separated states.

We suppose that the environment consists of a single photon, initially unentangled with the system. It is initially propagating rightwards from the direction $x = -\infty$, so that in the absence of any interaction it would reach $x = x_1$ at $t = t_1$ and $x = x_2$ at $t = t_2 = t_1 + (x_2 - x_1)$.

We take the photon-system interaction to have the effect of instantaneously reversing the photon's direction of travel, while leaving the system unaffected. (As noted above, we neglect the effect on the system: this violates conservation of momentum but simplifies the overall picture.)

Thus, for $t < t_1$, the state of the photon-system combination in our model is

$$\delta(X - x_1 - t + t_1)(a\psi_1^{\text{sys}}(Y) + b\psi_2^{\text{sys}}(Y)),$$

where X, Y are the position coordinates for the photon and system respectively.

For $t_1 < t < t_2$, the state is

$$\delta(X - x_1 + t - t_1)(a\psi_1^{\text{sys}}(Y)) + \delta(X - x_1 - t + t_1)(b\psi_2^{\text{sys}}(Y)).$$

For $t > t_2$, the state is

$$\delta(X - x_1 + t - t_1)(a\psi_1^{\text{sys}}(Y)) + \delta(X - x_2 + t - t_2)(b\psi_2^{\text{sys}}(Y)).$$

The possible outcomes of a (fictitious) stress-energy measurement at a late time $t = T \gg t_2$ are thus either finding the photon heading along the first ray $X = x_1 + t_1 - t$ and the system localized in the support of ψ_1^{sys}, or finding the photon heading along the second ray $X = x_2 + t_2 - t$ and the system localized in the support of ψ_2^{sys}.

Suppose, for example, we consider a real world defined by the first outcome. Our rules for constructing the system's beables imply that, for $t < 2t_1 - t_2$, and for $x = x_1$ or x_2, we condition on none of these outcomes, since all of them correspond to observations within the future light cone. Up to this time, then, the mass density beables for the system are distributed according to $|\psi^{\text{sys}}(Y)|^2$, with a proportion $|a|^2$ localized around $Y = x_1$ and a proportion $|b|^2$ localized around $Y = x_2$.

For $2t_1 - t_2 < t < t_1$, the observation of the photon on the first ray is outside the future light cone of the component of the system localized at x_2, but not outside the future light cone of the component localized at x_1. This gives us mass density beables distributing a proportion $|a|^2$ of the total system mass around x_1, but zero mass density beables around x_2.

For $t > t_1$, the observation is outside the future light cone of both localized components of the system. This gives us mass density beables distributing the full system mass around x_1, and zero around x_2.

In other words, in the picture given by the beables, the system is a combination of two mass clouds with appropriate Born rule weights initially, and "collapses" to a single cloud containing the full mass after $t > t_1$.

This quotation illustrates Kent's proposal well. To sum up: We suppose the real world is defined by the first outcome. That is: the photon that entered from the left (flying rightwards) reflects from x_1 at t_1, and registers on the given time slice $t = T$ (which is much later: $T \gg t_2 > t_1$). Given this outcome, there is zero-mass around x_2, in our frame, at all the times t for which the photon hits (while flying leftwards) the time slice $t = T$ outside the future light cone of (x_2, t). That is: at all times t later than $2t_1 - t_2$: a semi-infinite period. During the first part of this period—to be precise: for $t_1 > t > 2t_1 - t_2$—the photon hitting the $t = T$ time slice is still *inside* the future light cone of (x_1, t). It is only after t_1, i.e. when $t > t_1$, that hitting the much later slice is outside the future light cone of (x_1, t). Thus the 'collapse' to zero-mass around x_2 happens, in our frame, before the 'collapse' to full-mass around x_1 happens.

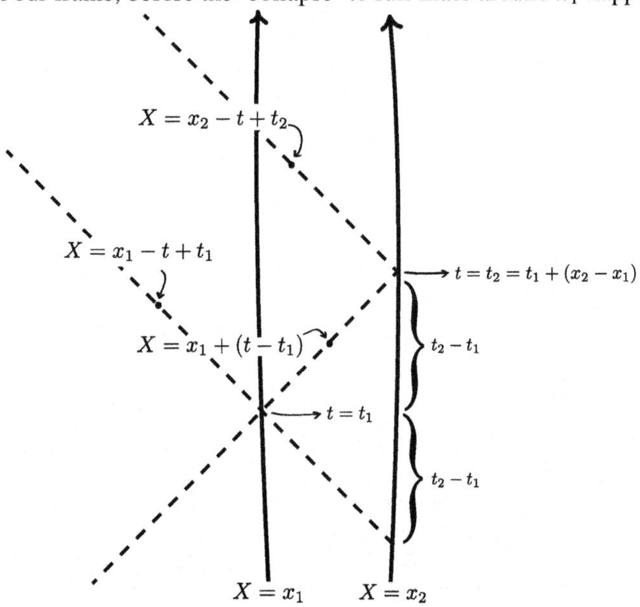

Finally, we must *beware* of beguiling words! Thus it is tempting to say things like: the photon registering on the late spacelike hypersurface records that it reflected from one peak of the superposition, rather than the other. If such statements are read *without* their usual temporal connotations, they are indeed innocent: they suggest that there is (in a timeless or 'block-universe' sense of 'is') an actual fact as to which reflection happens—and agreed, on Kent's proposal, there *is* such a fact. It is made true by the actual final condition: a final condition which is a matter of happenstance, of random selection (though not of an actual measurement). Thus in the preceding paragraph, my verbs like 'registers', 'hitting' and 'happens', are all to be read *without* temporal connotations: as what philosophers call 'tenseless' verbs, despite their syntactic form being the same as present-tensed verbs.[18]

But beware: such statements (especially words like 'registering', and 'records that it reflected') normally *do* carry temporal, indeed causal, connotations. So they suggest—not just that there is an actual fact as to which reflection happens (where 'happens' is tenseless!): which is true on Kent's proposal—but that:

(i) this actual fact is independent of the selection of the actual final condition; and even that

(ii) it *was* 'made true', or 'settled', before the time of the final condition.

[18]There is nothing suspect about such tenseless verbs. They are not a philosophers' fiction or contrivance: the verbs in proverbs, e.g. 'a stich in time saves nine', and in pure mathematics, e.g. '1 + 1 = 2', are tenseless.

And (i) and (ii) are, according to Kent, false. As I said in Sect. 3.2.3's second comment on the constraint (α): the randomly selected actual final condition *makes it true* that the beable has its value (the value it in fact has) at the earlier time. And in this last claim, the verbs are tenseless!

With this warning in mind, I submit that Kent's invoking a final boundary condition is conceptually unproblematic.

4.2 A Kentian Toy Model of a Bell Experiment

Let us now adapt the ideas of Sect. 4.1 to give a toy model of a Bell experiment. I will use the obvious analogy with pilot-wave descriptions of such experiments that use Stern–Gerlach magnets to measure spin. Recall from the summary in Sect. 2.2.3 that the pilot-wave theory makes precise the two possible outcomes—spin being 'up' or 'down' in the direction concerned—in a straightforward way: by the point-particle being 'up' or 'down' relative to the bifurcation plane.

Kent's proposal makes outcomes precise in a similar way. Agreed: his proposed beable is—not an always-existing (and continuously and deterministically evolving) point-particle position, but—roughly speaking, localized mass (or mass-energy) density. But this means that he can represent the two 'latent' outcomes of a spin measurement by a superposition of two separate localized states, $\psi_0^{\rm sys} = a\psi_1^{\rm sys} + b\psi_2^{\rm sys}$—just like the initial state of the massive system in Sect. 4.1. And the occurrence of a single definite outcome, in the one actual world, is to be represented by a photon registering on the late spacelike hypersurface (i.e. by the random selection of the actual final condition) and so recording that it reflected from one peak rather than the other.[19]

To spell this toy model out a little, I assume that we again idealize by using just one spatial dimension. So let the locations x_1 and x_2, as in Kent's model (so with $x_2 > x_1$), now be in the left-wing of the experiment. The representations of the two possible outcomes of a spin (or polarization) measurement on the massive quantum system entering the left wing ('the L-system') are the localization of mass density around x_1 and x_2 respectively. Again, the analogy is with how a Stern–Gerlach magnet makes the point-particle's position represent its spin. But I idealize by having just one spatial dimension, so that there is no bifurcation plane. And I will not try to represent alternative settings (parameters, components of spin) for the spin measurement: that will only become a topic in Sect. 5's discussion of Parameter Independence.

Of these two possible outcomes, one rather than the other occurs, in accordance with the actual final condition including a photon registering at a place on the late hypersurface $t = T$ that records a previous reflection off one peak rather than the other.

---

[19]Here again, the words 'registering', and 'recording that it reflected', are to be understood tenselessly, in line with the warning at the end of Sect. 4.1. That is: there is indeed an actual fact as to which reflection happens (tenselessly!). But this fact is *not* independent of, nor is it 'made true' or 'settled' before the time of, the actual final condition.

So much by way of describing the left-wing of the experiment. To describe the other wing, we assume there is also another massive system, the R-system, located far along the x-axis, and entangled with the the L-system. So: considering only spatial degrees of freedom (suppressing spin degrees of freedom), and ignoring how to treat the various possible settings (parameters) of measurement apparatuses, the initial joint quantum state can be written as (with Y_L, Y_R for the spatial coordinates of the left- and right-systems, respectively):

$$a\psi_1(Y_L)\psi_4(Y_R) + b\psi_2(Y_L)\psi_3(Y_R) \tag{6}$$

where x_3 and x_4 are located far along the x-axis, i.e. $x_3 >> x_1, x_2$ and $x_4 >> x_1, x_2$, and with $x_3 < x_4$; and where each factor wave-function ψ_i has support in a small neighbourhood of x_i. This state correlates the two systems' positions: there is Born-rule probability $|a|^2$ of their being in their respective outer positions (i.e. x_1 and x_4) and Born-rule probability $|b|^2$ of their being in their respective inner positions (i.e. x_2 and x_3). (This anti-correlation—i.e. the fact that one particle gets localized at its left alternative iff the other gets localized at its right alternative—is of course a simple analogue of the anti-correlation of spin results for parallel settings, on the singlet state of two spin-half particles.)

Now recall that Kent's first toy model, reviewed in Sect. 4.1, had one photon that (i) initially propagated rightwards from $x = -\infty$, but (ii) was supposed, by way of an example, to register 'left enough, soon enough' in the actual final condition so as to imply an earlier reflection at x_1 rather than x_2.

So in the Bell experiment, we can imagine, in a similar way, two photons:

(a) one initially propagating rightwards from $x = -\infty$, and as in Sect. 4.1, later registering 'left enough, soon enough' in the actual final condition so as to imply an earlier reflection at x_1 rather than x_2.

(b) one initially propagating leftwards from $x = +\infty$, and later registering 'right enough, soon enough' in the actual final condition so as to imply an earlier reflection at x_4 rather than x_3.

So (a) and (b) specify the joint quantum system's outcome as: each component system is localized in its outer position, i.e. the L-system gets localized at x_1 (its left alternative) and the R-system gets localized at x_4 (its right alternative).

So much by way of adapting the ideas of Kent's first toy model to give a (very!) toy model of a Bell experiment. Of course, other more complicated, less idealized models, are possible. But I have said enough to yield a verdict about Outcome Independence.

4.3 Outcome Independence at the 'Micro-Level', but not at the Observable Level

All the pieces are now in place. We only need to combine:

(a): Sect. 3.2.3's constraint (α), especially clause (iii), and constraint (β); with (b): Kent's postulated beables making precise, viz. as localizations of mass-energy, the outcomes of measurements in Sect. 4.2's toy model of a Bell experiment.

Combining (a) and (b), the analogy with the verdicts given by pilot-wave theory (Sect. 2.2.3) will be obvious.

4.3.1 The Micro-Level

Let us begin at the 'micro-level'. That is: we fix attention on the actual (but of course unknown) final condition $t_S(x)$ on the late hypersurface S in a universe where a Bell experiment is performed at a much earlier time, but after an initial hypersurface S_0. As stressed in Sect. 3.2.1: there is no claim that any measurement is made anywhere on the hypersurface S. The idea is rather that the actual final condition is part of the one real world: a part that by its orthodox quantum correlations (both strict and not strict) with earlier events, contributes to specifying the world—in particular, it specifies a quasiclassical history. Besides, it does so in a Lorentz-invariant way, thanks to Kent's calculational algorithm respecting the light-cone structure.

More precisely, in light of footnotes 11 and 13: There is not a single selected late hypersurface S; but rather, Kent postulates that:

(i) there is a well-defined limit to the distributions over all possible final conditions associated with an appropriate sequence of successively later hypersurfaces; and that

(ii) one element of the sample space on which the limiting distribution is defined—one 'outcome' in the jargon of probability theory (not our jargon!)—is actual; and this 'outcome', by its orthodox quantum correlations with various events throughout spacetime, specifies a quasiclassical history—including outcomes in *our* sense of macroscopic experimental results, such as pointer-readings.

But as in Sect. 3, I will set aside this subtlety, and talk of an actual final condition on the hypersurface S, not the less intuitive 'outcome' (ii) in a vast sample space.

So we wish to consider probabilities about events pertaining to the experiment, that are prescribed by the universal quantum state, but conditional on this actual final condition. In this endeavour, we will be guided by the evident analogy with the pilot-wave theory. Namely: Kent's actual final condition is like the pilot-wave theorist's actual possessed positions of all the point-particles; and we wish to calculate probabilities from, i.e. conditional on, both the orthodox quantum state and this extra 'micro-level' information. Besides, the future-*and-past* determinism of unitary dynamics means that, although the pilot-wave theory usually bases its description of time-evolution on an initial quantum state and initial particle positions, it could instead use the final quantum state and final particle positions—a 'temporal direction' of description like Kent's.

Thus, we recall the second paragraph of Sect. 3.2.1's quotation from Kent: 'our recipe is to take the expectation value for $T_{\mu\nu}(y)$ given that the initial state was $|\psi_0\rangle$ on S_0, conditioned on the [notional] measurement outcomes for $T_S(x)$ being [the actual] $t_S(x)$ for x outside the future light cone of y'. But we now need to amend the discussion so as to consider events, not at one spacetime point intermediate between S_0 and S (labelled y in Sect. 3.2.1), but at four.

So we replace the spacetime point labelled y, and its neighbourhood, by four appropriately chosen spacetime points and the regions around each of them. Their spatial coordinates will be given in Sect. 4.2's notation as $Y_L = x_1$, $Y_L = x_2$, $Y_R = x_3$, $Y_R = x_4$, where $Y_L(Y_R)$ is the position coordinate of the Bell experiment's L-system (R-system) respectively (cf. Eq. 6). And their temporal coordinates are chosen to match when (one run of!) the experiment is in fact performed. (So one expects the two left, respectively two right, spacetime points to be nearly simultaneous; and to arrange the pair of left points to be spacelike separated from the pair of right points.)

Let us suppose that in this run of the experiment the 'outer', x_1 and x_4, outcomes in fact occur in the quasiclassical history specified by the actual final condition $t_S(x)$. That is: $t_S(x)$ on S encodes that just after the run is completed, the L-system is localized (tenseless!) around $Y_L = x_1$, and the R-system is localized (again, tenseless) around $Y_R = x_4$; where this encoding is done by $t_S(x)$ describing photons registering in appropriate locations of S. To return to our example with one spatial dimension, described at the end of Sect. 4.2: the idea is that:

(a) a photon registers on S so far to the left as to imply that it earlier reflected at x_1 rather than at x_2, while

(b) another photon registers on S so far to the right as to imply that it earlier reflected at x_4 rather than at x_3.

So the idea is that if we condition orthodox quantum probabilities for experimental outcomes on all this (very rich!) information in $t_S(x)$, the probabilities will become trivial, i.e. 0 or 1. And so they will factorize. On our supposition, we shall obtain probability 1 for each of the two 'outer', x_1 and x_4, outcomes.

To use the language of hidden variable theories: *the theory is deterministic*. That is: Kent's proposal, as spelt out in this toy model—with its strict correlations between where on S a photon registers, and where the massive system has its mass-energy localized—is *past*-deterministic in the sense that in an individual run of the experiment, the later facts about photon registration, taken together with facts about the quantum state, imply with certainty what the earlier outcomes were. So *Outcome Independence is satisfied.* Again, the situation is analogous to pilot-wave theory, with its familiar, present-to-future, kind of determinism: in that theory, the earlier facts about particle positions, taken together with facts about the quantum state, imply with certainty what the later outcomes will be. Besides, as I noted five paragraphs above: the analogy can be strengthened. For the pilot-wave theory can instead use the final condition (of particle positions and quantum state), not the initial condition.

4.3.2 The Observable Level

I turn to the observable, i.e. experimental, level. You might say that Kent's proposal obeying Outcome Dependence (i.e. violation of Eq. 5 of Sect. 2.2.1) at this level is simply to be expected since, as we have seen (especially in Sects. 3.2.2 and 3.2.3), Kent's proposal is designed to match the experimental probabilities of quantum theory—which obey Outcome Dependence. But this dismissal would be too quick, for two reasons. (Both of them are about getting a better understanding of Kent's proposal.) First: Kent gives a precise meaning to 'outcome' in terms of localizations of mass-energy, and a calculational algorithm for probabilities of outcomes; so it is an important consistency check on his ideas to see that these lead to Outcome Dependence. Second: conducting this check provides an interesting comparison with the pilot-wave theory.

So I return to the earlier discussion (at the end of Sect. 3.2.3) about the fact that we of course do not know the final condition, and never will know it: and indeed, we will never know more than a tiny fraction of the actual single quasiclassical history.

I think Kent's proposal about this is natural; (compare clause (iii) of the first constraint, (α), in Sect. 3.2.3). Namely, he proposes that:

(1) one should average over the possible final mass-energy distributions $t_S(x)$ on S—and do this averaging with their Born-rule probabilities as prescribed by the final quantum state $|\psi_S\rangle$; and

(2) it is these averaged probabilities that are equal, or close enough, to the probabilities given by what I called the quantum textbook.

In other words: clause (iii) of constraint α is interpreted as: these textbook probabilities are (nearly) equal to a mixture of probabilities with (a) each component of the mixture conditioned on one of the various possible $t_S(x)$ (on the part of S outside the relevant point's future light cone); and with (b) each component weighted with the orthodox Born-rule probability, prescribed by $|\psi_S\rangle$, of its $t_S(x)$. To sum up: Kent proposes averaging over the possible final mass-energy distributions with their Born-rule probabilities, so as to recover the experimentally confirmed orthodox quantum probabilities.

Again, there is an evident analogy with the pilot-wave theory. It also recovers these probabilities by averaging over the values of its hidden variable, and by doing this averaging with the values' Born-rule probabilities. But agreed, there is also a difference: we have a much more detailed understanding of how the pilot-wave theory recovers orthodox quantum probabilities—indeed, so successfully that it is usually considered empirically equivalent to orthodoxy, at least for non-relativistic physics—than we do of how Kent's proposal does so.[20] Hence our answer to question (2) at the end of Sect. 3.2.3: that one must, as Kent does, investigate toy models.

The upshot as regards Outcome Independence (Eq. 5 of Sect. 2.2.1) is clear. In so far as we are confident that Kent's proposal recovers orthodox quantum probabilities at the observable level, it of course satisfies Outcome Dependence at that level. Besides, since Outcome Dependence is an inequation, not an equation: even if the recovery is not exact, i.e. there are systematic differences, so that Kent's proposal is an empirical rival to orthodox quantum theory, not an interpretative addition to it: nevertheless, Kent's proposal probably satisfies Outcome Dependence at the observable level.

---

[20]This difference implies no criticism of Kent: the pilot-wave theory is long-established, and most expositions of its recovery of orthodoxy do not involve any cosmological, and so inevitably speculative, considerations of the kind Kent's proposal must tangle with.

5 Parameter Independence?

So much for Outcome Dependence. What about Parameter Independence: does it hold in Kent's proposal?

The first point to make is that as we saw in Sect. 4.3, we should distinguish (as the pilot-wave theory does) between:

(i) the 'micro-level', i.e. probabilities conditioned on specific final boundary conditions of Kent's kind; and

(ii) an observable level, obtained by averaging over these boundary conditions: which is expected to match, or nearly match, orthodox quantum probabilities—though of course, this match should not just be assumed, but should be checked.

Assessing Parameter Independence, at either of these levels, is a substantive task: and not just because proving a universally quantified equation is harder than proving an existentially quantified inequation. Again, it is a matter of seeing how Kent's proposals for beables and their probabilities mesh with the formalism of standard quantum theory. Thus one needs to explicitly model in a Kentian fashion both:

(a) different choices of apparatus-setting (parameter) on one wing (the right wing in Eq. 4 of Sect. 2.2.1), and

(b) non-selective measurement, (i.e. defining the left-wing marginal probability by summing probabilities of the various possible right-wing outcomes).

A Kentian model of (a) will involve photons scattering off the apparatus' knob that sets which quantity gets measured, and registering a long time later on the late hypersurface; so that the actual, randomly selected, final condition encodes the setting. Then in order to model (b), the probabilities dependent on reflection from "which knob-setting" will need to be combined with Sect. 4.3's probabilities dependent on reflection from "which outcome, i.e. mass-energy lump".

As stressed in Sect. 2.2.1 after Eq. 2, and at the end of Sect. 4.1, there is nothing problematic or suspicious about a final condition encoding a knob-setting; nor about probabilities depending on such a setting. But writing down a model incorporating (a) and (b) *is* a substantive task. It is not obvious how such an explicit model would relate to standard quantum theory's no-signalling theorem, and more generally to the commutation of quantities. So I leave all this for another occasion!

But as announced in Sect. 1: recent theorems by Colbeck and Renner (made rigorous by Landsman), and by Leegwater, say (roughly speaking!) that, under some apparently natural extra assumptions: any theory that supplements orthodox quantum theory, in the sense of recovering orthodox Born-rule probabilities by averaging over other probabilities, must violate either a so-called 'no-conspiracy' assumption, or Parameter Independence. (Again: 'no-conspiracy' is an unfair label, since there need be nothing conspiratorial or problematic about a violation.) So Kent's proposal needs to be compared with these theorems, with a view to assessing which of their assumptions it satisfies, and which it violates (and maybe we will need to add: under which disambiguations). Besides, as Sect. 4.3's discussion brought out: Kent's proposal's probabilities at the micro-level, i.e. probabilities conditioned on a specific final boundary condition, are *not* equal to orthodox Born-rule probabilities. So indeed, his proposal supplements quantum theory in the relevant sense; and so the proposal must violate one or more of the assumptions of these theorems. And since Leegwater's theorem seems to dispense with the need for Colbeck and Renner's extra assumptions, we infer that Kent's proposal must violate either the no-conspiracy assumption, or Parameter Independence (at the micro-level), or both.

Properly working out which it violates must wait for another occasion. I will just, in Sect. 5.1, discuss the suggestion that it obeys the no-conspiracy assumption, and so *must* violate Parameter Independence. Then in Sect. 5.2, I make a final point, again based on these theorems: in effect, a peace-pipe to smoke with my dedicatee, Abner Shimony. Namely: these theorems, especially Leegwater's theorem, promote peaceful coexistence of the kind Shimony sought.

5.1 The Theorems of Colbeck and Renner, and of Leegwater: Parameter Dependence at the Micro-Level?

Colbeck and Renner claim to show that, under certain assumptions, 'no extension of quantum theory can have improved predictive power' (2011, 2012). The idea here of 'no improved predictive power' means, in the usual language of 'hidden variables', that the hidden variables are otiose, or trivial. A bit more precisely: if the hidden variable 'underpinning' of quantum theory is to recover orthodox Born-rule probabilities by probabilistic averaging over hidden variables using a measure μ_ψ dependent on the quantum state ψ, then (under certain assumptions) the hidden variables are trivial in the following sense. The probability prescribed by any hidden variable λ for any outcome for a measurement of any quantity (and in any context of simultaneously measuring some other quantities) is the same as the orthodox Born-rule probability—except perhaps for a set of hidden variables λ of zero μ_ψ-measure.

This is a stunning no-go result against a hidden variable underpinning of quantum probabilities, especially since the needed assumptions look natural enough; and indeed, Colbeck and Renner's proof-method is dazzlingly inventive. But their derivations are heuristic, rather than rigorous; and unsurprisingly, they prompted scrutiny, and even criticism. One concern was that their 'freedom of choice' assumption (about experimenters being free to choose what to measure) in fact encoded Parameter Independence. Indeed, anyone familiar with the pilot-wave theory will suspect that Parameter Independence is 'in there somewhere among' Colbeck and Renner's assumptions. For in the pilot-wave theory, the hidden variables, i.e. the point-particles' positions, are certainly *not* otiose in the above sense, and they violate Parameter Independence (cf. Sect. 2.2.3). Agreed, this concern might be met by simply admitting Parameter Independence as an explicit assumption.

But there were also other concerns about Colbeck and Renner's derivations:

(a) other assumptions, and the proof itself, needed to be formulated more rigorously;

(b) although these other assumptions (i) look natural enough, even when formulated rigorously, and (ii) are obeyed by quantum theory (in the sense of substituting the quantum state ψ for the schematic hidden variable λ, just as we did in Sect. 2.2.2), nevertheless: a hidden variable advocate might well doubt them—just as there is a noble precedent, viz. the pilot-wave theory, for denying Parameter Independence.

Of course, I cannot pursue all the topics that arise from the scrutiny of Colbeck and Renner's result. I will confine myself to reporting two main points: combining them will then lead directly to my conclusion about Kent.

First: Landsman has fully addressed concern (a) (while also emphasizing (b)). That is: he has given a mathematically rigorous statement of the assumptions, and a correspondingly rigorous proof of the theorem (2015; 2017, Chap. 6.6).

Second: [45] has a significantly different but also rigorous theorem that is a 'cousin' of the Colbeck and Renner theorem, as made rigorous by Landsman. Indeed, Leegwater manages to dispense with all the extra assumptions in (b) above, as clarified by Landsman, except for one: viz. the 'no-conspiracy' assumption, that the measure used to average over the hidden variables so as to recover Born-rule probabilities—while of course it can depend on ψ, as above—must be independent of which quantity, or quantities, are (chosen to be) measured. (We already saw this assumption way back in Sect. 2.2.1, in the discussion after Eq. 2.) Thus Leegwater proves that Parameter Independence and 'no-conspiracy', taken together, are enough to imply that the hidden variables must be otiose or trivial in the sense above (2016: Theorem 1, p. 21). This is a stunning no-go result.[21]

Thus, putting the two points together: while Colbeck and Renner gave a dazzlingly inventive but heuristic derivation, the papers by Landsman and Leegwater are also dazzling examples—of mathematical invention, as well as rigour.

What is the upshot as regards Kent? At first sight, it seems that his proposal *does* satisfy no-conspiracy. For he averages over the possible values of his hidden variables, i.e. over the possible final boundary conditions, with the Born-rule probabilities given by the final universal quantum state $|\psi_S\rangle$. And $|\psi_S\rangle$ is independent of which quantity, or quantities, were previously measured, even though it is the universal state. Agreed: there is a fact about which quantity got measured (and about what the outcome was), according to Kent's recovery of a quasiclassical world (Sects. 3.2.2 and 3.2.3). But these facts leave no 'mark' on—have no back-reaction on—the universal quantum state, which always evolves unitarily. Thus in the Schrödinger picture, $|\psi_S\rangle$ gets, at an intermediate time when an experiment is set up, components (non-zero amplitudes) for various possible choices of quantity (settings) and, soon thereafter, components for various possible outcomes. (Think, if you will, of an Everettian's description of the setting-up, and performance, of a run of an experiment: the components correspond

[21] Furthermore, Leegwater allows the measure over hidden variables to depend on more than just the quantum state (although not which quantity is measured). For example, it can depend on the specific method that was used to prepare the state. I stress that discerning the no-conspiracy assumption is not original to me: Landsman and Leegwater are perfectly clear that they make this assumption, even though it is not given a formal label or acronym ([43], pp. 122103–2, assumption CQ and its footnote 14; 2017, p. 221, Definition 6.20 and following text; [45], p. 21 footnote 7 and its preceding text).

to the Everettian's branches.) So it seems that Kent's proposal satisfies no-conspiracy; and so, thanks to Leegwater's theorem, must violate Parameter Independence at the micro-level—giving a clear analogy with the pilot-wave theory.[22]

But on reflection, it is *not* clear that Kent's proposal obeys no-conspiracy in the relevant sense, i.e. for the measure used in deriving a Bell inequality (cf. Sect. 2.2.1: here, I am indebted to Gijs Leegwater). For *a priori*, this measure need not be the Born-rule probabilities given by the final universal quantum state $|\psi_S\rangle$. In particular: since (we can presume) a Kentian toy model of a Bell experiment makes the final condition "rich enough" to determine which quantity was measured earlier, the relevant measures for different settings may have disjoint supports. In short: about Parameter Independence, the jury is still out.

5.2 Mandatory Peaceful Coexistence? A Peace-Pipe for Shimony

I end with one last, happy, point: in effect, a peace-pipe to smoke with my dedicatee, Abner Shimony. Namely: these theorems, especially Leegwater's theorem, promote peaceful coexistence of the kind Shimony sought—while leaving us plenty of work for the future, even if we join Shimony in favouring some process of dynamical reduction.

For recall that their gist is that Parameter Independence is enough to ban hidden variable supplementations. Thus Shimony's overall view of non-locality—accepting Outcome Dependence so as to avoid a Bell inequality, and seeing no reason *not* to endorse Parameter Independence—turns out, by dint of the hard work of these theorems, to lead to 'unsupplemented' quantum theory.

This is an important strengthening of the argument for peaceful coexistence. For the traditional Bell-Shimony argument is, essentially, that Parameter Independence and empirical adequacy for the Bell experiment imply Outcome Dependence: and we then worry about whether Outcome Dependence really satisfies the spirit of relativity (Bell himself thinking 'No'; cf. Sect. 1). But these new theorems imply that Parameter Independence and empirical adequacy (admittedly: in other nonlocality experiments—but ones where we are confident of quantum theory) imply 'unsupplemented' quantum theory: a much stronger conclusion that Outcome Dependence—which, in a relativistic world, amounts to mandatory peaceful coexistence.

Of course, this conclusion is not the last word. This new peaceful coexistence may be mandatory, given just Parameter Independence. But one naturally wants it to

[22]Note the obvious contrast with the dynamical reduction programme, i.e. with the view that the universal quantum state evolves *non*-unitarily, collapsing appropriately throughout history so as to yield a quasiclassical world (cf. footnote 2 and Sect. 5.2). On that view, the quasiclassical world no doubt includes which quantity, or quantities, are measured at all the various times, and so the final universal quantum state certainly *will* depend on such facts. So in such a universe, a Kentian algorithm applied to the final condition would violate no-conspiracy.

be spelt out in more detail: a project which, I argued in Sect. 2.3, eventually requires a solution, or at least a sketched solution, of the measurement problem.

To honour Shimony's memory, I shall end by briefly discussing this in relation to his preferred approach: dynamical reduction (i.e. non-unitary evolution for strictly isolated systems; cf. the references in footnote 2), rather than postulating extra values for quantities, as in the pilot-wave theory.

At first sight, this conclusion sits well with Shimony's preference. After all, dynamical reduction models do not try to recover the Born-rule probabilities of outcomes by probabilistically averaging over a hidden variable. But there is a subtlety, indeed work to be done, hereabouts. For one *can* think of such a model as recovering probabilities of outcomes by probabilistic averaging over realizations of the stochastic noise, i.e. over how the stochastic noise 'happens to go'. For recall that once given:

(i) an initial quantum state ψ, and

(ii) a choice of measured quantity Q, and more specifically

(iii) an apparatus in a specified state, and a total Hamiltonian for the joint system comprising micro-system and apparatus; then each realization of the noise leads to one definite measurement outcome.

So thinking of each realization as a hidden variable λ, these hidden variables are deterministic; so they are certainly not trivial in these theorems' sense. But the measure used in the probabilistic averaging over realizations is judiciously (brilliantly!) defined so as to be sensitive to ψ's Born-rule probabilities for values of Q, in just such a way as to recover the Born-rule probabilities for outcomes of the measurement process. Thus the measure depends on which quantity is chosen to be measured. So the point here is that the theorems of Colbeck, Renner and Leegwater are evaded in the sense that their 'no-conspiracy' assumption, that the measure over hidden variables be independent of the quantity measured, is violated.

But again, this is not the last word. Clearly, there is work to be done here, relating the notions of one's favoured dynamical reduction model to the notions and assumptions of the theorems. In particular: suppose one considers instead the measures over the noise that are independent of the quantity measured, and even independent of the quantum state—which are, after all, explicitly there in the formalism of dynamical reduction models. Yet it still seems reasonable to think of the model as underpinning orthodox quantum probabilities, with each realization of the noise being a (deterministic) hidden variable. So one faces the question: what assumptions of the theorems of Colbeck, Renner and Leegwater are now violated? Perhaps Parameter Independence? Work for the future!

6 Summary

My topic has been whether there can be 'peaceful coexistence' between quantum theory and special relativity; and in particular, Shimony's hope that Outcome Dependence would show a way—maybe the best or only way—to secure it.

In Sect. 2, I rehearsed the issues, and concluded sceptically that accepting Outcome Dependence is far from enough to get peaceful coexistence. In short, one needs a relativistic solution to the quantum measurement problem: given such a solution, one can then assess whether Outcome Independence fails—and agreed, one would expect it to fail.

Then I turned to one proposed solution: i.e. one realist one-world Lorentz-invariant interpretation of quantum theory—namely Kent's recent proposal (2014, 2015, 2017). I first reported his main ideas (Sect. 3). Then in Sect. 4, I spelt out how, with his proposed beable making precise what is an outcome of a measurement, his interpretation is like the pilot-wave theory, in that it satisfies Outcome Independence at the micro-level, and (presumably) Outcome Dependence at the observable level.

Then in Sect. 5, I reported recent remarkable theorems by Colbeck, Renner and Leegwater which bear on whether his interpretation obeys Parameter Independence at the micro-level. And I argued that these theorems make some kind of peaceful coexistence mandatory for someone who, like Shimony, endorses Parameter Independence.

So the upshot is that Shimony's hope for peaceful coexistence is alive and well. But to come true, it will need more than a judicious choice of which assumption of a Bell theorem to deny. It will probably need an agreed relativistic solution to the quantum measurement problem. In seeking such agreement, we need both to assess proposed solutions like Kent's, and to analyse the constraints on the solution implied by general theorems like those of Colbeck, Renner and Leegwater.

Acknowledgements Dedicated to the memory of Abner Shimony. As all who met him soon realized, it was a pleasure and a privilege to know him, both as a person and as an intellect. It is a pleasure to thank Masanao Ozawa for the invitation to the Nagoya conference; and to thank him, Francesco Buscemi and the other local organizers, for a very enjoyable and valuable meeting. I am very grateful to Adrian Kent for generous advice and encouragement; and to Bryan Roberts for the splendid diagram. For comments on a previous version, I thank an anonymous referee, audiences in Cambridge, Wayne Myrvold, James Read, Bryan Roberts and four Dutch wizards: Dennis Dieks, Fred Muller and especially Guido Bacciagaluppi and Gijs Leegwater.

References

1. Allahverdyan, A., Ballian R. and Nieuwenhuizen T.: Understanding quantum measurement from the solution of dynamical models. *Physics Reports* **525** 1–166 (2013).
2. Barrett, J.: *The Quantum Mechanics of Minds and Worlds*, Oxford University Press; (1999).
3. Bassi, A. and Ghirardi, G.C.: Dynamical reduction models. *Physics Reports* **379**, 257–426 (2003).
4. Bell, J: *Speakable and Unspeakable in Quantum Mechanics*, Cambridge: University Press; second edition(2004).
5. Bell, J. and Shimony, A., Horne, M. and Clauser, J.: An exchange on local beables, *Dialectica* **39**, 85–1110 (1985).
6. Berndl, K: Global existence and uniqueness of Bohmian trajectories, In Cushing J., Fine A. and Goldstein S. (eds), *Bohmian Mechanics and Quantum Theory: An Appraisal*, Kluwer Academic; arXiv:quant-ph/9509009 (1996).

7. Berndl, K., Dürr D., Goldstein S., Peruzzi G. and Zanghi, N: On the global existence of Bohmian mechanics, *Communications in Mathematical Physics* **173**, 647–673 (1995).
8. Bohm, D. and Hiley, B. *The Undivided Universe*, London: Routledge (1992).
9. Bricmont, J.: *Making Sense of Quantum Mechanics*, Springer (2016).
10. Brown, H. and Timpson, C.: Bell on Bell's theorem: the changing face of nonlocality, in M. Bell and Shan Gao (eds), *Quantum Nonlocality and Reality: Fifty Years of Bell's Theorem*, Cambridge University Press; arXiv:1501.03521 (2016).
11. Brown, H. and Wallace D.: Solving the measurement problem: de Broglie-Bohm loses out to Everett, *Foundations of Physics* **35**, 517–540 (2005).
12. Butterfield, J .: Bell's Theorem: what it takes, *British Journal for the Philosophy of Science*, **43**, 41–83 (1992).
13. Butterfield, J .: Stochastic Einstein Locality Revisited, *British Journal for the Philosophy of Science*, **58**, 805–867 (2007).
14. Butterfield, J .: Assessing the Montevideo interpretation of quantum mechanics, *Studies in the History and Philosophy of Modern Physics*, **52A**, 75–85; At: http://arXiv.org/abs/1406.4351; http://philsci-archive.pitt.edu/10761/ (2015).
15. d'Espagnat. B.: *Conceptual Foundations of Quantum Mechanics* Reading, Mass: Benjamin; second edition (1976).
16. Clauser, J. and Horne, M.: Experimental consequences of objective local theories, *Physical Review D* **10** 526–534 (1974).
17. Clifton, R. and Jones, M.: Against experimental metaphysics. In: French, P., Euling, T. and Wettstein, H. (eds.) *Midwest Studies in Philosophy*, volume 18: Philosophy of Science; Minneapolis: University of Minnesota Press; 295–316 (1993).
18. Colbeck, R. and Renner, R.: No extension of quantum theory can have improved predictive power, *Nature Communications* **2**, 411. https://doi.org/10.1038/ncomms1416 (2011).
19. Colbeck, R. and Renner, R.: The completeness of quantum theory for predicting measurement outcomes. arXiv:1208.4123 (2012).
20. Cushing, J. and McMullin, E. (eds): *Philosophical Lessons from Quantum Theory*, University of Notre Dame Press (1989).
21. Dewdney, C., Holland P. and Kyprianidis, A.: A causal account of non-local Einstein-Podolsky-Rosen spin correlations, *Journal of Physics A: Math. Gen.* **20**, 4717–4732 (1987).
22. Dieks, D and Vermaas, P. (eds.): *The Modal Interpretation of Quantum Mechanics*, Kluwer (1998).
23. Friederich, S.: *Interpreting Quantum Theory: a Therapeutic Approach*, Palgrave Macmillan (2015).
24. Ghirardi, G.: The interpretation of quantum mechanics: where do we stand?, *Journal of Physics: Conference Series* **174** (DICE 2008) 012013 (2009).
25. Ghirardi, G., Rimini, A. and Weber, T: A general argument against superluminal transmission through the quantum mechanical measurement process, *Lettere al Nuovo Cimento* **27**, 293–298 (1980).
26. Healey, R: Quantum theory: a pragmatist approach, *British Journal for the Philosophy of Science* **63**, 729D771 (2012).
27. Healey, R: How quantum theory helps us explain, *British Journal for the Philosophy of Science* **64**, 1–43; https://doi.org/10.1093/bjps/axt031 (2013).
28. Healey, R: Causality and chance in relativistic quantum field theories, *Studies in the History and Philosophy of Modern Physics* **48**, 156–167 (2014).
29. Healey, R.: *The Quantum Revolution in Philosophy*, Oxford University Press (2017).
30. Henson, J.: Non-separability does not relieve the problem of Bell's theorem, *Foundations of Physics* **43**, 1008–1038 (2013).
31. Holland, P.: *The Quantum Theory of Motion*, Cambridge: University Press (1993).
32. Howard, D.: Holism, separability and the metaphysical implications of the Bell experiments. In: Cushing and McMullin (eds.) (1989), pp. 224–253 (1989).
33. Isham, C: *Lectures on Quantum Theory*, London; Imperial College Press (1995).

34. Janssen, H.: Reconstructing Reality: Environment-Induced Decoherence, the Measurement Problem, and the Emergence of Definiteness in Quantum Mechanics. http://philsci-archive.pitt.edu/4224/ (2008).
35. Jarrett, J.: On the physical significance of the locality conditions in the Bell arguments, *Nous* **18**, 569–589 (1984).
36. Jarrett, J.: On the separability of physical systems, in Myrvold, W. and Christian, J. (eds.) (2009); pp. 105–124 (2009).
37. Jordan, T.: Quantum correlations do not transmit signals, *Physics Letters* **94A**, 264 (1983).
38. Kent, A.: Solution to the Lorentzian quantum reality problem, *Physical Review A* **90**, 012107; arXiv:1311.0249 (2014).
39. Kent, A.: Lorentzian quantum reality: postulates and toy models, *Philosophical Transactions of the Royal Society A* **373**, 20140241; arXiv:1411.2957 (2015).
40. Kent, A.: Quantum reality via late-time photodetection, *Physical Review A* **96**, 062121; arXiv:1608.04805 (2017).
41. Landsman, N.: Spontaneous symmetry breaking in quantum systems: emergence or reduction? *Studies in History and Philosophy of Modern Physics* **44(4)**, 379–394 (2013).
42. Landsman, N. and Reuvers, R: A flea on Schrödinger's cat. *Foundations of Physics* **43**, 373–407 (2013a).
43. Landsman, N.: The Colbeck-Renner theorem, *Journal of Mathematical Physics* **56**, 122103 (2015).
44. Landsman, N.: *Foundations of Quantum Theory*, Springer (2017). Open access: available at: https://doi.org/10.1007/978-3-319-51777-3
45. Leegwater, G.: An impossibility theorem for parameter independent hidden variable theories, *Studies in the History and Philosophy of Modern Physics*, **54** 18–34; http://philsci-archive.pitt.edu/12067/ (2016).
46. Leggett, A.: Probing Quantum Mechanics Towards the Everyday World: Where do we Stand?, *Physica Scripta* **T102**, 69–73 (2002).
47. Morganti, M. A new look at relational holism in quantum mechanics, *Philosophy of Science* **76**, 1027–1038 (2009).
48. Muller, F.: The locality scandal of quantum mechanics. In: Dalla Chiara, M. et al. (eds.), *Language, Quantum, Music*, Synthese Library volume 281, Dordrecht: Kluwer Academic, 241–248 (1999).
49. Myrvold, W.: On peaceful coexistence: is the collapse postulate incompatible with relativity? *Studies in the History and Philosophy of Modern Physics* **33**, 435–466 (2002).
50. Myrvold, W.: Relativistic quantum becoming, *British Journal for the Philosophy of Science* **54**, 475–500 (2003).
51. Myrvold, W.: Lessons of Bell's theorem: Nonlocality, yes; Action at a distance, not necessarily, in M. Bell and Shan Gao (eds), *Quantum Nonlocality and Reality: Fifty Years of Bell's Theorem*, Cambridge University Press. Available at: http://philsci-archive.pitt.edu/12382/ (2016).
52. Myrvold, W.: Ontology for collapse theories, forthcoming in Shan Gao (ed.) *Collapse of the Wave Function*, Cambridge University Press (2017).
53. Myrvold, W. and Christian, J. (eds.): *Quantum Reality, Relativistic Causality, and Closing the Epistemic Circle: Essays in Honor of Abner Shimony*, Springer (2009).
54. Norsen, T: Local causality and completeness: Bell vs. Jarrett, *Foundations of Physics* **39**, 273–294 (2009).
55. Norsen, T: John S. Bell's concept of local causality, *American Journal of Physics* **79**, 1261–1275 (2011).
56. Pearle, P.: How stands collapse II, in Myrvold, W. and Christian, J. (eds.) (2009); pp. 257–292 (2009).
57. Redhead, M: *Incompleteness, Nonlocality and Realism* Oxford: University Press (1987).
58. Schrödinger, E.: The Present Situation in Quantum Mechanics: A Translation of Schrödinger's 'Cat Paradox' Paper (trans: J D. Trimmer) *Proceedings of the American Philosophical Society*, **124**, (Oct. 10, 1980), pp. 323–338; American Philosophical Society; http://www.jstor.org/stable/986572 (1935).

59. Shimony, A.: Controllable and uncontrollable nonlocality. In: Kamefuchi , S. et al. (eds) *Foundations of Quantum Mechanics in the Light of New Technology*, Tokyo: Physical Society of Japan; reprinted in Shimony (1993), 130–139: page references to reprint (1984).
60. Shimony, A.: Events and processes in the quantum world. In: Penrose, R. and Isham C. (eds.) *Quantum Concepts in Space and Time* Oxford: University Press; reprinted in Shimony (1993), 140–162: page references to reprint (1986).
61. Shimony, A.: Desiderata for a modified quantum dynamics. In: (eds.) *PSA 1990* volume 2; Proceedings of the 1990 meeting of the Philosophy of Science Association; East Lansing, Michigan: Philosophy of Science Association; reprinted in Shimony (1993), 55–67 (1990).
62. Shimony, A.: *Search for a Naturalistic World View: Volume II: natural science and metaphysics*. Cambridge: University Press (1993).
63. Shimony, A.: Unfinished work, a bequest. In: Myrvold and Christian (eds.) (2009); pp. 479–491 (2009).
64. Shimony, A.: Bell's theorem, in *The Stanford Encyclopedia of Philosophy*. Available at: https://plato.stanford.edu/entries/bell-theorem/ (2009a).
65. Teller, P.: Relativity, relational holism and the Bell inequalities. In: Cushing and McMullin (eds.) (1989) pp. 208–223 (1989).
66. Valentini, A.: Signal locality in hidden variable theories, *Physics Letters A* **297**, 273–278; arXiv:quant-ph/0106098 (2002).
67. Valentini, A.: Signal locality and sub-quantum information in deterministic hidden variable theories, in T. Placek and J. Butterfield (eds.) *Non-Locality and Modality* NATO Science Series II: volume 64, Kluwer; arXiv:quant-ph/0112151 (2002a).
68. van Fraassen, B.: The charybdis of realism: epistemological implications of Bell's theorem, *Synthese* **52**; 25–38 (1982).
69. von Neumann, J.: *Mathematical Foundations of Quantum Mechanics*, Princeton: University Press (English translation 1955, reprinted in the Princeton Landmarks series 1996) (1932).
70. Wallace, D.: *The Emergent Multiverse*, Oxford University Press (2012).
71. Wiseman, H. and Cavalcanti, E.: *Causarum Investigatio* and the two Bell's theorems of John Bell, in *Quantum Unspeakables II*, ed. R. Bertlmann and A. Zeilinger, Springer; pp. 119–142; arXiv:1503.06413 (2017).
72. Zeh, H-D, Joos, E. et al.; *Decoherence and the Appearance of a Classical World in Quantum Theory*, second edition; Springer (2003).

Experimental Investigations of Uncertainty Relations Inherent in Successive 1/2–Spin Measurements

Yuji Hasegawa

Abstract There is no better way to study fundamental phenomena in quantum mechanics than the use of optical setup with matter-waves. In particular, neutrons in an interferometer or a polarimeter have been serving as an almost ideal tool for this sort of studies. Here, several experiments are described, which investigates the uncertainty relations appearing in successive quantum measurements: successive measurements of the neutron's 1/2-spin is carried out to evaluate the error of a measurement and the disturbance induced by that measurements. The results of the first experiment confirm the violation of Heisenberg's original reciprocal relation for measurement error and disturbance and the validity of the reformulated generally valid relation. Before this experiment, there have been no experimental study of the uncertainty relation inherent in the quantum measurement. Further experimental studies are carried out for extended relations, providing tight relations for pure as well as a generalized mixed input states: tightness of the relations for measurements on the two-level quantum system is demonstrated in the experiments.

Keywords Uncertainty relation · Spin · Neutron optics · Neutron polarimeter

1 Introduction

Quantum theory is one of the most successful theory developed in 20th century. From the early stage of its development, the peculiarities of this theory have fascinated on one hand side, but upset and confused not only the interested public but also physicists on the other hand side. It works well if one accepts the formalism of quantum mechanics merely to make statistical predictions; lots of its users do not need to reflect on what is really going on behind the theory. Nevertheless, since both implication of experiments studying most essential and vital phenomena in quantum mechanics and design of the devices on the ground of fundamental features

Y. Hasegawa (✉)
Atominsitut, TU-Wien, Stadionallee 2, Wien, Austria
e-mail: hasegawa@ati.ac.at

M. Ozawa et al. (eds.), *Reality and Measurement in Algebraic Quantum Theory*,
Springer Proceedings in Mathematics & Statistics 261,
https://doi.org/10.1007/978-981-13-2487-1_12

of quantum mechanics are becoming more and more complex and complicated, one can not emphasize the importance of an intellectual challenge to get deeper insight of quantum theory too much.

One of the essential consequence inherent in quantum theory is indeterminacy [1]: for instance, the path of a quantum particle propagating through a double-slit can not be revealed without washing out the interference fringes. Another kind of indeterminacy appears in the uncertainty principle: by using the famous γ-ray microscope thought experiment, Heisenberg formulated the uncertainty relation as a limitation of accuracies of position and momentum measurements [2]. He formulated the uncertain relation as $p_1 \cdot q_1 \sim h$, where q_1 and p_1 represent "the mean error" of the position measurement and the "the discontinuous change" of the momentum, respectively. He claimed that this relation is a straightforward mathematical consequence of the canonical commutation rule $qp - pq = i\hbar$ between position and momentum observables. Later, the uncertainty relation for the error of the position measurement $\varepsilon(q)$ and the disturbance of the momentum measurement $\eta(p)$ is reformulated as an inequality [3]

$$\varepsilon(q) \cdot \eta(p) \geq \frac{\hbar}{2}. \tag{1}$$

It is to be noted here that the error-disturbance uncertainty relation stems from physical consequence of an unavoidable and uncontrollable recoil through the interaction between the quantum object to be measured and the measurement apparatus.

Although the uncertainty relation in terms of standard deviation is more often mentioned as a *standard* uncertainty relation, this form for the position and momentum as $\Delta q \cdot \Delta p \geq \frac{\hbar}{2}$ was derived afterwards [4–6]. It should be emphasized here that this notion of uncertainty relation denotes only the statistical quantity, which is of significant only in repeated measurements, and is relevant to neither the error of the measurement nor the disturbance due to interactions in a quantum measurement. Later on, Robertson generalized the relation between standard deviations for arbitrary pairs of observables as

$$\Delta(A) \cdot \Delta(B) \geq \frac{1}{2}|\langle\psi|[A, B]|\psi\rangle| \tag{2}$$

where $[A, B] = AB - BA$ stands for the commutator between the two operators A and B. The physical content of this relation refers only to fact that the product of the root-mean-square half-widths of the two statistical distributions of (not a joint but) a single (repeated) measurement either A or B on lots of equivalently prepared states can never be less than $\hbar/2$ [7].

In the seventies, complete difference formulation of uncertainty relation in terms of entropy, which denotes bounds on the entropy of the measurement outcomes, was derived for position and momentum [8], followed by a generalization for any pairs of observables [9]. The improved version [10], which is proved in [11], has the form

$$H(A) \cdot H(B) \geq -\log 2 \max_{j,k} \| \langle a_j | b_k \rangle \| \tag{3}$$

where $H(A)$ $[H(B)]$ denotes the Shannon entropy of the probability distribution of the outcome and $|a_j\rangle$ $[|b_k\rangle]$ represent non-degenerating eigenstates of A [B]. Note that the uncertainty relation with entropy describes informational contents and does not refer to the interaction of quantum measurements.

A straightforward extension of the error-disturbance uncertainty relation for arbitrary pairs of observables A and B in a form

$$\varepsilon(A) \cdot \eta(B) \geq \frac{1}{2}|\langle\psi|[A, B]|\psi\rangle|. \tag{4}$$

seems to be satisfactory and reasonable at the first sight. Nevertheless, it was known that the validity of Heisenberg's original relation in this form is justified only under limited circumstances [12–15]. In order to quantify the accuracy of *joint* quantum measurements, it is essential to formulate a correct error-disturbance uncertainty relation.

2 First Experimental Investigation of the Error-Disturbance Uncertainty Relation

2.1 Theory

By applying rigorous and general theoretical treatments of quantum measurements, error-disturbance uncertainty relation was derived which is universally valid for arbitrary pairs of observables [16–19]. A new error-disturbance uncertainty relation (EDUR) is given in a form of

$$\varepsilon(A)\eta(B) + \varepsilon(A)\Delta(B) + \Delta(A)\eta(B) \geq C_{AB}, \;\; with \;\; C_{AB} = \frac{1}{2}|\langle\psi|[A, B]|\psi\rangle|. \tag{5}$$

Note that the additional second and third terms imply a new accuracy limitation. Here, the error of the measurement of an observable A and the thereby induced disturbance on the measurement of an observable B are defined as

$$\begin{aligned} \varepsilon(A) &= ||\left(A \otimes \mathbb{1} - U^\dagger(\mathbb{1} \otimes M)U\right)|\xi\rangle|\psi\rangle||, \\ \eta(B) &= ||\left(B \otimes \mathbb{1} - U^\dagger(B \otimes \mathbb{1})U\right)|\xi\rangle|\psi\rangle||, \end{aligned} \tag{6}$$

where $|\xi\rangle[|\psi\rangle]$ denotes the initial state of the apparatus[the system to be measured], M is the meter observable and U describes the unitary evolution of the composite object-apparatus system through the measurement.

For projective spin measurements, the error and the disturbance can be simplified as

$$\varepsilon(A)^2 = \langle(O_A - A)^2\rangle, \quad \eta(B)^2 = \langle(O_B - B)^2\rangle \tag{7}$$

between the observables actually measured O_A $[O_B]$ and the observables intended to be measured A $[B]$, where $\langle \cdots \rangle$ stands for the expectation value in the system state.

Here, the general results for successive 1/2-spin measurements of Pauli matrices are investigated. The operators are assigned for two 1/2-spin measurements in different directions **a** and **b**: $A = \mathbf{a} \cdot \boldsymbol{\sigma}$, $B = \mathbf{b} \cdot \boldsymbol{\sigma}$, where $\boldsymbol{\sigma} = (\sigma_x, \sigma_y, \sigma_z)^T$. The apparatus is actually supposed to perform a projective measurement along a distinct axis $\mathbf{o}_a$. The output operator O_A is then given by $O_A = \mathbf{o}_a \cdot \boldsymbol{\sigma}$. From these expressions, the error and the disturbance are expected to follow the relations

$$\varepsilon(A) = \sqrt{2 - 2\,\mathbf{a} \cdot \mathbf{o}_a}, \quad \eta(B) = \sqrt{2 - 2(\mathbf{b} \cdot \mathbf{o}_a)^2} \tag{8}$$

Both are thus independent of the input state, they are solely determined by the angle between the direction of the observable and the direction of the output operator. Due to the independence of the error and the disturbance from the input state, their product $\varepsilon(A)\eta(B)$ is simply calculated from the relative orientation of **a**, **b** and $\mathbf{o}_a$. The behavior of the error-disturbance product, appearing in Eq. (4), is depicted on the Bloch sphere for fixed as $A[B] = \sigma_x[\sigma_y]$ and $|\psi\rangle = |+z\rangle$ (see Fig. 1). In comparison, the three-terms sum of the new inequality given by Eq. 5 is plotted in Fig. 2 While the Heisenberg-term $\varepsilon(A)\eta(B)$ shows violation of the inequality in some regions, the three-terms sum appearing in Eq. 5 is always above the boundary value.

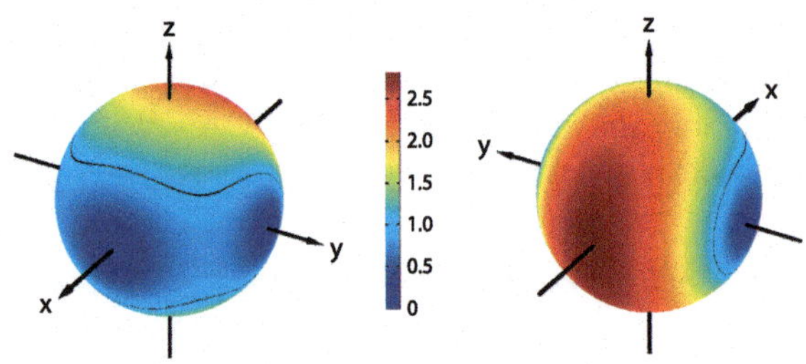

Fig. 1 The product $\varepsilon(A)\eta(B)$ of error and disturbance for $|\psi\rangle = |+z\rangle$, $A = \sigma_x$ and $B = \sigma_y$. Points on the Bloch sphere represent the direction of O_A. The bound of the inequality is indicated by the black solid line

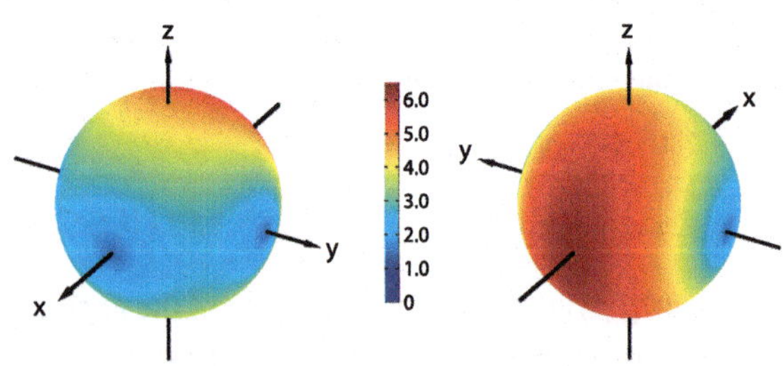

Fig. 2 The three-terms sum $\varepsilon(A)\eta(B) + \varepsilon(A)\sigma(B) + \sigma(A)\eta(B)$ for the same $|\psi\rangle$, A and B as in Fig. 1. All points on the Bloch sphere have the boundary value above $\frac{1}{2}\langle\psi|[A, B]|\psi\rangle = 1$

2.2 Experimental Scheme

Here, the validity of two forms of error-disturbance relations, Eqs. (4) and (5) are experimentally tested in neutron's successive 1/2-spin measurements. We applied so-called a three-state method: three independent states are sent to the apparatus to determine the error and the disturbance, which overcomes the operational obstacles. Note that, this three-state method is a kind of process tomograph, which resolves further extent of the characteristics of the measurement operators. For projective $\frac{1}{2}$-spin measurements, which is realized in the experiment, quadratic form of the error and the disturbance are given as the sum of expectation values of three different states:

$$\begin{aligned} \varepsilon(A)^2 &= 2 + \langle\psi|O_A|\psi\rangle + \langle\psi|AO_AA|\psi\rangle - \langle\psi|(A+\mathbb{1})O_A(A+\mathbb{1})|\psi\rangle\,, \\ \eta(B)^2 &= 2 + \langle\psi|BX_BB|\psi\rangle + \langle\psi|X_B|\psi\rangle - \langle\psi|(B+\mathbb{1})X_B(B+\mathbb{1})|\psi\rangle\,. \end{aligned} \tag{9}$$

These relations suggest that error and disturbance are solely determined by expectation values of experimentally accessible operators for various input states. For determination of the error $\varepsilon(A)$, three input sates, $|\psi\rangle$, $A|\psi\rangle$, and $(A+\mathbb{1})|\psi\rangle$ should be sent to the apparatus and expectation values are measured ; for the determination of the disturbance $\eta(B)$, other combination $|\psi\rangle$, $B|\psi\rangle$, and $(B+\mathbb{1})|\psi\rangle$ should be used.

The experimental scheme is depicted in Fig. 3. Three states are generated in a preparation stage; these are sent to a measurement apparatus M1 carrying out the O_A measurement and a second apparatus M2 performing the B measurement. The two projective measurement of O_A and B result in four possible outcomes. Four intensities ($I_{j,k}$ with $j,k=\pm$) are measured; The indices $(++)$, $(+-)$, $(-+)$, and $(--)$ of the output ports represent the direction of projections. Therefore, for instance, the expectation value of O_A in a state $|\psi\rangle$ is calculated from these four intensities at the four possible output ports to be

$$\langle\psi|O_A|\psi\rangle = \frac{(I_{++}+I_{+-})-(I_{-+}+I_{--})}{I_{++}+I_{+-}+I_{-+}+I_{--}} \tag{10}$$

Since the prior measurement of O_A modifies the measurement operator of M2 from B to X_B, the expectation values of X_B used for the determination of the disturbance

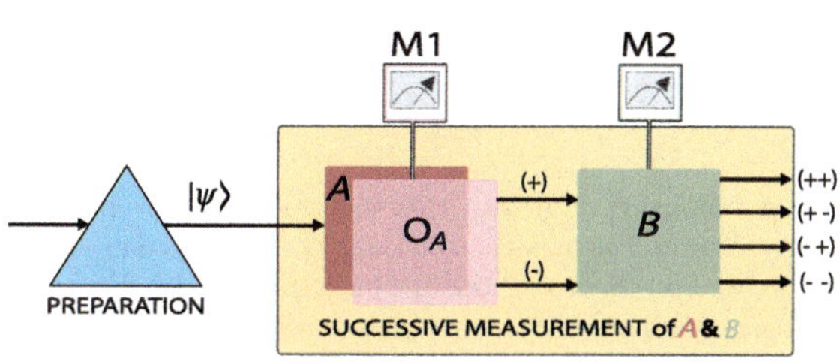

Fig. 3 Experimental scheme of the first test of the error-disturbance uncertain relation

are calculated as

$$\langle \psi | X_B | \psi \rangle = \frac{(I_{++} + I_{-+}) - (I_{+-} + I_{--})}{I_{++} + I_{+-} + I_{-+} + I_{--}} \tag{11}$$

All expectation values for the determination of error ε and disturbance η can thus be derived from the output intensities with the input states of $|\psi\rangle$, $A|\psi\rangle$, $(A + \mathbb{1})|\psi\rangle$, $B|\psi\rangle$, and $(B + \mathbb{1})|\psi\rangle$ to be sent to the joint measurement apparatuses M1 and M2. Note that, a different method for the experimental demonstration of the EDUR was proposed that exploits the weak-measurement technique [20].

The universally valid EDUR (Eq. 5) additionally contains the standard deviations of A and B in the state $|\psi\rangle$: in our measurements, these are easily calculated to be

$$\Delta(A)^2 = 1 - \langle \psi | A | \psi \rangle^2, \quad \Delta(B)^2 = 1 - \langle \psi | B | \psi \rangle^2. \tag{12}$$

That is, the standard deviation can be determined from the (single) expectation value of the measurement $A[B]$ for the input state $|\Psi\rangle$.

2.3 Experimental Results

The experiment was carried out at the research reactor facility TRIGA Mark II of the TU-Vienna. The schematic view of the experimental setup is depicted in Fig. 4. The monochromatic neutron beam is polarized through a super-mirror polarizer and two other super-mirrors are used as analyzers. The guide field together with four DC spin rotator allows state preparation and projective measurements of O_A in M1 and B in M2. Observables A and B are set as σ_x and σ_{ϕ_B} (an observable lying on the

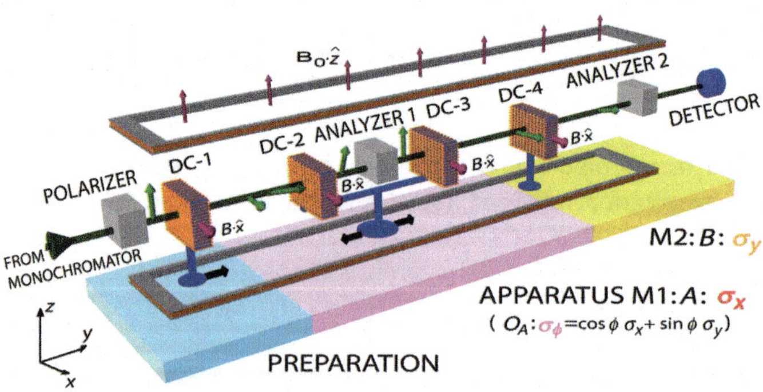

Fig. 4 Depiction of the experimental setup of the neutron optical test of error-disturbance uncertainty relation. The incident spin is set to be polarized to +z, followed by two measurements. While we set $A = \sigma_x$, $B = \sigma_y$ in the first measurement and $B = \sigma_x \cos(5\pi/6) + \sigma_y \sin(5\pi/6)$ in the second measurement. In the experiment, the observable of the first measurement is detuned by O_A

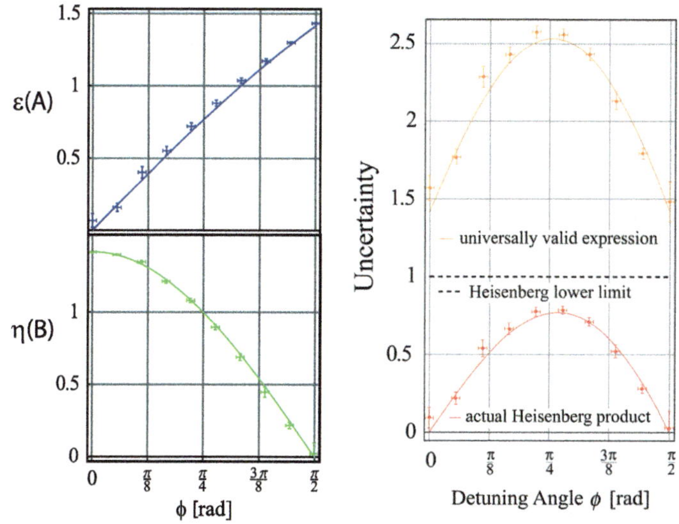

Fig. 5 Experimentally determined uncertainty. Trade-off relation between error and disturbance is clearly seen (left). Demonstration of the violation of the Heisenberg's uncertainty relation and the validity of the new relation given by Eq. 5 (right)

equator with the azimuthal angle ϕ_B of the Bloch sphere). The initial state $|\Psi\rangle$ is set to be +z spin state, $|+z\rangle$. In order to observe dependence of the error $\varepsilon(A)$ and the disturbance $\eta(B)$ on the output observable, $O_A = \sigma_x \cos\phi + \sigma_y \sin\phi$ (instead of exactly measuring $A = \sigma_x$), the apparatus M1 is designed to actually carry out measurements of adjustable observables. To test the new and old error-disturbance uncertainty relation, the error $\varepsilon(A)$, the disturbance $\eta(B)$ and the standard deviations $\Delta(A)$, $\Delta(B)$, are determined from the data. The error $\varepsilon(A)$ and the disturbance $\eta(B)$ are determined by successive projective measurements utilizing M1 and M2, while the measurements of the standard deviations $\Delta(A)$ and $\Delta(B)$ are carried out by single M1 and M2 measurements.

First, trade-off behavior of the error and the disturbance is investigated by setting the $B = \sigma_y$ and adjusting the detuning angle $\phi = [0, \frac{\pi}{2}]$. The result is depicted in Fig. 5 on the left side: clear trade-off relation between the error $\varepsilon(A)$ and the disturbance $\eta(B)$ is seen. Note that, the observable O_A coincides with A at $\phi = 0$, which results in the error of 0, and the zero disturbance at $\phi = \frac{\pi}{2}$ due to the fact that $O_A = B$. Next the validity of the new and the old EDUR is studied: the results is plotted in Fig. 5 on the right side. It is explicitly seen in this plot that, although the three-terms sum is always above the boundary value, the Heisenberg's product is below the boundary. This first experimental test of the EDUR is reported in [21].

Further, we proceed the measurements with more general parameters. Two cases of the results are shown in Fig. 6: (a) $B = \sigma_y$ and (b) $B = \sigma_x \cos(5\pi/6) + \sigma_y \sin(5\pi/6)$. The azimuthal angle of ϕ_{OA} of the output observable O_A is varied between 0 and 2π. The Heisenberg error-disturbance product $\varepsilon(A)\eta(B)$ and the three-terms

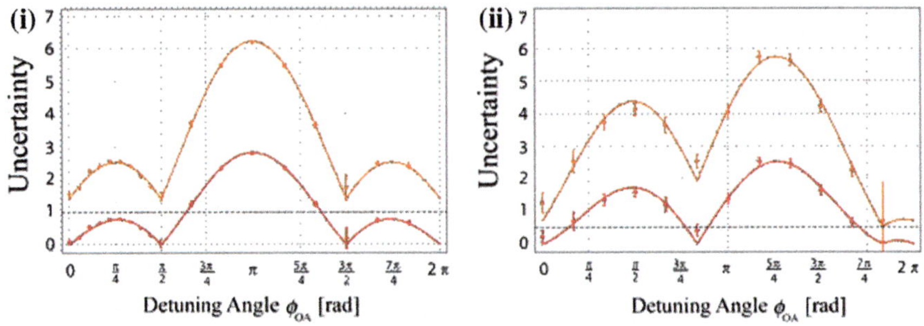

Fig. 6 Experimentally determined values of the universally valid uncertainty relation, (i) $\varepsilon(A)\eta(B) + \varepsilon(A)\Delta(B) + \Delta(A)\eta(B)$ (orange) and (ii) $\varepsilon(A)\eta(B)$ (red) as a function of the detuning angle ϕ. (a)$B = \sigma_y$ (left) and (b) $B = \sigma_x \cos(5\pi/6) + \sigma_y \sin(5\pi/6)$ (right)

sum $\varepsilon(A)\eta(B) + \varepsilon(A)\Delta(B) + \Delta(A)\eta(B)$ are plotted as a function of the detuned azimuthal angle ϕ_{OA}. These plots confirm the fact that the newly introduced three-rems sum is always larger than the boundary whereas the Heisenberg product is often below the limit. In Fig. 6 (b), the situation is observed where the universally expression actually touches the limit, which corresponds to the case where the equal sign of the inequality Eq. (5) really occurs. More detailed behavior of the error, disturbance, Heisenberg's product, and the three-terms sum is reported in our publications [22]. After the first experimental tests of the EDUR with neutrons, other investigations using photonic system appeared [23–25].

3 Experimental Studies of the Error-Disturbance Uncertainty Relation for Pure and Mixed States

3.1 Tight Relation for the Error-Disturbance

In pursuit of an improvement of EDUR, Branciard introduced a stronger inequality [26]

$$\varepsilon(A)^2 \Delta B^2 + \eta(B)^2 \Delta A^2 + 2\varepsilon(A)\eta(B)\sqrt{\Delta A^2 \Delta B^2 - C_{AB}^2} \geq C_{AB}^2. \qquad (13)$$

Experimental tests of this relation for pure input states were carried out by using photonic systems [27, 28] and neutrons [29]. An easy extension of the bound C_{AB} in the EDURs to $C'_{AB} = \frac{1}{2}|\mathrm{Tr}([A, B]\rho)|$ leads to a disappearance of uncertainty for totally mixed ensembles [30]; this extension for the mixed states turned out to be unsustainable. Next improvement of the bound was put forward by Ozawa [31]; C_{AB} in Eq. (13) can be replaced by a stronger quantity D_{AB} defined as $D_{AB} = \frac{1}{2}\mathrm{Tr}\left(|\sqrt{\rho}[A, B]\sqrt{\rho}|\right)$. This new parameter coincides with the Robertson's bound

C_{AB} when ρ is a pure state, but makes the EDUR in the form of Eq. (13) stronger for a mixed ensemble. This new relation is experimentally tested, which confirmed its validity: the experiments studied the residual character of the uncertainty for mixed ensemble [32].

For simplicity, we study here the case where spin-$\frac{1}{2}$ observables, represented by a set of Pauli operators, have been a major focus of investigations of EDURs. For binary measurements with $A^2 = B^2 = \mathbb{1}$ and $\langle A\rangle = \langle B\rangle = 0$, where $\langle\cdots\rangle$ stands for the expectation value in the system state, Eq. (13) can be strengthened to a stronger EDUR [26].

$$\hat{\varepsilon}_A^2 + \hat{\eta}_B^2 + 2\sqrt{1 - {C'_{AB}}^2}\hat{\varepsilon}_A\hat{\eta}_B \geq {C'_{AB}}^2, \tag{14}$$

where $\hat{\varepsilon}_A = \varepsilon(A)\sqrt{1-\varepsilon(A)^2/4}$ and $\hat{\eta}_B = \eta(B)\sqrt{1-\varepsilon(B)^2/4}$, Ozawa demonstrated that replacement of the bound C'_{AB} by D_{AB} improves the inequality in the binary case as well [31].

In the experiments, A and B are set σ_z and σ_y, respectively; a mixed ensemble $\rho_x(\alpha) = \frac{1}{2}(\mathbb{1} + \alpha\sigma_x)$ satisfying $\langle A\rangle = \langle B\rangle = 0$ is considered. In this case, the bound $D_{AB} = 1$ is constant and yields the tight relation [31]

$$\left(\varepsilon(A)^2 - 2\right)^2 + \left(\eta(B)^2 - 2\right)^2 \leq 4\,, \tag{15}$$

for any $\rho_x(\alpha)$ independent of the mixture of the state, while the bound C'_{AB} does depend on the parameter α.

3.2 Experimental

Based on the previous performance of the studies of the EDUR for pure states described in the previous section, we extend here the investigation by applying two procedures, i.e. the generation of mixed states and modification of the first measurement at apparatus M1 by adding correction procedure given by an unitary transformation of the output states. The former allows the study of the EDUR for mixed states and the latter enables to increase/decrease the disturbance. The polarimeter setup consists of three stages: (1) state preparation, (2) apparatus M1 performing a projective O_A measurement plus the correction procedure and (3) apparatus M2 performing the B measurement. The mixing of the state can be tuned by a noise magnetic field B_{noise}, which had been used in the former experiment [33]. In practice, $\pi/2$-rotations with noisy fields is realized by one DC-coil. In the correction stage, the influence of an unitary transformation U^{corr} on the output state $|O_A = \pm 1\rangle$ is studied. Thereby, an optimal (and anti-optimal) correction by adjusting U^{corr} is to be seen.

First, pure input states are generated and the detuning angle is fixed. Then, the eigenstate of O_A after apparatus M1 is unitarily transformed to the state $|\psi(\vartheta,\phi)\rangle = \left(\cos(\vartheta/2), e^{i\phi}\sin(\vartheta/2)\right)^T$: the minimal disturbance is realized. In con-

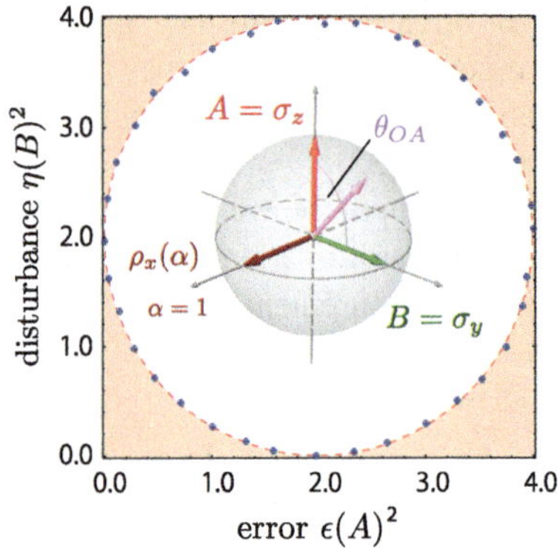

Fig. 7 Error-disturbance uncertainty relation as indicated by inequality Eq. (15)) measured with pure states. Not only the lower but also upper bounds of the disturbance are found: in squared plot, boundary is assigned on the circle

trast, the state after apparatus M1 is transformed to the orthogonal state of $|\psi(\vartheta, \phi)\rangle$, the maximum disturbance is realized. After determination of the disturbance-minimizing/maximizing unitary transformations, the EDUR given by Eq. (15) is analyzed. The experimentally determined error versus maximum and minimum disturbances together with the theoretically predicted bound are plotted in Fig. 7. The red shaded area represent the forbidden region. The lower and upper bound was measured by adjusting the detuning angle $\theta_{OA} = [0, \pi]$. At $\theta_{OA} = 0$, the error $\varepsilon(A)$ becomes 0 at which point the disturbance is unique. When $\theta_{OA} = \pi/2$ $(O_A = B)$, the disturbance reaches it's [maximum]minimum value, depending on the unitary [anti-]optimal correction transformation. When $\theta_{OA} = \pi$, $O_A = -A$ and the error is maximal and disturbance is independent of the transformation once again.

Next, the influence of the mixture of the input states is studied with applying the optimal correction procedure for minimal disturbances. The results are plotted in Fig. 8. Each plot exhibits optimal EDUR for a particular mixture with theoretical predictions by D_{AB} and C'_{AB}. It is immediately be seen that the error-disturbance uncertainty is insensitive to dephasing or amplitude damping of the input states; the bound is unchanged. The measured values always saturate inequality Eq. (15):

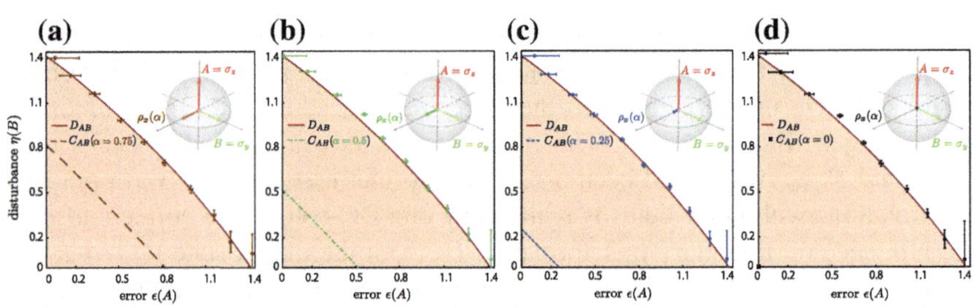

Fig. 8 Error $\varepsilon(A)$ versus disturbance $\eta(B)$ for the standard configuration $(A = \sigma_z, B = \sigma_y)$ with four different mixtures of the state $\rho_x(\alpha) = \frac{1}{2}(\mathbb{1} + \alpha\sigma_x)$: $\alpha = (a)0.75$, (b)0.5, (c)0.25 and (d)0. The red shaded areas are forbidden according to Eq. (15)

only the bound given by D_{AB} leads to saturation of the error-disturbance uncertainty relation. This statement is also true for different configurations of the observables A and B. Results for more arbitrary choices of the observables are reported in [32].

The experiment successfully confirms the validity of the tight bound D_{AB} and the non-tightness of the simply extended Robertson bound C'_{AB}. The independence of the EDUR on the mixture of the states is demonstrated for the case of dichotomic observables A, B with $\langle A \rangle = \langle B \rangle = 0$. Successive 1/2-spin measurement is implicated in the experiment, where the function of the correction procedure, i.e. realized by a unitary transformation of the output state of the first apparatus, is obviously seen in incorporation to the whole measurement. Note that, this kind of effect is unaccessible in the experiment reported in [27], due to insensibility of polarization[spin]-direction of the output state. Since the measurements for mixed states, in particular for states with lower purity, require high precision and accuracy of the instrument, we emphasize the use of the three-state tomography, which profitably satisfies these requirements.

4 Discussions

4.1 Ex-post Uncertainty Relations

The neutron's successive 1/2-spin measurement is the first experimental test of the EDUR. The validity of the new relation (Eq. (5)) proposed as a universally valid error-disturbance relation is confirmed. Furthermore the failure of the old relation as a reciprocal relation between the error and the disturbance is also demonstrated. This experiment stimulated advanced studies on the EDUR from the experimental as well as the theoretical view points. All measurements, other than ours, concern photon's polarization, which is described as a two-level quantum system in the same manner as the neutron's 1/2-spin. Now, more than 80 years after the publication of the first account of the uncertainty principle by Heisenberg, uncertainty relations have become again a hot topic in quantum physics. The uncertainty relations in terms of standard deviations is related with the state, which in turn accounts for the limitation of the precise preparation of a quantum system. In contrast, EDUR gives an explanation for the unavoidable influence of measuring instruments on quantum systems. It should be emphasized here that these two notations have physically completely different contents; two notations should not be mixed-up but be dealt individually.

In writing the paper, we actually regard it fair to say that *"our result demonstrates that the new relation solves a long-standing problem of describing the relation between measurement accuracy and disturbance, and sheds light on fundamental limitations of quantum measurements, for instance on the debate of the standard quantum limit for monitoring free-mass position"* [21]. Nevertheless, ex-post facto critical analysis appeared [34]: for instance, state-dependence of the error and the disturbance by Ozawa is claimed and state-independent definitions of error and disturbance are proposed to reconstruct the error-disturbance uncertainty relation in the same form as the original one proposed by Heisenberg [35]. For the alternatively defined error and disturbance, there are "state independent, each giving the worst-case estimates across all states". This allows overestimate of the measurement error and disturbance. This claim is immediately criticized [36, 37]: physical analysis in the former figures out the new definition as "disturbance power" and mathematical consideration in the latter reveals breakdowns of the new definition.

Furthermore, a paper is published which deal with "operational constraints" on the measures of the error and disturbance [38]. It is stated that, since *"only the change in the measurement statistics can be detected by the measurement," "a measurement cannot be treated as disturbed if its outcome statistics is identical to the one for the perfect measurement."* (underlines given by the author) Note that, the fact that this view does not accomplish its intended purpose is clearly seen in the first experimental test of EDUR: as we already emphasized as *"It is worth noting that the mean value of the observable A is correctly reproduced for any detuning angle ϕ, that is, $< +z|O_A| + z >=< +z|A| + z >$, so that the projective measurement of O_A reproduces the correct probability distribution of A, whereas we can detect the non-zero r.m.s. error $\varepsilon(A)$ for $\phi \neq 0$."* Reference [21], it is physically reasonable that the difference of the observable $O_A \neq A$ for $\phi \neq 0$, which is realized in the experiment, leads to the *error* of the measurement, even though the output statics of the measurements are identical. The author of the present paper often mentioned the case when one considers, for instance, an apparatus which (is broken and) always gives the results of the measurement as (+1) and (–1) with a fifty-fifty chance: can one regard this not a causal but an accidental coincidence as (physically) error-free? As far as *physical* consequences are concerned, causal differences, which can appear even in an operational form, are resources of the measurement error/disturbance. Modern quantum measurement schemes, i.e. process tomography or that in combination with weak-values, can actually reveal the operational difference. The *functional* differences, emerging only in the final results, can be considered as *informational* aspects of the measurements. Indeed, another form of noise-disturbance uncertainty relation in context of *information-theoretic approach* is derived [39, 40], where the correlations of the measurement results are considered as the resource. Let us clarify this treatment in more detail.

4.2 Information-Theoretic Noise-Disturbance Uncertainty Relation

It is very natural to seek a formulation of the uncertainty principle in terms of the information gained and lost due to measurement. Such a formulation was recently introduced by Buscemi et al. [39]. In their treatment, noise and disturbance are quantified not by a difference between a system observable and the quantity actually measured, but by the correlations between input states and measurement outcomes. A

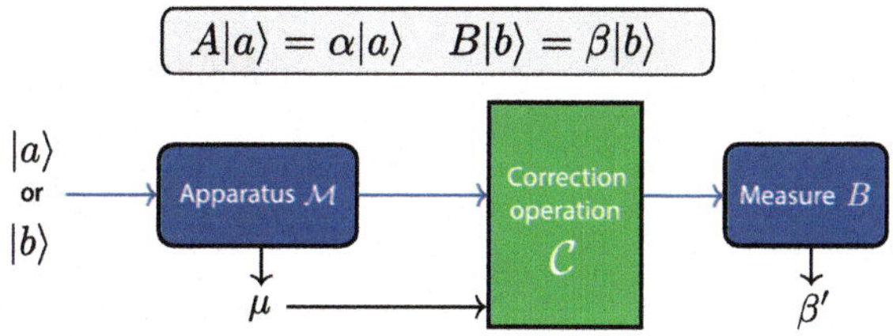

Fig. 9 The concept for determination of noise and disturbance. The information-theoretic noise $N(\mathcal{M}, A)$ and disturbance $D(\mathcal{M}, B)$ are calculated using the conditional probabilities $p(\alpha|\mu)$ and $p(\beta|\beta')$, respectively

tight uncertainty relation is derived for information-theoretic noise and disturbance, in the qubit case, and demonstrate its validity in a neutron polarimeter experiment [40].

Consider an observable A with eigenvalues α belonging to the non-degenerate eigenstates $|a\rangle$ and a measurement apparatus $\mathcal{M}$ representing a quantum instrument [41–43] with possible outcomes μ. All eigenstates $|a\rangle$ of A are now fed with equal probability into the apparatus, which is schematically illustrated in Fig. 9. The conditional probability $p(\alpha|\mu)$ for the input eigenstate $|a\rangle$ and a specific measurement outcome μ and the marginal probability $p(\mu)$ for occurrence of the specific outcome are used to define the information-theoretic noise $N(\mathcal{M}, A)$ as

$$N(\mathcal{M}, A) := -\sum_{\alpha,\mu} p(\mu)p(\alpha|\mu)\log p(\alpha|\mu) = H(\mathbb{A}|\mathbb{M}). \tag{16}$$

This equation represents the conditional entropy $H(\mathbb{A}|\mathbb{M})$, where $\mathbb{A}$ and $\mathbb{M}$ denote the classical random variables associated with input α and output μ. The information-theoretic noise thus quantifies how well the value of A can be inferred from the measurement outcome and only vanishes if an absolutely correct guess is possible.

The information-theoretic disturbance is defined in a similar manner as

$$D(\mathcal{M}, B) := -\sum_{\beta,\beta'} p(\beta')p(\beta|\beta')\log p(\beta|\beta') = H(\mathbb{B}|\mathbb{B}'). \tag{17}$$

Here uniformly distributed eigenstates $|b\rangle$ of an observable B are input to the apparatus $\mathcal{M}$, and a subsequent measurement of B is performed, with outcomes labeled by β' (see Fig. 9). The disturbance $D(\mathcal{M}, B)$ thus quantifies the correlation between the initial and final values of B, and is a measure of how much information about B is lost through the measurement $\mathcal{M}$.

In order to determine the *irreversible* loss of information about B, a correction operation C can be performed before the B-measurement to decrease the disturbance (see again Fig. 9), and consists of any completely positive, trace preserving map. Two cases are shown here; one is the uncorrected disturbance which we write as D_0 and the other is the optimally corrected disturbance denoted as D_{opt} corresponding to the

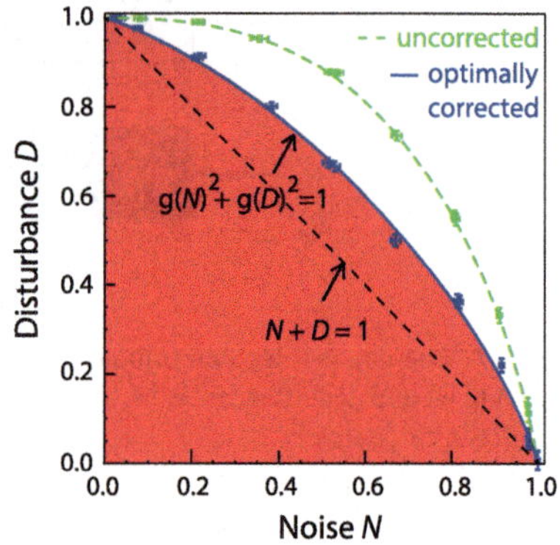

Fig. 10 Disturbance versus Noise with and without optimal correction procedure. The red shaded area marks the region which are prohibited according to Eq. (19)

correction operation that minimizes the disturbance. For any correction procedure the information theoretic noise and disturbance fulfil the following uncertainty relation [39]

$$N(\mathcal{M}, A) + D(\mathcal{M}, B) \geq c_{AB} := -\log \max |\langle a|b\rangle|^2, \tag{18}$$

where $|a\rangle$ and $|b\rangle$ denote the eigenstates of the observables A and B. For maximally incompatible qubit observables, represented by the Pauli matrices σ_z and σ_y, we have been able to significantly strengthen this relation to the tight relation

$$g[N(\mathcal{M}, \sigma_z)]^2 + g[D(\mathcal{M}, \sigma_y)]^2 \leq 1. \tag{19}$$

Here $g[x]$ denotes the inverse of the function $h(x)$ on the interval $x \in [0, 1]$ given by

$$h(x) := -\frac{1+x}{2}\log\frac{1+x}{2} - \frac{1-x}{2}\log\frac{1-x}{2}. \tag{20}$$

As the study of the uncertainty relations given by Eqs. (18) and (19) the noise and the disturbance determined in the experiments are plotted in Fig. 10. One immediately sees that the noise-disturbance uncertainty relation Eq. (18) is always fulfilled, but not saturated apart from extremal values, i.e. either N or D equal to 0. In contrast, the improved relation Eq. (19) valid for qubits provides a tight bound and is saturated if the optimal correction procedure is applied.

In this experiment, the validity of the information-theoretic formulation of noise-disturbance uncertainty relation is confirmed in the case of qubit measurements. Trade-off relation is again seen in this formulation: in particular, a completely reciprocal trade-off relation is observed for maximally incompatible combination of 1/2-spin observables. The obtained result is expected to stimulate the search for improved entropic uncertainty relations for observables of higher dimensional Hilbert spaces. Note that, in the scenario used in this experiment, content of the outcome statics is characterized: functional character of the measurement accounts for the uncertainty here. Two different considerations of the uncertainty are described in this article: one concerns error-disturbance and the other does noise-disturbance, of the uncertainty in a quantum measurement. The former characterize particularly the operational consequence of the quantum measurement and the latter does the functional. We hope, experimental investigations described here helps to clarify the physical contents of each uncertainty.

5 Concluding Remarks

Neutron optical investigations of the uncertainty relation as a fundamental limitation of quantum mechanics are presented here. Neutron's $1/2$-spin is exploited: high-reliable demonstrations are accomplished owing to manipulation and measurement of the spin with high efficiency and precision. No other experimental studies have been reported before the first experimental test of the error-disturbance uncertainty relation presented here. The experiment has provided the first evidence for the invalidity of the old (à la Heisenberg) and validity of the new (by Ozawa) uncertainty relation for measurements. Although, we thought, our result clarifies a long standing problem of describing the relation between measurement accuracy and disturbance, and sheds light on fundamental limitations of quantum measurements, there appeared not only the extension but also counter-arguments, some of which we mention and give a critique here. It would be fair to say that the experiment of ours opens up a new era of the study of uncertainty relations: our experiment activates this research field. More than four-decades after the Heisenberg's original publication, uncertainty in quantum mechanics is covered with new aroma and flavor and taken up again for hot discussions both from the theoretical and experimental point of view.

Acknowledgements It is great pleasure and honor of the author to present a review article on the recent experimental investigations of the uncertainty relation with neutrons in this issue of the journal. This research has been done in close cooperation between neutron optics group in Vienna and Prof. Masanao Ozawa. The author thanks all colleagues who were involved in carrying out the experiments presented here; in particular, we appreciate M. Ozawa, F. Busemi, B. Demirel, J. Erhart, M.J.W. Home, A. Hosoya, G. Sulyok and S. Sponar. This work was partially supported by the Austrian FWF (Fonds zur Föderung der Wissenschaftlichen Forschung) through grant numbers P27666-N20. Some parts of the results presented here have appeared in earlier publications.

References

1. J.A. Wheeler and W.H. Zurek (eds), *Quantum Theory and Measurement*, Princeton Univ. Press, (1983).
2. W. Heisenberg, Z. Phys. **43**, 172 (1927).
3. W. Heisenberg, *The Physical Principles of Quantum Mechanics* (University of Chicago Press, Chicago, IL, 1930).
4. E.H. Kennard, Z. Phys. **44**, 326 (1927).
5. H. P. Robertson, Phys. Rev. **34**, 163 (1929).

6. H.P. Robertson, *Sitzungsberichte der Preussischen Akademie der Wissenschaften* **14**, 296 (1930); *ibid.* Rev. Mod. Phys. **42**, 358 (1970).
7. L.E. Ballentine, *Quantum Mechanics*
8. I. Bialiniki-Birula and J. Mycielsky, Commun. Math. Phys.**44**, 129 (1975)
9. D. Deutsch, Phys. Rev. Lett. **50**,631 (1983).
10. K. Kraus, Phys. Rev. D **35**, 3070 (1987).
11. H. Maassen and J. B. M. Uffink, Phys. Rev. Lett. **60**, 1103 (1988)
12. E. Arthurs, and J.L. Kelly, Bell Syst. Tech. J. **44**, 725 (1965).
13. E. Arthurs, and M.S. Goodman, Phys. Rev. Lett. **60**, 2447 (1988).
14. S. Ishikawa, Rep. Math. Phys. **29**, 257 (1991).
15. M. Ozawa, (eds. C. Bendjaballah, O. Hirota, and S. Reynaud) (Lecture Notes in Physics, Vol. 378, 3, Springer, 1991).
16. M. Ozawa, Phys. Rev. A **67**, 042105 (2003).
17. M. Ozawa, Phys. Lett. A **318**, 21 (2003).
18. M. Ozawa, Ann. Phys. **311**, 350 (2004).
19. M. Ozawa, J. Opt. B **7**, S672 (2005).
20. A.P. Lund and H.M. Wiseman, New J. Phys. **12**, 093011 (2010).
21. J. Erhart, S. Sponar, G. Sulyok, G. Badurek, M. Ozawa and Y. Hasegawa, Nature Phys. **8**, 185 (2012).
22. G. Sulyok, S. Sponar, J. Erhart, G. Badurek, M. Ozawa and Y. Hasegawa, Phys. Rev. A **88**, 022110 (2013).
23. L.A. Rozema, A. Darabi, D.H. Mahler, A. Hayat, Y. Soudagar, and A.M. Steinberg, Phys. Rev. Lett. **109**, 100404 (2012).
24. S.Y. Baek, F. Kaneda, M. Ozawa, and K. Edamatsu, Sci. Rep. **3**, 2221 (2013).
25. M.M. Weston, M.J.W. Hall, M.S. Palsson, H.M. Wiseman, and G.J. Pryde, Phys. Rev. Lett. **110**, 220402 (2013).
26. C. Branciard, Proc. Natl. Acad. Sci. U.S.A. **110**, 6742 (2013).
27. M. Ringbauer, D.N. Biggerstaff, M.A. Broome, A. Fedrizzi, C. Branciard, and A.G. White, Phys. Rev. Lett. **112**, 020401 (2014).
28. F. Kaneda, S.Y. Baek, M. Ozawa, and K. Edamatsu, Phys. Rev. Lett. **112**, 020402 (2014).
29. S. Sponar, G. Sulyok, J. Erhart, and Y. Hasegawa, Adv. High Energy Phys. **44**, 36 (2015).
30. C. Branciard, Phys. Rev. A **89**, 022124 (2014).
31. M. Ozawa, arXiv:1404.3388v1 [quant-ph] (2014).
32. B. Demirel, S. Sponar, G. Sulyok, M. Ozawa, and Y. Hasegawa, Phys. Rev. Lett. **117**, 140402, (2016). https://doi.org/10.1103/PhysRevLett.117.140402.
33. For instance, J. Klepp, S. Sponar, S. Filipp, M. Lettner, G. Badurek and Y. Hasegawa, Phys. Rev. Lett. **101**, 150404 (2008).
34. for instance, see references in Lu X.-M. , Yu S., Fujikawa K., and Oh C. H., Phys. Rev. A **90**, 042113 (2014).
35. P. Busch, P. Lahti, and R.F. Werner, Phys. Rev. Lett. **111**, 160405 (2013).
36. L.A. Rozema, D.H. Mahler, A. Hayat, and A.M. Steinberg, arXiv:1307.3604 (2013).
37. M. Ozawa, arXiv:1308.3540 (2013).
38. K. Korzekwa, D. Jennings, and T. Rudolph, Phys. Rev. A **89**, 052108 (2014).
39. F. Buscemi, M.J.W. Hall, M. Ozawa, and M.M. Wilde, Phys. Rev. Lett. **112**, 050401 (2014).
40. G. Sulyok, S. Sponar, B. Demirel, F. Busemi, M.J.W. Hall, M. Ozawa, and Y. Hasegawa, Phys. Rev. Lett. **115** 030401 (2015).
41. E. B. Davies and J. T. Lewis, Communications in Mathematical Physics **17**, 239 (1970).
42. E. B. Davies, *Quantum theory of open systems* (Academic Press London, New York).
43. M. Ozawa, J. Math. Phys. **25**, 79 (1984).

Obituary for a Flea

Jasper van Heugten and Sander Wolters

Abstract The Landsman–Reuvers proposal to solve the measurement problem from within quantum theory is extensively analysed. In favor of proposals of this kind, it is shown that the standard reasoning behind objections to solving the measurement problem from within quantum theory rely on counterfactual reasoning or mathematical idealisations. Subsequently, a list of objections/challenges to the proposal are made. Part of these objections are equally important for all attempts at solving the measurement problem, such as the problem of interpreting small numbers in the density matrix, the problem of reproducing the Born rule, the use of pure states as a tool to alleviate the interpretational issues of quantum states, and the necessity of introducing classical certainties which are not strictly present in quantum theory. The additional objections that are particular to the proposal, such as the physical interpretation/origin of the flea perturbation, the use of potentials to solve a dynamical problem, slow collapse times, the inability to handle unequal probabilities, and the dictatorial role of the flea perturbation, lead us to believe that the Landsman–Reuvers proposal is lacking in both physical grounding and theoretical promise. Finally, an overview is given of the challenges that were encountered in this attempt to solve the measurement problem from within quantum theory.

Keywords Measurement problem · Quantum measurement · Bohrification

The research in this chapter is part of the project Experimental Tests of Quantum Reality, funded by the Templeton World Charity Foundation.

J. van Heugten · S. Wolters (✉)
Institute for Mathematics, Astrophysics and Particle Physics, Radboud Universiteit Nijmegen, Heyendaalseweg 135, 6525AJ Nijmegen, The Netherlands
e-mail: sretlowrednas@gmail.com

J. van Heugten
e-mail: jvanheugten@gmail.com

M. Ozawa et al. (eds.), *Reality and Measurement in Algebraic Quantum Theory*, Springer Proceedings in Mathematics & Statistics 261,
https://doi.org/10.1007/978-981-13-2487-1_13

1 Introduction

In 2013, Landsman, and his student Reuvers, proposed a reformulation of the measurement problem of quantum mechanics [13]. Using the sensitivity of bound states with respect to small perturbations in the semi-classical limit, it was claimed that this reformulation may hold the possibility of resolving the measurement problem within the formalism of quantum theory. We shall refer to this idea and the (simple) mathematical model expressing this idea as the *flea model* or *flea approach*. Later [10], Landsman embedded his treatment of the measurement problem within a larger framework regarding, but not limited to, the relations between classical and quantum physics. Following the contribution of Landsman and Lindenhovius in this volume, we refer to this framework as *asymptotic Bohrification*.

Our goals are twofold:

1. To investigate the feasibility of the flea model for dynamical collapse.
2. To analyse to what extent asymptotic Bohrification captures the measurement problem.

Apart from this introduction, the final discussion and the appendices, the text is divided into two parts. Each part deals with one of the two goals. The introductory section only contains a single subsection, where a rather general formulation of the measurement problem is recalled. This subsection contains no new insights, but helps us to provide something to compare the measurement problem of the flea approach to.

In Sect. 2, we consider the reformulation of the measurement problem within asymptotic Bohrification. Section 2.1 introduces the measurement problem and highlights some of the underlying ideas. The asymptotic Bohrification formulation of the measurement problem requires a more liberal notion of outcome state for measurements. The lack of empirical grounding for such states is briefly treated in Sect. 2.2. Next, in Sect. 2.3, we consider the pressing problem of obtaining the correct Born probabilities for the outcomes in the flea model. The main problem here is that getting the Born probabilities correct may require us to abandon independence between the initial system state and the measurement apparatus. Finally, in Sect. 2.4 we attempt to relate the flea model to a conventional version of the measurement problem. The flea approach is incomplete in the sense that it is silent about setting up system-pointer correlations and gives no physical background on the flea as well as the wave function which is to be collapsed.

Subsequently, in Sect. 3 we consider the effectiveness of the collapse mechanism of the flea model. In Sect. 3.1 we consider the sizable obstables in generalising the collapse mechanism of the flea model to more outcomes, and unequal Born probabilities. In Sect. 3.2 we investigate the time scale of the collapse. There it turns out that the collapse of the wave function takes an unrealistically long time.

We conclude regretfully that as a model for dynamical collapse the flea model as it now stands is not feasible. Although the underlying idea looks good statically, there is no reason to believe that it would work dynamically. But even if the collapse were effective, it is not clear what this would say about measurements. The model is silent on the system-pointer correlation and the physics behind the flea and the potential, and seems far removed from existing models of quantum measurement. And even if a working model of quantum measurement, based on the flea model, could be found, the problem of replicating the Born probabilities may still require the addition of some sort of conspiracy theory or superdeterminism. If this is the case, then it is not clear what more the flea model has to offer, with respect to conventional formulations of the measurement problem.

1.1 The Measurement Problem

We briefly consider (a simple version of) the measurement problem as formulated in [2]. This formulation is of interest because it uses a sufficiently liberal notion of outcome, and because it does not rely on idealisations, making it relevant to the flea model. We ask to what extent there is still room for the flea model to circumvent the measurement problem posed here.

Consider a two-level system $\mathscr{S}$. Let $|+\rangle$ and $|-\rangle$ be an orthonormal basis of the Hilbert space $\mathscr{H}_{\mathscr{S}}$. The post-measurement states are written in the form $|A, \alpha\rangle$. The label A denotes a value of a certain macroscopic variable, which we call the pointer variable. With respect to the state $|A, \alpha\rangle$, the pointer is assigned the value A. The second index, α, refers to all other degrees of freedom we deem relevant. These might belong to the system $\mathscr{S}$, or be environmental, and may even include the whole universe (apart from the pointer variable).

Let V_A denote the set of all states $|A, \alpha\rangle$, for which we agree that it is sensible to say that the pointer variable has value A. Suppose that A and B are two macroscopically distinguishable values for the pointer variable. We shall not assume that the sets V_A and V_B are closed linear subspaces of the relevant Hilbert space, or, even stronger, that there are associated projections which are orthogonal. We instead consider a weaker relation between the elements of V_A and V_B. This weaker relation allows for the possibility of states in which the pointer has an outcome, but which display some form of tail with respect to the pointer variable. Recall that if $|\Phi\rangle$ and $|\Psi\rangle$ are orthogonal vectors, then $\||\Phi\rangle - |\psi\rangle\|^2 = 2$. We assume that elements of V_A and V_B are close to orthogonal in the sense that there exists a number $\eta \ll 1$ such that for all $|A, \alpha\rangle$ in V_A and all $|B, \beta\rangle$ in V_B we have,

$$\||A, \alpha\rangle - |B, \beta\rangle\| \geq \sqrt{2} - \eta. \tag{1}$$

Alternatively, this assumption follows from assuming that for macroscopically distinguishable pointer values A and B, there exists a positive number $\varepsilon \ll 1$, such that the transition probability satisfies

$$p(|A, \alpha\rangle \mid |B, \beta\rangle) = |\langle A, \alpha|B, \beta\rangle| \leq \varepsilon, \tag{2}$$

for all states in V_A paired with states in V_B. The initial pre-measurement state is assumed to be factorised

$$(a|+\rangle + b|-\rangle) \otimes |\Phi\rangle, \tag{3}$$

where $|\Phi\rangle$ is an appropriate initial state of the measurement apparatus and the environment. We do not need to further specify this state. The post-measurement state is not assumed to be factorised. Time evolution is assumed to be given by a unitary transformation, and is in particular linear. An initial state of the form $|+\rangle \otimes |\Phi\rangle$ evolves to a state $|P, \alpha\rangle$, with pointer value P. We assume that initial states of the form $|-\rangle \otimes |\Phi\rangle$ evolve into states $|M, \beta\rangle$ for which we attribute the value M to the pointer.

The measurement problem arises when the initial system state is taken to be

$$|\psi\rangle = \frac{1}{\sqrt{2}} (|+\rangle + |-\rangle) . \tag{4}$$

By linearity the measurement interaction yields the evolved state

$$|\psi\rangle \otimes |\Phi\rangle \mapsto U(t_0, t_f)|\psi\rangle \otimes |\Phi\rangle = \frac{1}{\sqrt{2}} (|P, \alpha\rangle + |M, \beta\rangle) . \tag{5}$$

We assume that the pointer values M and P are macroscopically distinguishable. As a result $|P, \alpha\rangle \in V_P$ and $|M, \beta\rangle \in V_M$ are almost orthogonal in the sense of the inequality (1). But from (5) we deduce that with respect to the norm metric, the distance of the evolved state to both V_P and V_M must at least be large enough that the post-measurement state is forbidden to be in V_P, and forbidden to be in V_M. As a consequence, the post-measurement state is neither an element of V_P, nor an element of V_M, and hence we are unable to assign any outcome to the evolved state.

This formulation of the measurement problem does not seem to depend on any idealisation. In anticipation of later sections, the usage of a *pure* initial state may itself be seen as an idealisation, but this has no impact on the issue at hand. The use of the same unitary operator for different runs of the experiment (in particular, when considering different initial states $|\phi\rangle \otimes |\Phi\rangle$ of the measurement, where $\phi \in \{+, -, \psi\}$) may be viewed as an idealisation. In fact, for the flea model, we shall see that different runs of an experiment use slightly different hamiltonians. In a similar vein, van Wezel [21] adds a small non-hermitian term to the hamiltonian in order to create a dynamical collapse. Regardless of the origin of the variation, we assume that (possibly after tracing out many environmental degrees of freedom) the different unitary operators $U(t_0, t_f)$ are sufficiently close in operator norm with respect to each other. Under this assumption, the previous conclusion that the post-measurement state is neither in V_P, nor in V_M, remains valid.

As the flea model aims to solve the measurement problem from within quantum mechanics. the above formulation of the measurement problem seems to provide a serious obstacle for the flea model. Before considering the reformulation of the measurement problem for the flea mechanism, we consider whether there are still some loopholes left. There is at least one loophole, but it comes at a hefty price. We assumed the initial state to be factorised between system and the rest, as $|\phi\rangle \otimes |\Phi\rangle$. For different runs we assumed that $|\Phi\rangle$ was the same each time. Effectively, we were assuming that the system and apparatus states are independent. This independence is needed to derive the measurement problem. However, exploiting this independence loophole immediately puts us on thin ice, as it questions the very notion of measurement.

2 Bohrification and the Measurement Problem

The flea model provides a new formulation for the measurement problem, as part of a programme called asymptotic Bohrification, as well as a collapse mechanism for the wave function within this formulation. In this section we concentrate on the reformulation of the measurement problem, whereas the next section focuses on the collapse mechanism. In Sect. 2.1 we introduce asymptotic Bohrification, wherein lies the formulation of the measurement problem which the flea model aims to solve. This formulation of the measurement problem requires us to rely on an approximate version of outcome states, unfortunately without providing physical grounding of how to understand such states, as discussed in Sect. 2.2. However, such approximate outcome states are typical of approaches to the measurement problem which take the formalism of quantum mechanics seriously, as elaborated in Appendix 6 and should not (in itself) be seen as a weakness of the approach. Next, in Sect. 2.3 we briefly consider the pressing problem of replicating the Born rule in the flea model. Finally, in Sect. 2.4 we look at the connection between the asymptotic Bohrification formulation of the measurement problem, and more conventional formulations of this problem, such as in Sect. 1.1.

2.1 Asymptotic Bohrification

Consider the following incarnation of the measurement problem, adapted from [15]. According to Maudlin, the measurement problem is the incompatibility of the following three assumptions:

1. Quantum mechanical pure states are complete in the sense that they specify all physical properties of a system.
2. Time-evolution of the states is described by a linear unitary operator.
3. Measurements always (or at least usually) have single outcomes, i.e. at the end of the measurement, the measuring device indicates a definite physical state.

Indeed, in the previous section we saw this incompatibility formulated in a general setting, without any apparent reliance on idealisations. The only loophole mentioned in that section was the independence implicit in the form of the initial state, factorised between system and environment. Note that this assumption is not listed by Maudlin, in particular since denial of this assumption challenges the very notion of measurement.

The various approaches to the measurement problem differ in which of the above three assumptions are challenged. In hidden variable models the first assumption is denied. For dynamical collapse theories such as the GRW models, the second assumption is rejected. For the many-worlds interpretations the third assumption gets a new perspective. But how would the practitioner of the Copenhagen interpretation consider this problem? He (or she) would most likely frown at the first assumption. Properties are classical concepts and have no place in quantum theory. For now, let us ignore this issue. At this point, the Copenhagenist may claim that there is no measurement problem, in the sense of a contradiction. Indeed. assumptions (1) and (2) are about the quantum mechanical formalism whereas assumption (3) deals with the purely classical notions of outcomes and measurements. If we are to connect these assumptions, then we need to use classical approximations at some point of the description. It is exactly because of the need of these approximations that the irreducible probabilities of quantum theory arise. There is nothing mysterious about these probabilities, in the sense that in a purely quantum mechanical description probabilities need not arise. However, aside from putting us in the awkward position that a theory which supposedly generalises classical physics, also needs classical physics for its formulation, the Copenhagen move teaches us little, if anything, about measurements in quantum theory.

And so we ask: why would we *need* classical approximations in the first place? Consider Bohr's doctrine of classical concepts:

> However far the phenomena transcend the scope of classical physical explanation, the account of all evidence must be expressed in classical terms. (?) The argument is simply that by the word experiment we refer to a situation where we can tell others what we have done and what we have learned and that, therefore, the account of the experimental arrangements and of the results of the observations must be expressed in unambiguous language with suitable application of the terminology of classical physics.

Although this strikes us as common sense, there are different ways in which we can implement this philosophy with respect to the measurement problem. Note that if we accept the need to use the *language* of, and experience from, classical physics, this does not automatically entail that we need to replace approximations from the formalism of classical mechanics in studying the measurement problem. Just to be clear, we do not see wisdom in attempting to describe any realistic measurement apparatus (and environment) completely at the level of the standard model, and to use this as a model of measurement. However, we could endorse Bohr's doctrine of classical concepts, without making the jump to the rest of the Copenhagen interpretation. That is we could take the stance that the language of classical physics is essential to the measurement problem, but, unlike the Copenhagen interpretation not a priori assuming that the occurrence of an outcome cannot be explained by using a suitable quantum-mechanical model.

This brings us to the *Bohrification* approach to the measurement problem, proposed by Landsman. In this approach, Bohr's doctrine of classical concepts is given the following mathematical interpretation: study non-commutative C*-algebras, such as the algebra of all bounded operators on a Hilbert space, by means of commutative C*-algebras. There are two different ways in which this is done. The first approach, called *exact Bohrification* replaces a non-commutative C*-algebra by its partially ordered set of commutative C*-subalgebras, where the order is given by inclusion. Exact Bohrification is the central theme of the contribution of Landsman and Lindenhovius to this volume. Since this approach has not been applied to the measurement problem, we shall not consider it any further. In the second approach, called *asymptotic Bohrification*, the commutative and non-commutative C*-algebras, no longer related by an inclusion relation, are glued together in a bundle called a countinuous field of C*-algebras. Rather than discuss this approach in full generality, we first consider the motivating example of the flea approach to the measurement problem, as introduced by Landsman and his student Reuvers in [13], and later embedded by Landsman in the asymptotic Bohrification programme in [10]. The terminology which we adopt here was introduced by Landsman in [12].

Consider the following simplistic model used to reformulate the measurement problem. The relevant Hilbert space is $\mathscr{H} = L^2(\mathbb{R})$. The dynamics is generated by the hamiltonian

$$\hat{H}_\hbar = -\frac{\hbar^2}{2m}\frac{d^2}{dx^2} + \frac{\lambda}{8}\left(x^2 - a^2\right)^2 =: \frac{\hat{p}^2}{2m} + V(x), \tag{6}$$

using a symmetric double well potential $V(x)$, with barrier height $V_b = \lambda a^4/8$. The subscript $\hbar$ was added because we consider different values of $\hbar$, and are in particular interested in the limit $\hbar \to 0$. In this model, the value of $\hbar$ is used to represent a scale for macroscopicity, where smaller values of $\hbar$ correspond to more macroscopic situations. The reader who is uncomfortable about varying a constant of nature may instead consider the limit $\lambda \to \infty$ where the potential energy becomes steeper, or the limit $m \to \infty$.

The state for the model on which we initially focus is the (non-degenerate) ground state ψ_0 for this hamiltonian. The characteristic length of the problem, determining the scale on which the wavefunctions vary, is $l = (\hbar^2/mk)^{1/4}$. Here $k = \lambda a^2$ is the spring constant as determined by the quadratic approximation at the well minima $x = \pm a$. Consider the case with fixed λ, a but variable $\xi = \sqrt{\hbar^2/m}$. We are interested in the setting where $l \ll a$, i.e., large mass or small $\hbar$, where the lowest energy eigenstates are localised within the two wells, see Fig. 1.

The ground state ψ_0 is to represent the state of a pointer or some other macroscopic variable at some point of time during the measurement interaction. Supposedly, there is a system $\mathscr{S}$ which was initially in a Schrödinger cat state. Through its interaction with the measurement apparatus, this Schrödinger cat state was passed on to the pointer. The two wells in the potential correspond to two macroscopically distinct values that the pointer variable may take. The x-variable need not correspond to a physical distance; we only identify the two wells with two pointer values. The

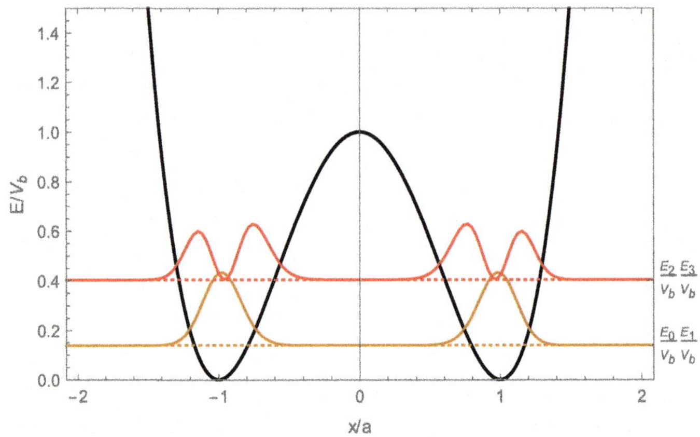

Fig. 1 The first four eigenfunctions as $|\psi(x)|^2$ for the double well potential with $l \ll a$. The first and second, and the third and fourth, wavefunctions overlap due to the symmetry

system $\mathscr{S}$ itself is not visible in this simple model. Neither does the model include an environment $\mathscr{E}$, which contains the uncontrollable degrees of freedom of the apparatus and any other degrees of freedom we may consider to be of interest. The environment plays an important conceptual role in the flea approach to the measurement problem, but the environmental degrees of freedom do not themselves appear in the model.

It may strike the reader as odd to use a bound state, the textbook example of a stable state, as the of the model to focus on. But keep in mind that this state of the model is not the initial state of the measurement. The initial apparatus state may very well have been metastable, as is typical for models of quantum measurement. In addition, we shall see that the flea approach uses a time-dependent hamiltonian, so the initial state does not remain a bound state. Regardless, we may still be sceptical whether the setting of a bound state sporting a macroscopic superposition actually occurs during *any* measurement. But we postpone further discussion with regard to the justification of the model until the end of this section.

At this point we can discuss the reformulation of the measurement problem. The ground state $\psi_0(x)$ has a $\mathbb{Z}_2$-symmetry, through reflection in the y-axis, a symmetry inherited from the invariance of $\hat{H}_\hbar$ under $x \mapsto -x$. Without any further additions or modifications (such as the flea) the wave function does not change over time and retains this symmetry, even when it becomes the post-measurement state. This happens for all non-zero values of $\hbar$. In the limit $\hbar \to 0$, the state - which we now think of as a post-measurement state - $\psi_0(x)$ converges, in a sense we will make precise mathematically in a moment, to the following classical mixed state

$$\rho_0^{(0)} = \frac{1}{2}\left(\rho_0^+ + \rho_0^-\right), \tag{7}$$

where the two phase space points

$$\rho_0^{\pm} = (p = 0, q = \pm a)$$

are the two classical ground states of the (classical) hamiltonian

$$h_0(p, q) = \frac{p^2}{2m} + \frac{\lambda}{8}\left(x^2 - a^2\right)^2. \tag{8}$$

From the point of view of asymptotic Bohrification, the measurement problem is that the quantum-mechanical post-measurement pure state converges to a classical mixed state. Or, alternatively stated, it is not the case that for sufficiently large mass m or small $\hbar$, the post-measurement state approximates a classical pure state.

So from this perspective the measurement problem can be seen as the incompatibility of the following three assumptions:

1. Measurements and their outcomes are notions from classical physics.
2. In many cases of interest, the transition from quantum physics to classical physics can be described by a limit such as $\hbar \to 0$ (or, to be briefly considered at the end of the following section, $N \to \infty$, where N is the number of degrees of freedom in the model).
3. Whenever such a limit is applicable, any physical effect in classical physics must be foreshadowed in quantum physics.

Let us consider the third assumption, which is vital. Even though the notion of outcome only has a meaning in the limiting classical theory, the classical limit itself is an idealisation and any phenomenon cannot be counted as genuinely physical if it only appears in this idealised limitting theory. The philosophy adopted in asymptotic Bohrification, telling us that outcomes should have approximate quantum-mechanical counterparts, is partly captured by *Earman's principle* [5]:

> While idealizations are useful and, perhaps, even essential to progress in physics, a sound principle of interpretation would seem to be that no effect can be counted as a genuine physical effect if it disappears when the idealizations are removed.

The rest is captured by *Butterfield's Principle* [3]:

> there is a weaker, yet still vivid, novel and robust behaviour that occurs before we get to the limit, i.e. for finite N. And it is this weaker behaviour which is physically real.

Thus the measurement problem, as it arises in the simple double well model, amounts to the problem that for small non-zero values of $\hbar$, the post-measurement state does not *approximate* one of the classical pure states $(p = 0, q = \pm a)$, which we identify as outcomes in this model. Mathematically, the term *approximate* has a precise meaning in asymptotic Bohrification, expressed in the language of continuous fields of C*-algebras. The definition can be found in Appendix 5.

It is crucial to understand the way in which the post-measurement state should approximate an outcome. By assumption, the very notion of outcome is a classical one. Yet, in following Butterfield's principle we should concentrate on quantum mechanics for small values of $\hbar$ (or in another suitable limit). It is the notion of convergence that tells us in which way the quantum mechanical states should be close to classical outcomes. Therefore, in addition to a precise mathematical formulation, we need a physical grounding for convergence of states. The following quote, taken from [11] on p. 9, should help in providing insight into the physical grounding of convergence of states, as it explains the way that the doctrine of classical concepts is understood in the operator algebraic setting of asymptotic Bohrification. The quantisation map $Q_\hbar$ used in this quote is defined in Appendix 5.

> The map $Q_\hbar$ is the quantization map at value $\hbar$ of Planck's constant; we feel it is the most precise formulation of Heisenberg's original *Umdeutung* of classical observables known to date. It has the same interpretation as the heuristic symbol $Q_\hbar$ used so far: the operator $Q_\hbar(f)$ is the quantum-mechanical observable whose classical counterpart is f.

Thus, Classical physics enters the measurement problem through the theory of quantization rather than through classical approximations, as used in the Copenhagen interpretation. The Umdeutung of the quotation is central to asymptotic Bohrification. Because of this, one might think that when a salesman, selling asymptotic Bohrification, is at the door, Heisenberg, rather than Bohr, is the one who knocks.

Although this gives some physical underpinning for the quantum mechanical approximate outcomes, we will argue in what follows that this understanding is incomplete, at least for solving the measurement problem.

To make the discussion more concrete, briefly consider the flea proposal [13], intended to alleviate the measurement problem. The time-independent hamiltonian of (6), which, from now on, we denote as $\hat{H}_{\hbar,0}$ is replaced by a time-dependent one $\hat{H}_\hbar(t) = \hat{H}_{\hbar,0} + f(t)\delta V$, where δV is a small perturbation of the potential, localised in one of the two wells, and $f(t)$ is some scalar function changing the hamiltonian from $\hat{H}_{\hbar,0}$ initially, to $\hat{H}_{\hbar,0} + \delta V$. The motivation for this move can be found in the work of Jona-Lasinio et al. [8] regarding the sensitivity of ground states with respect to small perturbations in the semi-classical setting. Provided that $\hbar$ is taken to be small enough, even for a minute perturbation, the ground state of $\hat{H}_{\hbar,0} + \delta V$ is highly concentrated in one of the two wells, as in Fig. 2.

More physically, there exists a scale for the size of the flea perturbation, dependent on $\xi = \sqrt{\hbar^2/m}$ and the shape of the flea, such that if the flea is larger than this scale it causes the groundstate to be almost completely localized in a single well, as shown in Fig. 3. This scale can be used to verify the static prediction of the flea model, namely that a small symmetry-breaking perturbation of the potential can cause the groundstate to be localized, by performing many experiments determining the groundstate of a double well for several sizes of a flea perturbation. Note that for larger masses m (smaller ξ), we need a smaller flea to localize the wavefunction!

This result of the (static) flea model thus begs to be applied to the (dynamical) measurement problem. The idea is that for a suitable dynamical introduction of the flea, the wave function, initially expressing a Schrödinger cat state, could evolve to the ground state with the perturbed potential resulting in a post-measurement state which is highly concentrated in one of the two wells. In Sect. 3 we explore the feasibility of such an approach. For the remainder of this section, however, we

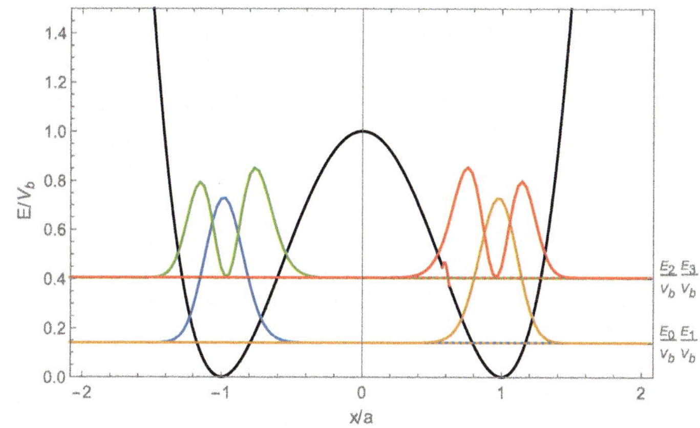

Fig. 2 The first four eigenfunctions as $|\psi(x)|^2$ for the double well potential with flea perturbation (marked red). From left to right, bottom to top, the maxima give the first (blue), second (yellow), third (green) and fourth (red) wavefunctions

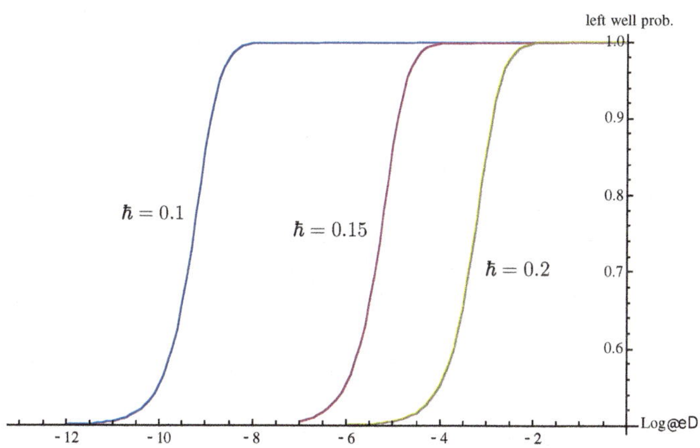

Fig. 3 Left-well probability of the groundstate of a double-well potential as a function of flea size (flea is scaled by shrinking parameter ε), for $m = 1$ and $\hbar = \xi \in \{0.1, 0.15, 0.2\}$ and a parabolic flea such as that shown in Fig. 2. The shrinking parameter ε is scaled logarithmically with base 10. This shows that there exist a scale for the size of the perturbation, dependent on $\xi = \sqrt{\hbar^2/m}$, above which the flea causes the groundstate to be almost completely located in a single well. As will be discussed in Sect. 3.1.2, however, it also shows that adding an asymmetry (interpreted as a flea) to the double well to allow for unequal initial probabilities relies critically on the size of the asymmetry

ask a different question. Starting from the assumption that the flea can (be made to) provide an effective collapse mechanism, how much closer would that bring us to solving the measurement problem?

2.2 Generalised Outcomes

Asymptotic Bohrification relies on a more generalized notion of outcome state, which have non-zero Born probabilities associated to macroscopically distinct pointer values. Even though these probabilites vanish in the semi-classical limit, it is the states *before* taking the limit which count as physically real, as made explicit by Butterfield's principle. This raises the question of how to understand such states as outcomes. In this subsection we argue that we lack a physical grounding for understanding post-measurement states of the flea model as outcomes. However, we do not consider this to be weakness of asymptotic Bohrification, but rather a challenge for any approach to the measurement problem, at least when the approach sticks within quantum mechanics.

Let us start with the second point: As argued in Appendix 6, in any physically realistic setting, the fundamental uncertainties of quantum mechanics entail that we cannot prepare a pure state for a system. Instead of an eigenstate such as $|+\rangle$ as in Sect. 1 of the measurement problem, it is more realistic to have an entangled state such as

$$\sqrt{1-\varepsilon^2}\,|+\rangle \otimes |\Phi\rangle + \varepsilon e^{i\phi}\,|-\rangle \otimes |\Psi\rangle,$$

with $\varepsilon > 0$.

The pragmatic physicist can of course safely ignore this and, for example, trade in the mathematically cumbersome initial states where the system is entangled with the environment, for a state such that the reduced system state is assumed pure. For a well designed preparation method, the difference between the eigenstate and the exact state is negligible as far as the statistics of outcomes of the subsequent measurements are concerned. The Born rule provides physical grounding for dismissing the small 'ε' terms in the state. However, when targeting the specific foundational issue of explaining the occurrence and statistics of outcomes, such a move may very well throw the baby out with the bath water.

As with the initial state in the previous example, the Born rule would also provide empirical grounding to understand the ε-states which arise as outcomes in the flea model. However, we should be cautious about a priori assuming the Born rule. In particular, as argued in the following subsection, this is because the Born rule dictates the statistics of the flea perturbations, thus challenging the independence assumption between system and pointer.

A different attempt at defining (approximate) outcomes can be found in the many worlds interpretation, or Everrettian quantum mechanics, where the ε-states obtain an even grander status than that of approximate outcomes, namely distinct branching worlds, and the Born rule can then allegedly be derived.[1] The discrete branching ontology of Everettian quantum mechanics is not exactly realised by decoherence, but only approximated and results in ubiquitous generation of ε-states. The problem

[1]The derivation of the Born rule has faced much criticism, such as its reliance on decoherence, which produces the ε states, through which the derivation becomes circular, as noted by various authors such as Zurek [22, 23], Baker [1], and Kent [9].

thus remains to bridge the gap between the ε-states and the exact branching ontology without smuggling in new concepts, such as the Born rule. Smallness of ε by itself does not provide any justification, especially if we are not clear on what these numbers mean. According to Zurek [22, 23], to pass from this effective description to the setting where classical decision theory can be used, we need to assume the Born rule. Proponents of Everettian quantum mechanics have, of course, been aware of such circularity objections. Wallace [20], for example, might dismiss the circularity objection by denying that the last step is needed. According to Wallace, the discrete branching structure of Everettian quantum mechanics is understood as a *robust* yet *emergent* feature of reality. But as is typical for discussions surrounding questions of emergence and reduction, it is hard to understand what this robustness exactly means. In a critique of such robustness, Dawid and Thébault [4] argue that the only robustness that holds explanatory power is *empirically grounded* robustness. In their words:

> The first crucial distinction that can be made is between a notion of robustness that is empirically grounded and one that is not. By this we mean some qualification such that whether a structure within the formalism of a theory is taken to be robust is dependent upon some interpretational connection between that structure and empirical phenomenology.

Similarly, the GRW dynamical collapse theories [2], although not within the framework of standard quantum theory, suffer from the same problem of epsilonics, in their case called “the tail problem”. The collapsed states are approximately Gaussian and therefore the associated wavefunctions are non-zero throughout the whole space. From the point of view of standard quantum theory, this once again raises the question of how such states can be understood as outcomes. Ghirardi, Grassi and Benatti, reply that the tails do not constitute a problem, as these can be seemingly defined away in the matter density formulation of the theory [6]. The tails of a collapsed state correspond to low-density matter regions of the wavefunction, which are deemed not relevant because they are inaccessible. But what does it mean for a matter distribution to be accessible or inaccessible? If “inaccessible” means that any observer is unable to measure it, then we share the worries expressed in [14, 19] that we should not rely on observers before the tail problem is resolved. Thus the problem of connecting ε-states to physical reality without a priori assuming the Born rule again rears its head.

In short, ε-states, such as the generalised outcomes of asymptotic Bohrification are ubiquitous in the foundations of quantum mechanics, and the status of such states (whenever assuming the Born rule is out of the question) is a source of controversy. However, we leave this challenge of understanding outcome states for now, and move on to more pressing problems for asymptotic Bohrification.

2.3 Independence and Idealizations

If a flea-type perturbation is effective in collapsing the wave function, then the location of the flea determines to which well the state collapses. In order to be empirically consistent with quantum theory, the statistics of the flea locations should match the Born probabilities of the initial system state under consideration. Not much is said about the origin of the flea, other than it being of environmental nature. If the flea is thought to be some random perturbation, present simply because we do not want the physics of our model to hinge on mathematical idealizations, then indeed we expect it to be independent of the initial system state. But if that is the case, nothing short of a conspiracy theory would be needed to replicate the Born rule. This question is therefore; how could the flea be dependent, in the appropriate way, on the initial system state?

What makes this question particularly hard to answer, is that the flea model is not explicit on what role the system plays in the model. It may very well be that the problem of independence in replicating the Born rule simply amounts to the need to violate the independence between the system and the other degrees of freedom: as discussed at the end of Sect. 1.1, where this was stated as a loophole in avoiding the measurement problem. However, starting out with dependencies between pointer and system, it becomes unclear in what sense the flea model describes measurements. This is a pressing problem as it challenges the relevance of the flea model to the measurement problem.

In order to address the issue of independence, we need a physical underpinning for the flea. Related to the point of interpreting the flea, is the question of what the potentials represent. Recall that the flea proposal is based on the observation that two different potentials, one with a flea and the other without, behave quite differently when the parameter ξ is changed. This is, in itself, a completely static argument, void of any dynamical considerations. Note that the proposal relies heavily on potentials, which, according to quantum theory, are themselves always to be seen as an approximation of some dynamic quantum fields. For example, the possible energy of a photon emitted by an atom can be understood in terms of the energy levels of the electromagnetic potential governing the electrons. However, to understand the dynamics of the emission, the electrons must be coupled with an electromagnetic field. This is because the dynamics depends not only on the system's state but also on the state of the environment, in this case the electromagnetic field, whose interplay determines the speed of the changes in the system. This means that to connect the two different potentials, the flea must be dynamically added, and thus it must be treated as a dynamic quantum field.

In short, the flea model needs further work in making the system explicit and the providing the flea with a physical underpinning, before it can be considered as an actual model of quantum measurement. And even then it remains to be seen if the problem of independence can be resolved in such a way that the model has something to offer to the measurement problem without resorting to a version of superdeterminism.

2.4 Interpreting the Ground State

How is it possible that in the setting of quantum measurements, where entanglement plays such an important role, the flea model describes the dynamics of the pointer in terms of a single pure state?

The flea model assumes that, from the moment at which the flea is introduced, up to the end of the measurement, the degrees of freedom which are explicitly modelled are described by a pure state. It is tempting to think of the wave-function under investigation as representing the state of the pointer variable, but things cannot be that simple. Certainly, there is a relation between the pointer variable and the wave function since the two wells of the potential correspond to the two possible values of the pointer variable. However, if we think of the flea model as being obtained from a more complete measurement model by tracing out the system and environmental degrees of freedom, then we would expect entanglement between the system and the pointer variable to result in a non-pure state. This is crucial to the flea model since it relies on properties of bound states.

For the sake of concreteness, consider the time evolution for a von Neumann ideal measurement

$$\frac{1}{\sqrt{2}}\left(|m_1\rangle + |m_2\rangle\right) \otimes |r\rangle \mapsto \frac{1}{\sqrt{2}}\left(|m_1\rangle \otimes |E_1\rangle + |m_2\rangle \otimes |E_2\rangle\right). \tag{9}$$

Here $|r\rangle$ is the initial environmental state, and the environmental states $|E_i\rangle$ are close to orthogonal. When we trace out everything but the pointer, the entanglement causes the end result to be a non-pure state.

We could, as in Sect. 1.1, argue that different runs of the measurement correspond to different initial environmental states. However, if (still as in Sect. 1.1) after tracing out all environmental degrees of freedom, this would only lead to a small variation in the time-evolution operator, then this would have no impact on the above conclusion. The final state would still be close enough to (9) to yield a non-pure state. By the previous reasoning, the wave function of the flea model cannot represent the state of the pointer variable in a straightforward way. But then, what does it actually describe?

More pressingly: do asymptotic Bohrification and the flea model even deal with measurements? The flea model concentrates solely on collapsing a wave function, but much more is needed for any model of measurement. The post-measurement state not only needs to assign a value to the pointer variable, but also to the system's measured observable in such a way that these values are correlated. In this sense the flea model, concentrating only on the collapse of what we presume to be the pointer variable, is far from complete.

It should be noted that *we assumed* we should apply the flea perturbation solely to try to collapse the *pointer* to a definite outcome, which goes against the recommendations given to us by Landsman himself. In his view, the collapse should occur on the level of the combined pointer and observable.

Aside from the issues presented above, there is at least one other good reason for not wanting to identify the wave function with the pointer. To illustrate this, suppose we wish to include something akin to a flea perturbation in existing quantum measurement models. As an example, consider the model by Haake and Spehner [18], who use a double well for the pointer and an appropriate interaction to correlate the pointer's position with a system's observable (z-component of a spin-1/2 system). The obvious place in the Haake–Spehner model to apply the flea is to the potential of the pointer. However, since the original hamiltonian of the model commutes with the observable, and the model is only modified at the level of the pointer, the Born probabilities for the observable are unaffected by the introduction of the flea. Even if the flea is effective in causing a collapse of the pointer wave function, the correlation between pointer and observable disappears. As a side note, introducing the flea model to the Haake–Spehner model is not as straightforward as one might think from the previous remarks. For the model contains two separate potentials which are heavily-slanted double wells depending on the spin component, making it quite different from the flea model.

Thinking of entanglement and in keeping the observable/pointer correlation, one naturally thinks one should consider Landsman's proposal of the flea model as representing the combined observable and pointer. However, it is not clear to the authors what is meant by this statement, and, more concretely, how to apply this idea to any physical model. In the case where the collapse occurs on the level of the pointer, it is intuitively clear how to construct a model. Namely: x can, for example, describe the center-of-mass position of all particles in a gauge on the measurement device (pointer) that is subject to a electromagnetic potential due to its interaction with the particles in the measurement device and the wavefunction is the center-of-mass wavefunction of the particles in the gauge. The flea can be imagined to be a result of some change in the electromagnetic potential, although whether its cause should come from outside or inside the measurement device is unclear. But what if the collapsing wave function somehow represents the state of both pointer and observable? Where does the potential for this model come from, and how should it be interpreted? What would be the physical significance of x? What model can we build to give rise to a potential in x? It seems that the variable x is now much harder to interpret. Without knowing what x might stand for, it is hard to answer questions such as: why is it energetically unfavourable to have large x or $x = 0$ for the pointer+observable? The values of x are clear in the case of only a pointer (as in the Haake–Spehner model). But what does it mean in the case of a pointer and observable: does $x > 0$ correspond to spin up and $x < 0$ to spin down or only in the minima? What does $x = 0$ mean for the observable?

We end up in the situation where we do not have a physical interpretation for the potential, the wavefunction, or the flea, and where we are unable to connect with existing models of quantum measurement. Although this state of affairs is already troubling, it becomes more so after the next section where we conclude that the flea model is unable to perform its task of collapsing the wave function effectively.

To sum up the situation: How do you salvage a model if it is so incomplete at an interpretational level (i.e., what does the wave function mean, where does the flea come from, what does the potential represent) and at the same time so far removed from any other model of quantum measurement (it would need a model containing a bound state representing the combined observable and pointer state, and which is completely quantum mechanical)?

3 The Collapse Mechanism

In the previous section we looked at the measurement problem through the eyes of asymptotic Bohrification, without considering the effectiveness of the collapse mechanism. The collapse mechanism is the central topic of this section. In Sect. 3.1 we consider obstacles to generalising the collapse mechanism of the flea model to more outcomes, and unequal Born probabilities. In Sect. 3.2 we investigate the time scale of the collapse.

First, let us recall the setting of Sect. 2.1. The Hilbert space is $\mathscr{H} = L^2(\mathbb{R})$, and the initial state is the symmetric ground state of the hamiltonian in (6), i.e.,

$$\hat{H}_\hbar = -\frac{\hbar^2}{2m}\frac{d^2}{dx^2} + \frac{\lambda}{8}\left(x^2 - a^2\right)^2 =: \frac{\hat{p}^2}{2m} + V(x), \tag{10}$$

The two wells of the potential represents two distinct values which the pointer variable of the measurement device can assume, and the symmetric ground state represents a Schrödinger cat like state, presumably transferred from the superposed state of some quantum two-level system. Thus the starting point of the flea model is supposed to take place near the end of a measurement interaction. The flea is a small (in sup-norm) asymmetric potential $W(x)$ which is localised near the bottom of one of the two wells of the symmetric double well potential $V(x)$. For a small value of $\hbar$, the ground state of the perturbed hamiltonian $\hat{H}_\hbar + W$ is well-nigh completely localised in a single well.

Ultimately, however, one of the main challenges of applying the flea model to the measurement problem is to find a time dependence for the flea, e.g. a function $t \mapsto f(t)$, in such a way that the initial symmetric wave function evolves to the localised ground state of the perturbed setting, as the hamiltonian evolves as $t \mapsto \hat{H}_\hbar + f(t)W$.

3.1 Generalisations

We consider two generalisations of the flea proposal; measurements with more than two possible outcomes, and system preparations with unequal associated Born probabilities.

We could have chosen to generalise in a different direction. For example, we could trade in the Hilbert space $L^2(\mathbb{R})$, the hamiltonian with a double well potential, and the semi-classical limit $\hbar \to 0$ for the spin chain Hilbert space $\mathcal{H}_N = \otimes^N \mathbb{C}^2$, the quantum Curie–Weisz hamiltonian,

$$H_N = -\sum_{-\frac{N}{2}}^{\frac{N}{2}-1} \sigma_i^z \sigma_{i+1}^z - B \sum_{i=1}^{N} \sigma_i^z$$

and the limit $N \to \infty$, where N is the number of sites on the lattice. This example was treated in the setting of asymptotic Bohrification in [10]. The only thing that we need to add is a flea, for example in the shape of a matrix $W = \text{diag}(0, 0, \ldots, 0, \varepsilon, 0, \ldots, 0)$, where ε is some small number and the matrix representation is relative to the basis of $\mathcal{H}_N$ generated by the eigenstates of the operators σ_i^z. However, for the sake of brevity we will not discuss such examples any further, at the risk of giving the impression that the idea of the flea model is limited to n-well potentials and the semi-classical limit ($\hbar \to 0$).

3.1.1 More Outcomes

The two wells of the potential correspond to the two values of the observable. How should we generalise the model to observables with n distinct values, where $n > 2$? One strategy is to consider the groundstate of an n-well potential with periodic boundaries. As an example we use $V(x) = V_b \cos^2(\pi x/2a)$ on $x \in [-na, na]$. This choice has the advantage that the energy eigenstates are known, namely they are the Mathieu functions, shown in Fig. 4.

A flea perturbation is added to the potential and the eigenfunctions are determined numerically. From Fig. 4 it is seen that the groundstate need not localise in a single well when only a single flea perturbation is added. In the figure, the flea perturbation is parabolic with finite support, although these details have no impact on the discussion. Since a single flea does not provide a fully localised ground state, one should consider adding multiple flea perturbations in the different wells. These multiple fleas must be chosen in such a way that the resulting potential no longer has any symmetry. If many fleas are added, and their locations, sizes and shapes are chosen largely at random, then this should be no problem.

A potential cause of problems for this generalisation is the increase in collapse times, especially when there are many wells, providing many barriers through which parts of the wave function have to tunnel. In Sect. 3.2 it will become clear that collapse times are already problematic for the double well setting. Therefore we shall not explore the increase of collapse times due to the added wells.

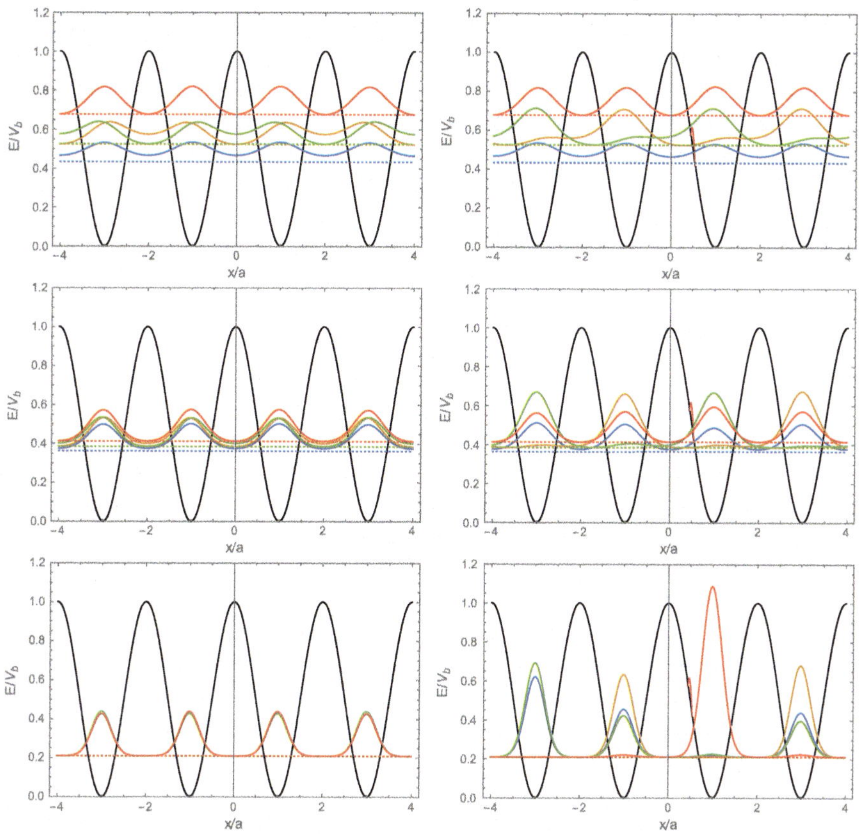

Fig. 4 The first four eigenfunctions as $|\psi(x)|^2$ of the periodic potential without (left) and with (right) a flea for $\xi = \{0.6, 0.4, 0.2\}$ from top to bottom, respectively. Clearly, without a flea the solutions are symmetric and will remain present in all wells as ξ becomes smaller. With the flea the symmetry is broken, and the groundstate (blue) will move out from the well with the Flea

3.1.2 Unequal Born Probabilities

How can we adapt the flea model to deal with other Born probabilities than the 50/50 case? At least two options are available. The first option is to keep using a symmetric potential, but to add additional wells. Suppose we consider a two level system, and that the Born probability assigned to one of the two values of the observable can be expressed as a rational number p/q, where we assume that 1 is the only common divisor of natural numbers p and q. We can consider q wells, p of which correspond to the value with the Born probability p/q, and the other $q - p$ correspond to the other value. Then we can proceed as before. This option will typically involve many wells, making it complicated and providing potential additional problems with regards to collapse times. In addition, this option looks far removed from the idea that the system state becomes correlated to the pointer variable. Since the connection between the system's measured observable and pointer is already so weak in the flea model, we consider this to be a serious drawback.

A second option, which does not share the previous drawbacks, is to replace the symmetric ground state of the symmetric double well potential by an asymmetric potential yielding a ground state such that the probabilities assigned to the wells are equal to the desired Born probabilities. However, since we are working in the semi-classical limit, the asymmetry of the potential should shrink with the value of $\hbar$. Otherwise, the asymmetry localises the wave function, just as the flea would.

In addition to the symmetric potential V there are three additional asymmetric terms added to the potential:

$$V(x) + W_0(x) + W_b(x) + W_f(t). \tag{11}$$

There is some noise W_0 which is too small (in sup-norm) to affect localisation, and which may be time dependent. We only add it to emphasise that the initial potential need not be completely symmetric in order to start out with a (sufficiently) symmetric ground state. Otherwise, we would need to worry whether the flea model itself conflicts with Earman's principle. Then there is a contribution W_b which ensures that the initial state of the flea model has the desired Born probabilities. Since it affects the Born probabilities, it is larger (in sup-norm) than the noise, but must be smaller than the flea in order to avoid localisation. The final contribution is the flea $W_f(t)$ which localises the wave function.

Consider again Fig. 3 in Sect. 2.1 where, for three different values of $\hbar$, we considered a parabolic flea W. We shrunk this flea as εW, where ε is a number ranging from 10^{-1} to 10^{-12}. As the flea shrinks we consider the probability assigned to the left well. The sensitivity of the ground state relative to small perturbations is seen as the curves shift to the right, when $\hbar$ decreases. Now suppose we are given a value of $\hbar$ and some perturbation W. If W is to be part of the noise, then there is an upper bound such that if we shrink W below this bound, then it will not affect the Born probabilities. Likewise, if W is intended as the flea, then there is a lower bound, and if W is, relative to the sup norm, larger than this boundary, it will allow localisation. If W is needed to set the initial Born probabilities, then there is only a single value of ε_0, such that $\varepsilon_0 W$ provides the right Born probabilities. Any deviation from this value affects the Born probabilities. Note that the variations around ε_0, such that the Born probabilities do not change significantly, decrease in order of magnitude as $\hbar$ decreases. More importantly, in all cases these asymmetries that we are fine-tuning are necessarily smaller than the flea perturbation. The flea is supposed to represent the result of an extremely small environmental fluctuation. Yet, we are fine-tuning even smaller fluctuations, just to get the right initial state.

In addition to the previous fine-tuning problem, the problem of independence also plays a role here. For recall that it is the location of the flea that determines in which well the final state is localised! The Born probabilities, which we have just modeled using a finely tuned asymmetry, do not influence how a flea is chosen. In [13], the flea perturbation was compared to a hung parliament, where a small political party acquired influence far exceeding its relative size. We feel that the flea is more like a dictator, determining what is going to happen, unfettered by any democratic constraint. The Born probabilities of the initial wave function, motivated by the correlation between observable and pointer variable, do not play any active role in the model.

We can only conclude that none of the options presented here provides a satisfactory extension of the flea model to arbitrary Born probabilities. But of how much interest can a solution to the measurement problem be, if it can only be applied to the special 50/50 case?

3.2 Problem of Collapse Times

Next, we consider the time scales of the collapse for the flea model. For a macroscopic device, modelled by a small value for $\hbar$, we find that a collapse takes an unrealistically long time, because we are considering a tunnelling problem with respect to a relatively large potential barrier.

Recall the work on Schrödinger operators by Jona-Lasinio et al. [8], which plays a key role in the flea approach. In the words of Landsman and Reuvers [13]:

> ...the ground state of a symmetric double-well Hamiltonian (which is paradigmatically of Schrodinger's Cat type) becomes exponentially sensitive to tiny perturbations of the potential as $\hbar \to 0$.

This phenomenon may have relevance for the quantum to classical transition. However, being a relation between bound states for slightly different potentials, it is a *static* phenomenon that has no obvious connection to the *dynamical* problem of the collapse of the wave function. In fact, as we argue below for the double well, when we decrease $\hbar$, we are effectively increasing the potential barrier between the wells, and the time needed for symmetric wave function to localise increases at least exponentially in the semi-classical limit, regardless of the way in which the flea is introduced. Since smaller values of $\hbar$ are supposed to represent a larger degree of macroscopicity, such an increase in collapse times undermines the flea model.

How should we choose the time dependence of the flea? Since there is presently no physics explaining the emergence of the flea we treat this to a large extent as a mathematical problem, rather than a physical one. The only physical restriction is that the collapse times should decrease as the scale $\hbar$ decreases, since this parameter is used to represent the degree of macroscopicity. Landsman and Reuvers [13] consider various ways of dynamically introducing a flea term $W(x)$ to the potential. The first option is to introduce it as a 'quench',

$$H(t) = H_0 + \theta(t)\varepsilon W, \tag{12}$$

where $\theta(t)$ is the Heaviside step function, and $\varepsilon > 0$ is a real number. Depending on the size of ε, introducing the flea either results in a wave function which oscillates between the two wells, or the original symmetric wave function barely changes at

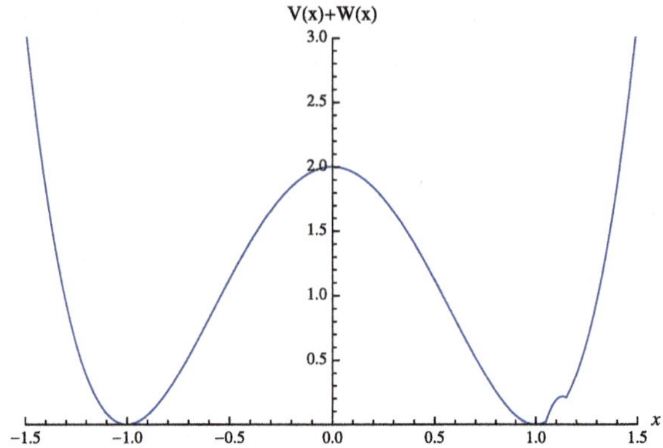

Fig. 5 Potential with flea

all. For no value of ε however, do we obtain a wave function which is localised in a single well.

Other attempts, such as adding white noise or Poisson noise did not lead to a localisation either. In all these attempts both the ground state and the first excited state of the perturbed setting contribute significantly to the wave function. In order to avoid this, Landsman and Reuvers consider introducing the flea in the adiabatic limit, ensuring that we end up with the ground state of the perturbed setting. As it turns out, the combination of the adiabatic limit in quantum mechanics and the semi-classical limit poses a problem of too large collapse times.

For the sake of concreteness, consider the time-dependent Hamiltonian:

$$H(t) = H(0) + \sin\left(\frac{\pi t}{2T}\right) W, \quad \text{for } t \leq T \tag{13}$$

and $H(t) = H(0) + W$ for $t > T$. For the potential V of $H(0)$ we use the symmetric double well potential as in (6) and the flea W is a parabolic shape, as shown in Fig. 5.

Let $(\psi_n)_{n\in\mathbb{N}}$ be a (time-dependent) orthonormal basis of eigenfunctions of $H(t)$, where the eigenvalues $E_n(t)$ are ordered in increasing size. The wave function can be expressed as:

$$\Psi(t) = \sum_{i=1}^{\infty} c_n(t)\psi_n(t) \exp\left(\frac{i}{\hbar}\int_0^t ds\, E_n(s)\right), \tag{14}$$

where the x-dependence is suppressed in the notation. We start out in the ground state of $H(0)$, so $c_0(0) = 1$ and $c_n(0) = 0$ for all $n > 0$. The time dependence of $c_n(t)$ is given by

$$\dot{c}_n = c_n\langle\psi_n|\dot{\psi}_n\rangle - \sum_{m\neq n} c_m \frac{\langle\psi_n|\dot{H}|\psi_m\rangle}{E_m - E_n} \exp\left(-\frac{i}{\hbar}\theta_{mn}(t)\right), \tag{15}$$

$$\theta_{mn}(t) = \int_0^t ds(E_m(s) - E_n(s)). \tag{16}$$

In what follows we concentrate on $|\dot{c}_1(0)|$, for several reasons. As already noted, the first excited state is localised in the other well with respect to the ground state, therefore, $|\dot{c}_1(t)|$ is of interest. In addition, consider the energy splitting $\Delta(t) = |E_0(t) - E_1(t)|$, which rapidly decreases in the semi-classical limit. During the collapse, the splitting $\Delta(t)$ takes on its smallest value for the unperturbed setting $t = 0$. It is also at this time that we expect the overlap $|\langle\psi_1(t)|W|\psi_0(t)\rangle|$, appearing in the matrix element $|\langle\psi_1(t)|\dot{H}|\psi_0(t)\rangle|$, to be at its largest. In other words, if $|\dot{c}_1(0)|$ turns out to be sufficiently small, then this may help yield a final state which is close to the ground state of the perturbed potential. Thus, we ask for which order of magnitude of T, does the quantity

$$|\dot{c}_1(0)| = \frac{\pi}{2T\Delta(0)}|\langle\psi_1(0)|W|\psi_0(0)\rangle| \tag{17}$$

become sufficiently small. As in [13] on p. 8 as $\hbar \to 0$, $\Delta(0)$ decreases as

$$\Delta(0) \approx \frac{2\hbar a\sqrt{\lambda}}{\sqrt{e\pi}} e^{-\frac{d_V}{\hbar}}. \tag{18}$$

Unless $|\langle\psi_1(0)|W|\psi_0(0)\rangle|$ decreases with a rate of at least $e^{-d_V/\hbar}$ in the semi-classical limit, the time $T \to +\infty$ must increase exponentially fast in that same limit, in order to keep $T\Delta(0)$ finite. Here d_V denotes the WKB-factor.

In itself, the limit $T \to +\infty$ as $\hbar \to 0$ need not be problematic. As $\hbar$ decreases, we can shrink the flea W, and consequently shrink $|\langle\psi_1(0)|W|\psi_0(0)\rangle|$, whilst retaining a localised perturbed ground state. Consider the quantity:

$$\Gamma = \frac{|\langle\psi_1(0)|W|\psi_0(0)\rangle|}{\Delta(0)}, \tag{19}$$

which we use to quantify the rate at which T needs to increase. Figure 6 shows how $\mathrm{Log}(\Gamma)$ varies with $\hbar$ for the fixed parabolic flea of Fig. 5. Note the exponential growth, indicating that T needs to increase at a rate $x \mapsto e^{e^x}$. In the range of the graph, if we halve the value of $\hbar$, then Γ increases roughly 15 orders of magnitude.

How much can we shrink the flea $W(x) \mapsto 10^{-n}W(x)$ in order to compensate for this effect. For any $\hbar$ in the range of Fig. 6, if $n \geq 12$, the perturbed ground state is no longer localised in a single well. In other words, the collapse times rapidly increase as $\hbar \to 0$ even if we shrink the flea W as much as possible along the way. Why do we expect this behaviour to hold in general? Instead of $\hbar$, consider $\lambda := 1/\hbar$. Regardless of the details of the flea, we are considering situations where part of the wave function has to tunnel through a barrier of increasing height λV_0.

If the flea is introduced too fast, then the first excited state becomes occupied and we retain the superposition. If the flea is introduced slower, to ensure that we

Fig. 6 Collapse times

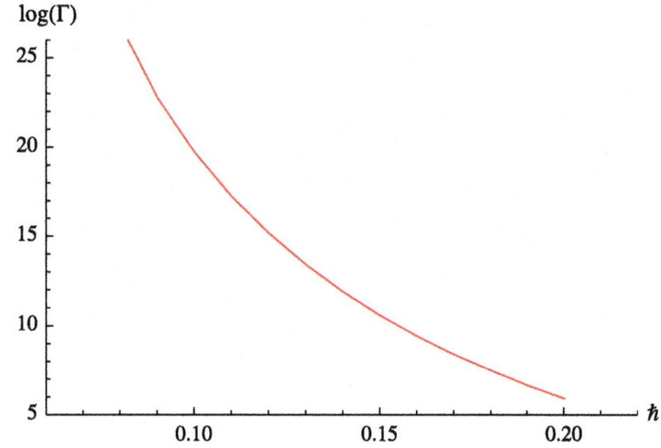

remain in the ground state, then the collapse times increase dramatically as the macroscopicity scale parameter $\hbar$ decreases. The cause of the increase of collapse times is the rapid decrease of the initial splitting $\Delta(0)$. Yet, this same smallness of the splitting with regard to the semi-classical limit is important to the asymptotic Bohrification programme, as argued in [10] on p. 12:

> The essential point is that in our models, the energy difference $\Delta E_\bullet = E_\bullet^{(1)} - E_\bullet^{(0)}$, between $\Psi_\bullet^{(1)}$ and $\Psi_\bullet^{(0)}$ vanishes exponentially as $\Delta E_N \sim exp(-C \cdot N)$ for $N \to \infty$, or as $\Delta E_\hbar \sim exp(-C'/\hbar)$ for $\hbar \to 0$ respectively. This means that asymptotically any linear combination of $\Psi_\bullet^{(1)}$ and $\Psi_\bullet^{(0)}$ is almost an energy eigenstate.

The flea model hinges on the precept that first a macroscopic superposition for the pointer variable is formed, and subsequently this superposition is broken down dynamically. As argued in this subsection, this is an open invitation to problems about tunnelling times. However, conceptually it is suspicious as well. Already from decoherence we know that setting up a macroscopic superposition is highly non-trivial.

4 Conclusion

The most important physical results of the flea model is that in the static setting it gives a scale for the size of the flea, dependent on $\hbar^2/m$, above which the groundstate is almost fully located in one well, as was shown in Fig. 3, and that with increasing mass a smaller perturbation is needed. This static property of potentials could, in principle, be confirmed in experiments by determining the shape of the groundstate of a double well with varying sizes of an artificial flea perturbation. The drastic effect of such small perturbations on the form of the groundstate remains a remarkable feature of quantum theory.

Although the static picture looks appealing, when we consider the collapse using a time-dependent flea, we run into various problems:

- Only adiabatic introduction of the flea provides a collapse, but it comes with unrealistically long collapse times. For faster introduction of the flea, we find that the first excited state becomes relevant and no collapse is observed.
- We cannot generalise to system states with unequal associated Born probabilities.
- It is unclear how to scale to more than two outcomes.

When considering the statistics of outcomes in addition to individual runs, the following problem of independence is added:

- It is only the statistics of the flea distribution that determines the statistics of outcomes. Without any relation between the flea and the initial system state, there is no reason why the Born probabilities (apart from the special 50/50 case) should be replicated by the flea model.

In some sense the wave function, representing the system and pointer state, and the flea, which is assumed to be of environmental origin, need to be related. As the flea model, as so far formulated, is incomplete concerning the interpretation of the flea and the dynamics between system and pointer, it is hard to understand the exact ramifications of imposing relations between flea and system state. However, it would not be too surprising if, after further working out the flea model, it does turn out that replicating the Born probabilities simply means denying independence between the system state and the degrees of freedom of the measuring apparatus. If that is the case, we need to ask in what sense we are still considering a measurement.

With regard to interpretation of the flea model we are left with the following issues:

- The problem of potentials: The idea behind the flea proposal comes from comparing two separate potentials and thus has no bearing on dynamics. True dynamics only follows from a treatment in terms of two coupled quantum systems.
- Physical interpretational issues: The origin of both the potential and flea is unclear; which makes a realistic physical model almost impossible.
- Where is the system in the flea description, and how does it become correlated with the pointer? What kind of model would yield the double well ground state as an intermediate state for the combined system and pointer?

These are more than enough questions. But with the flea model being far removed from known models of quantum measurement, it is unclear how to construct an actual model which provides physical grounding for the potentials, the flea and/or the wave function. As a final point we mention that the generalised notion of outcome state in the flea approach still lacks an empirical grounding.

In short, the flea model as it now stands is not able to provide an effective collapse. Although the underlying idea looks good statically, there is no reason why it would work dynamically. Generalising the model to system states with unequal Born probabilities appears to provide a sizable obstacle as well. But even if the collapse were effective, it is not clear what this would say about measurements. The model is silent on the system-pointer correlation and the physics behind the flea and the potential. Thus, to the authors it is not clear that we are even dealing with measurements. And if a working model of quantum measurement, underlying the flea model, could be found, the problem of independence may still require the addition of some sort of conspiracy theory or superdeterminism. If this is the case, then it is not clear what more the flea model has to offer, about the measurement problem as formulated in Sect. 1.1.

Acknowledgements The research in this paper is part of the project *Experimental Tests of Quantum Reality*, funded by the *Templeton World Charity Foundation*. The authors would like to thank Klaas Landsman for his investment in this project. We gratefully acknowledge the helpful discussions with Andrew Briggs, Hans Halvorson, Andrew Steane and various members of the Oxford Materials groups. The authors would also like to thank the two anonymous referees that greatly helped this paper.

5 Appendix 1: Convergence of States

For concreteness, let $\hbar$ take values in the unit interval $[0, 1]$. To each strictly positive $\hbar > 0$, associate the non-commutative algebra $\mathfrak{A}_\hbar = \mathscr{K}(L^2(\mathbb{R}))$ of compact operators acting on the Hilbert space of square-integrable functions. To $\hbar = 0$ associate the commutative algebra $\mathfrak{A}_0 = C_0(\mathbb{R}^2)$ of continuous real-valued functions on the phase space $\mathbb{R}^2$, which vanish at infinity. Through their disjoint union $\mathfrak{A} = \coprod_{\hbar\in[0,1]}\mathfrak{A}_\hbar$ these algebras combine in a single algebra fibred over the unit interval, $\mathfrak{A} \to [0, 1]$. Dual to this bundle there is the bundle of state spaces $\mathscr{S} \to [0, 1]$ where $\mathscr{S} = \coprod_{\hbar\in[0,1]}\mathscr{S}_\hbar$, and $\mathscr{S}_\hbar$ denotes the state space of $\mathfrak{A}_\hbar$. For $\hbar > 0$ the states are density operators acting on $L^2(\mathbb{R})$, and for $\hbar = 0$ the states are probability measures on the phase space $\mathbb{R}^2$.

Next, we could consider the algebraic and topological aspects of these bundles. But since we are only concerned with the measurement problem, we refer the reader to [10] and proceed directly to the our main question; how is convergence of states defined in this scheme? More precise, when does a family of density operators $(\rho_\hbar)_{\hbar\in(0,1]}$ converge to a classical state $\mu_0 \in \mathscr{S}_0$? To define convergence, note that each density operator $\rho_\hbar$ defines a probability measure $\mu_\hbar$ on $\mathbb{R}^2$, through

$$\int_{\mathbb{R}^2} d\mu_\hbar f := Tr\left(\rho_\hbar Q_\hbar(f)\right), \quad \forall f \in C_0(\mathbb{R}^2) \tag{20}$$

where $Q_\hbar(f)$ is the compact operator acting on $L^2(\mathbb{R})$, called the 'Berezin quantisation' of f. The Berezin quantisation map $Q_\hbar$ is defined as

$$Q_\hbar(f) = \int_{\mathbb{R}^2} \frac{dpdq}{2\pi\hbar} f(p, q)|\Phi_\hbar^{(p,q)}\rangle\langle\Phi_\hbar^{(p,q)}|, \tag{21}$$

through the coherent states $\Phi_{\hbar}^{(p,q)} \in L^2(\mathbb{R})$;

$$\Phi_{\hbar}^{(p,q)}(x) = (\pi\hbar)^{-1/4} e^{-ipq/2\hbar} e^{ipx/\hbar} e^{-(x-q)^2/2\hbar}. \tag{22}$$

The states $\rho_{\hbar}$ converge to the classical (possibly mixed) state μ_0 iff the probability measures $\mu_{\hbar}$ converge weakly to μ_0 in the sense

$$\lim_{\hbar \to 0} \int_{\mathbb{R}^2} d\mu_{\hbar} f = \int_{\mathbb{R}^2} d\mu_0 f,$$

for each $f \in C_0(\mathbb{R}^2)$ with compact support.

For the double-well model, as $\hbar \to 0$, the ground state $\psi_0(x)$, or rather its associated density operator, converges to the classical mixed state (7), a convex combination of probability distributions with support in the two different wells. As emphasised in the main paper, this is a classical state which does not qualify as an outcome.

6 Appendix 2: Practical Necessities

The fundamental uncertainties of quantum theory prohibit the preparation of a pure state, let alone an eigenstate with respect to a fixed basis. For most purposes this is irrelevant since we can get close enough in terms of Born probabilities; but for the measurement problem, the distinction may matter. We illustrate the point using the simple example of preparing an initial state using a Stern–Gerlach experiment. The point will be that for any *fully* quantum mechanical treatment, the Born probabilities associated to both eigenvalues of the observable are always non-zero. If we started with an n-level system, the same would hold for all the eigenvalues of any observable.

Consider the Stern–Gerlach experiment where spin-$1/2$ particles are sent through an inhomogeneous magnetic field. The textbook view is that due to the spin-magnetic-field interaction the spin along the magnetic field gets correlated with the position of the particle. In this simple view of the device, the incoming particle is described by a wavepacket $\psi(\mathbf{x}, t)$ and then after some interaction time the position of the wavepacket is correlated to the spin in the z-direction

$$\begin{aligned}|\Psi(t=0)\rangle &= (\alpha\,|\uparrow\rangle + \beta\,|\downarrow\rangle) \int \psi(\mathbf{x}, 0)\,|\mathbf{x}\rangle\,\mathrm{d}\mathbf{x} \\ \to |\Psi(t)\rangle &= \alpha\,|\uparrow\rangle \int \psi_+(\mathbf{x}, t)\,|\mathbf{x}\rangle\,\mathrm{d}\mathbf{x} + \beta\,|\downarrow\rangle \int \psi_-(\mathbf{x}, t)\,|\mathbf{x}\rangle\,\mathrm{d}\mathbf{x},\end{aligned}$$

where $\psi_\pm(\mathbf{x}, t)$ indicate the wavepackets as they leave the Stern–Gerlach apparatus. If the initial wavepacket is Gaussian, the wavepackets $\psi_\pm$ will also be approximately Gaussian but shifted upwards or downwards along the z-axis. For a derivation of the typical form of such wavepackets see [7, 16, 17].

For the Stern–Gerlach apparatus to serve its purpose, to distinguish spin states based on the position of the particle, the conditions of the experiment should be such that most particles will be detected at one of the two well-separated positions on a detector screen. Some of the dominant parameters, which determine the separation of the final wavepackets, are the initial wavepacket width, the strength and inhomogeneity of the magnetic field, and the time of flight during and after the interaction.

These parameters are varied by the experimenter when designing and testing the experiment until the overlap between the wavepackets $\int \psi_+^*(\mathbf{x}, t)\psi_-(\mathbf{x}, t)\mathrm{d}\mathbf{x}$ becomes extremely small such that for all practical purposes the wavepackets seem to be separated in space. However, according to quantum theory the overlap will in general always be non-zero. In other words, spin is not perfectly correlated with position on the detector screen.

If a small slit is made in the detector in the region we identify with "spin-up", the state immediately after the slit will be given by

$$\alpha\,|\!\uparrow\rangle \int \psi_+(\mathbf{x}, t)\,|\mathbf{x}\rangle\,\mathrm{d}\mathbf{x} + \beta\,|\!\downarrow\rangle \int \psi_-(\mathbf{x}, t)\,|\mathbf{x}\rangle\,\mathrm{d}\mathbf{x},$$

where now the integration is restricted to the small slit. If the experiment is well designed, one of the terms will be (exponentially) smaller than the other. In principle, we should also allow for an extremely small contribution where the particle tunnels through the detector screen.[2]

In practice, when the Stern–Gerlach apparatus is used as a preparation device, the smaller term will be discarded as the parameters of the setup were tuned specifically for reproducibility, i.e., it is tuned such that the smaller term is experimentally inaccessible to subsequent verification (using another device) due to the finite statistics and the resolution of any experiment. This leads to the erroneous conclusion that a pure state in spin-space can be obtained by application of the Stern–Gerlach apparatus. Theoretically, after the slit the following density matrix in the $\uparrow, \downarrow$-basis is obtained

$$\rho_s = \int \begin{pmatrix} |\alpha\psi_+|^2 & \alpha^*\psi_+^*\beta\psi_- \\ \alpha\psi_+\beta^*\psi_-^* & |\beta\psi_-|^2 \end{pmatrix} \mathrm{d}\mathbf{x}. \tag{23}$$

Experimentally, the factors in the density matrix can be tuned more-or-less continuously by the above-mentioned parameters, however, they will never be strictly equal to one or zero unless the exact initial spin-state was known, i.e., the exact value of α and β is known beforehand.

A further fundamental complication is that the magnetic field must have zero divergence, which implies that it cannot have a gradient in the field in only one direction [16]. Therefore, as the wavepacket has a finite width in space, each part of it couples to its local direction of the magnetic field; and these local directions are not precisely aligned with the single z-axis that is considered theoretically. Thus particles

[2]Similarly, in the two-slit experiment a photon is said to travel through both slits at once; however, in principle there is also a contribution that it tunnels through the screen itself, which is dependent on the thickness of the screen.

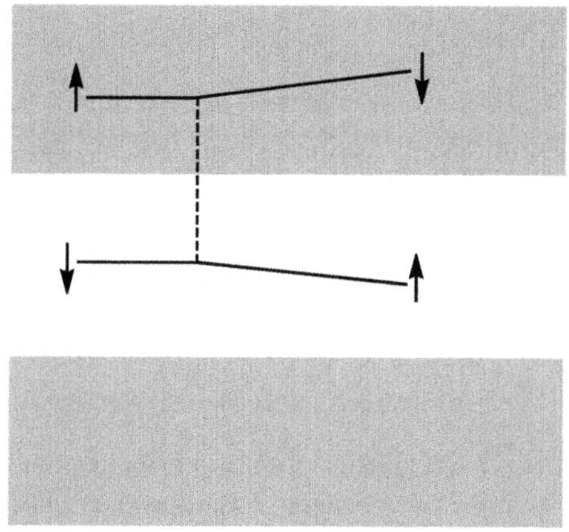

Fig. 7 A spin-flip process in the Stern–Gerlach experiment

with initially the same spin state along the quantization axis can, nevertheless, deviate according to that of the opposite spin.

Another important point is that the magnetic field and magnet were presumed to be classical. If the electromagnetic field is treated quantum mechanically as mediating interactions between the spin-1/2 test-particle and the particles in the Stern–Gerlach magnets, it would result in entanglement between the test-particle's position and those of the charge carriers in the coils of the Stern–Gerlach magnets. Namely, the charge carriers in the magnet would undergo a momentum increase, and thereby change their position, along the z-direction depending on the spin-component of the test-particle's wavefunction. After tracing out the states of the magnet, a density matrix is obtained similar to Eq. (23). Also, as spin exchange processes between the test-particle and the magnet's particles are always possible, there are again contributions which cause incorrect deflections to occur. Such processes are easy to visualize in the path-integral picture, which sums the amplitudes over all possible paths and interactions, see Fig. 7. Agreed, the suppression of spin-flips can be argued to be to be very strong under typical circumstances due to Pauli blocking of transitions to already occupied electronic states in the magnet: whereby we normally assume classical properties to the magnet.

Summarizing, the Stern–Gerlach experiment cannot be used as an ideal and reliable preparation device of spin states, even in principle, as there is no perfect one-to-one correspondence with position and spin. Note that the objections to this experiment creating pure states in spin are of a fundamental nature, namely they lie in the divergencelessness of the magnetic field, or the entanglement with the magnet with which it necessarily interacts, or the spatial extent of the wavefunctions.

References

1. D.J. Baker: Measurement outcomes and probability in Everettian quantum mechanics. Studies in History and Philosophy of Science Part B: Studies in History and Philosophy of Modern Physics 38, n1. 153–169. 2007.
2. A. Bassi, and G.C. Ghirardi: Dynamical reduction models. Phys. Rep. 379, 257–426. 2003. arXiv:quant-ph/0302164v2.
3. J. Butterfield: Less is different: Emergence and reduction reconciled. Found. Phys. 41, 1065. 2011. arXiv:1106.0702v1.
4. R. Dawid, and K. Thébault: Many worlds: decoherent or incoherent? Synthese 192(5), 1559–1580. 2015.
5. J. Earman: Curie's principle and spontaneous symmetry breaking. Int. Stud. Phil. Sci. 18, 173. 2004.
6. G.C. Ghirardi, R. Grassi, and F. Benatti: Describing the macroscopic world: closing the circle within the dynamical reduction program. Found. Phys. 25, 5–38. 1995.
7. G. Vandegrift: Accelerating wave packet solution to Schrödinger's equation. American Journal of Physics 68(6), 576–577. 2000.
8. G. Jona-Lasinio, F. Martinelli, and E. Scoppola: New approach to the semiclassical limit of quantum mechanics. Commun. Math. Phys. 80, 223–254. 1981.
9. A. Kent: One world versus many: The inadequacy of Everettian accounts of evolution, probability, and scientific confirmation. In Many Worlds? Everett, Quantum Theory, and Reality. S. Saunders, J. Barrett, D. Wallace and A. Kent, eds. ch 10. 307–355. Oxford University Press. 2010. arXiv:0905.0624v3.
10. N.P. Landsman: Spontaneous symmetry breaking in quantum systems: emergence or reduction? Stud. Hist. Phil. Mod. Phys. 44(4), 379–394. 2013. arXiv:1305.4473v2.
11. N.P. Landsman: Between classical and quantum, Handbook of the Philosophy of Science Vol. 2: Philosophy of Physics, Ed. J. Butterfield and J. Earman, 417–553. 2007. arXiv:quant-ph/0506082v2.
12. N.P. Landsman: Bohrification: From classical concepts to commutative algebras, to appear in Niels Bohr in the 21st Century, eds. J. Faye, J. Folse. arXiv:1601.02794v2.
13. N.P. Landsman, and R. Reuvers: A Flea on Schrödinger's Cat. Found. Phys. 43(3), 373–407. 2013. arXiv:1210.2353v2.
14. K.J. McQueen: Four tails problems for dynamical collapse theories. Studies in History and Philosophy of Modern Physics 29, 10–18. 2015. arXiv:1501.05778v1.
15. T. Maudlin: Three measurement problems. Topoi 14, 7–15. 1995.
16. G. Potel, F. Barranco, S. Cruz-Barrios, and J. Gmez-Camacho: Quantum mechanical description of Stern–Gerlach experiments. Physical Review A 71(5). 2005. arXiv:quant-ph/0409206v1.
17. G.B. Roston, M. Casas, A. Plastino, and A.R. Plastino: Quantum entanglement, spin-1/2 and the Stern–Gerlach experiment. European Journal of Physics 26(4), 657–672. 2005.
18. D. Spehner, and F. Haake: Quantum measurements without macroscopic superpositions. Physical Review A 77, 052114. 2008. arXiv:0711.0943v2.
19. R. Tumulka: Paradoxes and primitive ontology in collapse theories of quantum mechanics. arXiv:1102.5767. 2011.
20. D. Wallace: The Emergent Multiverse. Oxford University Press. 2012. arXiv:1102.5767v2.
21. J. van Wezel: An instability of unitary quantum dynamics. arXiv:1502.07527v1. 2015.
22. W.H. Zurek: Probabilities from entanglement, Born's rule $p_k = |\psi_k|^2$ from envariance. Phys. Rev. A71, 052105. 2005. arXiv:quant-ph/0405161v2.
23. W.H. Zurek: Quantum jumps, Born's rule, and objective reality. In Many Worlds? Everett, Quantum Theory, and Reality. S. Saunders, sJ. Barrett, D. Wallace and A. Kent, eds. ch 10. 307–355. Oxford University Press. 2010.

Measuring Processes and the Heisenberg Picture

Kazuya Okamura

Abstract In this paper, we attempt to establish quantum measurement theory in the Heisenberg picture. First, we review foundations of quantum measurement theory, that is usually based on the Schrödinger picture. The concept of instrument is introduced there. Next, we define the concept of system of measurement correlations and that of measuring process. The former is the exact counterpart of instrument in the (generalized) Heisenberg picture. In quantum mechanical systems, we then show a one-to-one correspondence between systems of measurement correlations and measuring processes up to complete equivalence. This is nothing but a unitary dilation theorem of systems of measurement correlations. Furthermore, from the viewpoint of the statistical approach to quantum measurement theory, we focus on the extendability of instruments to systems of measurement correlations. It is shown that all completely positive (CP) instruments are extended into systems of measurement correlations. Lastly, we study the approximate realizability of CP instruments by measuring processes within arbitrarily given error limits.

Keywords Quantum measurement theory · System of measurement correlations · CP instrument

1 Introduction

In this paper, we mathematically investigate measuring processes in the Heisenberg picture. We aim to extend the framework of quantum measurement theory and to apply the method in this paper not only to quantum systems of finite degrees of freedom but also to those with infinite degrees of freedom.

K. Okamura (✉)
Graduate School of Informatics, Nagoya University, Chikusa-ku, Nagoya 464-8601, Japan
e-mail: okamura@math.cm.is.nagoya-u.ac.jp

M. Ozawa et al. (eds.), *Reality and Measurement in Algebraic Quantum Theory*,
Springer Proceedings in Mathematics & Statistics 261,
https://doi.org/10.1007/978-981-13-2487-1_14

It is well-known that correlation functions are essential for the description of systems in quantum theory and in quantum probability theory. Typical examples are Wightman functions in (axiomatic) quantum field theory [45] and several (algebraic, noncommutative) independence in quantum probability theory (see [19, 25] and references therein), which are characterized by behaviors of correlation functions. In the famous paper by Accardi, Frigerio and Lewis [1], general classes of quantum stochastic processes including quantum Markov processes were characterized in terms of correlation functions. The main result of [1] is a noncommutative version of Kolmogorov's theorem stating that every quantum stochastic process can be reconstructed from a family of correlation functions up to equivalence. The proof of this result is made by the efficient use of positive-definiteness of a family of correlation functions. Later Belavkin [6] formulated the theory of operator-valued correlation functions, which is more flexible than the original formulation in [1] and gives an opportunity for reconsidering the standard formulation of qunatum theory. And he extended the main result of [1]. We apply Belavkin's theory, with some modifications, to a systematic characterization of measurement correlations herein.

Measurements are described by the notion of instrument introduced Davies and Lewis [11]. An instrument $\mathcal{I}$ for $(\mathcal{M}, S)$ is defined as a $P(\mathcal{M}_*)$-valued measure $\mathcal{F} \ni \Delta \mapsto \mathcal{I}(\Delta) \in P(\mathcal{M}_*)$, where $\mathcal{M}$ is a von Neumann algebra with predual $\mathcal{M}_*$, $P(\mathcal{M}_*)$ is the set of positive linear map of $\mathcal{M}$ and $(S, \mathcal{M})$ is a measurable space. The statistical description of measurements in terms of instruments can be regarded as a kind of quantum dynamical process based on the so-called Schrödinger picture. As widely accepted, the Schrödinger picture stands on describing states as time-dependent variables and observables as constants with respect to time, i.e., time-independent variables while we treat states as constants with respect to time and observables as time-dependent variables in the Heisenberg picture. To the author's knowledge, the Schrödinger picture is matched with an operational approach to quantum theory concerning probability distributions of observables and of output variables of apparatuses [9–11, 34]. On the other hand, no systematic treatment of measurements in the Heisenberg picture, which can compare with the theory of instruments, has been investigated. In contrast to the Schrödinger picture, the Heisenberg picture focuses on dynamical changes of observables and can naturally treat correlation functions of observables at different times, so that enables us to examine the dynamical nature of the system under consideration itself in detail. The Heisenberg picture is better than the Schrödinger picture at this point. Therefore, inspired by the previous investigations on quantum stochastic processes and correlation functions [1, 6], we define the concept of system of measurement correlations. This is the exact counterpart of instrument in a "generalized" Heisenberg picture and defined as a family of "operator-valued" multilinear maps satisfying "positive-definiteness", "σ-additivity" and other conditions. An instrument induced by a system of measurement correlations is always completely positive. In addition, we redefine measuring process (Definition 9) in order that it becomes consistent with the definition of system of measurement correlations. In the quantum mechanical case, we show that every system of measurement correlations is defined by a measuring process. It is, however, difficult to extend this result to general von Neumann algebras. Therefore, we develop another aspect of measurements, which is deeply analyzed for the first time in this paper. From the statistical viewpoint as the starting point of quantum measurement theory, we discuss the extendability of CP instruments to systems of measurement correlations and the realizability of CP instruments by measuring processes. In physically relevant cases, we show that both are possible within arbitrary accuracy.

The purpose of this paper is to mathematically develop the unitary dilation theory of systems of measurement correlations and of CP instruments. Dilation theory is one of main topics in functional analysis and enables us to apply representation theory and harmonic analysis to operators or to operator algebras. Especially, the unitary dilation theory of contractions on Hilbert space [18, 26] and the dilation theory of completely positive maps [4, 44] have been studied in many investigations (see [4, 13, 18, 20, 21, 26, 39, 42–44] and references therein). A representation theorem of CP instruments on the set $\mathbf{B}(\mathscr{H})$ of bounded operators on a Hilbert space $\mathscr{H}$ [31, Theorem 5.1] (Theorem 1) follows from these results, which shows the existence of unitary dilations of CP instruments. The proof of this theorem is based on the theory of CP-measure space [28, 29]. In the case of CP instruments, a unitary dilation of a CP instrument is nothing but a measuring process which realizes it. We generalize this representation theorem to systems of measurement correlations defined on $\mathbf{B}(\mathscr{H})$ in terms of Kolmogorov's theorem. It should be remarked that CP instruments defined on general von Neumann algebras do not always admit unitary dilations (see Examples 1 and 2). Next, we consider the extendability of CP instruments to systems of measurement correlations. It will be shown that all CP instruments can be extended into systems of measurement correlations. Furthermore, we show that every CP instrument defined on general von Neumann algebras can be approximated by measuring processes within arbitrarily given error limits $\varepsilon > 0$. If von Neumann algebras are injective or injective factors, measuring processes approximating a CP instrument can be chosen to be faithful or inner, respectively.

Preliminaries are given in Sect. 2; Foundations of quantum measurement theory, kernels and their Kolmogorov decompositions are explained. We introduce a system of measurement correlations and prove a representation theorem of systems of measurement correlations in Sect. 3. In Sect. 4, we define measuring processes and their complete equivalence, and in the case of $\mathbf{B}(\mathscr{H})$ we show a unitary dilation theorem of systems of measurement correlations establishing a one-to-one correspondence between systems of measurement correlations and complete equivalence classes of measuring processes. In Sect. 5, we discuss a generalization of the main result in Sect. 4 to arbitrary von Neumann algebras, and the extendability of CP instruments to systems of measurement correlations. We show that for any CP instruments there always exists a systems of measurement correlations which defines a given CP instrument. In Sect. 6, we explore the existence of measuring processes which approximately realizes a given CP instrument. We show several approximate realization theorems of CP instruments by measuring processes.

2 Preliminaries

In this paper, we assume that von Neumann algebras $\mathcal{M}$ are σ-finite. However, only in the case of $\mathcal{M} = \mathbf{B}(\mathcal{H})$, the von Neumann algebra of bounded operators on a Hilbert space $\mathcal{H}$, this assumption is ignored.

2.1 Foundations of Quantum Measurement Theory

We introduce foundations of quantum measurement theory herein. To precisely understand the theory of quantum measurement and its mathematics, the most important thing is to know how measurements physically realizable in experimental settings are described as physical processes consistent with statistical characterization of measurements. We refer the reader to [29, 35, 37] for detailed introductory expositions of quantum measurement theory.

The history of quantum measurement theory is long as much as those of quantum theory, but the modern theory of quantum measurement began with the mathematical study of the notion of instruments introduced by Davies and Lewis [11]. They proposed that we should abandon *the repeatability hypothesis* [11, 27, 38] as a general principle and employ an operational approach to quantum measurement, which is based on the mathematical description of measurements in terms of instruments defined as follows. Let **S** be a system whose observables and states are described by self-adjoint operators affiliated to a von Neumann algebra $\mathcal{M}$ on a Hilbert space $\mathcal{H}$ and by normal states on $\mathcal{M}$, respectively. $\mathcal{M}_*$ denotes the predual of $\mathcal{M}$, i.e., the set of ultraweakly continuous linear functionals on $\mathcal{M}$, $\mathcal{S}_n(\mathcal{M})$ does that of normal states on $\mathcal{M}$ and $P(\mathcal{M}_*)$ does that of positive linear maps on $\mathcal{M}_*$.

Definition 1 (*Instrument, Davies-Lewis* [11, p.243, ll.21–26]) Let $\mathcal{M}$ be a von Neumann algebra on a Hilbert space $\mathcal{H}$ and $(S, \mathcal{F})$ a measurable space. A map $\mathcal{I} : \mathcal{F} \to P(\mathcal{M}_*)$ is called an instrument for $(\mathcal{M}, S)$ if it satisfies the following conditions:

(1) $\langle \mathcal{I}(S)\rho, 1\rangle = \langle \rho, 1\rangle$ for all $\rho \in \mathcal{M}_*$, where $\langle \cdot, \cdot\rangle$ denotes the duality pairing of $\mathcal{M}_*$ and $\mathcal{M}$;
(2) For every $M \in \mathcal{M}$, $\rho \in \mathcal{M}_*$ and mutually disjoint sequence $\{\Delta_j\}$ of $\mathcal{F}$,

$$\langle \mathcal{I}(\cup_j \Delta_j)\rho, M\rangle = \sum_j \langle \mathcal{I}(\Delta_j)\rho, M\rangle. \tag{1}$$

We define the dual map $\mathcal{I}^*$ of an instrument $\mathcal{I}$ by $\langle \rho, \mathcal{I}^*(\Delta)M\rangle = \langle \mathcal{I}(\Delta)\rho, M\rangle$ and use the notation $\mathcal{I}(M, \Delta) = \mathcal{I}^*(\Delta)M$ for all $\Delta \in \mathcal{F}$ and $M \in \mathcal{M}$.

It is obvious, by the definition, that every instrument describes the weighted state changes caused by the measurement. The dual map $\mathcal{I} : \mathcal{M} \times \mathcal{F} \to \mathcal{M}$ of an instrument $\mathcal{I}$ for $(\mathcal{M}, S)$ is characterized by the following conditions:

(i) For every $\Delta \in \mathscr{F}$, the map $\mathscr{M} \ni M \mapsto \mathscr{I}(M, \Delta) \in \mathscr{M}$ is normal positive linear map of $\mathscr{M}$;
(ii) $\mathscr{I}(1, S) = 1$;
(iii) For every $M \in \mathscr{M}$, $\rho \in \mathscr{M}_*$ and mutually disjoint sequence $\{\Delta_j\}$ of $\mathscr{F}$,

$$\langle \rho, \mathscr{I}(M, \cup_j \Delta_j)\rangle = \sum_j \langle \rho, \mathscr{I}(M, \Delta_j)\rangle. \tag{2}$$

Since every map $\mathscr{I} : \mathscr{M} \times \mathscr{F} \to \mathscr{M}$ satisfying the above conditions is always the dual map of an instrument for $(\mathscr{M}, S)$, we also call the map $\mathscr{I}$ an instrument for $(\mathscr{M}, S)$.

Davies and Lewis claimed that experimentally and statistically accessible ingredients via measurements by a given measuring apparatus should be specified by instruments as follows:

Davies-Lewis proposal *For every apparatus* $\mathbf{A}(\mathbf{x})$ *measuring* $\mathbf{S}$*, where* $\mathbf{x}$ *is the output variable of* $\mathbf{A}(\mathbf{x})$ *taking values in a measurable space* $(S, \mathscr{F})$*, there always exists an instrument* $\mathscr{I}$ *for* $(\mathscr{M}, S)$ *corresponding to* $\mathbf{A}(\mathbf{x})$ *in the following sense. For every input state* ρ*, the probability distribution* $\Pr\{\mathbf{x} \in \Delta \| \rho\}$ *of* $\mathbf{x}$ *in* ρ *is given by*

$$\Pr\{\mathbf{x} \in \Delta \| \rho\} = \|\mathscr{I}(\Delta)\rho\| = \langle \mathscr{I}(\Delta)\rho, 1\rangle \tag{3}$$

for all $\Delta \in \mathscr{F}$*, and the state* $\rho_{\{\mathbf{x}\in\Delta\}}$ *after the measurement under the condition that* ρ *is the prepared state and the outcome* $\mathbf{x} \in \Delta$ *is given by*

$$\rho_{\{\mathbf{x}\in\Delta\}} = \frac{\mathscr{I}(\Delta)\rho}{\|\mathscr{I}(\Delta)\rho\|} \tag{4}$$

if $\Pr\{\mathbf{x} \in \Delta \| \rho\} > 0$*, and* $\rho_{\{\mathbf{x}\in\Delta\}}$ *is indefinite if* $\Pr\{\mathbf{x} \in \Delta \| \rho\} = 0$.

Although this proposal is very general, it was not evident at that time that how this is related to the standard formulation of quantum theory. In the 1980s, Ozawa [30, 31] introduced both completely positive (CP) instruments and measuring processes. Following this investigation, the standpoint of the above proposal in quantum mechanics was settled and the circumstances changed at all. An instrument $\mathscr{I}$ for $(\mathscr{M}, S)$ is said to be *completely positive* if $\mathscr{I}(\Delta)$ (or $\mathscr{I}(\cdot, \Delta)$, equivalently) is completely positive for every $\Delta \in \mathscr{F}$. $\mathrm{CPInst}(\mathscr{M}, S)$ denotes the set of CP instruments for $(\mathscr{M}, S)$. The notion of measuring process is defined as a quantum mechanical modeling of an apparatus as a physical system, of the meter of the apparatus, and of the measuring interaction between the system and the apparatus. Let $\mathscr{H}$ and $\mathscr{K}$ be Hilbert spaces. $\mathbf{B}(\mathscr{H})$ denotes the set of bounded linear operators on $\mathscr{H}$ and $\mathbf{B}(\mathscr{H}, \mathscr{K})$ does the set of bounded linear operators of $\mathscr{H}$ to $\mathscr{K}$. Let $\mathscr{M}$ and $\mathscr{N}$ be von Neumann algebras on $\mathscr{H}$ and $\mathscr{K}$, respectively. $\mathscr{M}\overline{\otimes}\mathscr{N}$ denotes the W*-tensor product of $\mathscr{M}$ and $\mathscr{N}$. For every $\sigma \in \mathscr{N}_*$, a linear map $\mathrm{id} \otimes \sigma : \mathscr{M}\overline{\otimes}\mathscr{N} \to \mathscr{M}$ is defined by $\langle \rho, (\mathrm{id} \otimes \sigma)X\rangle = \langle \rho \otimes \sigma, X\rangle$ for all $X \in \mathscr{M}\overline{\otimes}\mathscr{N}$ and $\rho \in \mathscr{M}_*$. The following is the mathematical definition of measuring processes:

Definition 2 (*Measuring process* [29, Definition 3.2]) A measuring process for $(\mathcal{M}, S)$ is a 4-tuple $\mathbb{M} = (\mathcal{K}, \sigma, E, U)$ of a Hilbert space $\mathcal{K}$, a normal state σ on $\mathbf{B}(\mathcal{K})$, a PVM $E : \mathcal{F} \to \mathbf{B}(\mathcal{K})$ and a unitary operator U on $\mathcal{H} \otimes \mathcal{K}$ satisfying

$$(\mathrm{id} \otimes \sigma)[U^*(M \otimes E(\Delta))U] \in \mathcal{M} \tag{5}$$

for all $M \in \mathcal{M}$ and $\Delta \in \mathcal{F}$.

Let $\mathbb{M} = (\mathcal{K}, \sigma, E, U)$ be a measuring process for $(\mathcal{M}, S)$. Then a CP instrument $\mathcal{I}_{\mathbb{M}}$ for $(\mathcal{M}, S)$ is defined by

$$\mathcal{I}_{\mathbb{M}}(M, \Delta) = (\mathrm{id} \otimes \sigma)[U^*(M \otimes E(\Delta))U] \tag{6}$$

for $M \in \mathcal{M}$ and $\Delta \in \mathcal{F}$. The most important example of meauring processes is a von Neumann model $(L^2(\mathbb{R}), \omega_\alpha, E^Q, e^{-iA\otimes P/\hbar})$ of measurement of an observable A, a self-adjoint operator affiliated with $\mathcal{M}$, where α is a unit vector of $L^2(\mathbb{R})$, ω_α is defined by $\omega_\alpha(M) = \langle \alpha | M\alpha \rangle$ for all $M \in \mathbf{B}(L^2(\mathbb{R}))$, and $Q = \int_{\mathbb{R}} q \, dE^Q(q)$ and P are self-adjoint operators defined on dense linear subspaces of $L^2(\mathbb{R})$ such that $\overline{[Q, P]} = i\hbar 1$. Here, E^X denotes the spectral measure of a self-adjoint operator X densely defined on a Hilbert space. Quantum mechanical modeling of appratuses began with this model [27, 33].

Two measuring processes are *statistically equivalent* if they define an identical instrument. As seen above, a measuring process $\mathbb{M}$ for $(\mathcal{M}, S)$ defines a CP instrument $\mathcal{I}_{\mathbb{M}}$ for $(\mathcal{M}, S)$. In the case of $\mathcal{M} = \mathbf{B}(\mathcal{H})$, the following theorem, a unitary dilation theorem of CP instruments for $(\mathbf{B}(\mathcal{H}), S)$, is known to hold.

Theorem 1 ([30], [31, Theorem 5.1], [29, Theorem 3.6]) *For every CP instrument $\mathcal{I}$ for $(\mathbf{B}(\mathcal{H}), S)$, there uniquely exists a statistical equivalence class of measuring processes $\mathbb{M} = (\mathcal{K}, \sigma, E, U)$ for $(\mathbf{B}(\mathcal{H}), S)$ such that $\mathcal{I}(M, \Delta) = \mathcal{I}_{\mathbb{M}}(M, \Delta)$ for all $M \in \mathbf{B}(\mathcal{H})$ and $\Delta \in \mathcal{F}$. Conversely, every statistical equivalence class of measuring processes for $(\mathbf{B}(\mathcal{H}), S)$ defines a unique CP instrument $\mathcal{I}$ for $(\mathbf{B}(\mathcal{H}), S)$.*

A generalization of this theorem to arbitrary von Neumann algebras is shown to hold not for all CP instruments but for those with the normal extension property (NEP) in [29] (see Theorem 2). Let $(S, \mathcal{F}, \mu)$ be a measure space. $\mathcal{L}(S, \mu)$ denotes the *-algebra of complex-valued μ-measurable functions on S. A μ-measurable function f is said to be μ-negligible if $f(s) = 0$ for μ-a.e. $s \in S$. $\mathcal{N}(S, \mu)$ denotes the set of μ-negligible functions on S and $M^\infty(S, \mu)$ does the *-algebra of bounded μ-measurable functions on S. It is obvious that $M^\infty(S, \mu) \subset \mathcal{L}(S, \mu)$ as *-algebra. For any $1 \leq p < \infty$, $L^p(S, \mu)$ denotes the Banach space of p-integrable functions on S with respect to μ modulo the μ-negligible functions. $[f]$ denotes the μ-negligible equivalence class of $f \in \mathcal{L}(S, \mu)$ and $L^\infty(S, \mu)$ does the commutative von Neumann algebra on $L^2(S, \mu)$. $L^\infty(S, \mathcal{I})$ denotes the W*-algebra of essentially bounded $\mathcal{I}$-measurable functions on S modulo the $\mathcal{I}$-negligible functions.

Definition 3 (*Normal extension property* [29, Definition 3.3]) Let $\mathcal{M}$ be a von Neumann algebra on a Hilbert space $\mathcal{H}$ and $(S, \mathcal{F})$ a measurable space. Let $\mathcal{I}$ be a CP instrument for $(\mathcal{M}, S)$ and $\Psi_{\mathcal{I}} : \mathcal{M} \otimes_{\min} L^{\infty}(S, \mathcal{I}) \to \mathcal{M}$ the corresponding unital binormal CP map, i.e., $\Psi_{\mathcal{I}}$ is normal on $\mathcal{M}$ and $L^{\infty}(S, \mathcal{I})$ and satisfies $\Psi_{\mathcal{I}}(M \otimes [\chi_{\Delta}]) = \mathcal{I}(M, \Delta)$ for all $M \in \mathcal{M}$ and $\Delta \in \mathcal{F}$ [29, Proposition 3.3]. $\mathcal{I}$ is said to have the normal extension property (NEP) if there exists a unital normal CP map $\widetilde{\Psi_{\mathcal{I}}} : \mathcal{M}\overline{\otimes}L^{\infty}(S, \mathcal{I}) \to \mathcal{M}$ such that $\widetilde{\Psi_{\mathcal{I}}}|_{\mathcal{M} \otimes_{\min} L^{\infty}(S,\mathcal{I})} = \Psi_{\mathcal{I}}$. $\mathrm{CPInst}_{\mathrm{NE}}(\mathcal{M}, S)$ denotes the set of CP instruments for $(\mathcal{M}, S)$ with the NEP.

We then have the following theorem, a generalization of Theorem 1.

Theorem 2 ([29, Theorem 3.4]) *For a CP instrument $\mathcal{I}$ for $(\mathcal{M}, S)$, the following conditions are equivalent:*
(i) $\mathcal{I}$ has the NEP.
(ii) There exists CP instrument $\mathcal{I}$ for $(\mathbf{B}(\mathcal{H}), S)$ such that $\widetilde{\mathcal{I}}(M, \Delta) = \mathcal{I}(M, \Delta)$ for all $M \in \mathcal{M}$ and $\Delta \in \mathcal{F}$.
(iii) There exists a measuring process $\mathbb{M} = (\mathcal{K}, \sigma, E, U)$ for $(\mathcal{M}, S)$ such that $\mathcal{I}(M, \Delta) = \mathcal{I}_{\mathbb{M}}(M, \Delta)$ for all $M \in \mathcal{M}$ and $\Delta \in \mathcal{F}$.

It is also shown that all CP instruments defined on a von Neumann algebra $\mathcal{M}$ have the NEP, i.e., $\mathrm{CPInst}(\mathcal{M}, S) = \mathrm{CPInst}_{\mathrm{NE}}(\mathcal{M}, S)$ if $\mathcal{M}$ is atomic [29, Theorem 4.1]. We should remember the famous fact that a von Neumann algebra $\mathcal{M}$ on a Hilbert space $\mathcal{H}$ is atomic if and only if there exists a normal conditional expectation $\mathcal{E} : \mathbf{B}(\mathcal{H}) \to \mathcal{M}$ [46, Chapter V, Section 2, Excercise 8]. Then the following question naturally arises.

Question 1 Let $\mathcal{M}$ be a non-atomic von Neumann algebra on a Hilbert space $\mathcal{H}$ and $(S, \mathcal{F})$ a measurable space. Are there CP instruments for $(\mathcal{M}, S)$ without the NEP? For any CP instrument $\mathcal{I}$ for $(\mathcal{M}, S)$, does there exist a measuring process $\mathbb{M}$ for $(\mathcal{M}, S)$ which realizes $\mathcal{I}$ within arbitrarily given error limits $\varepsilon > 0$?

A CP instrument $\mathcal{I}$ for $(\mathcal{M}, S)$ is said to have *the approximately normal extension property* (ANEP) if there is a net $\{\mathcal{I}_{\alpha}\}$ of CP instruments with the NEP such that $\mathcal{I}_{\alpha}(M, \Delta)$ is ultraweakly converges to $\mathcal{I}(M, \Delta)$ for all $M \in \mathcal{M}$ and $\Delta \in \mathcal{F}$. $\mathrm{CPInst}_{\mathrm{AN}}(\mathcal{M}, S)$ denotes the set of CP instruments for $(\mathcal{M}, S)$ with the ANEP.

Contrary to physicists' expectations, Question 1 was positively resolved in [29, Section V] for non-atomic but injective von Neumann algebras.

Definition 4 ([29, Definition 5.3]) (1) An instrument $\mathcal{I}$ for $(\mathcal{M}, S)$ is called repeatable if $\mathcal{I}(\Delta_2)\mathcal{I}(\Delta_1) = \mathcal{I}(\Delta_2 \cap \Delta_1)$ for all $\Delta_1, \Delta_2 \in \mathcal{F}$.
(2) An instrument $\mathcal{I}$ for $(\mathcal{M}, S)$ is called weakly repeatable if $\mathcal{I}(\mathcal{I}(1, \Delta_2), \Delta_1) = \mathcal{I}(1, \Delta_2 \cap \Delta_1)$ for all $\Delta_1, \Delta_2 \in \mathcal{F}$.
(3) An instrument $\mathcal{I}$ for $(\mathcal{M}, S)$ is called discrete if there exist a countable subset S_0 of S and a map $T : S_0 \to P(\mathcal{M}_*)$ such that

$$\mathcal{I}(\Delta) = \sum_{s\in\Delta} T(s) \tag{7}$$

for all $\Delta \in \mathcal{F}$.

Proposition 1 ([29, Proposition 5.9]) *Let $\mathscr{M}$ be a von Neumann algebra on a Hilbert space $\mathscr{H}$ and $(S, \mathscr{F})$ a measurable space. Every discrete CP instrument $\mathscr{I}$ for $(\mathscr{M}, S)$ has the NEP.*

Theorem 3 ([29, Theorem 5.10]) *Let $\mathscr{M}$ be a von Neumann algebra on a Hilbert space $\mathscr{H}$ and $(S, \mathscr{F})$ a standard Borel space. A weakly repeatable CP instrument $\mathscr{I}$ for $(\mathscr{M}, S)$ is discrete if and only if it has the NEP.*

In the case where $\mathscr{M}$ is non-atomic, there exist CP instruments for $(\mathscr{M}, S)$ without the NEP. The following two CP instruments are such examples.

Example 1 ([32, pp. 292–293], [29, Example 5.1]) Let m be Lebesgue measure on $[0, 1]$. A CP instrument $\mathscr{I}_m$ for $(L^\infty([0, 1], m), [0, 1])$ is defined by $\mathscr{I}_m(f, \Delta) = [\chi_\Delta]f$ for all $\Delta \in \mathscr{B}([0, 1])$ and $f \in L^\infty([0, 1], m)$.

A von Neumann algebra $\mathscr{M}$ is said to be approximately finite-dimensional (AFD) if there is an increasing net $\{\mathscr{M}_\alpha\}_{\alpha\in A}$ of finite-dimensional von Neumann subalgebras of $\mathscr{M}$ such that

$$\mathscr{M} = \overline{\bigcup_{\alpha\in A} \mathscr{M}_\alpha}^{uw}. \tag{8}$$

Example 2 ([29, Example 5.2]) Let $\mathscr{M}$ be an AFD von Neumann algebra of type II_1 on a separable Hilbert space $\mathscr{H}$. Let $A = \int_{\mathbb{R}} a\, dE^A(a)$ be a self-adjoint operator with continuous spectrum affiliated with $\mathscr{M}$ and $\mathscr{E}$ a (normal) conditional expectation of $\mathscr{M}$ onto $\{A\}' \cap \mathscr{M}$ (the existence of $\mathscr{E}$ was first found by [48, Theorem 1]), where $\{A\}' = \{E^A(\Delta) \mid \Delta \in \mathscr{B}(\mathbb{R})\}'$. A CP instrument $\mathscr{I}_A$ for $(\mathscr{M}, \mathbb{R})$ is defined by

$$\mathscr{I}_A(M, \Delta) = \mathscr{E}(M)E^A(\Delta) \tag{9}$$

for all $M \in \mathscr{M}$ and $\Delta \in \mathscr{B}(\mathbb{R})$.

By Theorem 3, the weak repeatability and the non-discreteness of $\mathscr{I}_m$ and $\mathscr{I}_A$ imply the non-existence of measuring processes which define them. These examples are very important for the dilation theory of CP maps since they revealed the existence of families of CP maps which do not admit unitary dilations.

The following theorem holds for general σ-finite von Neumann algebras without assuming any other conditions.

Theorem 4 *Let $\mathscr{M}$ be a von Neumann algebra on a Hilbert space $\mathscr{H}$ and $(S, \mathscr{F})$ a measurable space. For every CP instrument $\mathscr{I}$ for $(\mathscr{M}, S)$, $n \in \mathbb{N}$, $\rho_1, \ldots, \rho_n \in \mathscr{S}_n(\mathscr{M})$, $M_1, \ldots, M_n \in \mathscr{M}$ and $\Delta_1, \ldots, \Delta_n \in \mathscr{F}$, there exists a measuring process $\mathbb{M} = (\mathscr{K}, \sigma, E, U)$ for $(\mathscr{M}, S)$ such that*

$$\langle \rho_j, \mathscr{I}(M_j, \Delta_j)\rangle = \langle \rho_j, \mathscr{I}_{\mathbb{M}}(M_j, \Delta_j)\rangle \tag{10}$$

for all $j = 1, \ldots, n$.

Proof Let $n \in \mathbb{N}$, $\rho_1, \ldots, \rho_n \in \mathcal{S}_n(\mathcal{M})$, $M_1, \ldots, M_n \in \mathcal{M}\backslash\{0\}$ and $\Delta_1, \ldots, \Delta_n \in \mathcal{F}\backslash\{\emptyset\}$. Let $\mathcal{F}'$ be a σ-subfield of $\mathcal{F}$ generated by $\Delta_1, \ldots, \Delta_n, S$. Let $\{\Gamma_i\}_{i=1}^m \subset \mathcal{F}'\backslash\{\emptyset\}$ be a maximal partition of S, i.e., $\{\Gamma_i\}_{i=1}^m$ satisfies the following conditions:
(1) For every $i = 1, \ldots, m$, if $\Delta \in \mathcal{F}'$ satisfies $\Delta \subset \Gamma_i$, then Δ is Γ_i or $\emptyset$;
(2) $\cup_{i=1}^m \Gamma_i = S$;
(3) $\Gamma_i \cap \Gamma_j = \emptyset$ if $i \neq j$.
We fix $s_1, \ldots .s_m \in S$ such that $s_i \in \Gamma_i$ for all $i = 1, \ldots, m$. We define a discrete CP instrument $\mathcal{I}'$ for $(\mathcal{M}, S)$ by

$$\mathcal{I}'(M, \Delta) = \sum_{j=1}^{m} \delta_{s_j}(\Delta)\mathcal{I}(M, \Gamma_j) \tag{11}$$

for all $M \in \mathcal{M}$ and $\Delta \in \mathcal{F}$. It is then obvious that $\mathcal{I}'$ satisfies

$$\langle \rho_j, \mathcal{I}(M_j, \Delta_j)\rangle = \langle \rho_j, \mathcal{I}'(M_j, \Delta_j)\rangle \tag{12}$$

for all $j = 1, \ldots, n$. By Proposition 1, there exists a measuring process $\mathbb{M} = (\mathcal{K}, \sigma, E, U)$ for $(\mathcal{M}, S)$ such that $\mathcal{I}'(M, \Delta) = \mathcal{I}_{\mathbb{M}}(M, \Delta)$ for all $M \in \mathcal{M}$ and $\Delta \in \mathcal{F}$. The proof is complete.

Corollary 1 *Let $\mathcal{M}$ be a von Neumann algebra on a Hilbert space $\mathcal{H}$ and $(S, \mathcal{F})$ a measurable space. Then we have*

$$\mathrm{CPInst}_{\mathrm{AN}}(\mathcal{M}, S) = \mathrm{CPInst}(\mathcal{M}, S). \tag{13}$$

In the case where $\mathcal{M}$ is injective, the result stronger than Theorem 4 is shown in [29, Theorem 4.2]: for every CP instrument $\mathcal{I}$ for $(\mathcal{M}, S)$, $\varepsilon > 0$, $n \in \mathbb{N}$, $\{\rho_i\}_{i=1}^n \subset \mathcal{S}_n(\mathcal{M})$, $\{\Delta_i\}_{i=1}^n \subset \mathcal{F}$ and $\{M_i\}_{i=1}^n \subset \mathcal{M}$, there exists a measuring process $\mathbb{M}$ for $(\mathcal{M}, S)$ such that

$$|\langle \mathcal{I}(\Delta_i)\rho_i, M_i\rangle - \langle \mathcal{I}_{\mathbb{M}}(\Delta_i)\rho_i, M_i\rangle| < \varepsilon \tag{14}$$

for all $i = 1, 2, \ldots, n$, and that $\mathcal{I}(1, \Delta) = \mathcal{I}_{\mathbb{M}}(1, \Delta)$ for all $\Delta \in \mathcal{F}$. In physically relevant cases, it is known that every von Neumann algebra $\mathcal{M}$ describing the observable algebra of a quantum system acts on a separable Hilbert space and is AFD. For example, it is shown in [7] that von Neumann algebras of local observables in quantum field theory are AFD and acts on a separable Hilbert space under natural postulates, e.g., the Wightman axioms, the nuclearity condition and the asymptotic scale invariance. For every von Neumann algebra $\mathcal{M}$ on a separable Hilbert space (or with separable dual, equivalently), $\mathcal{M}$ is AFD if and only if it is injective, furthermore, if and only if it is amenable [8, 47]. Hence the assumption of the injectivity for von Neumann algebras is very natural.

In quantum mechanics, complete positivity of instruments is physically justified in [35, 37] by considering a natural extendability, called *the trivial extendability*, of an instrument $\mathcal{I}$ on the system $\mathbf{S}$ to that $\mathcal{I}'$ on the composite system $\mathbf{S}+\mathbf{S}'$

containing the original one **S**, where **S**$'$ is an arbitrary system not interacting with **S** nor **A**(**x**). This justification of complete positivity is obtained as a part of an axiomatic characterization of physically realizable measurements [35, 37]. Then Theorem 1 enables us to regard the Davies-Lewis proposal restricted to CP instruments as a statement that is consistent with the standard formulation of quantum mechanics and hence acceptable for physicists. The above discussion is summarized as follows.

Davies-Lewis-Ozawa criterion *For every apparatus* **A**(**x**) *measuring* **S**, *where* **x** *is the output variable of* **A**(**x**) *taking values in a measurable space* $(S, \mathscr{F})$, *there always exists a CP instrument* $\mathscr{I}$ *for* $(\mathscr{M}, S)$ *corresponding to* **A**(**x**) *in the sense of the Davies-Lewis proposal, i.e., for every input state* ρ *and outcome* $\Delta \in \mathscr{F}$ *both the probability distribution* $\Pr\{\mathbf{x} \in \Delta \| \rho\}$ *of* **x** *and the state* $\rho_{\{\mathbf{x}\in\Delta\}}$ *after the measurement are obtained from* $\mathscr{I}$.

2.2 Kernels

Here, we briefly summerize the theory of kernels. We refer the reader to [13, 14, 43] for standard references.

Definition 5 (*Kernel* [14, p.11, ll.1–3]) Let C be a set and $\mathscr{H}$ a Hilbert space. A map $K : C \times C \to \mathbf{B}(\mathscr{H})$ is called a kernel of C on $\mathscr{H}$. $\mathbb{K}(C; \mathscr{H})$ denotes the set of kernels of C on $\mathscr{H}$.

It should be noted that $\mathbb{K}(C; \mathscr{H})$ has a natural $\mathbf{B}(\mathscr{H})$-bimodule structure.

Definition 6 ([14, Definition 1.1]) Let $K \in \mathbb{K}(C; \mathscr{H})$. K is said to be positive definite if

$$\sum_{i,j=1}^{n} \langle \xi_i | K(c_i, c_j)\xi_j \rangle \geq 0 \tag{15}$$

for every $n \in \mathbb{N}$, $c_1, c_2, \ldots, c_n \in C$ and $\xi_1, \xi_2, \ldots, \xi_n \in \mathscr{H}$. $\mathbb{K}(C; \mathscr{H})_+$ denotes the set of positive definite kernels of C on $\mathscr{H}$.

Definition 7 (*Kolmogorov decomposition* [14, Definition 1.3]) Let $K \in \mathbb{K}(C; \mathscr{H})$. A pair $(\mathscr{K}, \Lambda)$ of a Hilbert space $\mathscr{K}$ and a map $\Lambda : C \to \mathbf{B}(\mathscr{H}, \mathscr{K})$ is called a Kolmogorov decomposition of K if it satisfies

$$K(c, c') = \Lambda(c)^* \Lambda(c') \tag{16}$$

for all $c, c' \in C$. A Kolmogorov decomposition $(\mathscr{K}, \Lambda)$ of K is said to be minimal if $\mathscr{K} = \overline{\mathrm{span}}(\Lambda(C)\mathscr{H})$.

The following representation theorem holds for kernels.

Theorem 5 ([14, Lemma 1.4, Theorems 1.8 and 1.9]) *Let C be a set and $\mathscr{H}$ a Hilbert space. For every $K \in \mathbb{K}(C; \mathscr{H})$, K admits a Kolmogorov decomposition if and only if it is an element of $\mathbb{K}(C; \mathscr{H})_+$. For every $K \in \mathbb{K}(C; \mathscr{H})_+$, there exists a minimal Kolmogorov decomposition $(\mathscr{K}, \Lambda)$ of K, which is unique up to unitary equivalence.*

This theorem is a key to the proof of the main theorem of the paper. The famous Stinespring representation theorem is regarded as a corollary of this theorem.

Theorem 6 (Arveson [4, Theorem 1.3.1], [39, Theorem 12.7]) *Let $\mathscr{H}$, $\mathscr{K}$ be Hilbert spaces, $\mathscr{B}$ a unital C^*-subalgebra of $\mathbf{B}(\mathscr{K})$ and V an element of $\mathbf{B}(\mathscr{H}, \mathscr{K})$ such that $\mathscr{K} = \overline{\text{span}}(\mathscr{B}V\mathscr{H})$. For every $A \in (V^*\mathscr{B}V)'$, there exists a unique $A_1 \in \mathscr{B}'$ such that $VA = A_1V$. Furthermore, the map $\pi' : A \in (V^*\mathscr{B}V)' \ni A \mapsto A_1 \in \mathscr{B}' \cap \{VV^*\}'$ is an ultraweakly continuous surjective $*$-homomorphism.*

The following theorem holds as a corollary of [12, Part I, Chapter 4, Theorem 3], [46, Chapter IV, Theorem 5.5]:

Theorem 7 *Let $\mathscr{H}_1$ and $\mathscr{H}_2$ be Hilbert spaces. If π is a normal representation of $\mathbf{B}(\mathscr{H}_1)$ on $\mathscr{H}_2$, there exist a Hilbert space $\mathscr{K}$ and a unitary operator U of $\mathscr{H}_1 \otimes \mathscr{K}$ onto $\mathscr{H}_2$ such that*

$$\pi(X) = U(X \otimes 1_{\mathscr{K}})U^* \tag{17}$$

for all $X \in \mathbf{B}(\mathscr{H}_1)$.

This theorem is also a key to the proof of the main theorem of the paper.

3 Systems of Measurement Correlations

In this section, we introduce the concept of system of measurement correlations, which is a natural, multivariate version of instrument and is defined as a family of multilinear maps satisfying "positive-definiteness", "σ-additivity" and other conditions. This is an appropriate abstraction of measurement correlations in the context of quantum stochastic processes [1]. It is known that the representation theory of CP instruments contributed to quantum measurement theory [29–31]. Hence we adopt a representation-theoretical approach to system of measurement correlations. The "positive-definiteness" of systems of measurement correlations enables us to apply the (minimal) Kolmogorov decomposition to them, so that provides them with representation-theoretical structures. As a result, a representation theorem (Theorem 8) similar to that for CP instruments [31, Proposition 4.2] will be shown to hold for systems of measurement correlations defined on an arbitrary von Neumann algebra. To precisely understand physics described by systems of measurement correlations we need a generalization of the Heisenberg picture which is introduced after the proof of Theorem 8 and is called the generalized Heisenberg picture. The introduction of this new picture is motivated also by the present circumstances that the understanding of the (usual) Heisenberg picture has not been deepened in contrast to the Schrödinger picture. It should be stressed that the circumstances are never restricted to quantum measurement theory.

We adopt the following notations.

Notation 3.1 *Let $\mathscr{T}^{(1)}$ be a set. We define a set $\mathscr{T}$ by $\mathscr{T} = \cup_{j=1}^{\infty}(\mathscr{T}^{(1)})^j$.*
(i) For each $T \in \mathscr{T}$, $|T|$ denotes the natural number n such that $T \in (\mathscr{T}^{(1)})^n$.
(ii) For each $T = (t_1, t_2, \dots, t_{n-1}, t_n) \in \mathscr{T}$, we define $T^{\#} \in \mathscr{T}$ by $T^{\#} = (t_n, t_{n-1}, \dots, t_2, t_1)$.
(iii) For any $T = (t_{1,1}, \dots, t_{1,m}), T_2 = (t_{2,1}, \dots, t_{2,n}) \in \mathscr{T}$, the product $T_1 \times T_2$ is defined by

$$T_1 \times T_2 = (t_{1,1}, \dots, t_{1,m}, t_{2,1}, \dots, t_{2,n}). \tag{18}$$

Since it holds that $T_1 \times (T_2 \times T_3) = (T_1 \times T_2) \times T_3$, $T_1 \times (T_2 \times T_3)$ is written as $T_1 \times T_2 \times T_3$.
(iv) For any $n \in \mathbb{N}$ and $\overrightarrow{M} = (M_1, M_2, \dots, M_{n-1}, M_n) \in \mathcal{M}^n$, we define $\overrightarrow{M}^{\#} \in \mathcal{M}^n$ by

$$\overrightarrow{M}^{\#} = (M_n^*, M_{n-1}^*, \dots, M_2^*, M_1^*). \tag{19}$$

(v) For any $m, n \in \mathbb{N}$, $\overrightarrow{M}_1 = (M_{1,1}, \dots, M_{1,m}) \in \mathcal{M}^m$ and $\overrightarrow{M}_2 = (M_{2,1}, \dots, M_{2,n}) \in \mathcal{M}^n$, the product $\overrightarrow{M}_1 \times \overrightarrow{M}_2 \in \mathcal{M}^{m+n}$ is defined by

$$\overrightarrow{M}_1 \times \overrightarrow{M}_2 = (M_{1,1}, \dots, M_{1,m}, M_{2,1}, \dots, M_{2,n}). \tag{20}$$

Since it holds that $\overrightarrow{M}_1 \times (\overrightarrow{M}_2 \times \overrightarrow{M}_3) = (\overrightarrow{M}_1 \times \overrightarrow{M}_2) \times \overrightarrow{M}_3$, $\overrightarrow{M}_1 \times (\overrightarrow{M}_2 \times \overrightarrow{M}_3)$ is written as $\overrightarrow{M}_1 \times \overrightarrow{M}_2 \times \overrightarrow{M}_3$.

In addition, for every family $\{\Pi_t\}_{t\in\mathscr{T}^{(1)}}$ of representations of $\mathcal{M}$ on a Hilbert space $\mathscr{L}$, we adopt the notation

$$\Pi_T(\overrightarrow{M}) = \Pi_{t_1}(M_1)\dots\Pi_{t_{|T|}}(M_{|T|}) \tag{21}$$

for all $T = (t_1, \dots, t_{|T|}) \in \mathscr{T}$ and $\overrightarrow{M} = (M_1, \dots, M_{|T|}) \in \mathcal{M}^{|T|}$.

Let $(S, \mathscr{F})$ be a measurable space. We define a set $\mathscr{T}_S$ by

$$\mathscr{T}_S = \cup_{j=1}^{\infty} (\mathscr{T}_S^{(1)})^j, \tag{22}$$

$$\mathscr{T}_S^{(1)} = \{in\} \cup \mathscr{F}, \tag{23}$$

where in is a symbol.

We shall define the notion of system of measurement correlations, which is a modified version of projective system of multikernels analyzed in the previous investigations [1, 6]. We define and analyze only the case that systems of measurement correlations do not have explicit time-dependence for simplicity herein.

Definition 8 (*System of measurement correlations*) A family $\{W_T\}_{T\in\mathscr{T}}$ of maps $W_T : \mathscr{M}^{|T|} = \overbrace{\mathscr{M} \times \cdots \times \mathscr{M}}^{|T|} \to \mathscr{M}$ is called a system of measurement correlations for $(\mathscr{M}, S)$ if it satisfies $\mathscr{T}^{(1)} = \mathscr{T}_S^{(1)}$ and the following six conditions:
(MC1) For any $T \in \mathscr{T}$, $W_T(M_1, \ldots, M_{|T|})$ is separately linear and ultraweakly continuous in each variable $M_1, \ldots, M_{|T|} \in \mathscr{M}$.
(MC2) For any $n \in \mathbb{N}$, $(T_1, \overrightarrow{M}_1), \ldots, (T_n, \overrightarrow{M}_n) \in \cup_{T\in\mathscr{T}}(\{T\} \times \mathscr{M}^{|T|})$, and $\xi_1, \ldots, \xi_n \in \mathscr{H}$,

$$\sum_{i,j=1}^{n} \langle \xi_i | W_{T_i^\# \times T_j}(\overrightarrow{M}_i^\# \times \overrightarrow{M}_j)\xi_j\rangle \geq 0. \tag{24}$$

(MC3) For any $T = (t_1, \ldots, t_{|T|}) \in \mathscr{T}$, $\overrightarrow{M} = (M_1, \ldots, M_{|T|}) \in \mathscr{M}^{|T|}$ and $M \in \mathscr{M}$,

$$M W_T(\overrightarrow{M}) = W_{(in)\times T}((M) \times \overrightarrow{M}), \tag{25}$$

$$W_T(\overrightarrow{M})M = W_{T\times(in)}(\overrightarrow{M} \times (M)). \tag{26}$$

(MC4) Let $T = (t_1, \ldots, t_{|T|}) \in \mathscr{T}$. If $t_k = t_{k+1} = in$ or $t_k, t_{k+1} \in \mathscr{F}$ for some $1 \leq k \leq |T| - 1$,

$$W_T(M_1, \ldots, M_k, M_{k+1}, \ldots, M_{|T|}) = W_{T'}(M_1, \ldots, M_k M_{k+1}, \ldots, M_{|T|}) \tag{27}$$

for all $(M_1, \ldots, M_{|T|}) \in \mathscr{M}^{|T|}$, where $T' = (t_1, \ldots, t_{k-1}, t_k \cap t_{k+1}, t_{k+2} \ldots, t_{|T|})$ and

$$t_k \cap t_{k+1} = \begin{cases} in, & (\text{if } t_k = t_{k+1} = in) \\ t_k \cap t_{k+1}, & (\text{if } t_k, t_{k+1} \in \mathscr{F}). \end{cases}$$

(MC5) For any $T = (t_1, \ldots, t_{|T|}) \in \mathscr{T}$ with $t_k = in$ or S, and $(M_1, \ldots, M_{|T|}) \in \mathscr{M}^{|T|}$ with $M_k = 1$,

$$W_T(M_1, \ldots, M_{k-1}, 1, M_{k+1}, \ldots, M_{|T|}) = W_{\hat{k}T}(M_1, \ldots, M_{k-1}, M_{k+1}, \ldots, M_{|T|}), \tag{28}$$

where $\hat{k}T = (t_1, \ldots, t_{k-1}, t_{k+1}, t_{k+2} \ldots, t_{|T|})$. In addition,

$$W_{in}(1) = W_S(1) = 1. \tag{29}$$

(MC6) For any $n \in \mathbb{N}$, $1 \leq k \leq n$, $t_1, \ldots, t_{k-1}, t_{k+1}, \ldots, t_n \in \mathscr{T}^{(1)}$, mutually disjoint sequence $\{t_{k,j}\}_j \subset \mathscr{F}$, $\overrightarrow{M} \in \mathscr{M}^n$ and $\rho \in \mathscr{M}_*$,

$$\langle \rho, W_{(t_1,\ldots,t_{k-1},\cup_j t_{k,j},t_{k+1},\ldots,t_n)}(\overrightarrow{M})\rangle = \sum_j \langle \rho, W_{(t_1,\ldots,t_{k-1},t_{k,j},t_{k+1},\ldots,t_n)}(\overrightarrow{M})\rangle. \tag{30}$$

in and $\Delta \in \mathcal{F}$ are subscripts that specify the time before the measurement and the time after the measurement, respectively. In $W_T(\overrightarrow{M})$, components of $\overrightarrow{M}$ indexed by in and those of $\overrightarrow{M}$ indexed by $\Delta \in \mathcal{F}$, describe observables before the measurement and those after the measurement, respectively, for each $T \in \mathcal{T}$ and $\overrightarrow{M} \in \mathcal{M}^{|T|}$. Especially, the latter represents observables of the system after the measurement in the situation that values of the output variable of the measuring appratus are restricted to $\Delta \in \mathcal{F}$. The discussion in Sect. 4 will support this interpretation.

It is easy to generalize systems of measurement correlations to the case that they have explicit time-dependence by modifying the definition. For this purpose, $\mathcal{T}_S^{(1)}$ is replaced by $\mathcal{T}_{G,S}^{(1)} = \{in\} \cup (G \times \mathcal{F})$, where G is the set representing time and is usually assumed to be a subset of $\mathbb{R}$, and, for instance, the condition (MC4) is replaced by (MC4′) Let $T = (t_1, \ldots, t_{|T|}) \in \mathcal{T}$. If $t_k = t_{k+1} = in$ or $t_k = (g, \Delta_k), t_{k+1} = (g, \Delta_{k+1}) \in G \times \mathcal{F}$ for some $1 \le k \le |T| - 1$, then

$$W_T(M_1, \ldots, M_k, M_{k+1}, \ldots, M_{|T|}) = W_{T'}(M_1, \ldots, M_k M_{k+1}, \ldots, M_{|T|}) \tag{31}$$

for all $(M_1, \ldots, M_{|T|}) \in \mathcal{M}^{|T|}$, where $T' = (t_1, \ldots, t_{k-1}, t_k \cap t_{k+1}, t_{k+2} \ldots, t_{|T|})$ and

$$t_k \cap t_{k+1} = \begin{cases} in, & (\text{if } t_k = t_{k+1} = in) \\ (g, \Delta_k \cap \Delta_{k+1}), & (\text{if } t_k = (g, \Delta_k), t_{k+1} = (g, \Delta_{k+1}) \in G \times \mathcal{F}). \end{cases}$$

Other conditions are also modified in the same manner.

When a system $\{W_T\}_{T \in \mathcal{T}}$ of measurement correlations for $(\mathcal{M}, S)$ is given, an instrument $\mathcal{I}_W$ for $(\mathcal{M}, S)$ is defined by

$$\mathcal{I}_W(M, \Delta) = W_\Delta(M) \tag{32}$$

for all $\Delta \in \mathcal{F}$ and $M \in \mathcal{M}$, which is seen to be completely positive by the condition (MC2).

Every system of measurement correlations admits the following representation theorem.

Theorem 8 *Let $\mathcal{M}$ be a von Neumann algebra on a Hilbert space $\mathcal{H}$ and $(S, \mathcal{F})$ a measurable space. For any systems $\{W_T\}_{T \in \mathcal{T}}$ of measurement correlations for $(\mathcal{M}, S)$, there exist a Hilbert space $\mathcal{L}$, a family $\{\Pi_t\}_{t \in \mathcal{T}^{(1)}}$ of normal (*-) representations of $\mathcal{M}$ on $\mathcal{L}$ and an isometry V from $\mathcal{H}$ to $\mathcal{L}$ such that*

$$\Pi_{in}(M)V = VM \tag{33}$$

for all $M \in \mathcal{M}$, and that

$$W_T(\overrightarrow{M}) = V^* \Pi_T(\overrightarrow{M}) V \tag{34}$$

for all $T \in \mathcal{T}$ and $\overrightarrow{M} \in \mathcal{M}^{|T|}$.

Proof Let $\{W_T\}_{T\in\mathscr{T}}$ be a system of measurement correlations for $(\mathcal{M}, S)$. We set $\mathscr{C} = \cup_{T\in\mathscr{T}}(\{T\} \times \mathcal{M}^{|T|})$. We define a kernel $K : \mathscr{C} \times \mathscr{C} \to \mathcal{M}$ by

$$K(\mathbf{a}, \mathbf{b}) = W_{T_1^{\#}\times T_2}(\overrightarrow{M}_1^{\#} \times \overrightarrow{M}_2) \tag{35}$$

for all $\mathbf{a} = (T_1, \overrightarrow{M}_1)$, $\mathbf{b} = (T_2, \overrightarrow{M}_2) \in \mathscr{C}$. By the definition of a system of measurement correlations, K is positive definite. By Theorem 5, there exists the minimal Kolmogorov decomposition $(\mathscr{L}, \Lambda)$ of K such that

$$K(\mathbf{a}, \mathbf{b}) = \Lambda(\mathbf{a})^*\Lambda(\mathbf{b}) \tag{36}$$

for all $\mathbf{a} = (T_1, \overrightarrow{M}_1)$, $\mathbf{b} = (T_2, \overrightarrow{M}_2) \in \mathscr{C}$. We remark that we use the fact that $\mathrm{span}(\Lambda(\mathscr{C})\mathscr{H})$ is dense in $\mathscr{L}$ many times in this proof.

For each $t \in \mathscr{T}^{(1)}$ and $M \in \mathcal{M}$, we define a map $\Pi_t(M)$ on span $\Lambda(\mathscr{C})\mathscr{H}$ by

$$\Pi_t(M)\Lambda(\mathbf{a})\xi = \Lambda(\mathbf{a}')\xi \tag{37}$$

for all $\mathbf{a} = (T, \overrightarrow{M}) = ((t_1, \ldots, t_{|T|}), (M_1, \ldots, M_{|T|})) \in \mathscr{C}$ and $\xi \in \mathscr{H}$, where

$$\mathbf{a}' = ((t) \times T, (M) \times \overrightarrow{M}) = ((t, t_1, \ldots, t_{|T|}), (M, M_1, \ldots, M_{|T|})). \tag{38}$$

For all $t \in \mathscr{T}^{(1)}$, we show that $\Pi_t : M \mapsto \Pi_t(M)$ is a normal *-representation of $\mathcal{M}$. By the condition (MC1), it holds that

$$\begin{aligned}
&\langle \Lambda(\mathbf{a})\xi_1 | \Pi_t(\alpha M + \beta N)\Lambda(\mathbf{b})\xi_2\rangle \\
&= \langle \xi_1 | \Lambda(\mathbf{a})^*\Pi_t(\alpha M + \beta N)\Lambda(\mathbf{b})\xi_2\rangle \\
&= \langle \xi_1 | W_{T_1^{\#}\times(t)\times T_2}(\overrightarrow{M}_1^{\#} \times (\alpha M + \beta N) \times \overrightarrow{M}_2)\xi_2\rangle \\
&= \alpha\langle \xi_1 | W_{T_1^{\#}\times(t)\times T_2}(\overrightarrow{M}_1^{\#} \times (M) \times \overrightarrow{M}_2)\xi_2\rangle \\
&\quad + \beta\langle \xi_1 | W_{T_1^{\#}\times(t)\times T_2}(\overrightarrow{M}_1^{\#} \times (N) \times \overrightarrow{M}_2)\xi_2\rangle \\
&= \alpha\langle \xi_1 | \Lambda(\mathbf{a})^*\Pi_t(M)\Lambda(\mathbf{b})\xi_2\rangle + \beta\langle \xi_1 | \Lambda(\mathbf{a})^*\Pi_t(N)\Lambda(\mathbf{b})\xi_2\rangle \\
&= \langle \xi_1 | \Lambda(\mathbf{a})^*(\alpha\Pi_t(M) + \beta\Pi_t(N))\Lambda(\mathbf{b})\xi_2\rangle \\
&= \langle \Lambda(\mathbf{a})\xi_1 | (\alpha\Pi_t(M) + \beta\Pi_t(N))\Lambda(\mathbf{b})\xi_2\rangle
\end{aligned} \tag{39}$$

for any $t \in \mathscr{T}^{(1)}$, $\alpha, \beta \in \mathbb{C}$, $M, N \in \mathcal{M}$, $\mathbf{a} = (T_1, \overrightarrow{M}_1)$, $\mathbf{b} = (T_2, \overrightarrow{M}_2) \in \mathscr{C}$ and $\xi_1, \xi_2 \in \mathscr{H}$, so that $\Pi_t(\alpha M + \beta N) = \alpha\Pi_t(M) + \beta\Pi_t(N)$ for all $t \in \mathscr{T}^{(1)}$, $\alpha, \beta \in \mathbb{C}$ and $M, N \in \mathcal{M}$.

Similarly, by the condition (MC4) it holds that

$$\begin{aligned}
&\langle \Lambda(\mathbf{a})\xi_1 | \Pi_t(M)\Pi_t(N)\Lambda(\mathbf{b})\xi_2\rangle = \langle \xi_1 | \Lambda(\mathbf{a})^* \Pi_t(M)\Pi_t(N)\Lambda(\mathbf{b})\xi_2\rangle \\
&= \langle \xi_1 | W_{T_1^\# \times (t,t) \times T_2}(\overrightarrow{M}_1^\# \times (M, N) \times \overrightarrow{M}_2)\xi_2\rangle \\
&= \langle \xi_1 | W_{T_1^\# \times (t) \times T_2}(\overrightarrow{M}_1^\# \times (MN) \times \overrightarrow{M}_2)\xi_2\rangle \\
&= \langle \xi_1 | \Lambda(\mathbf{a})^* \Pi_t(MN)\Lambda(\mathbf{b})\xi_2\rangle = \langle \Lambda(\mathbf{a})\xi_1 | \Pi_t(MN)\Lambda(\mathbf{b})\xi_2\rangle
\end{aligned} \tag{40}$$

for any $t \in \mathscr{T}^{(1)}$, $M, N \in \mathscr{M}$, $\mathbf{a} = (T_1, \overrightarrow{M}_1)$, $\mathbf{b} = (T_2, \overrightarrow{M}_2) \in \mathscr{C}$ and $\xi_1, \xi_2 \in \mathscr{H}$, so that $\Pi_t(MN) = \Pi_t(M)\Pi_t(N)$ for all $t \in \mathscr{T}^{(1)}$ and $M, N \in \mathscr{M}$.

For any $t \in \mathscr{T}^{(1)}$, $n \in \mathbb{N}$, $\mathbf{a}_1 = (T_1, \overrightarrow{M}_1)$, $\mathbf{a}_2 = (T_2, \overrightarrow{M}_2), \ldots, \mathbf{a}_n = (T_n, \overrightarrow{M}_n) \in \mathscr{C}$ and $\xi_1, \xi_2, \ldots, \xi_n \in \mathscr{H}$, the map

$$\mathscr{M} \ni M \mapsto \sum_{i,j=1}^{n} \langle \Lambda(\mathbf{a}_i)\xi_i | \Pi_t(M)\Lambda(\mathbf{a}_j)\xi_j\rangle \in \mathbb{C} \tag{41}$$

is normal linear functional on $\mathscr{M}$, which is also positive since it holds by the conditions (MC2) and (MC4) that

$$\begin{aligned}
&\sum_{i,j=1}^{n} \langle \Lambda(\mathbf{a}_i)\xi_i | \Pi_t(M)\Lambda(\mathbf{a}_j)\xi_j\rangle = \sum_{i,j=1}^{n} \langle \xi_i | W_{T_i^\# \times (t) \times T_j}(\overrightarrow{M}_i^\# \times (M) \times \overrightarrow{M}_j)\xi_j\rangle \\
&= \sum_{i,j=1}^{n} \langle \xi_i | W_{T_i^\# \times (t,t) \times T_j}(\overrightarrow{M}_i^\# \times (\sqrt{M}, \sqrt{M}) \times \overrightarrow{M}_j)\xi_j\rangle \\
&= \sum_{i,j=1}^{n} \langle \xi_i | W_{((t)\times T_i)^\# \times ((t) \times T_j)}(((\sqrt{M}) \times \overrightarrow{M}_i)^\# \times ((\sqrt{M}) \times \overrightarrow{M}_j))\xi_j\rangle \geq 0
\end{aligned} \tag{42}$$

for all $M \in \mathscr{M}_+ = \{M \in \mathscr{M} \mid M \geq 0\}$, $t \in \mathscr{T}^{(1)}$, $n \in \mathbb{N}$, $\mathbf{a}_1 = (T_1, \overrightarrow{M}_1)$, $\mathbf{a}_2 = (T_2, \overrightarrow{M}_2), \ldots, \mathbf{a}_n = (T_n, \overrightarrow{M}_n) \in \mathscr{C}$ and $\xi_1, \xi_2, \ldots, \xi_n \in \mathscr{H}$. Thus, for any $t \in \mathscr{T}^{(1)}$, $M \in \mathscr{M}$, $n \in \mathbb{N}$, $\mathbf{a}_1 = (T_1, \overrightarrow{M}_1)$, $\mathbf{a}_2 = (T_2, \overrightarrow{M}_2), \ldots, \mathbf{a}_n = (T_n, \overrightarrow{M}_n) \in \mathscr{C}$ and $\xi_1, \xi_2, \ldots, \xi_n \in \mathscr{H}$ we have

$$\left\| \sum_{i=1}^{n} \Pi_t(M)\Lambda(\mathbf{a}_i)\xi_i \right\| \leq \|M\| \cdot \left\| \sum_{i=1}^{n} \Lambda(\mathbf{a}_i)\xi_i \right\|. \tag{43}$$

For every $t \in \mathscr{T}^{(1)}$ and $M \in \mathscr{M}$, $\Pi_t(M)$ is a bounded operator on $\mathscr{L}$. In addition, for all $t \in \mathscr{T}^{(1)}$, $M \in \mathscr{M}$ and $\mathbf{a} = (T_1, \overrightarrow{M}_1)$, $\mathbf{b} = (T_2, \overrightarrow{M}_2) \in \mathscr{C}$,

$$\begin{aligned}&\langle \Lambda(\mathbf{a})\xi_1|\Pi_t(M)^*\Lambda(\mathbf{b})\xi_2\rangle = \langle \xi_1|(\Pi_t(M)\Lambda(\mathbf{a}))^*\Lambda(\mathbf{b})\xi_2\rangle\\ &= \langle \xi_1|W_{((t)\times T_1)^\#\times T_2}(((M)\times \overrightarrow{M}_1)^\# \times \overrightarrow{M}_2)\xi_2\rangle\\ &= \langle \xi_1|W_{T_1^\#\times(t)\times T_2}(\overrightarrow{M}_1^\# \times (M^*)\times \overrightarrow{M}_2)\xi_2\rangle\\ &= \langle \xi_1|\Lambda(\mathbf{a})^*\Pi_t(M^*)\Lambda(\mathbf{b})\xi_2\rangle = \langle \Lambda(\mathbf{a})\xi_1|\Pi_t(M^*)\Lambda(\mathbf{b})\xi_2\rangle. \end{aligned} \tag{44}$$

Thus, for every $t \in \mathscr{T}^{(1)}$ Π_t is a normal *-representation of $\mathscr{M}$ on $\mathscr{L}$. By the condition (MC5), Π_{in} and Π_S are nondegenerate, i.e., $\Pi_{in}(1) = \Pi_S(1) = 1_{\mathbf{B}(\mathscr{L})}$.

For every $t \in \mathscr{T}$ and and $\overrightarrow{M} \in \mathscr{M}^{|T|}$, it then holds that

$$\begin{aligned} V^*\Pi_T(\overrightarrow{M})V &= \Lambda((in,1))^*\Pi_T(\overrightarrow{M})\Lambda((in,1))\\ &= W_{(in)\times T\times(in)}((1)\times \overrightarrow{M}\times(1)) = W_T(\overrightarrow{M}). \end{aligned} \tag{45}$$

By the above relation and the condition (MC3), we have $V^*\Pi_{in}(M)V = M$ for all $M \in \mathscr{M}$, and

$$\begin{aligned}&(\Pi_{in}(M)V - VM)^*(\Pi_{in}(M)V - VM)\\ &= V^*\Pi_{in}(M)^*\Pi_{in}(M)V - V^*\Pi_{in}(M)^*VM - M^*V^*\Pi_{in}(M)V + M^*V^*VM\\ &= V^*\Pi_{in}(M^*M)V - V^*\Pi_{in}(M^*)VM - M^*V^*\Pi_{in}(M)V + M^*V^*VM\\ &= M^*M - M^*M - M^*M + M^*M = 0 \end{aligned} \tag{46}$$

for all $M \in \mathscr{M}$, which implies $\Pi_{in}(M)V = VM$ for all $M \in \mathscr{M}$.

Remark 1 By the above proof, we see the following. For every family $\{W_T\}_{T\in\mathscr{T}}$ of maps $W_T : \mathscr{M}^{|T|} \to \mathscr{M}$ satisfying the conditions (MC1), (MC2), (MC4), (MC5) and (MC6), there exist a Hilbert space $\mathscr{L}$, a family $\{\Pi_t\}_{t\in\mathscr{T}^{(1)}}$ of normal (*-)representations of $\mathscr{M}$ on $\mathscr{L}$ and an isometry V from $\mathscr{H}$ to $\mathscr{L}$ such that

$$W_T(\overrightarrow{M}) = V^*\Pi_T(\overrightarrow{M})V \tag{47}$$

for all $T \in \mathscr{T}$ and $\overrightarrow{M} \in \mathscr{M}^{|T|}$. Equation (47) then implies

$$W_T(\overrightarrow{M})^* = W_{T^\#}(\overrightarrow{M}^\#) \tag{48}$$

for all $T \in \mathscr{T}$ and $\overrightarrow{M} \in \mathscr{M}^{|T|}$.

As seen the proof of Theorem 8, the following fact holds, which will be used in the next section.

Corollary 2 *Let $\mathscr{M}$ be a von Neumann algebra on a Hilbert space $\mathscr{H}$ and $(S, \mathscr{F})$ a measurable space. For any systems $\{W_T\}_{T\in\mathscr{T}}$ of measurement correlations for $(\mathscr{M}, S)$, let $(\mathscr{L}, \{\Pi_t\}_{t\in\mathscr{T}^{(1)}}, V)$ be a triplet in Theorem 8. Then the map $\mathscr{F} \ni \Delta \mapsto \Pi_\Delta(1) \in \mathbf{B}(\mathscr{L})$ is a projection-valued measure (PVM).*

Proof The proof can be easily done in terms of the conditions (MC4), (MC5) and (MC6).

In [1], a (noncommutative) stochastic process over a C*-algebra $\mathscr{B}$, indexed by a set $\mathbb{T}$, is defined by a pair $(\mathscr{A}, \{j_t\}_{t\in\mathbb{T}})$ of a C*-algebra $\mathscr{A}$ and a family $\{j_t\}_{t\in\mathbb{T}}$ of *-homomorphisms from $\mathscr{B}$ into $\mathscr{A}$. Obviously, a pair $(\mathbf{B}(\mathscr{L}), \{\Pi_t\}_{t\in\mathscr{T}^{(1)}})$ in Theorem 8 is nothing but a stochastic process over a von Neumann algebra $\mathscr{M}$ indexed by $\mathscr{T}^{(1)}$ in this sense.

Let $\mathscr{M}$ be a von Neumann algebra on a Hilbert space. Let $\mathbb{T}$ be a set. We set $\mathscr{T}_{\mathbb{T}} = \cup_{j=1}^{\infty}(\{in\} \cup \mathbb{T})^j$. Let $(\mathscr{L}, \{\Pi_t\}_{t\in\{in\}\cup\mathbb{T}}, V)$ be a triplet consisting of a Hilbert space $\mathscr{L}$, a family $\{\Pi_t\}_{t\in\{in\}\cup\mathbb{T}}$ of normal representations of $\mathscr{M}$ on $\mathscr{L}$ and V an isometry from $\mathscr{H}$ to $\mathscr{L}$ such that $\Pi_{in}(M)V = VM$ for all $M \in \mathscr{M}$ and that $V^*\Pi_T(\overrightarrow{M})V \in \mathscr{M}$ for all $T \in \mathscr{T}_{\mathbb{T}}$ and $\overrightarrow{M} \in \mathscr{M}^{|T|}$. The generalized Heisenberg picture is then formulated by this triple $(\mathscr{L}, \{\Pi_t\}_{t\in\{in\}\cup\mathbb{T}}, V)$, which enables us to compare the situtation before the change, specified by a representation Π_{in}, with the situtation after the change, specified by $\{\Pi_t\}_{t\in\mathbb{T}}$. This interpretation naturally follows from the intertwining relation $\Pi_{in}(M)V = VM$ for all $M \in \mathscr{M}$ and from the generation of correlation functions $\mathscr{W}_T(\overrightarrow{M}) = V^*\Pi_T(\overrightarrow{M})V$ for all $T \in \mathscr{T}_{\mathbb{T}}$ and $\overrightarrow{M} \in \mathscr{M}^{|T|}$. For example, in a triplet $(\mathscr{L}, \{\Pi_t\}_{t\in\mathscr{T}^{(1)}}, V)$ in Theorem 8, Π_{in} and $\{\Pi_t\}_{t\in\mathscr{F}}$ correspond to a representation before the measurement and those after the measurement, respectively. The author believes that the generalized Heisenberg picture introduced here gives a right extension of the description of dynamical processes in the standard formulation of quantum mechanics since it succeeds to the advantage of the (usual) Heisenberg picture that we can calculate correlation functions of observables at different times. This topic will be discussed in detail in the succeeding paper of the author.

4 Unitary Dilation Theorem

As previously mentioned, the introduction of the concept of measuring process was cruicial for the progress of the theory of quantum measurement and of instruments. Measuring processes redefined as follows also play the central role in quantum measurement theory based on the generalized Heisenberg picture.

Definition 9 A measuring process $\mathbb{M}$ for $(\mathscr{M}, S)$ is a 4-tuple $\mathbb{M} = (\mathscr{K}, \sigma, E, U)$ which consists of a Hilbert space $\mathscr{K}$, a normal state σ on $\mathbf{B}(\mathscr{K})$, a spectral measure $E : \mathscr{F} \to \mathbf{B}(\mathscr{K})$, and a unitary operator U on $\mathscr{H} \otimes \mathscr{K}$ and defines a system of measurement correlations $\{W_T^{\mathbb{M}}\}_{T\in\mathscr{T}_S}$ for $(\mathscr{M}, S)$ as follows: We define a representation π_{in} of $\mathscr{M}$ and a family $\{\pi_\Delta\}_{\Delta\in\mathscr{F}}$ of those of $\mathscr{M}$ on $\mathscr{H} \otimes \mathscr{K}$ by

$$\pi_{in}(M) = M \otimes 1_{\mathscr{K}}, \qquad \pi_\Delta(M) = U^*(M \otimes E(\Delta))U \tag{49}$$

for all $M \in \mathscr{M}$ and $\Delta \in \mathscr{F}$, respectively. We use the notation

$$\pi_T(\overrightarrow{M}) = \pi_{t_1}(M_1)\ldots\pi_{t_{|T|}}(M_{|T|}) \tag{50}$$

for all $T = (t_1, \ldots, t_{|T|}) \in \mathscr{T}$ and $\overrightarrow{M} = (M_1, \ldots, M_{|T|}) \in \mathscr{M}^{|T|}$. For each $T \in \mathscr{T}_S$, $W_T^{\mathbb{M}} : \mathscr{M}^{|T|} \to \mathscr{M}$ is defined by

$$W_T^{\mathbb{M}}(\overrightarrow{M}) = (\mathrm{id} \otimes \sigma)(\pi_T(\overrightarrow{M})) \tag{51}$$

for all $\overrightarrow{M} \in \mathscr{M}^{|T|}$.

It is easily seen that two definitions of measuring processes for $(\mathbf{B}(\mathscr{H}), S)$ are equivalent.

We say that a CP instrument $\mathscr{I}$ for $(\mathscr{M}, S)$ is *realized* by a measuring process $\mathbb{M}$ for $(\mathscr{M}, S)$ in the sense of Definition 9, or $\mathbb{M}$ *realizes* $\mathscr{I}$ if $\mathscr{I} = \mathscr{I}_{\mathbb{M}}$. $\mathrm{CPInst}_{\mathrm{RE}}(\mathscr{M}, S)$ denotes the set of CP instruments for $(\mathscr{M}, S)$ realized by measuring processes for $(\mathscr{M}, S)$ in the sense of Definition 9. Then we have $\mathrm{CPInst}_{\mathrm{RE}}(\mathscr{M}, S) \subseteq \mathrm{CPInst}_{\mathrm{NE}}(\mathscr{M}, S)$. It will be shown in Sect. 6 that

$$\mathrm{CPInst}_{\mathrm{RE}}(\mathscr{M}, S) = \mathrm{CPInst}_{\mathrm{NE}}(\mathscr{M}, S). \tag{52}$$

Definition 10 Let $n \in \mathbb{N}$. Two measuring processes $\mathbb{M}_1$ and $\mathbb{M}_2$ for $(\mathscr{M}, S)$ are said to be n-equivalent if $W_T^{\mathbb{M}_1} = W_T^{\mathbb{M}_2}$ for all $T \in \mathscr{T}$ such that $|T| \leq n$. Two measuring processes $\mathbb{M}_1$ and $\mathbb{M}_2$ for $(\mathscr{M}, S)$ are said to be completely equivalent if they are n-equivalent for all $n \in \mathbb{N}$.

The n-equivalence class of a measuring process $\mathbb{M}$ for $(\mathscr{M}, S)$ is nothing but the set of measuring processes $\mathbb{M}'$ for $(\mathscr{M}, S)$ whose correlation functions of order less or equal to n are identical to those defined by $\mathbb{M}$, i.e., $W_T^{\mathbb{M}} = W_T^{\mathbb{M}'}$ for all $T \in \mathscr{T}$ such that $|T| \leq n$. Since a measuring process $\mathbb{M}$ for $(\mathscr{M}, S)$ in the sense of Definition 9 is also that in the sense of Definition 2, the statistical equivalence works for the former. Of course, the 2-equivalence is the same as the statistical equivalence. In practical situations, dynamical aspects of physical systems are usually analyzed in terms of correlation functions of finite order. Thus it is natural to consider that the classification of measuring processes by the n-equivalence for not so large $n \in \mathbb{N}$ is valid in the same way. It should be stressed here that causal relations cannot be verified without using correlation functions (of observables at different times) and that situations concerned with measurements are not the exception. A successful example of causal relations in the context of measurement has been already given by the notion of perfect correlation introduced in [36], which uses correlation functions of order 2. One may consider that the complete equivalence of measuring processes is unrealistic and useless, but we believe that it is much useful since the following theorem holds.

Theorem 9 *Let $\mathscr{H}$ be a Hilbert space and $(S, \mathscr{F})$ a measurable space. Then there is a one-to-one correpondence between complete equivalence classes of measuring processes $\mathbb{M} = (\mathscr{K}, \sigma, E, U)$ for $(\mathbf{B}(\mathscr{H}), S)$ and systems $\{W_T\}_{T\in\mathscr{T}}$ of measurement correlations for $(\mathbf{B}(\mathscr{H}), S)$, which is given by the relation*

$$W_T(\overrightarrow{M}) = W_T^{\mathbb{M}}(\overrightarrow{M}) \tag{53}$$

for all $T \in \mathscr{T}$ *and* $\overrightarrow{M} \in \mathscr{M}^{|T|}$.

Let $\mathscr{H}_1$ and $\mathscr{H}_2$ be Hilbert spaces. For each $\eta \in \mathscr{H}_2$, we define a linear map $V_\eta : \mathscr{H}_1 \to \mathscr{H}_1 \otimes \mathscr{H}_2$ by $V_\eta \xi = \xi \otimes \eta$ for all $\xi \in \mathscr{H}_1$. It is easily seen that, for each $\eta \in \mathscr{H}_2$, V_η satisfies $(X \otimes 1)V_\eta = V_\eta X$ for all $X \in \mathbf{B}(\mathscr{H}_1)$. For any $x \in \mathscr{H}_1 \backslash \{0\}$, P_x denotes the projection from $\mathscr{H}_1$ onto the linear subspace $\mathbb{C}x$ of $\mathscr{H}_1$ linearly spanned by x. For any $x, y \in \mathscr{H}_1$, we define $|y\rangle\langle x| \in \mathbf{B}(\mathscr{H}_1)$ by $|y\rangle\langle x|z = \langle x|z\rangle y$ for all $z \in \mathscr{H}_1$.

Lemma 1 *Let* $\mathscr{H}_1$ *and* $\mathscr{H}_2$ *be Hilbert spaces. Let* V *be an isometry from* $\mathscr{H}_1$ *to* $\mathscr{H}_1 \otimes \mathscr{H}_2$. *If* V *satisfies* $(X \otimes 1)V = VX$ *for all* $X \in \mathbf{B}(\mathscr{H}_1)$, *then there exists* $\eta \in \mathscr{H}_2$ *such that* $V = V_\eta$.

Proof Let $x \in \mathscr{H}_1 \backslash \{0\}$. Since $(P_x \otimes 1)Vx = VP_x x = Vx$, it holds that $Vx \in \mathbb{C}x \otimes \mathscr{H}_2$. Hence, for any $x \in \mathscr{H}_1 \backslash \{0\}$, there is $\eta_x \in \mathscr{H}_2$ such that $Vx = x \otimes \eta_x$ and that $\|\eta_x\| = 1$.

For any $x, y \in \mathscr{H}_1 \backslash \{0\}$,

$$\begin{aligned}\langle x|x\rangle y \otimes \eta_y &= \langle x|x\rangle Vy = V(\langle x|x\rangle y) = V(|y\rangle\langle x|x) \\ &= (|y\rangle\langle x| \otimes 1)Vx = (|y\rangle\langle x| \otimes 1)(x \otimes \eta_x) = \langle x|x\rangle y \otimes \eta_x\end{aligned} \tag{54}$$

Thus $\eta_x = \eta_y$ for all $x, y \in \mathscr{H}_1 \backslash \{0\}$. This fact implies that the range of the map $\mathscr{H}_1 \backslash \{0\} \ni x \mapsto \eta_x \in \mathscr{H}_2$ is one point. We put $\eta = \eta_x$ for some $x \in \mathscr{H}_1 \backslash \{0\}$. By the linearity of V, $V0 = 0 = 0 \otimes \eta$. Thus we have $V = V_\eta$.

Proof (*Proof of Theorem* 9) Let $\{W_T\}_{T \in \mathscr{T}}$ be a system of measurement correlations for $(\mathbf{B}(\mathscr{H}), S)$. By Theorem 8, there exist a Hilbert space $\mathscr{L}_0$, a family $\{\Pi_t\}_{t \in \mathscr{T}^{(1)}}$ of normal representations of $\mathbf{B}(\mathscr{H})$ on $\mathscr{L}_0$ and an isometry V_0 from $\mathscr{H}$ to $\mathscr{L}_0$ such that

$$W_T(\overrightarrow{M}) = V_0^* \Pi_T(\overrightarrow{M}) V_0 \tag{55}$$

for all $T \in \mathscr{T}$ and $\overrightarrow{M} \in \mathbf{B}(\mathscr{H})^{|T|}$.

By Theorem 7, there exist Hilbert space $\mathscr{L}_1$ and a unitary operator $U_1 : \mathscr{L}_0 \to \mathscr{H} \otimes \mathscr{L}_1$ such that

$$\Pi_{in}(M) = U_1^*(M \otimes 1)U_1 \tag{56}$$

for all $M \in \mathbf{B}(\mathscr{H})$. Similarly, by Theorem 7, there exist a Hilbert space $\mathscr{L}_2$ and a unitary operator $U_2 : \mathscr{L}_0 \to \mathscr{H} \otimes \mathscr{L}_2$ such that

$$\Pi_S(M) = U_2^*(M \otimes 1)U_2 \tag{57}$$

for all $M \in \mathbf{B}(\mathscr{H})$, and by Theorem 6 there exist a PVM $E_0 : \mathscr{F} \to \mathbf{B}(\mathscr{L}_2)$ such that

$$\Pi_\Delta(1) = U_2^*(1 \otimes E_0(\Delta))U_2 \tag{58}$$

for all $\Delta \in \mathscr{F}$.

We define a linear map $V : \mathscr{H} \to \mathscr{H} \otimes \mathscr{L}_1$ by $V = U_1 V_0$, which is obviously seen to be an isometry. Here, it holds that $V^*(M \otimes 1)V = M$ for all $M \in \mathbf{B}(\mathscr{H})$. For all $M \in \mathbf{B}(\mathscr{H})$,

$$\begin{aligned}
&((M \otimes 1)V - VM)^*((M \otimes 1)V - VM) \\
&= V^*(M^*M \otimes 1)V - V^*(M^* \otimes 1)VM - M^*V^*(M \otimes 1)V + M^*V^*VM \\
&= M^*M - M^* \cdot M - M^* \cdot M + M^*M = 0.
\end{aligned} \tag{59}$$

Thus we have $(M \otimes 1)V = VM$ for all $M \in \mathbf{B}(\mathscr{H})$. By Lemma 1, there is $\eta_1 \in \mathscr{L}_1$ such that $V = V_{\eta_1}$.

Let $\eta_2 \in \mathscr{L}_2$ such that $\|\eta_2\| = 1$. Let ζ be an isomorphism from $\mathscr{L}_1 \otimes \mathscr{L}_2$ to $\mathscr{L}_2 \otimes \mathscr{L}_1$ defined by $\zeta(\xi_1 \otimes \xi_2) = \xi_2 \otimes \xi_1$ for all $\xi_1 \in \mathscr{L}_1$ and $\xi_2 \in \mathscr{L}_2$. We define a unitary operator U_3 from $\mathscr{H} \otimes \mathscr{L}_1 \otimes \mathbb{C}\eta_2$ to $\mathscr{H} \otimes \mathbb{C}\eta_1 \otimes \mathscr{L}_2$ by

$$U_3(\xi \otimes \eta_2) = (1 \otimes \zeta)(U_2 U_1^* \xi \otimes \eta_1) \tag{60}$$

for all $\xi \in \mathscr{H} \otimes \mathscr{L}_1$. We define a unitary operator U_5 from $\mathbb{C}\eta_2$ to $\mathbb{C}\eta_1$ by $U_5 x = \langle \eta_2 | x \rangle \eta_1$ for all $x \in \mathbb{C}\eta_2$. Then U_3 has the following form:

$$U_3 = (1 \otimes \zeta)(U_2 U_1^* \otimes U_5). \tag{61}$$

Since both $\mathscr{H} \otimes \mathscr{L}_1 \otimes \mathbb{C}\eta_2$ and $\mathscr{H} \otimes \mathbb{C}\eta_1 \otimes \mathscr{L}_2$ are subspaces of $\mathscr{H} \otimes \mathscr{L}_1 \otimes \mathscr{L}_2$ and satisfies $\dim(\mathscr{H} \otimes \mathscr{L}_1 \otimes \mathbb{C}\eta_2) = \dim(\mathscr{H} \otimes \mathbb{C}\eta_1 \otimes \mathscr{L}_2)$ by the above observation, it holds that

$$\dim((\mathscr{H} \otimes \mathscr{L}_1 \otimes \mathbb{C}\eta_2)^\perp) = \dim((\mathscr{H} \otimes \mathbb{C}\eta_1 \otimes \mathscr{L}_2)^\perp). \tag{62}$$

This fact implies that there is a unitary operator U_4 from $(\mathscr{H} \otimes \mathscr{L}_1 \otimes \mathbb{C}\eta_2)^\perp$ to $(\mathscr{H} \otimes \mathbb{C}\eta_1 \otimes \mathscr{L}_2)^\perp$. Let Q be a projection operator from $\mathscr{H} \otimes \mathscr{L}_1 \otimes \mathscr{L}_2$ onto $\mathscr{H} \otimes \mathscr{L}_1 \otimes \mathbb{C}\eta_2$, i.e., $Q = 1 \otimes 1 \otimes P_{\eta_2}$. Let R be a projection operator from $\mathscr{H} \otimes \mathscr{L}_1 \otimes \mathscr{L}_2$ onto $\mathscr{H} \otimes \mathbb{C}\eta_1 \otimes \mathscr{L}_2$, i.e., $R = 1 \otimes P_{\eta_1} \otimes 1$. We then define a unitary operator U on $\mathscr{H}$ by $U = U_3 Q + U_4(1 - Q)$. It is obvious that U satisfies $UQ = U_3 Q = RU_3 = RU$.

We define a Hilbert space $\mathscr{K}$ by $\mathscr{K} = \mathscr{L}_1 \otimes \mathscr{L}_2$, a normal state σ on $\mathbf{B}(\mathscr{K})$ by

$$\sigma(Y) = \langle \eta_1 \otimes \eta_2 | Y(\eta_1 \otimes \eta_2) \rangle \tag{63}$$

for all $Y \in \mathbf{B}(\mathscr{K})$, and a spectral measure $E : \mathscr{F} \to \mathbf{B}(\mathscr{K})$ by

$$E(\Delta) = 1 \otimes E_0(\Delta) \tag{64}$$

for all $\Delta \in \mathscr{F}$.

We show that the 4-tuple $\mathbb{M} := (\mathscr{K}, \sigma, E, U)$ is a measuring process for $(\mathbf{B}(\mathscr{H}), S)$ such that

$$W_T(\overrightarrow{M}) = W_T^{\mathbb{M}}(\overrightarrow{M}) \tag{65}$$

for all $T \in \mathscr{T}$ and $\overrightarrow{M} \in \mathbf{B}(\mathscr{H})^{|T|}$. Since $Q = 1 \otimes 1 \otimes P_{\eta_2}$ and

$$\pi_{in}(M) = U_1 \Pi_{in}(M) U_1^* \otimes 1_{\mathbf{B}(\mathscr{L}_2)} \tag{66}$$

for all $M \in \mathbf{B}(\mathscr{H})$, we have

$$\pi_{in}(M) Q = Q \pi_{in}(M) \tag{67}$$

for all $M \in \mathbf{B}(\mathscr{H})$. Similarly, we have

$$\begin{aligned}
\pi_\Delta(M) Q &= U^*(M \otimes E(\Delta)) U Q \\
&= U^*(M \otimes E(\Delta)) R U_3 Q \\
&= U^* R (M \otimes E(\Delta)) U_3 Q \\
&= U_3^* R (M \otimes E(\Delta)) U_3 Q \\
&= U_3^* (M \otimes E(\Delta)) U_3 Q \\
&= ((1 \otimes \zeta)(U_2 U_1^* \otimes U_5))^* (M \otimes E(\Delta))((1 \otimes \zeta)(U_2 U_1^* \otimes U_5)) \\
&= (U_1 U_2^* \otimes U_5^*)(M \otimes \zeta^* E(\Delta) \zeta)(U_2 U_1^* \otimes U_5) Q \\
&= (U_1 U_2^* \otimes U_5^*)(M \otimes E_0(\Delta) \otimes 1_{\mathbf{B}(\mathscr{L}_1)})(U_2 U_1^* \otimes U_5)) Q \\
&= (U_1 \Pi_\Delta(M) U_1^* \otimes P_{\eta_2}) Q \\
&= (U_1 \Pi_\Delta(M) U_1^* \otimes 1_{\mathbf{B}(\mathscr{L}_2)}) Q,
\end{aligned} \tag{68}$$

and

$$\pi_\Delta(M) Q = Q \pi_\Delta(M) \tag{69}$$

for all $M \in \mathbf{B}(\mathscr{H})$ and $\Delta \in \mathscr{F}$. By Eqs. (67) and (69), it holds that

$$Q \pi_T(\overrightarrow{M}) Q = Q(U_1 \Pi_{t_1}(M_1) \ldots \Pi_{t_{|T|}}(M_{|T|}) U_1^* \otimes 1_{\mathbf{B}(\mathscr{L}_2)}) Q \tag{70}$$

for all $T = (t_1, \ldots, t_{|T|}) \in \mathscr{T}$ and $\overrightarrow{M} = (M_1, \ldots, M_{|T|}) \in \mathbf{B}(\mathscr{H})^{|T|}$.

For all $\xi \in \mathscr{H}$, $T = (t_1, \ldots, t_{|T|}) \in \mathscr{T}$ and $\overrightarrow{M} = (M_1, \ldots, M_{|T|}) \in \mathbf{B}(\mathscr{H})^{|T|}$.

$$\begin{aligned}
\langle \xi | W_T^{\mathbb{M}}(\overrightarrow{M}) \xi \rangle &= \langle \xi | (\mathrm{id} \otimes \sigma)(\pi_T(\overrightarrow{M})) \xi \rangle \\
&= \langle \xi \otimes \eta_1 \otimes \eta_2 | \pi_T(\overrightarrow{M})(\xi \otimes \eta_1 \otimes \eta_2) \rangle \\
&= \langle Q(\xi \otimes \eta_1 \otimes \eta_2) | \pi_T(\overrightarrow{M}) Q(\xi \otimes \eta_1 \otimes \eta_2) \rangle \\
&= \langle V\xi \otimes \eta_2 | Q \pi_T(\overrightarrow{M}) Q(V\xi \otimes \eta_2) \rangle \\
&= \langle V\xi \otimes \eta_2 | Q(U_1 \Pi_{t_1}(M_1) \ldots \Pi_{t_{|T|}}(M_{|T|}) U_1^* \otimes 1_{\mathbf{B}(\mathscr{L}_2)}) Q(V\xi \otimes \eta_2) \rangle
\end{aligned}$$

$$= \langle V\xi \otimes \eta_2 | (U_1 \Pi_{t_1}(M_1) \ldots \Pi_{t_{|T|}}(M_{|T|}) U_1^* \otimes 1_{\mathbf{B}(\mathscr{L}_2)})(V\xi \otimes \eta_2) \rangle$$
$$= \langle V\xi | U_1 \Pi_{t_1}(M_1) \ldots \Pi_{t_{|T|}}(M_{|T|}) U_1^* V\xi \rangle$$
$$= \langle \xi | V^* U_1 \Pi_{t_1}(M_1) \ldots \Pi_{t_{|T|}}(M_{|T|}) U_1^* V\xi \rangle$$
$$= \langle \xi | V_0^* \Pi_{t_1}(M_1) \ldots \Pi_{t_{|T|}}(M_{|T|}) V_0 \xi \rangle$$
$$= \langle \xi | W_T(\overrightarrow{M}) \xi \rangle, \tag{71}$$

which completes the proof.

Remark 2 We adopt here the same notations as in the proof of the above theorem. Suppose that $\mathscr{H}$ is separable and $(S, \mathscr{F})$ is a standard Borel space. Let $\{\Delta_n\}_{n\in\mathbb{N}}$ be a countable generator of $\mathscr{F}$, $\{M_n\}_{n\in\mathbb{N}}$ a dense subset of $\mathbf{B}(\mathscr{H})$ in the strong topology, and $\{\xi_n\}_{n\in\mathbb{N}}$ a dense subset of $\mathscr{H}$. Let $\{C_n\}_{n\in\mathbb{N}}$ be a well-ordering of the countable set $\{\Pi_{in}(M_n) \mid n \in \mathbb{N}\} \cup \{\Pi_{\Delta_m}(M_n) \mid m, n \in \mathbb{N}\}$. $\mathscr{L}_0$ has the increasing sequence $\{\mathscr{L}_{0,n}\}_{n\in\mathbb{N}}$ of separable closed subspaces, defined by

$$\mathscr{L}_{0,n} = \overline{\text{span}}\{C_{f(1)} \ldots C_{f(n)} V_0 \xi_k \mid f \in \mathbb{N}^{\{1,\ldots,n\}}, k \in \mathbb{N}\} \tag{72}$$

for all $n \in \mathbb{N}$, such that $\mathscr{L}_0 = \overline{\text{span}}(\cup_n \mathscr{L}_{0,n})$, where $\mathbb{N}^{\{1,\ldots,n\}}$ is the set of maps from $\{1, \ldots, n\}$ to $\mathbb{N}$. Hence, $\mathscr{L}_0$ is separable because we have

$$\mathscr{L}_0 = \overline{\text{span}} \left\{ \bigoplus_{n=1}^{\infty} (\mathscr{L}_{0,n-1})^{\perp} \cap \mathscr{L}_{0,n} \right\}, \tag{73}$$

where $\mathscr{L}_{0,0} = \{0\}$. It is immediately seen that $\mathscr{L}_1$, $\mathscr{L}_2$ and $\mathscr{K} = \mathscr{L}_1 \otimes \mathscr{L}_2$ are also separable.

5 Extendability of CP Instruments to Systems of Measurement Correlations

To begin with, the following theorem similar to [29, Theorem 3.4] holds for arbitrary von Neumann algebras $\mathscr{M}$.

Corollary 3 *Let $\mathscr{M}$ be a von Neumann algebra on a Hilbert space $\mathscr{H}$ and $(S, \mathscr{F})$ a measurable space. For a system $\{W_T\}_{T\in\mathscr{T}}$ of measurement correlations for $(\mathscr{M}, S)$, the following conditions are equivalent:*
(1) *There is a system $\{\widetilde{W}_T\}_{T\in\mathscr{T}}$ of measurement correlations for $(\mathbf{B}(\mathscr{H}), S)$ such that*

$$W_T(\overrightarrow{M}) = \widetilde{W}_T(\overrightarrow{M}) \tag{74}$$

for all $T \in \mathscr{T}$ and $\overrightarrow{M} \in \mathscr{M}^{|T|}$.
(2) *There is a measuring process $\mathbb{M} = (\mathscr{K}, \sigma, E, U)$ for $(\mathscr{M}, S)$ such that*

$$W_T(\overrightarrow{M}) = W_T^{\mathbb{M}}(\overrightarrow{M}) \tag{75}$$

for all $T \in \mathscr{T}$ *and* $\overrightarrow{M} \in \mathscr{M}^{|T|}$.

The proof of this corollary is obvious by Theorem 9. It is not known that how large the set of systems of measurement correlations for $(\mathscr{M}, S)$ satisfying the above equivalent conditions in the set of systems of measurement correlations for $(\mathscr{M}, S)$ at the present time.

Going back to the starting point of quantum measurement theory, we do not have to rack our brain to resolve the above difficulty. This is because we should recall that each CP instrument statistically corresponds to an appratus measuring the system under consideration in the sense of the Davies-Lewis proposal. In addition, the introduction of systems of measurement correlations was motivated by the necessity of the counterpart of CP instruments in the (generalized) Heisenberg picture in order to systematically treat measurement correlations. Hence it is natural to consider that an instrument $\mathscr{I}$ for $(\mathscr{M}, S)$ describing a physically realizable measurement should be defined by a system of measurement correlations $\{W_T\}_{T\in\mathscr{T}_S}$ for $(\mathscr{M}, S)$, i.e.,

$$\mathscr{I}(M, \Delta) = \mathscr{I}_{\mathrm{W}}(M, \Delta) \tag{76}$$

for all $M \in \mathscr{M}$ and $\Delta \in \mathscr{F}$.

Question 2 Let $\mathscr{M}$ be a von Neumann algebra on a Hilbert space $\mathscr{H}$ and $(S, \mathscr{F})$ a measurable space. For any CP instrument $\mathscr{I}$ for $(\mathscr{M}, S)$, does there exist a system of measurement correlations $\{W_T\}_{T\in\mathscr{T}_S}$ for $(\mathscr{M}, S)$ which defines $\mathscr{I}$?

In the case of $\mathbf{B}(\mathscr{H})$, this question is already affirmatively answered by the existence of measuring processes for $(\mathbf{B}(\mathscr{H}), S)$ for every CP instrument for $(\mathbf{B}(\mathscr{H}), S)$ (Theorem 1). Surprisingly, Question 2 is affirmatively resolved for all CP instruments defined on arbitrarily given von Neumann algebras.

Theorem 10 *Let* $\mathscr{M}$ *be a von Neumann algebra on a Hilbert space* $\mathscr{H}$ *and* $(S, \mathscr{F})$ *a measurable space. For every CP instrument* $\mathscr{I}$ *for* $(\mathscr{M}, S)$*, there exists a system of measurement correlations* $\{W_T\}_{T\in\mathscr{T}_S}$ *for* $(\mathscr{M}, S)$ *such that*

$$\mathscr{I}(M, \Delta) = \mathscr{I}_{\mathrm{W}}(M, \Delta) \tag{77}$$

for all $M \in \mathscr{M}$ *and* $\Delta \in \mathscr{F}$.

Proof By [31, Proposition 4.2] (or [29, Proposition 3.2]), there exist a Hilbert space $\mathscr{K}$, a normal representation π_0 of $\mathscr{M}$ on $\mathscr{K}$, a PVM $E_0 : \mathscr{F} \to \mathbf{B}(\mathscr{K})$ and an isometry $V : \mathscr{H} \to \mathscr{K}$ such that

$$\mathscr{I}(M, \Delta) = V^*\pi_0(M)E_0(\Delta)V, \tag{78}$$

$$\pi_0(M)E_0(\Delta) = E_0(\Delta)\pi_0(M) \tag{79}$$

for all $M \in \mathscr{M}$ and $\Delta \in \mathscr{F}$, and that $\mathscr{K} = \overline{\mathrm{span}}(\pi_0(\mathscr{M})E_0(\mathscr{F})V\mathscr{H})$.

We follow the identification

$$\mathbf{B}(\mathscr{H} \oplus \mathscr{K}) = \begin{pmatrix} \mathbf{B}(\mathscr{H}) & \mathbf{B}(\mathscr{K}, \mathscr{H}) \\ \mathbf{B}(\mathscr{H}, \mathscr{K}) & \mathbf{B}(\mathscr{K}) \end{pmatrix} \tag{80}$$

with multiplication and involution compatible with the usual matrix operations. We define a normal represetation Π_{in} of $\mathscr{M}$ on $\mathscr{H} \oplus \mathscr{K}$ by

$$\Pi_{in}(M) = \begin{pmatrix} M & 0 \\ 0 & \pi_0(M) \end{pmatrix} \tag{81}$$

for all $M \in \mathscr{M}$, a PVM $E : \mathscr{F} \to \mathbf{B}(\mathscr{H} \oplus \mathscr{K})$ by

$$E(\Delta) = \begin{pmatrix} \delta_s(\Delta)1 & 0 \\ 0 & E_0(\Delta) \end{pmatrix} \tag{82}$$

for all $\Delta \in \mathscr{F}$, where $s \in S$ and δ_s is a delta measure on $(S, \mathscr{F})$ concentrated on s, and a unitary operator U of $\mathscr{H} \oplus \mathscr{K}$ by

$$U = \begin{pmatrix} 0 & -V^* \\ V & Q \end{pmatrix}, \tag{83}$$

where $Q = 1 - VV^*$. For every $\Delta \in \mathscr{F}$, we define a representation Π_Δ of $\mathscr{M}$ on $\mathscr{H} \oplus \mathscr{K}$ by

$$\begin{aligned} \Pi_\Delta(M) &= U^*\Pi_{in}(M)E(\Delta)U \\ &= \begin{pmatrix} \mathscr{I}(M, \Delta) & -V^*\pi_0(M)E_0(\Delta)Q \\ -Q\pi_0(M)E_0(\Delta)V & \delta_s(\Delta)VMV^* + Q\pi_0(M)E_0(\Delta)Q \end{pmatrix} \end{aligned} \tag{84}$$

for all $M \in \mathscr{M}$. We define a unital normal CP linear map $P_{11} : \mathbf{B}(\mathscr{H} \oplus \mathscr{K}) \to \mathbf{B}(\mathscr{H})$ by

$$\begin{pmatrix} \mathbf{B}(\mathscr{H}) & \mathbf{B}(\mathscr{K}, \mathscr{H}) \\ \mathbf{B}(\mathscr{H}, \mathscr{K}) & \mathbf{B}(\mathscr{K}) \end{pmatrix} \ni \begin{pmatrix} X_{11} & X_{12} \\ X_{21} & X_{22} \end{pmatrix} \mapsto X_{11} \in \mathbf{B}(\mathscr{H}). \tag{85}$$

For every $T = (t_1, \ldots, t_{|T|}) \in \mathscr{T}_S$, we define a map $W_T : \mathscr{M}^{|T|} \to \mathbf{B}(\mathscr{H})$ by

$$W_T(\overrightarrow{M}) = P_{11}[\Pi_T(\overrightarrow{M})] = P_{11}[\Pi_{t_1}(M_1) \ldots \Pi_{t_{|T|}}(M_{|T|})] \tag{86}$$

for all $\overrightarrow{M} = (M_1, \ldots, M_{|T|}) \in \mathscr{M}^{|T|}$.

We show that the family $\{W_T\}_{T\in\mathcal{T}_S}$ is a system of measurement correlations for $(\mathcal{M}, S)$ such that

$$\mathcal{I}(M, \Delta) = W_\Delta(M) \tag{87}$$

for all $M \in \mathcal{M}$ and $\Delta \in \mathcal{F}$. For this purpose, it suffices to show $W_T(\overrightarrow{M}) \in \mathcal{M}$ for all $T \in \mathcal{T}_S$ and $\overrightarrow{M} \in \mathcal{M}^{|T|}$. Then the set

$$\mathcal{D} = \begin{pmatrix} \mathcal{M} & \mathrm{span}(\mathcal{M} V^* \mathcal{A}) \\ \mathrm{span}(\mathcal{A} V \mathcal{M}) & \mathcal{A} \end{pmatrix} \tag{88}$$

is a *-subalgebra of $\mathbf{B}(\mathcal{H} \oplus \mathcal{K})$, where $\mathcal{A}$ is a *-subalgebra of $\mathbf{B}(\mathcal{K})$ algebraically generated by $V\mathcal{M}V^*$, $\pi_0(\mathcal{M})E_0(\mathcal{F})$ and $Q = 1 - VV^*$. This fact follows from the usual matrix operations and $V^*\mathcal{A}V \subset \mathcal{M}$.[1] Since it is obvious that $\Pi_{in}(M), \Pi_\Delta(M') \in \mathcal{D}$ for all $M, M' \in \mathcal{M}$ and $\Delta \in \mathcal{F}$, we have $\Pi_T(\overrightarrow{M}) \in \mathcal{D}$ for all $T \in \mathcal{T}_S$ and $\overrightarrow{M} \in \mathcal{M}^{|T|}$. Therefore, for every $T \in \mathcal{T}_S$ and $\overrightarrow{M} \in \mathcal{M}^{|T|}$, the $(1, 1)$-component of $\Pi_T(\overrightarrow{M})$ is also an element of $\mathcal{M}$, which completes the proof.

Remark 3 In the case of an atomic von Neumann algebra $\mathcal{M}$ on a Hilbert space $\mathcal{H}$, we have another construction of a system of measurement correlations which defines a given CP instrument.

Let $\mathcal{E} : \mathbf{B}(\mathcal{H}) \to \mathcal{M}$ be a normal conditional expectation. We define a CP instrument $\widetilde{\mathcal{I}}$ for $(\mathbf{B}(\mathcal{H}), S)$ by

$$\widetilde{\mathcal{I}}(X, \Delta) = \mathcal{I}(\mathcal{E}(X), \Delta) \tag{89}$$

for all $X \in \mathbf{B}(\mathcal{H})$ and $\Delta \in \mathcal{F}$. By Theorem 1, there exists a measuring process $\mathbb{M} = (\mathcal{K}, \sigma, E, U)$ for $(\mathbf{B}(\mathcal{H}), S)$ such that

$$\widetilde{\mathcal{I}}(X, \Delta) = \mathcal{I}_{\mathbb{M}}(X, \Delta) \tag{90}$$

for all $X \in \mathbf{B}(\mathcal{H})$ and $\Delta \in \mathcal{F}$. A system of measurement correlations $\{W_T\}_{T\in\mathcal{T}_S}$ for $(\mathcal{M}, S)$ is defined by

$$W_T(\overrightarrow{M}) = \mathcal{E}(W_T^{\mathbb{M}}(\overrightarrow{M})) \tag{91}$$

for all $T \in \mathcal{T}$ and $\overrightarrow{M} \in \mathcal{M}^{|T|}$. Then $\mathcal{I}_W$ satisfies

$$\mathcal{I}_W(M, \Delta) = \mathcal{E}(W_\Delta^{\mathbb{M}}(M)) = \mathcal{E}(\widetilde{\mathcal{I}}(M, \Delta)) = \mathcal{E}(\mathcal{I}(\mathcal{E}(M), \Delta)) = \mathcal{I}(M, \Delta) \tag{92}$$

for all $M \in \mathcal{M}$ and $\Delta \in \mathcal{F}$.

We should remark that the above construction does not show the existence of measuring processes for $(\mathcal{M}, S)$ for every CP instrument for $(\mathcal{M}, S)$.

[1] To show this, we use $Q = 1 - VV^*$ and Eq. (78).

6 Approximate Realization of CP Instruments by Measuring Processes

We discuss the realizability of CP instruments by measuring processes in this section. Here, we shall start from the following question similar to Question 2.

Question 3 Let $\mathcal{M}$ be a von Neumann algebra on a Hilbert space $\mathcal{H}$ and $(S, \mathcal{F})$ a measurable space. For any CP instrument $\mathcal{I}$ for $(\mathcal{M}, S)$, does there exist a measuring process $\mathbb{M}$ for $(\mathcal{M}, S)$ which realizes $\mathcal{I}$ within arbitrarily given error limits $\varepsilon > 0$?

We say that a CP instrument $\mathcal{I}$ for $(\mathcal{M}, S)$ is *approximately realized* by a net of measuring processes $\{\mathbb{M}_\alpha\}_{\alpha\in A}$ for $(\mathcal{M}, S)$, or $\{\mathbb{M}_\alpha\}_{\alpha\in A}$ *approximately realizes* $\mathcal{I}$ if, for every $\varepsilon > 0$, $n \in \mathbb{N}$, $\rho_1, \ldots, \rho_n \in \mathcal{S}_n(\mathcal{M})$, $\Delta_1, \ldots, \Delta_n \in \mathcal{F}$ and $M_1, \ldots, M_n \in \mathcal{M}$, there is $\alpha \in A$ such that $|\langle \rho_i, \mathcal{I}(M_i, \Delta_i)\rangle - \langle \rho_i, \mathcal{I}_{\mathbb{M}_\alpha}(M_i, \Delta_i)\rangle| < \varepsilon$ for all $i = 1, \ldots, n$. $\mathrm{CPInst}_{\mathrm{AR}}(\mathcal{M}, S)$ denotes the set of CP instruments for $(\mathcal{M}, S)$ approximately realized by nets of measuring processes for $(\mathcal{M}, S)$.

Before answering to Question 3, we shall extend the program, advocated and developed by many researchers [15, 20, 22, 41], which states that physical processes should be described by (inner) CP maps usually called operations [15] or effects [22].

Definition 11 ([3, 23]) Let $\mathcal{M}$ be a von Neumann algebra on a Hilbert space $\mathcal{H}$.
(1) A positive linear map Ψ of $\mathcal{M}$ is said to be finitely inner if there is a finite sequence $\{V_j\}_{j=1,\ldots,m}$ of $\mathcal{M}$ such that

$$\Psi(M) = \sum_{j=1}^{m} V_j^* M V_j \tag{93}$$

for all $M \in \mathcal{M}$.
(2) A positive linear map Ψ of $\mathcal{M}$ is said to be inner if there is a sequence $\{V_j\}_{j\in\mathbb{N}}$ of $\mathcal{M}$ such that

$$\Psi(M) = \sum_{j=1}^{\infty} V_j^* M V_j \tag{94}$$

for all $M \in \mathcal{M}$, where the convergence is ultraweak.
(3) A positive linear map Ψ of $\mathcal{M}$ is said to be approximately inner if it is the pointwise ultraweak limit of a net $\{\Psi_\alpha\}_{\alpha\in A}$ of finitely inner positive linear maps such that $\Psi_\alpha(1) \le \Psi(1)$ for all $\alpha \in A$.

In [3], finite innerness and approximate innerness of CP maps are called factorization through the identity map $\mathrm{id}_{\mathcal{M}} : \mathcal{M} \to \mathcal{M}$ and approximate factorization through $\mathrm{id}_{\mathcal{M}}$, respectively. We refer the reader to [2, 3, 23, 24] for more detailed discussions. It is obvious that every finitely inner positive linear map Ψ of $\mathcal{M}$ is inner. Similarly, every inner positive linear map $\Psi(M) = \sum_{j=1}^{\infty} V_j^* M V_j$, $M \in \mathcal{M}$, is approximately inner since it is the ultraweak limit of a sequence $\{\Psi_j\}$ of finitely inner positive maps

$\Psi_j(M) = \sum_{k=1}^{j} V_k^* M V_k$, $M \in \mathcal{M}$, such that $\Psi_j(1) = \sum_{k=1}^{j} V_k^* V_k \leq \Psi(1)$ for all $j \in \mathbb{N}$. Every approximately inner positive linear map Ψ of $\mathcal{M}$ is always completely positive.

Definition 12 An instrument $\mathcal{I}$ for $(\mathcal{M}, S)$ is said to be finitely inner [inner, or approximately inner, respectively] if $\mathcal{I}(\cdot, \Delta)$ is finitely inner [inner, or approximately inner, respectively] for every $\Delta \in \mathcal{F}$. $\mathrm{CPInst}_{\mathrm{FI}}(\mathcal{M}, S)$ [$\mathrm{CPInst}_{\mathrm{IN}}(\mathcal{M}, S)$, or $\mathrm{CPInst}_{\mathrm{AI}}(\mathcal{M}, S)$, respectively] denotes the set of finitely inner [inner, or approximately inner, respectively] CP instruments for $(\mathcal{M}, S)$.

The following relation holds.

$$\mathrm{CPInst}_{\mathrm{FI}}(\mathcal{M}, S) \subset \mathrm{CPInst}_{\mathrm{IN}}(\mathcal{M}, S) \subset \mathrm{CPInst}_{\mathrm{AI}}(\mathcal{M}, S). \tag{95}$$

Definition 13 (*Inner measuring process*) A measuring process $\mathbb{M} = (\mathcal{K}, \sigma, E, U)$ for $(\mathcal{M}, S)$ is said to be inner if U is contained in $\mathcal{M}\overline{\otimes}\mathbf{B}(\mathcal{K})$.

We then have the following theorem.

Theorem 11 *Let $\mathcal{M}$ be a von Neumann algebra on a Hilbert space $\mathcal{H}$ and $(S, \mathcal{F})$ a measurable space. For every approximately inner (hence CP) instrument $\mathcal{I}$ for $(\mathcal{M}, S)$, $\varepsilon > 0$, $n \in \mathbb{N}$, $\rho_1, \ldots, \rho_n \in \mathcal{S}_n(\mathcal{M})$, $M_1, \ldots, M_n \in \mathcal{M}$ and $\Delta_1, \ldots, \Delta_n \in \mathcal{F}$, there exists an inner measuring process $\mathbb{M} = (\mathcal{K}, \sigma, E, U)$ for $(\mathcal{M}, S)$ in the sense of Definition 9 such that*

$$|\langle \rho_j, \mathcal{I}(M_j, \Delta_j)\rangle - \langle \rho_j, \mathcal{I}_{\mathbb{M}}(M_j, \Delta_j)\rangle| < \varepsilon \tag{96}$$

for all $j = 1, \ldots, n$.

Corollary 4 *Let $\mathcal{M}$ be a von Neumann algebra on a Hilbert space $\mathcal{H}$ and $(S, \mathcal{F})$ a measurable space. We have*

$$\mathrm{CPInst}_{\mathrm{AI}}(\mathcal{M}, S) \subset \mathrm{CPInst}_{\mathrm{AR}}(\mathcal{M}, S). \tag{97}$$

Only for injective factors the following holds as a corollary of the above theorem.

Theorem 12 *Let $\mathcal{M}$ be an injective factor on a Hilbert space $\mathcal{H}$ and $(S, \mathcal{F})$ a measurable space. For every CP instrument $\mathcal{I}$ for $(\mathcal{M}, S)$, $\varepsilon > 0$, $n \in \mathbb{N}$, $\rho_1, \ldots, \rho_n \in \mathcal{S}_n(\mathcal{M})$, $M_1, \ldots, M_n \in \mathcal{M}$ and $\Delta_1, \ldots, \Delta_n \in \mathcal{F}$, there exists an inner measuring process $\mathbb{M} = (\mathcal{K}, \sigma, E, U)$ for $(\mathcal{M}, S)$ in the sense of Definition 9 such that*

$$|\langle \rho_j, \mathcal{I}(M_j, \Delta_j)\rangle - \langle \rho_j, \mathcal{I}_{\mathbb{M}}(M_j, \Delta_j)\rangle| < \varepsilon \tag{98}$$

for all $j = 1, \ldots, n$.

Corollary 5 *Let $\mathcal{M}$ be an injective factor on a Hilbert space $\mathcal{H}$ and $(S, \mathcal{F})$ a measurable space. Then we have*

$$\mathrm{CPInst}_{\mathrm{AI}}(\mathcal{M}, S) = \mathrm{CPInst}_{\mathrm{AR}}(\mathcal{M}, S) = \mathrm{CPInst}(\mathcal{M}, S). \tag{99}$$

Theorem 12 is a stronger result than [29, Theorem 4.4] for factors, so that Question 3 is affirmatively resolved for injective factors. We use the following proposition for the proof of Theorem 12.

Proposition 2 (Anantharaman-Delaroche and Havet [3, Lemma 2.2, Remarks 5.4]) *Let $\mathcal{M}$ be an injective factor on on a Hilbert space $\mathcal{H}$. Every CP map Ψ of $\mathcal{M}$ is approximately inner,* i.e., *it is the pointwise ultraweak limit of a net of CP maps $\{\Psi_\theta\}_{\theta\in\Theta}$ of the form $\Psi_\theta(M) = \sum_{j=1}^{n_\theta} V_{\theta,j}^* M V_{\theta,j}$, $M \in \mathcal{M}$, with $n_\theta \in \mathbb{N}$, $V_{\theta,1}, \ldots, V_{\theta,n_\theta} \in \mathcal{M}$ such that $\Psi_\theta(1) = \sum_{j=1}^{n_\theta} V_{\theta,j}^* V_{\theta,j} \le \Psi(1)$ for all $\theta \in \Theta$.*

The following proof is inspired by [40].

Proof (*Proof of Theorem* 11) Let $n \in \mathbb{N}$, $\rho_1, \ldots, \rho_n \in \mathcal{S}_n(\mathcal{M})$, $M_1, \ldots, M_n \in \mathcal{M}\backslash\{0\}$ and $\Delta_1, \ldots, \Delta_n \in \mathcal{F}\backslash\{\emptyset\}$. Let $\mathcal{F}'$ be a σ-subfield of $\mathcal{F}$ generated by $\Delta_1, \ldots, \Delta_n, S$. Let $\{\Gamma_i\}_{i=1}^m \subset \mathcal{F}'\backslash\{\emptyset\}$ be a maximal partition of $\cup_{i=1}^n \Delta_i$, i.e., $\{\Gamma_i\}_{i=1}^m$ satisfies the following conditions:

(1) For every $i = 1, \ldots, m$, if $\Delta \in \mathcal{F}'$ satisfies $\Delta \subset \Gamma_i$, then Δ is Γ_i or $\emptyset$;
(2) $\cup_{i=1}^m \Gamma_i = \cup_{i=1}^n \Delta_i$;
(3) $\Gamma_i \cap \Gamma_j = \emptyset$ if $i \neq j$.

For every $i = 1, \ldots, m$, there is a net of finitely inner CP maps $\{\Psi_{\theta_i}\}_{\theta_i\in\Theta_i}$ of the form $\Psi_{i,\theta_i}(M) = \sum_{j=1}^{n_{\theta_i}} V_{i,\theta_i,j}^* M V_{i,\theta_i,j}$, $M \in \mathcal{M}$, with $n_{\theta_i} \in \mathbb{N}$, $V_{i,\theta_i,1}, \ldots, V_{i,\theta_i,n_{\theta_i}} \in \mathcal{M}$ such that is pointwisely convergent to $\mathcal{I}(\cdot, \Gamma_i)$ in the ultraweak topology and $\Psi_{i,\theta_i}(1) \le \mathcal{I}(1, \Gamma_i)$.

We fix $s_0, s_1, \ldots .s_m \in S$ such that $s_i \in \Gamma_i$ for all $i = 1, \ldots, m$. For every $\boldsymbol{\theta} = (\theta_1, \ldots, \theta_m) \in \Theta_1 \times \ldots \times \Theta_m$, we define a finitely inner CP instrument $\mathcal{I}_{\boldsymbol{\theta}}$ for $(\mathcal{M}, S)$ by

$$\mathcal{I}_{\boldsymbol{\theta}}(M, \Delta) = \sum_{i=1}^m \delta_{s_i}(\Delta)\Psi_{\theta_i}(M) + \delta_{s_m}(\Delta) L_{\boldsymbol{\theta}} M L_{\boldsymbol{\theta}} \tag{100}$$

for all $M \in \mathcal{M}$ and $\Delta \in \mathcal{F}$, where δ_s is a delta measure on $(S, \mathcal{F})$ concentrated on s and $L_{\boldsymbol{\theta}} = \sqrt{1 - \sum_{i=1}^m \Psi_{i,\theta_i}(1)}$.

Let ε be a positive real number. For every $i = 1, \ldots, m$, there is $\bar{\theta}_i \in \Theta_i$ such that

$$|\langle \rho_j, \mathcal{I}(M_j, \Gamma_i)\rangle - \langle \rho_j, \Psi_{i,\bar{\theta}_i}(M_j)\rangle| < \frac{\varepsilon}{2m} \tag{101}$$

for all $j = 1, \ldots, n$, and that

$$\langle \rho_j, \mathcal{I}(1, \Gamma_i) - \Psi_{i,\bar{\theta}_i}(1)\rangle < \frac{\varepsilon}{2m \sum_{k=1}^n \|M_k\|} \tag{102}$$

for all $j = 1, \ldots, n$.

By Eqs. (101), (102), we have

$$
\begin{aligned}
&|\langle \rho_j, \mathscr{I}(M_j, \Delta_j)\rangle - \sum_{i=1}^{m} \delta_{s_i}(\Delta_j)\langle \rho_j, \Psi_{i,\bar{\theta}_i}(M_j)\rangle| \\
=&|\sum_{i=1}^{m} \delta_{s_i}(\Delta_j)\langle \rho_j, \mathscr{I}(M_j, \Gamma_i)\rangle - \sum_{i=1}^{m} \delta_{s_i}(\Delta_j)\langle \rho_j, \Psi_{i,\bar{\theta}_i}(M_j)\rangle| \\
\leq& \sum_{i=1}^{m} \delta_{s_i}(\Delta_j)|\langle \rho_j, \mathscr{I}(M_j, \Gamma_i)\rangle - \langle \rho_j, \Psi_{i,\bar{\theta}_i}(M_j)\rangle| < \sum_{i=1}^{m} \delta_{s_i}(\Delta_j) \cdot \frac{\varepsilon}{2m} \leq \frac{\varepsilon}{2}
\end{aligned} \tag{103}
$$

for all $j = 1, \ldots, n$, and

$$
\begin{aligned}
|\rho_j(L_{\bar{\theta}} M_j L_{\bar{\theta}})| \leq \|M_j\|\rho_j(L_{\bar{\theta}}^2) &= \|M_j\| \cdot \langle \rho_j, \sum_{i=1}^{m} \left(\mathscr{I}(1, \Gamma_i) - \Psi_{i,\bar{\theta}_i}(1)\right)\rangle \\
&= \|M_j\| \cdot \sum_{i=1}^{m} \langle \rho_j, \mathscr{I}(1, \Gamma_i) - \Psi_{i,\bar{\theta}_i}(1)\rangle \\
&< \|M_j\| \cdot m \cdot \frac{\varepsilon}{2m \sum_{k=1}^{n} \|M_k\|} \leq \frac{\varepsilon}{2}.
\end{aligned} \tag{104}
$$

for all $j = 1, \ldots, n$. Then the CP instrument $\mathscr{I}_{\bar{\theta}}$ for $(\mathscr{M}, S)$ with $\bar{\boldsymbol{\theta}} = (\bar{\theta}_1, \ldots, \bar{\theta}_m)$ satisfies

$$
|\langle \rho_j, \mathscr{I}(M_j, \Delta_j)\rangle - \langle \rho_j, \mathscr{I}_{\bar{\theta}}(M_j, \Delta_j)\rangle| < \varepsilon \tag{105}
$$

for all $j = 1, \ldots, n$.

Next, we shall define an inner measuring process $\mathbb{M} = (\mathscr{K}, \sigma, E, U)$ for $(\mathscr{M}, S)$ that realizes $\mathscr{I}_{\bar{\theta}}$. Let $\eta = \{\eta_j\}_{j=0,1,\ldots,\|\bar{\theta}\|+1}$ be a complete orthonormal system of $\mathbb{C}^{\|\bar{\theta}\|+2}$. A partial isometry $V : \mathscr{H} \otimes \mathbb{C}^{\|\bar{\theta}\|+2} \to \mathscr{H} \otimes \mathbb{C}^{\|\bar{\theta}\|+2}$ is defined by

$$
V = \sum_{i=1}^{m} \sum_{j=\sum_{k=1}^{i-1}\bar{\theta}_k+1}^{\sum_{k=1}^{i}\bar{\theta}_k} V_{i,\bar{\theta}_i,j} \otimes |\eta_j\rangle\langle \eta_0| + L_{\bar{\theta}} \otimes |\eta_{\|\bar{\theta}\|+1}\rangle\langle \eta_0| \tag{106}
$$

It is obvious that V satisfies $V^*V = 1 \otimes |\eta_0\rangle\langle \eta_0|$. We define a PVM $E : \mathscr{F} \to M_{\|\bar{\theta}\|+2}(\mathbb{C})$ by

$$
E_\eta(\Delta) = \delta_{s_0}(\Delta)|\eta_0\rangle\langle \eta_0| + \sum_{i=1}^{m} \delta_{s_i}(\Delta) \sum_{j=\sum_{k=1}^{i-1}\bar{\theta}_k+1}^{\sum_{k=1}^{i}\bar{\theta}_k} |\eta_j\rangle\langle \eta_j| + \delta_{s_m}(\Delta)|\eta_{\|\bar{\theta}\|+1}\rangle\langle \eta_{\|\bar{\theta}\|+1}| \tag{107}
$$

for all $\Delta \in \mathscr{F}$.

We define a Hilbert space $\mathscr{K} = \mathbb{C}^{\|\bar{\theta}\|+2} \otimes \mathbb{C}^2$, a normal state σ on $\mathbf{B}(\mathscr{K}) = M_{\|\bar{\theta}\|+2}(\mathbb{C}) \otimes M_2(\mathbb{C})$, a PVM $E : \mathscr{F} \to \mathbf{B}(\mathscr{K})$ and a unitary operator U on $\mathscr{H} \otimes \mathscr{K}$ by

$$\sigma(X) = \mathrm{Tr}\,[X(|\eta_0\rangle\langle\eta_0| \otimes G_{11})], \qquad X \in \mathbf{B}(\mathscr{K}), \tag{108}$$

$$E(\Delta) = E_\eta(\Delta) \otimes 1, \qquad \Delta \in \mathscr{F}, \tag{109}$$

$$U = V \otimes G_{11} + (1 - VV^*) \otimes G_{12} + (1 - V^*V) \otimes G_{12}^* - V^* \otimes G_{22}, \tag{110}$$

respectively, where Tr is the trace on $M_{\|\bar{\theta}\|+2}(\mathbb{C}) \otimes M_2(\mathbb{C})$ and

$$G_{11} = \begin{pmatrix} 1 & 0 \\ 0 & 0 \end{pmatrix}, \quad G_{12} = \begin{pmatrix} 0 & 1 \\ 0 & 0 \end{pmatrix}, \quad G_{22} = \begin{pmatrix} 0 & 0 \\ 0 & 1 \end{pmatrix}. \tag{111}$$

Since $U \in \mathscr{M} \overline{\otimes} M_{\|\bar{\theta}\|+2}(\mathbb{C}) \overline{\otimes} M_2(\mathbb{C})$, the 4-tuple $\mathbb{M} = (\mathscr{K}, \sigma, E, U)$ is an inner measuring process for $(\mathscr{M}, S)$ satisfying

$$|\langle\rho_j, \mathscr{I}(M_j, \Delta_j)\rangle - \langle\rho_j, \mathscr{I}_{\mathbb{M}}(M_j, \Delta_j)\rangle| < \varepsilon \tag{112}$$

for all $j = 1, \ldots, n$.

Proof (*Proof of Theorem* 12) Let $\mathscr{I}$ be a CP instrument for $(\mathscr{M}, S)$. Since $\mathscr{M}$ is an injective factor, $\mathscr{I}(\cdot, \Delta)$ is approximately inner for every $\Delta \in \mathscr{F}$ by Proposition 2. Thus the proof of Theorem 11 works.

Remark 4 We use the same notations as in the proof of Theorem 11. In the case where $\mathscr{M}$ is factor, we have another construction of a measuring process $\mathbb{M}$ for $(\mathscr{M}, S)$ such that $\mathscr{I}_{\bar{\theta}}(M, \Delta) = \mathscr{I}_{\mathbb{M}}(M, \Delta)$ for all $M \in \mathscr{M}$ and $\Delta \in \mathscr{F}$.

Let $\mathscr{N}$ be an AFD type III factor on a separable Hilbert space $\mathscr{L}$. Let Y be a partial isometry of $\mathscr{N}$ such that $Y^*Y \neq 1$ and $YY^* \neq 1$. There then exists a partial isometry W of $\mathscr{M} \overline{\otimes} M_{\|\bar{\theta}\|+2}(\mathbb{C}) \overline{\otimes} \mathscr{N}$ such that $W^*W = 1 - V^*V \otimes Y^*Y = 1 - 1 \otimes |\eta_0\rangle\langle\eta_0| \otimes Y^*Y$ and $WW^* = 1 - VV^* \otimes YY^*$. We define a unitary operator U of $\mathscr{M} \overline{\otimes} M_{\|\bar{\theta}\|+2}(\mathbb{C}) \overline{\otimes} \mathscr{N}$ by

$$U = V \otimes Y + W, \tag{113}$$

and a PVM $E : \mathscr{F} \to M_{\|\bar{\theta}\|+2}(\mathbb{C}) \overline{\otimes} \mathscr{N}$ by

$$E(\Delta) = E_\eta(\Delta) \otimes 1_{\mathscr{L}} \tag{114}$$

for all $\Delta \in \mathscr{F}$. Let ψ be a unit vector of $\mathscr{L}$ such that $Y^*Y\psi = \psi$. Then we have $W(\xi \otimes \eta_0 \otimes \psi) = 0$ for all $\xi \in \mathscr{H}$. We define a normal state σ on $M_{\|\bar{\theta}\|+2}(\mathbb{C}) \overline{\otimes} \mathscr{N}$ by

$$\sigma(X) = \langle\eta_0 \otimes \psi | X(\eta_0 \otimes \psi)\rangle \tag{115}$$

for all $X \in M_{\|\bar{\theta}\|+2}(\mathbb{C}) \overline{\otimes} \mathscr{N}$.

A Hilbert space $\mathscr{K}$ is then defined by $\mathscr{K} = \mathbb{C}^{\|\bar{\theta}\|+2} \otimes \mathscr{L}$. Since $U \in \mathscr{M}\overline{\otimes} M_{\|\bar{\theta}\|+2}(\mathbb{C})\overline{\otimes}\mathscr{N}$, the 4-tuple $\mathbb{M} = (\mathscr{K}, \sigma, E, U)$ is an inner measuring process for $(\mathscr{M}, S)$ satisfying the desired property.

Not only for factors $\mathscr{M}$, we have the following theorem affirmatively resolving Question 3 for physically relevant cases.

Definition 14 A measuring process $\mathbb{M} = (\mathscr{K}, \sigma, E, U)$ for $(\mathscr{M}, S)$ is said to be faithful if there exists a normal faithul representation $\widetilde{E} : L^\infty(S, \mathscr{I}_{\mathbb{M}}) \to \mathbf{B}(\mathscr{K})$ such that $\widetilde{E}([\chi_\Delta]) = E(\Delta)$ for all $\Delta \in \mathscr{F}$.

This definition is the same as [29, Definition 3.4] except that the definition of measuring process for $(\mathscr{M}, S)$ is different.

Theorem 13 *Let $\mathscr{M}$ be an injective von Neumann algebra on a Hilbert space $\mathscr{H}$ and $(S, \mathscr{F})$ a measurable space. For every CP instrument $\mathscr{I}$ for $(\mathscr{M}, S)$, $\varepsilon > 0$, $n \in \mathbb{N}$, $\rho_1, \ldots, \rho_n \in \mathscr{S}_n(\mathscr{M})$, $M_1, \ldots, M_n \in \mathscr{M}$ and $\Delta_1, \ldots, \Delta_n \in \mathscr{F}$, there exists a faithful measuring process $\mathbb{M} = (\mathscr{K}, \sigma, E, U)$ for $(\mathscr{M}, S)$ in the sense of Definition 9 such that*

$$|\langle \rho_j, \mathscr{I}(M_j, \Delta_j)\rangle - \langle \rho_j, \mathscr{I}_{\mathbb{M}}(M_j, \Delta_j)\rangle| < \varepsilon \tag{116}$$

for all $j = 1, \ldots, n$, and that

$$\mathscr{I}(1, \Delta) = \mathscr{I}_{\mathbb{M}}(1, \Delta) \tag{117}$$

for all $\Delta \in \mathscr{F}$.

Proof Suppose that $\mathscr{M}$ is in a standard form without loss of generality. Then there is a norm one projection $\mathscr{E} : \mathbf{B}(\mathscr{H}) \to \mathscr{M}$ and a net $\{\Phi_\alpha\}_{\alpha \in A}$ of unital CP maps such that $\Phi_\alpha(X) \to^{uw} \mathscr{E}(X)$ for all $X \in \mathbf{B}(\mathscr{H})$ by [2, Corollary 3.9], (or [29, Proposition 4.2]). For every $\alpha \in A$, a CP instrument $\mathscr{I}_\alpha$ for $(\mathbf{B}(\mathscr{H}), S)$ is defined by

$$\mathscr{I}_\alpha(X, \Delta) = \mathscr{I}(\Phi_\alpha(X), \Delta) \tag{118}$$

for all $X \in \mathbf{B}(\mathscr{H})$ and $\Delta \in \mathscr{F}$. For every $\alpha \in A$, $\mathscr{I}_\alpha$ satisfies

$$\mathscr{I}_\alpha(1, \Delta) = \mathscr{I}(\Phi_\alpha(1), \Delta) = \mathscr{I}(1, \Delta) \tag{119}$$

for all $\Delta \in \mathscr{F}$.

Let $\varepsilon > 0$, $n \in \mathbb{N}$, $\rho_1, \ldots, \rho_n \in \mathscr{S}_n(\mathscr{M})$, $M_1, \ldots, M_n \in \mathscr{M}$ and $\Delta_1, \ldots, \Delta_n \in \mathscr{F}$. There exists $\alpha_0 \in A$ such that

$$|\langle \rho_i, \mathscr{I}(M_i, \Delta_i)\rangle - \langle \rho_i, \mathscr{I}_{\alpha_0}(M_i, \Delta_i)\rangle| < \varepsilon \tag{120}$$

for every $i = 1, \ldots, n$.

By [29, Proposition 3.2] and Theorem 7, there exist a Hilbert space $\mathscr{L}_1$, a normal faithful representation $E_1 : L^\infty(S, \mathscr{I}_{\alpha_0}) \to \mathbf{B}(\mathscr{L}_1)$ and an isometry $V : \mathscr{H} \to \mathscr{H} \otimes \mathscr{L}_1$ such that

$$\mathscr{I}_{\alpha_0}(X, \Delta) = V^*(X \otimes E_1([\chi_\Delta]))V \tag{121}$$

for all $X \in \mathbf{B}(\mathscr{H})$ and $\Delta \in \mathscr{F}$.

Because the discussion below is not needed in the case of $\dim(\mathscr{L}_1) = 1$, we assume that $\dim(\mathscr{L}_1) \geq 2$. Let η_1 be a unit vector of $\mathscr{L}_1$. Let $\mathscr{N}$ be an AFD type III factor on a separable Hilbert space $\mathscr{L}_2$. We define a partial isometry $U_1 : \mathscr{H} \otimes \mathscr{L}_1 \otimes \mathscr{L}_2 \to \mathscr{H} \otimes \mathscr{L}_1 \otimes \mathscr{L}_2$ by

$$U_1(x \otimes \xi \otimes \psi) = \langle \eta_1 | \xi \rangle V x \otimes \psi \tag{122}$$

for all $x \in \mathscr{H}$, $\xi \in \mathscr{L}_1$ and $\psi \in \mathscr{L}_2$. Let U_2 be an isometry of $\mathbf{B}(\mathscr{L}_1)\overline{\otimes}\mathscr{N}$ such that $U_2 U_2^* = |\eta_1\rangle\langle\eta_1| \otimes 1$. We define an isometry U_3 of $\mathbf{B}(\mathscr{H} \otimes \mathscr{L}_1)\overline{\otimes}\mathscr{N}$ by $U_3 = 1 \otimes U_2$. We then define a unitary operator U of $\mathscr{H} \otimes \mathscr{L}_1 \otimes \mathscr{L}_2 \otimes \mathbb{C}^2$ by

$$\begin{aligned} U &= U_1 U_3 \otimes G_{11} + [1 - (U_1 U_3)(U_1 U_3)^*] \otimes G_{12} \\ &\quad + [1 - (U_1 U_3)^*(U_1 U_3)] \otimes G_{12}^* - (U_1 U_3)^* \otimes G_{22} \\ &= U_1 U_3 \otimes G_{11} + (1 - U_1 U_1^*) \otimes G_{12} - (U_1 U_3)^* \otimes G_{22}, \end{aligned} \tag{123}$$

where

$$G_{11} = \begin{pmatrix} 1 & 0 \\ 0 & 0 \end{pmatrix}, \quad G_{12} = \begin{pmatrix} 0 & 1 \\ 0 & 0 \end{pmatrix}, \quad G_{22} = \begin{pmatrix} 0 & 0 \\ 0 & 1 \end{pmatrix}. \tag{124}$$

Let η_2 be a unit vector of $\mathscr{L}_2$. We define a Hilbert space $\mathscr{K} = \mathscr{L}_1 \otimes \mathscr{L}_2 \otimes \mathbb{C}^2$, a normal state σ on $\mathbf{B}(\mathscr{K})$ by

$$\sigma(X) = \mathrm{Tr}\,[X(|\eta_1 \otimes \eta_2\rangle\langle\eta_1 \otimes \eta_2| \otimes G_{11})] \tag{125}$$

for all $X \in \mathbf{B}(\mathscr{K})$, and a PVM $E : \mathscr{F} \to \mathbf{B}(\mathscr{K})$ by

$$E(\Delta) = E_1([\chi_\Delta]) \otimes 1_{\mathscr{N}\overline{\otimes}M_2(\mathbb{C})} \tag{126}$$

for all $\Delta \in \mathscr{F}$, respectively, where Tr is the trace on $\mathbf{B}(\mathscr{L}_1 \otimes \mathscr{L}_2)\overline{\otimes}M_2(\mathbb{C})$.

The 4-tuple $\mathbb{M} = (\mathscr{K}, \sigma, E, U)$ is then a faithful measuring process for $(\mathscr{M}, S)$ that realizes $\mathscr{I}_{\alpha_0}$ and that satisfies

$$|\langle \rho_j, \mathscr{I}(M_j, \Delta_j)\rangle - \langle \rho_j, \mathscr{I}_{\mathbb{M}}(M_j, \Delta_j)\rangle| < \varepsilon \tag{127}$$

for all $j = 1, \ldots, n$, and

$$\mathscr{I}(1, \Delta) = \mathscr{I}_{\mathbb{M}}(1, \Delta) \tag{128}$$

for all $\Delta \in \mathscr{F}$.

By the proof of Theorem 13 and facts in Sect. 2, we have the following corollaries.

Corollary 6 *Let $\mathscr{M}$ be a von Neumann algebra on a Hilbert space $\mathscr{H}$ and $(S, \mathscr{F})$ a measurable space. Then we have*

$$\mathrm{CPInst}_{\mathrm{RE}}(\mathscr{M}, S) = \mathrm{CPInst}_{\mathrm{NE}}(\mathscr{M}, S). \tag{129}$$

Proof Use [29, Theorem 3.4 (iii)].

Corollary 7 *Let $\mathscr{M}$ be an atomic von Neumann algebra on a Hilbert space $\mathscr{H}$ and $(S, \mathscr{F})$ a measurable space. Then we have*

$$\mathrm{CPInst}_{\mathrm{RE}}(\mathscr{M}, S) = \mathrm{CPInst}(\mathscr{M}, S). \tag{130}$$

Corollary 8 *Let $\mathscr{M}$ be a von Neumann algebra on a Hilbert space $\mathscr{H}$ and $(S, \mathscr{F})$ a measurable space. Then we have*

$$\mathrm{CPInst}_{\mathrm{AR}}(\mathscr{M}, S) = \mathrm{CPInst}(\mathscr{M}, S). \tag{131}$$

Following these results, Question 3 is affirmatively resolved for general σ-finite von Neumann algebras.

Throughout the present paper, we have developed the dilation theory of systems of measurement correlations and CP instruments, and established many unitary dilation theorems of them. In the succeeding paper, we systematically develop successive and continuous measurements in the generalized Heisenberg picture. The author believes that the approach to quantum measurement theory given in the present and succeeding papers contributes to the categorical (re-)formulation of quantum theory. On the other hand, though we do not know how it is related to the topic of the paper at the present time, the future task is to find the connection with the results of Haagerup and Musat [16, 17], which develop the asymptotic factorizability of CP maps on finite von Neumann algebras.

Acknowledgements The author would like to thank Professor Masanao Ozawa for his useful comments and warm encouragement. This work was supported by the John Templeton Foundations, No. 35771 and by the JSPS KAKENHI, No. 26247016, and No. 16K17641.

References

1. L. Accardi, A. Frigerio and J.T. Lewis, Quantum stochastic processes, Publ. Res. Inst. Math. Sci. **18**, 97–133 (1982).
2. C. Anantharaman-Delaroche, Amenable correspondences and approximation properties for von Neumann algebras, Pacific J. Math. **171**, 309–341 (1995).
3. C. Anantharaman-Delaroche and J.F. Havet, On approximate factorizations of completely positive maps, J. Func. Anal. **90**, 411–428 (1990).
4. W. Arveson, Subalgebras of C*-algebras, Acta Math. **123**, 141–224 (1969).
5. A. Barchielli and V.P. Belavkin, Measurements continuous in time and a posteriori states in quantum mechanics, J. Phys. A: Math. Gen. **24**, 1495–1514 (1991).

6. V.P. Belavkin, Reconstruction Theorem for Quantum Stochastic Processes, Theoret. Math. Phys. **3**, 409–431 (1985), arXiv:math/0512410.
7. D. Buchholz, C. D'Antoni and K. Fredenhagen, The universal structure of local algebras, Commun. Math. Phys. **111**, 123–135 (1987).
8. A. Connes, *Noncommutative Geometry*, (Academic Press, San Diego, CA, 1994).
9. P. Busch, M. Grabowski and P.J. Lahti, *Operational quantum physics*, (Springer, Berlin, 1995).
10. E.B. Davies, *Quantum Theory of Open Systems*, (Academic Press, London, 1976).
11. E.B. Davies and J.T. Lewis, An operational approach to quantum probability, Commun. Math. Phys. **17**, 239–260 (1970).
12. J. Dixmier, *Von Neumann Algebras*, (North-Holland, Amsterdam, 1981).
13. E.C. Lance, *Hilbert C*-modules: a toolkit for operator algebraists*, (Cambridge UP, Cambridge, 1995).
14. D.E. Evans and J.T. Lewis, *Dilations of irreversible evolutions in algebraic quantum theory*, Comm. Dublin Inst. Adv. Studies Ser. A **24**, (1977).
15. R. Haag and D. Kastler, An algebraic approach to quantum field theory, J. Math. Phys. **5**, 848–861 (1964).
16. U. Haagerup and M. Musat, Factorization and dilation problems for completely positive maps on von Neumann algebras, Commun. Math. Phys. **303**, 555–594 (2011).
17. U. Haagerup and M. Musat, An asymptotic property of factorizable completely positive maps and the Connes embedding problem, Commun. Math. Phys. **338**, 721–752 (2015).
18. P.R. Halmos, *A Hilbert space problem book*, 2nd Ed., (Springer, New York, 1982).
19. A. Hora and N. Obata, *Quantum probability and spectral analysis of graphs*, (Springer, Berlin, 2007).
20. K. Kraus, General state changes in quantum theory, Ann. Phys. **64**, 311–335 (1971).
21. K. Kraus, *States, Effects, and Operations: Fundamental Notions of Quantum Theory*, Lecture Notes Phys. **190**, (Springer, Belin, 1983).
22. G. Ludwig, Attempt of an axiomatic foundation of quantum mechanics and more general theories, II, Commun. Math. Phys. **4**, 331–348 (1967); Attempt of an axiomatic foundation of quantum mechanics and more general theories, III, *ibid.* **9**, 1–12 (1968).
23. J.A. Mingo, The correspondence associated to an inner completely positive map, Math. Ann. **284**, 121–135 (1989).
24. J.A. Mingo, Weak containment of correspondences and approximate factorization of completely positive maps, J. Func. Anal. **89**, 90–105 (1990).
25. N. Muraki, A simple proof of the classification theorem for positive natural products, Prob. Math. Stat. **33**, 315–326 (2013).
26. B. Sz.-Nagy, C. Foias, H. Bercovici and L. Kerchy, *Harmonic analysis of operators on Hilbert space*, 2nd Ed., (Springer, New York, 2010).
27. J. von Neumann, *Mathematische Grundlagen der Quantenmechanik*, (Springer, Berlin, 1932); *Mathematical Foundations of Quantum Mechanics*, (Princeton UP, Princeton, 1955).
28. I. Ojima, K. Okamura and H. Saigo, Local state and sector theory in local quantum physics, Lett. Math. Phys. **106**, 741–763 (2016).
29. K. Okamura and M. Ozawa, Measurement theory in local quantum physics, J. Math. Phys. **57**, 015209 (2016).
30. M. Ozawa, Conditional expectation and repeated measurements of continuous quantum observables, In; *Probability Theory and Mathematical Statistics*, (eds. K. Ito and J.V. Prohorov), Lecture Notes Math. **1021**, pp.518–525 (Springer, Berlin, 1983).
31. M. Ozawa, Quantum measuring processes of continuous observables, J. Math. Phys. **25**, 79–87 (1984).
32. M. Ozawa, Conditional probability and a posteriori states in quantum mechanics, Publ. Res. Inst. Math. Sci. **21**, 279–295 (1985).
33. M. Ozawa, Canonical approximate quantum measurements, J. Math. Phys. **34**, 5596–5624 (1993).
34. M. Ozawa, An Operational Approach to Quantum State Reduction, Ann. Phys. (N.Y.) **259**, 121–137 (1997).

35. M. Ozawa, Uncertainty relations for noise and disturbance in generalized quantum measurements, Ann. Phys. (N.Y.) **331**, 350–416 (2004).
36. M. Ozawa, Quantum perfect correlations, Ann. Phys. (N.Y.) **321**, 744–769 (2006).
37. M. Ozawa, Mathematical foundations of quantum information: Measurement and foundations, Sugaku Expositions **27**, 195–221 (2014).
38. M. Ozawa, Heisenberg's original derivation of the uncertainty principle and its universally valid reformulations, Current Science **10**, 2006–2016 (2015).
39. V. Paulsen, *Completely bounded maps and operator algebras*, (Cambridge UP, Cambridge, 2002).
40. J.-P. Pellonpää and M. Tukiainen, Minimal normal measurement models of quantum instruments, (2015), arXiv:1509.08886 [quant-ph].
41. J. Schwinger, The algebra of microscopic measurement, Proc. Nat. Acad. Sci. U.S. **45**, 1542–1554 (1959); The geometry of quantum states, *ibid.* **46**, 257–265 (1960).
42. M. Skeide. Generalized matrix C*-algebras and representations of Hilbert modules, Math. Proc. Royal Irish Academy, **100A**, 11–38 (2000).
43. M. Skeide, *Hilbert modules and applications in quantum probability*, Habilitationsschrift, (Cottbus, 2001).
44. W.F. Stinespring, Positive functions on C*-algebras, Proc. Amer. Math. Soc. **6**, 211-216 (1955).
45. R.F. Streater and A.S. Wightman, *PCT, spin and statistics, and all that*, (Princeton UP, Princeton, 2000).
46. M. Takesaki, *Theory of Operator Algebras I*, (Springer, Berlin, 1979).
47. M. Takesaki, *Theory of Operator Algebras III*, (Springer, Berlin, 2002).
48. H. Umegaki, Conditional expectation in an operator algebra, Tohoku Math. J. **6**, 177–181 (1954).

Printed by Printforce, the Netherlands